International
REVIEW OF
Neurobiology

Volume 50

Neurobiology

OF

Diabetic Neuropathy

International

REVIEW OF

Neurobiology

Volume 50

SERIES EDITORS

RONALD J. BRADLEY

Department of Psychiatry, School of Medicine
Louisiana State University Medical Center
Shreveport, Louisiana, USA

R. ADRON HARRIS

Waggoner Center for Alcohol and Drug Addiction Research
The University of Texas at Austin
Austin, Texas, USA

PETER JENNER

Division of Pharmacology and Therapeutics
GKT School of Biomedical Sciences
King's College, London, UK

EDITORIAL BOARD

Neurobiology OF *Diabetic Neuropathy*

EDITED BY

DAVID TOMLINSON

Neuroscience Division
University of Manchester
School of Biological Sciences
Manchester, United Kingdom

ACADEMIC PRESS

An imprint of Elsevier Science

Amsterdam   Boston   London   New York   Oxford   Paris
San Diego   San Francisco   Singapore   Sydney   Tokyo

Academic Press
An imprint of Elsevier Science
525 B Street, Suite 1900, San Diego, California 92101-4495, USA
http://www.academicpress.com

Academic Press
84 Theobolds Road, London WC1X 8RR, UK
http://www.academicpress.com

International Standard Book Number: 0-12-366850-6

PRINTED IN THE UNITED STATES OF AMERICA
02 03 04 05 06 07 MB 9 8 7 6 5 4 3 2 1

DEDICATION

In 1979 there appeared in the journal *Diabetes*, a review entitled "Diabetic Neuropathy: New Concepts of its Etiology." The author was Rex S. Clements, Jr. Two years later I met Rex when he was on sabbatical leave at the University of Nottingham, where I was a junior academic. Rex suggested a collaboration on axonal transport in experimental diabetes. He taught me about diabetic rats, *myo*-inositol, the polyol pathway, and other things. I taught him how to knots (around nerves). It was a fair swap?

Rex's review gave force and direction to an approach to this problem, which demanded the integration of biochemistry, neurophysiology, and microanatomy with accurate clinical observation. This volume reviews much of the culmination of that force.

Rex died of cancer in May 1999 and we dedicate this book with fondness and respect to his memory.

David Tomlinson

CONTENTS

PART I

Primary Mechanisms

How Does Glucose Generate Oxidative Stress In Peripheral Nerve?

IRINA G. OBROSOVA

Glycation in Diabetic Neuropathy: Characteristics, Consequences, Causes, and Therapeutic Options

PAUL J. THORNALLEY

PART II

Secondary Changes

Protein Kinase C Changes in Diabetes: Is the Concept Relevant to Neuropathy?

JOSEPH EICHBERG

Are Mitogen-Activated Protein Kinases Glucose Transducers for Diabetic Neuropathies?

TERTIA D. PURVES AND DAVID R. TOMLINSON

Neurofilaments in Diabetic Neuropathy

PAUL FERNYHOUGH AND ROBERT E. SCHMIDT

Apoptosis in Diabetic Neuropathy

AVIVA TOLKOVSKY

Nerve and Ganglion Blood Flow in Diabetes:
An Appraisal

DOUGLAS W. ZOCHODNE

PART III

Manifestations

Potential Mechanisms of Neuropathic Pain in Diabetes

Nigel A. Calcutt

Electrophysiologic Measures of Diabetic Neuropathy: Mechanism and Meaning

Joseph C. Arezzo and Elena Zotova

Neuropathology and Pathogenesis of Diabetic Autonomic Neuropathy

Robert E. Schmidt

Role of the Schwann Cell in Diabetic Neuropathy

LUKE ECKERSLEY

PART IV

Potential Treatment

Polyol Pathway and Diabetic Peripheral Neuropathy

PETER J. OATES

Nerve Growth Factor for the Treatment of Diabetic Neuropathy: What Went Wrong, What Went Right, and What Does the Future Hold?

STUART C. APFEL

Angiotensin-Converting Enzyme Inhibitors: Are there Credible Mechanisms for Beneficial Effects in Diabetic Neuropathy?

RAYAZ A. MALIK AND DAVID R. TOMLINSON

Clinical Trials for Drugs Against Diabetic Neuropathy: Can We Combine Scientific Needs With Clinical Practicalities?

DAN ZIEGLER AND DIETER LUFT

CONTRIBUTORS

Numbers in parentheses indicate the pages on which the authors' contributions begin.

Stuart C. Apfel (393), Albert Einstein College of Medicine, Bronx, New York 10461

Joseph C. Arezzo (229), Albert Einstein College of Medicine, Bronx, New York 10461

Nigel A. Calcutt (205), Department of Pathology, University of California, San Diego, La Jolla, California 92093

Luke Eckersley (293), Neuroscience Division, University of Manchester, School of Biological Sciences, Manchester M13 9PT, United Kingdom

Joseph Eichberg (61), Department of Biology and Biochemistry, University of Houston, Houston, Texas 77204

Paul Fernyhough (115), School of Biological Sciences, University of Manchester, Manchester M13 9PT, United Kingdom

Dieter Luft (431), 4th Medical Department, Eberhard Karis University, 72076 Tübingen, Germany

Rayaz A. Malik (415), Department of Medicine, Manchester Royal Infirmary, Manchester M13 9WL, United Kingdom

Peter J. Oates (325), Department of Cardiovascular and Metabolic Diseases, Pfizer Global Research and Development, Groton, Connecticut 06340

Irina G. Obrosova (3), Department of Internal Medicine, Division of Endocrinology on Metabolism, University of Michigan Medical Center, Ann Arbor, Michigan 48109

Tertia D. Purves (83), Neuroscience Division, University of Manchester, School of Biological Sciences, Manchester M13 9PT, United Kingdom

Robert E. Schmidt (115, 257), Department of Pathology and Immunology, Washington University School of Medicine, St. Louis, Missouri 63110

Paul J. Thornalley (37), Department of Biological Sciences, University of Essex, Colchester, Essex CO4 3SQ, United Kingdom

Aviva Tolkovsky (145), Department of Biochemistry, University of Cambridge, Cambridge CB2 1QW, United Kingdom

David R. Tomlinson (83, 415), Neuroscience Division, University of Manchester, School of Biological Sciences, Manchester M13 9PT, United Kingdom

Dan Ziegler (431), German Diabetes Research Institute at the Heinrich Heine University, German Diabetes Clinic, 40225 Düsseldorf, Germany

Douglas W. Zochodne (161), Department of Clinical Neurosciences, University of Calgary, Calgary, Alberta, Canada T2N 4N1
Elena Zotova (229), Albert Einstein College of Medicine, Bronx, New York 10461

FOREWORD

This volume, entitled *Neurobiology of Diabetic Neuropathy*, reviews an evolving and complex, medically important neurological discipline marked by vigorous debate and healthy disagreement rather than by consensus; the volume might have just as appropriately been entitled "Neurodichotomies of Diabetic Neuropathy," because this is indeed a dichotomous discipline. Diabetic neuropathy severely burdens individuals suffering with or caring for type 1 or type 2 diabetes, yet proven and accepted specific disease-modifying treatment (save for improved metabolic control) is virtually nonexistent. The current scientific climate of this field is defined by ubiquitously positive therapeutic intervention studies in animals followed by dismal treatment failures in elaborate and expensive human clinical trials. The anatomically dispersed and cellularly heterogeneous peripheral nervous system comprising neuronal, glial, vascular, and connective tissues, leaves scientific opinion hotly divided as to the cellular and anatomical origins of diabetic neuropathy: Does the disorder arise peripherally and spread centrally, or visa versa; does it arise in the neural or vascular components of peripheral nerve? Although these arguments may be rendered intrinsically specious by the dispersed anatomy and interlocking cellular complexity of the peripheral nervous system, they nevertheless channel the quest for early biomarkers and eventual surrogates for disease prevention, stabilization, amelioration and/or cure. The absence of validated biomarkers and surrogates, combined with the relative inaccessibility of human nerve tissue, greatly complicates the direct testing of specific animal-derived hypotheses in the clinic short of full-scale clinical treatment trials.

On the positive side, two large government-supported clinical trials, the Diabetes Control and Complications Trial (DCCT) and the United Kingdom Prospective Diabetes Study (UKPDS), have clearly and unequivocally implicated the altered metabolic milieu in diabetic patients in development of clinically-detectable diabetic neuropathy. This conceptual framework should enable the rapidly advancing fields of neurobiology and molecular biology to define more focused and testable scientific hypotheses, whose proof-of-concept testing in patients should be facilitated by new noninvasive imaging technologies. The reader should view the current scientific debates encompassed within this volume in this context.

Part I of the volume explores plausible links between the altered metabolism of the diabetic milieu and the initiating biochemical insults within various components of peripheral nerve. This discussion focuses exclusively on hyperglycemia to the exclusion of accompanying lipid, electrolyte, and amino acid metabolic aberrations.

Dr. Obrosova elegantly reviews animal and *in vitro* studies that implicate reactive free radicals that accumulate consequent to enzymatic metabolism of excess glucose through the aldose reductase (AR) or "sorbitol" pathway (Fig. 1). A cascade of secondary biochemical phenomena generates excess free radicals and/or impairs oxidative defense mechanisms. She invokes an incompletely understood, relationship between increased glucose flux through AR and depletion of total (oxidized plus reduced) glutathione, a key buffer against free radical damage.

In contrast, Dr. Thornalley's chapter emphasizes nonenzymatic rather than enzymatic glucose metabolism to explain similar free radical-related phenomena, and additional structural modifications of long-lived proteins. By this construct, nonenzymatically-derived metabolites of glucose undergo transformation to "advanced glycosylation end-products" (AGE's), which form irreversible protein adducts and cross-links that impair structure and function of long-lived peripheral nerve proteins (e.g., myelin proteins, neurofilaments, and neurotubules). Glucose-derived AGE's also interact with specific cytokine-like "AGE" receptors (RAGE), thereby activating a pathological cascade of signaling pathway abnormalities, oxidative stress, and free radical formation, that plausibly culminate in neuronal apoptosis and/or vascular dysfunction and other functional defects in diabetic peripheral nerve.

Part II focuses on downstream pathophysiological steps triggered by these more proximal consequences of hyperglycemia. These "secondary" aberrations involve protein kinase C (PKC); the family of mitogen-activated protein (MAP) kinases, ERK, JNK, and P38; neurofilament phosphorylation and axonal transport; apoptosis; and nerve and ganglionic blood flow.

Dr. Eichberg reviews the dichotomous roles of PKC in neural and non-neural diabetic complications. Accumulation of diacylglycerol (DAG), the activator of "classic" PKC isoforms, and increased PKCβ activation are uniformly associated with non-neural microvascular and macrovascular complications of diabetes, whereas diabetic neuropathy is marked by diverse directionally and qualitatively different patterns of PKC perturbations. These patterns appear to differ between the neural and vascular components of peripheral nerve and between different neuronal sub-populations. Under varying experimental conditions, PKC exhibits dichotomous relationships with important neurophysiological mediators such as the activity of the (Na,K)-ATPase. Thus, the conventional wisdom linking DAG accumulation and PKCβ activation to non-neural complications may not apply directly to diabetic neuropathy.

In the next chapter, Drs. Purves and Tomlinson link hyperglycemia to a pattern of widespread activation of peripheral nerve ERK, JNK, and P38 kinases that normally play important roles in normal cell homeostasis, neurite outgrowth and stress-response. The authors link MAP kinase activation to nonenzymatic AGE formation, glucose-induced osmotic stress, mitochondrial, and cytosolic enzymatic pathways of glucose metabolism, oxygen and nitrogen free radical generation, and PKC activation. Glucose-induced MAP kinase activation may also function as a sensitizer to, rather than a consequence of, glucose-induced free radical toxicity. Glucose-induced ERK, JNK, and P38 kinase activation patterns may be sufficient to explain neuronal cell apoptosis in diabetic neuropathy.

Drs. Fernyhough and Schmidt implicate the neurofilament-(NF)-based axonal transport system unique to neurons as a key target for glucose toxicity in diabetic somatic and autonomic neuropathy. Glucose- and stress-mediated activation of MAP kinases and other kinases in peripheral nerve are thought to lead to aberrant NF phosphorylation, trafficking and function, leading to axonal dysfunction, dystrophy, atrophy and degeneration in diabetic nerve.

Dr. Tolkovsky reviews the current controversy regarding the possible role of components of the apoptotic pathway in diabetic peripheral neurons, including both the mitochondrial and Fas-activated pathways. She minimizes the prominence of classic apoptosis with attendant rapid clearing of damaged cells in diabetic neuropathy, but does not exclude the thesis of Eva Feldman and Philip Low and colleagues that a more stately neuronal degenerative process (''apoptosis lente'') involving elements of the apoptotic cascade (e.g., positive TUNEL staining, DNA fragmentation and caspase activation) may chronically damage axons in diabetes. She notes that diabetic mouse models deficient in components of the apoptotic pathway may help define this process more rigorously.

Dr. Zochodne's comprehensive review of conflicting experimental evidence and interpretation surrounding the impact of experimental diabetes on various measurements of peripheral nerve and ganglion blood flow clearly highlights and delineates the full scope of the problem. It would appear that these dichotomous results and interpretations might reflect technical considerations such as instrumentation of the nerve and exposure to temperatures lower than body temperature, and the measurement techniques themselves. Dr. Zochodne's thoughtful discussion defines the relevant physiological dimensions of the issue, and sets a standard for future progress in the field. These discussions define glucose-related pathogenetic pathways and their derivative physiologic and cellular defects that constitute potential disease-modifying therapeutic targets.

Part III extrapolates these neurobiological defects as the basis of the clinical manifestations of diabetic neuropathy. The disabling neuropathic pain, which afflicts about 10% of diabetic neuropaths, combines hyperalgesia and allodynia with objectively diminished peripheral sensory function. New experimental animal models lead Dr. Calcutt to invoke augmented aberrant spinal amplification secondary to diminished peripheral sensory input as a partial explanation of this paradox.

Dr. Arezzo elucidates the role of traditional and more modern computer-assisted electrophysiological techniques to characterize and quantify the structural and functional involvement of large and small peripheral nerve fibers in the peripheral injury accompanying human diabetic neuropathy, pointing out techniques such as distribution of velocities and area of depolarization to assess small-fiber integrity, and the use of repetitive stimulation to measure the status of axonal (Na,K)-ATPase (altered in animal models) in human diabetic neuropathy.

Dr. Schmidt identifies noradrenergic sympathetic neuroaxonal dystrophy and distal nerve terminal degeneration rather than neuronal loss as a hallmark of diabetic autonomic neuropathy, but considers the underlying toxic insult to be either multifactorial, or as yet unidentified.

Rather than implicating the Schwann cell in axonal dysfunction in diabetes, Dr. Eckersley's chapter delineates the normal developmental, molecular, cellular and neurophysiological interactions between axons and Schwann cells, and describes a series of diabetes-induced Schwann cell metabolic abnormalities without drawing any definitive causative conclusions.

The profound and durable effects of improved metabolic control on the development and progression of diabetic neuropathy have been established by the DCCT, the UKPDS, and the long-term follow-up of the DCCT cohort. However, specific mechanism-based disease modifying therapy for diabetic neuropathy remains an unfulfilled promise.

Dr. Oates' review encompasses a detailed analysis of the full breadth of animal and human experiments with pharmacological inhibitors of the enzymes of the polyol pathway, aldose reductase and sorbitol dehydrogenase. He implicates metabolic flux through the pathway rather than accumulation of pathway intermediates as an important factor in the pathogenesis of diabetic neuropathy, and concludes that at least 10-fold more potent inhibitors suitable for human use will be required to confirm this hypothesis and provide therapeutic benefit.

Dr. Apfel reviews the evidence implicating reduced neurotrophism in the pathogenesis of diabetic neuropathy in animals, and the conflicting results of two positive "phase II" clinical trials in diabetic neuropathy and HIV-associated neuropathy, followed by a negative larger "phase III" trial in diabetic neuropathy. He attributes the discrepant results to either the intervention vs prevention paradigm in human trials, changes in formulation of recombinant NGF between the phase II and phase III studies, unmasking in the smaller "phase II" trials, or insufficient dose of NGF.

Drs. Malik and Tomlinson review a small clinical trial demonstrating improved nerve function in diabetic neuropathic patients treated with an angiotensin converting enzyme (ACE) inhibitor. Their review of possible underlying mechanisms includes vasodilatation and improved nerve blood flow, and theoretical non-vascular mechanisms possibly involving Schwann cell and neuronal angiotensin II receptor expression, or effects related to bradykinin. The paradoxical lack of effect of ACE inhibitor treatment on neuropathy in the large Appropriate Blood Pressure Control in Diabetes [ABCD] trial, presents yet another dichotomy.

Drs. Ziegler and Luft review the history of randomized controlled clinical trials in diabetic neuropathy, and the striking limitation in their consistency and interpretability. They underline the extreme rigor necessary to document therapeutic disease-modifying benefit in patients with clinically manifest diabetic neuropathy, but do not comment on well-validated clinical methods to demonstrate prevention of neuropathy in non-neuropathic study subjects (e.g., DCCT, and the long-term follow-up of the DCCT cohort in the Epidemiology of Diabetic Complications Trial).

Thus, this volume comprehensively reviews and updates the underlying basic neuroscience of diabetic neuropathy, its clinical correlates, and the current status of disease-modifying therapeutics for diabetic neuropathy, effectively delineating the present state and present gaps in our knowledge. The value of the volume is primarily derived not from the conclusions it draws, but from the conclusions it cannot draw. It serves to focus and challenge scientists in the field to vigorously pursue the definitive therapeutic targets that will inevitably eliminate disabling peripheral neuropathy, neuropathic foot ulceration and amputation, from the specter of diabetes.

Douglas A. Greene
Merck Research Laboratories
Rahway, New Jersey

PREFACE

This book is unique and the title is not an affectation. In this volume you will not find clinical descriptions of neuropathies, accounts of staging, or epidemiology. Such matters are found elsewhere. This is a book by scientists, written for scientists and clinicians who are prepared to address the scientific aspects of their specialty. Most of the authors are active in aspects of neuroscience that are broader than diabetes-derived problems, indeed some do not work on diabetic neuropathy at all. Therefore, they bring knowledge and perspectives from these other fields, which is the only way that this clinical problem can be beaten.

The book is ordered to examine by stages the manifestations of disordered plasma glucose regulation on the peripheral nervous system. The first two chapters examine the immediate disturbances that follow from persistently raised glucose. Consideration of the polyol pathway could, of course, have appeared in this section of the book also, but it was decided to place this alongside consideration of potential treatments toward the end of the book, but it does not matter because you are all clever people who can move around a book with ease.

What then follows is a section on the next stages of biochemical and physiological transduction of the damage pathways. I believe that the critical stage in the development of diabetes complications occurs when reversible biochemical derangements bring about changes in cellular phenotype. This is as true in peripheral nerves as it is in the retina and the kidney. Once these altered phenotypes become established, they provoke changes that become progressively less reversible and, hence, a chronic disease becomes established. These five chapters, grouped together as ''Secondary Changes,'' consider aspects of this process.

We then reach the stage where secondary and tertiary changes manifest themselves as signs and symptoms. The following four chapters consider aspects of this, concentrating on mechanistics of pain and conduction changes, plus an attempt to rationalize autonomic neuropathy—the Cinderella of the diabetic neuropathies—perhaps an inappropriate metaphor given that autonomic degeneration does indeed get to the balls.

Finally, we explore hope and despair by considering treatments. Inevitably, aldose reductase inhibitors are covered in great depth. This is entirely appropriate, even though their success has been limited or, as some would say, nonexistent. Nevertheless, the development of these agents has provided incentive and much-needed funds for a substantial proportion of the work in this field over the past 20 years. These agents have also taught some lessons. Paramount among these is the clear and solid fact that we do not yet know how to subject potential new drugs for diabetic neuropathy to clinical trials. Thus, we explore the lessons from the nerve growth factor trial and, in the final chapter, consider how the process might be rationalized, given enough patience and enough money.

This book was written very briskly by a keen band of authors, which shows in the freshness of the approach. Some of these chapter topics have not yet appeared as reviews in journals. The text will not go out of date quickly because the topics themselves form essential substrate for future work, and the book, therefore, serves as a perfect primer for researchers and those seeking to learn from them.

David Tomlinson

ACKNOWLEDGMENTS

This book would not have been written but for the efforts of our students and postdocs and the funding from the grant bodies and the pharmaceutical industry. It is, therefore, appropriate to acknowledge these anonymous sources of inspiration and substrate. There are also authors who have had a profound impact on this field, who could have been contributors, but are not because one simply cannot ask everybody. In particular we must acknowledge Peter Dyck and P.K. Thomas, virtually the fathers of this subject clinically; Doug Greene and Phil Low, who set a pattern for related clinical and laboratory exploration; and Anders Sima for the application of neuropathological techniques to clinical and animal material. In industry, Pik Dvornik can be considered to be the father of aldose reductase inhibitors and, as such, sired an enormous body of work.

To these bodies, and these and other nameless individuals, we give thanks.

PART I
PRIMARY MECHANISMS

HOW DOES GLUCOSE GENERATE OXIDATIVE STRESS IN PERIPHERAL NERVE?

Irina G. Obrosova

Department of Internal Medicine, Division of Endocrinology and Metabolism
University of Michigan Medical Center, Ann Arbor, Michigan 48109

Diabetes-associated oxidative stress is clearly manifest in peripheral nerve, dorsal root, and sympathetic ganglia of the peripheral nervous system and endothelial cells and is implicated in nerve blood flow and conduction deficits, impaired neurotrophic support, changes in signal transduction and metabolism, and morphological abnormalities characteristic of peripheral diabetic neuropathy (diabetic peripheral neuropathy). Hyperglycemia has a key role in oxidative stress in diabetic nerve, whereas the contribution of other factors, such as endoneurial hypoxia, transition metal imbalance, and hyperlipidemia, has not been rigorously proven. It has been suggested that oxidative stress, particularly mitochondrial superoxide production, is responsible for sorbitol pathway hyperactivity, nonenzymatic glycation/glycooxidation, and activation of protein kinase C. However, this concept is not supported by *in vivo* studies demonstrating the lack of any inhibition of the sorbitol pathway activity in peripheral nerve, retina, and lens by antioxidants, including potent super-oxide scavengers. It has been also hypothesized that aldose reductase (AR) detoxifies lipid peroxidation products, and therefore, the enzyme inhibition in diabetes is detrimental rather than beneficial. However, the role for AR in lipid peroxidation product metabolism has never been demonstrated *in vivo*, and the effects of aldose reductase inhibitors and antioxidants on diabetic peripheral neuropathy are unidirectional, i.e., both classes of agents prevent and correct functional, metabolic, neurotrophic, and morphological changes in diabetic nerve. Growing evidence indicates that AR has a key role in oxidative stress in the peripheral nerve and contributes to superoxide production by the vascular endothelium. The potential mechanisms of this phenomenon are discussed. © 2002, Elsevier Science (USA).

I. Manifestations of Diabetes-Associated Oxidative Stress in the Peripheral Nervous System (PNS)

Diabetic distal symmetric sensorimotor polyneuropathy, the most common peripheral neuropathy in developed countries, affects up to 60–70% of diabetic patients (National Institutes of Diabetes and Digestive and Kidney Diseases, 1995) and is the leading cause of foot amputation (Brand, 1982). Improved blood glucose control substantially reduces the risk of developing diabetic polyneuropathy in insulin-dependent (type 1) (Diabetes Control and Complications Trial Research Group, 1993) and noninsulin-dependent (type 2) diabetes (UK Prospective Diabetes Study Group, 1998), thereby strongly implicating hyperglycemia as a causative factor. One of the important consequences of chronic hyperglycemia in PNS and *vasa nervorum* is enhanced oxidative stress resulting from an imbalance between the production and the neutralization of reactive oxygen species (ROS). These comprise highly reactive hydroxyl radicals, as well as superoxide anion and peroxyl radicals, singlet oxygen, peroxynitrite, and hydrogen peroxide. Reliable, sensitive, and specific techniques for the detection of free radicals in biological tissues are either unavailable or under development and are rather difficult to perform (Luo and Lehotay, 1997; Acworth *et al.*, 1999). The "footprints" of diabetes-associated free radical injury are of two types: (1) accumulation of lipid peroxidation products, malondialdehyde (Lowitt *et al.*, 1995) and 4-hydroxyalkenals (Obrosova *et al.*, 2000d) and conjugated dienes (Kihara *et al.*, 1991; Low and Nickander, 1991), and (2) disruption of the antioxidative defense mechanisms, as listed, starting with depletion of nonenzymatic antioxidants.

1. Depletion of reduced glutathione (GSH) (Nagamatsu *et al.*, 1995; Obrosova *et al.*, 1999b).
2. Depletion of ascorbate (Obrosova *et al.*, 2002a).
3. Depletion of taurine (Pop-Busui *et al.*, 2001; Obrosova *et al.*, 2002a).
4. Increase in oxidized glutathione (GSSG)/GSH ratio (Nagamatsu *et al.*, 1995; Stevens *et al.*, 2000).
5. Increased dehydroascorbate/ascorbate (DHAA/AA) ratio (Obrosova *et al.*, 2002a).
6. Downregulation of superoxide dismutase (SOD) (Low and Nickander, 1991; Obrosova *et al.*, 2000c; Stevens *et al.*, 2000).
7. Impaired activity of catalase (Obrosova *et al.*, 2000c; Stevens *et al.*, 2000).
8. Impaired activity of quinone reductase (Obrosova *et al.*, 2000c).

All of these phenomena have been reported in diabetes in peripheral nerve, dorsal root, and sympathetic ganglia of the PNS, as well as endothelial cells (Giardino *et al.*, 1996; Paget *et al.*, 1998). Glutathione peroxidase was found downregulated in the diabetic mouse (Romero *et al.*, 1999) and tended

TABLE I

MANIFESTATIONS OF DIABETES-INDUCED OXIDATIVE STRESS IN PERIPHERAL NERVE[a]

Index	Source
Lipid peroxidation	
Malondialdehyde — no change (3 weeks)	Obrosova *et al.* (2000d)
↑ Malondialdehyde + 4-hydroxyalkenals (3 and 6 weeks)	Obrosova *et al.* (2000c,d, 2002a)
Nonenzymatic antioxidants	
↓GSH (3 and 6 weeks)	Obrosova *et al.* (1999b, 2000c, 2002a); Stevens *et al.* (2000)
GSSG — no change (6 weeks)	Obrosova *et al.* (2002a)
↑GSSG/GSH (6 weeks)	Stevens *et al.* (2000)
↓Total AA (6 weeks)	Obrosova *et al.* (2002a)
↓Free ascorbate (6 weeks)	Obrosova *et al.* (2002a)
DHAA — no change (6 weeks)	Obrosova *et al.* (2002a)
↑DHAA/AA	Obrosova *et al.* (2002a)
↓Taurine (3 and 6 weeks)	Pop-Busui *et al.* (2001); Obrosova *et al.* (2002a)
Antioxidative enzymes	
↓Superoxide dismutase (6 weeks)[b]	Obrosova *et al.* (2000c); Stevens *et al.* (2000)
↓Catalase (6 weeks)[b]	Stevens *et al.* (2000)
↓Quinone reductase (6 weeks)[b]	Stevens *et al.* (2000)

[a]Findings of our group only. The duration of diabetes is given in parentheses.
[b]None of the listed antioxidative defense enzymes demonstrated any downregulation in rats with a 3 week duration of diabetes.

to decrease in diabetic rat nerve, but the difference between diabetic and nondiabetic groups did not achieve statistical significance (9.8 ± 4.6 vs 13.7 ± 3.6 nmol/min mg protein in controls; Obrosova *et al.*, 2000c). Diabetes-induced changes in lipid peroxidation, GSH, AA, taurine, and the glutathione and ascorbate redox states, as well as antioxidative defense enzyme activities identified in the studies of our laboratory (Obrosova *et al.*, 1999b, 2000c,d, 2002a; Stevens *et al.*, 2000), are summarized in Table I.

II. Role for Oxidative Stress in Peripheral Diabetic Neuropathy

Numerous reports implicate ROS in the pathogenesis of diabetic peripheral neuropathy (Fig. 1). The role for oxidative stress in diabetes-induced neurovascular dysfunction and nerve conduction deficits has been demonstrated in studies of (1) glutathione (Bravenboer *et al.*, 1992) and the precursor for glutathione biosynthesis, *N*-acetyl-L-cysteine (Love *et al.*, 1996); (2) lipid-soluble antioxidants, probucol (Cameron *et al.*, 1994a; Karasu *et al.*, 1995), butylated hydroxytoluene (Cameron *et al.*, 1993), and vitamin E (Karasu *et al.*, 1995; Love *et al.*, 1996); (3) metal chelators, deferoxamine, trientine, and extracellular high molecular weight

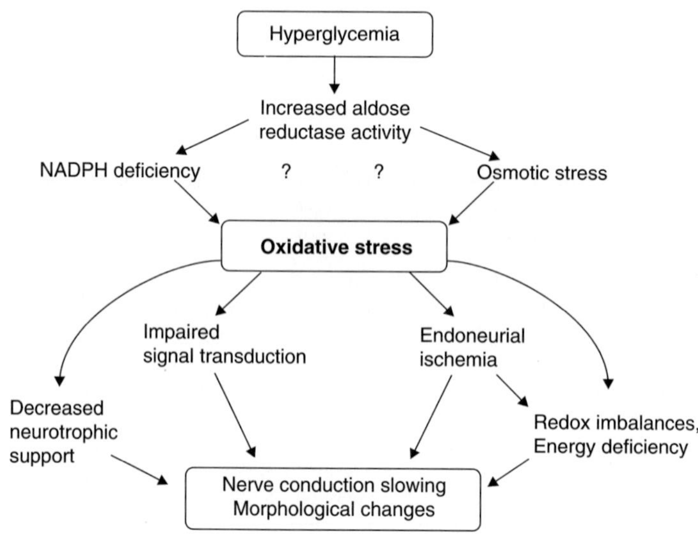

FIG. 1. Role for hyperglycemia-induced oxidative stress in the pathogenesis of peripheral diabetic neuropathy.

hydroxyethyl starch-deferoxamine (Cameron and Cotter, 1995; Love *et al.*, 1996; Cameron and Cotter, 2001); (4) prooxidant primaquine (Cameron *et al.*, 1994a; Hounsom *et al.*, 2001); and (5) the potent "universal" antioxidant DL-α-lipoic acid (Nagamatsu *et al.*, 1995; Low *et al.*, 1997; Cameron *et al.*, 1998; Obrosova *et al.*, 2000c; Stevens *et al.*, 2000), which combines free radical scavenging and metal chelator properties with an ability (after conversion to dihydrolipoic acid) to regenerate levels of other nonenzymatic (GSH, ascorbate, α-tocopherol) and enzymatic (catalase, glutathione peroxidase) antioxidants (Maitra *et al.*, 1995; Packer *et al.*, 1995). Oxidative stress has an important role in the diabetes-induced impairment of neurotrophic support (Garrett *et al.*, 1997; Tomlinson *et al.*, 1996, 1997; Hounsom *et al.*, 1998, 2001), which is closely associated with Schwann cell injury (Kalichman *et al.*, 1998; Mizisin *et al.*, 1998). Hounsom *et al.*(2001), employing (1) γ–linolenic acid and α-lipoic acid diester for the correction of enhanced oxidative stress in streptozotocin-diabetic rats and (2) prooxidant primaquine or a vitamin E-deficient diet for modeling ROS-induced breakdown of the neuronal phenotype in control rats, have generated compelling evidence of the important role of oxidative stress in diabetes–associated deficits of nerve growth factor (NGF) and NGF-regulated peptides, i.e., substance P and neuropeptide Y. The conclusions of the aforementioned study are supported by experiments from our laboratory (Obrosova *et al.*, 2002a; Fig. 2) demonstrating a partial prevention of the diabetes-induced NGF deficit in the peripheral nerve of streptozotocin-diabetic rats by dietary 1% taurine supplementation, i.e., the treatment that partially arrested diabetes-induced lipid peroxidation. Numerous studies indicate that ROS are powerful activators of three subfamilies of mitogen-activated protein kinases (MAPKs): stress-activated protein kinase/c-Jun-terminal kinases (SAPKs/JNKs), extracellularly responsive kinases (ERKs), and p38-MAPK (Clerk *et al.*, 1998; Maulik *et al.*, 1998; Adler *et al.*, 1999), glucose transducers for diabetic complications (Tomlinson, 1999), which have been implicated in axonopathy (Fernyhough *et al.*, 1999) and neuropathic pain (Calcutt *et al.*, 2000). Studies in both diabetic and nondiabetic models of oxidative stress have revealed that, in addition to MAPKs, oxidative stress affects multiple signal transduction pathways, such as the arachidonic acid cascade (Whisler *et al.*, 1994; Jennings, 2000), phosphoinositide (Li *et al.*, 1998; Servitja *et al.*, 2000; Halstead *et al.*, 2001), and Ca^{2+} signaling (Goldhaber and Qayyum, 2000; Lounsbury *et al.*, 2000; Okabe *et al.*, 2000), as well as neurotransmission (Langeveld *et al.*, 1995). Hydroxyl- and superoxide anion radical and peroxynitrite-induced DNA single strand breakage activates poly(ADP-ribosyl)ation (Soldatenkov and Smulson, 2000), which in turn leads to NAD depletion and energy failure (Schraufstatter *et al.*, 1986; Thies and Autor, 1991; Plaschke *et al.*, 2000;

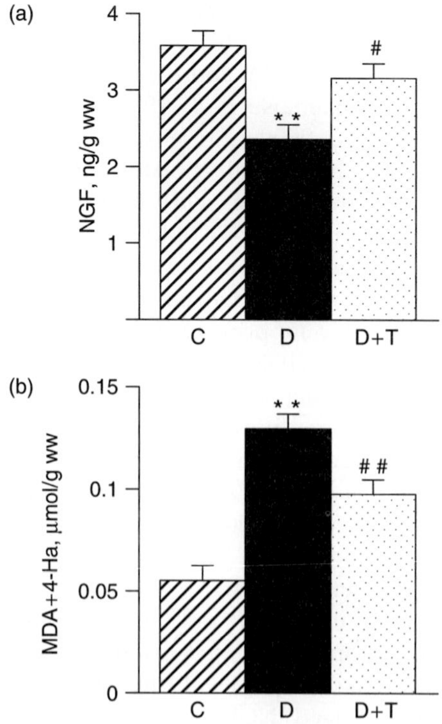

FIG. 2. Effect of dietary 1% taurine supplementation on NGF (a) and malondialdehyde + 4-hydroxyalkenal concentrations (b) in the sciatic nerve of rats with a 6-week duration of strepotozotocin diabetes (mean ±SEM, $n = 7$–10). From Obrosova *et al.* (2002a).

Soriano *et al.*, 2001), alters gene expression (Soldatenkov and Smulson, 2000), and is essential for the execution of apoptosis (Simbulan-Rosenthal *et al.*, 1998). Poly(ADP-ribose)synthetase (PARS) inhibition protects against oxidant-induced apoptosis in retinal pericytes (Shojanee *et al.*, 1999), and no apoptosis was detected in cells from PARS knockout ($-/-$) mice subjected to anti-Fas treatment (Simbulan-Rosenthal *et al.*, 1998). PARS has been identified in Schwann cells of the PNS (Berciano *et al.*, 1999), as well as in endothelial cells (Walisser *et al.*, 1999; Cuzzocrea *et al.*, 2000), and its role in diabetes-induced oxidative injury in peripheral nerve and *vasa nervorum* still needs to be explored. Finally, oxidative stress has been implicated in myelinated fiber atrophy and other morphological changes characteristic for advanced diabetic peripheral neuropathy (Sagara *et al.*, 1996).

III. Origin of Diabetes-Induced Oxidative Stress in PNS

Numerous findings indicate that hyperglycemia is a key causative factor in oxidative stress in tissue sites for diabetic complications. Some investigators (Nagamatsu *et al.*, 1996; Nickander *et al.*, 1996; Low *et al.*, 1997; Sasaki *et al.*, 1997) suggest that diabetes-associated oxidative stress in the PNS has a composite origin; however, specific studies are needed to sort out the role of potential contributing factors independent from and additive to hyperglycemia, such as endoneurial hypoxia, hyperlipidemia, and increased free fatty acid abundance, as well as transition metal imbalance (Qian *et al.*, 1998; Qian and Eaton, 2000; Cameron *et al.*, 2001; Cameron and Cotter, 2001). It is important to note that diabetes-induced lipid peroxidation is completely arrested by an aldose reductase inhibitor, presumably as a result of correction of exaggerated flux through the sorbitol pathway in some compartment of peripheral nerve (Obrosova *et al.*, 2002; Fig. 3). There are those who suggest that all effects of aldose reductase inhibitors are secondary to correction of impaired endoneurial blood flow (Cameron *et al.*, 1994b), although others contest this assertion (Tomlinson *et al.*, 1998). However, the fact that diabetes-induced lipid peroxidation is unaffected by the α_1-adrenoceptor antagonist prazosin (Obrosova *et al.*, 2000d; Fig. 4), which normalizes nerve blood flow without affecting biochemical parameters (Cameron *et al.*, 1991), demonstrates that the role of endoneurial hypoxia in diabetes-associated oxidative stress in PNS is fairly minor. This conclusion is supported by two other studies (Cameron *et al.*, 1999; Hohman *et al.*, 2000) indicating that the protein kinase C (PKC) inhibitor and the vasodilators, ATP-sensitive (K^+) channel openers, celikalim, and WAY135201 correct nerve blood flow, but not the peripheral nerve GSH deficit in diabetic rats. It should be noted that severe hypoxia creates enhances lipid peroxidation in PNS (Nagamatsu *et al.*, 1996), as it does in other tissues (Yoshida *et al.*, 2000; Mackenson *et al.*, 2001). The aforementioned study of Nagamatsu *et al.*, (1996), however, was performed in the model of ischemia caused by ligation of the supplying arteries to the sciatic-tibial nerve. Such an approach creates a far more profound hypoxia than the one present in streptozotocin-diabetic rats in which mean endoneurial oxygen tensions are only 40–50% lower than in an equivalent nondiabetic (normoxic) group (Tuck *et al.*, 1984; Cameron *et al.*, 1994b; Cameron and Cotter, 1994). The role for hyperlipidemia and free fatty acid oxidation in diabetes-associated oxidative stress still remains to be explored. Increased free fatty acid concentrations produce oxidative stress in endothelial cells through the protein kinase C-dependent activation of NADH oxidase (Inoguchi *et al.*, 2000). The transition metal imbalance and associated glucose autoxidation have important roles in oxidative stress in *vasa nervorum* (Love *et al.*,

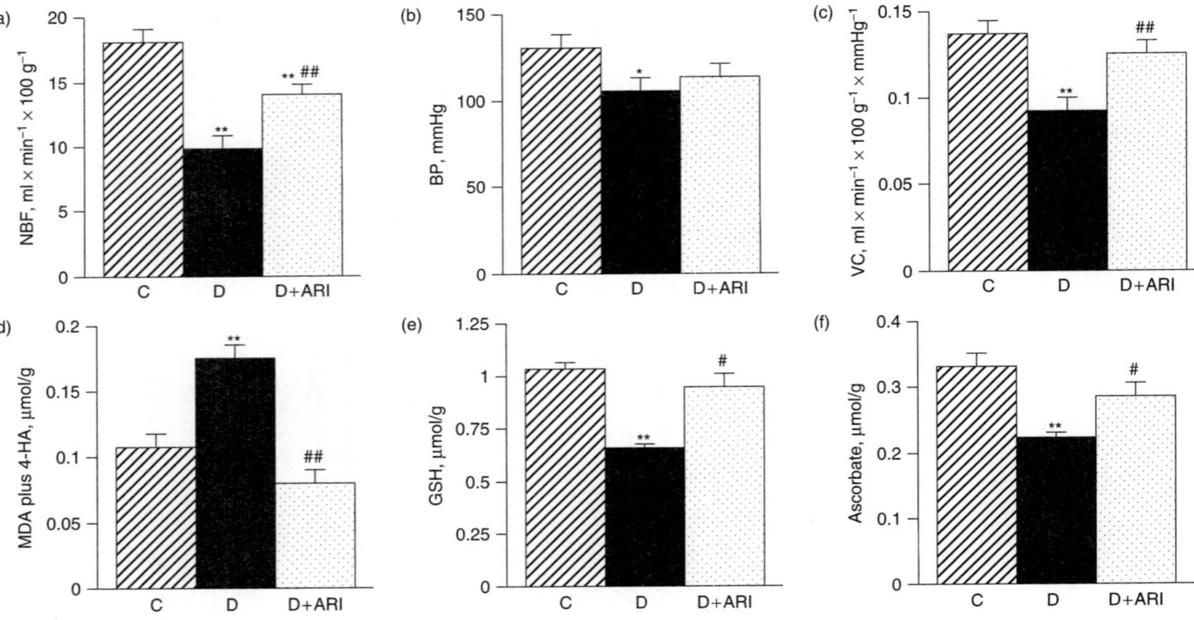

FIG. 3. Effect of the ARI sorbinil ($65 \text{ mg/kg}^{-1} \text{ day}^{-1}$ in the diet for 2 weeks after 4 weeks of untreated diabetes) on neurovascular dysfunction — (a) NBF, sciatic endoneurial nutritive blood flow; (b) BP, blood pressure; and (c) VC, endoneurial vascular conductance (mean $\pm$ SEM, $n = 10$–12) — and on variables of lipid peroxidation and antioxidative defense in the sciatic nerve — (d) MDA + 4-HA, lipid peroxidation products; (e) GSH, reduced glutathione; and (f) Total ascorbate (mean $\pm$ SEM, $n = 6$–10). C, controls; D, untreated diabetic group; D + ARI, diabetic group treated with ARI. Significantly different vs controls ($p < 0.05^*$ and $<0.01^{**}$). Significantly different vs untreated diabetic group ($p < 0.05^\#$ and $< 0.01^{\#\#}$).

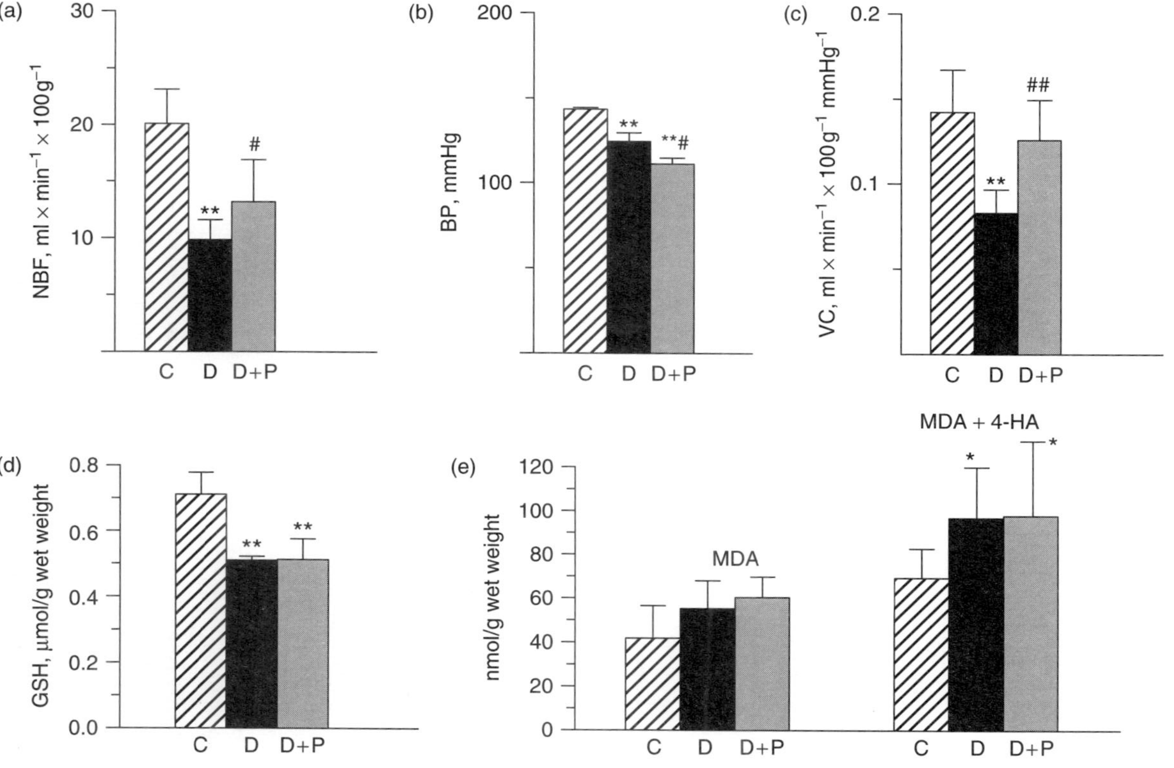

Fig. 4. Effect of the vasodilator prazosin ($5 \text{ mg/kg}^{-1} \text{ day}^{-1}$ in the drinking water for 3 weeks) on neurovascular dysfunction — (a) NBF, sciatic endoneurial nutritive blood flow; (b) BP, blood pressure; and (c) VC, endoneurial vascular conductance (mean $\pm$ SD, $n = 12$–15) — and on variables of lipid peroxidation and antioxidative defense in the sciatic nerve — (d) GSH, reduced glutathione; and (e) lipid peroxidation products (mean $\pm$ D, $n = 12$–15). C, controls; D, untreated diabetic group; D + P, diabetic group treated with prazosin. Significantly different vs controls ($p < 0.05*$ and $< 0.01**$). Significantly different vs untreated diabetic group ($p < 0.05^{\#}$ and $< 0.01^{\#\#}$).

11

1996; Cameron and Cotter, 2001), whereas their importance in neural components of PNS has not been elucidated. It should be noted that the relation among oxidative stress and a number of diabetes-associated signal transduction, metabolic, and neurotrophic imbalances in PNS is extremely complex. For example, NGF contributes to the neutralization of superoxide anion radicals and hydrogen peroxide by inducing expression of the superoxide dismutase and catalase genes (Mattson *et al.*, 1995; Li *et al.*, 1998). Therefore, the ROS-induced deficiency of the nerve NGF concentration in diabetes may, in turn, further disrupt the antioxidative defense. It has been reported that NGF deprivation almost doubles free radical abundance in sympathetic neurons (Nair *et al.*, 2000). In a similar fashion, MAPK activation and poly(ADP-tibosyl)ation, originally identified as consequences but not the cause of oxidative injury (Clerk *et al.*, 1998; Maulik *et al.*, 1998; Adler *et al.*, 1999; Szabo, 1998; Soldatenkov and Smulson, 2000), have been found to exacerbate ROS production (Myhre and Fonnum, 2001) and lipid peroxidation (Zingarelli *et al.*, 1999).

IV. Interactions between Oxidative Stress and Other Hyperglycemia-Initiated Factors in Pathogenesis of Diabetic Peripheral Neuropathy

The continuing debate about a "primary mechanism" of diabetic complications has centered around oxidative stress and its relationship with other hyperglycemia-initiated factors (Tomlinson *et al.*, 1999). The effects of antioxidants and metal chelators (Karasu *et al.*, 1995; Nagamatsu *et al.*, 1995; Love *et al.*, 1996; Sagara *et al.*, 1996; Low *et al.*, 1997; Tomlinson, 1999; Stevens *et al.*, 2000; Cameron and Cotter, 2001; Coppey *et al.*, 2001; Obrosova *et al.*, 2002a) on indices of diabetic peripheral neuropathy are unidirectional with those of inhibitors of aldose reductase (Tomlinson *et al.*, 1982, 1998; Hotta *et al.*, 1986; Diemel *et al.*, 1992; Calcutt *et al.*, 1994; Cameron *et al.*, 1997; Ohi *et al.*, 1998; Mizuno *et al.*, 1999; Oates and Mylari, 1999; Kato *et al.*, 2000), nonenzymatic glycation (Kihara *et al.*, 1991; Cameron and Cotter, 1996; Dewhurst *et al.*, 1997), and protein kinase C (Cameron *et al.*, 1999; Nakamura *et al.*, 1999), which implies that the major pathways implicated in diabetic complications, including diabetic peripheral neuropathy, are interrelated. It has been suggested that increased sorbitol pathway activity, glycation/glycoxidation, and PKC activation originate from oxidative stress and, in particular, production of superoxide anion radicals in mitochondria (Nishikawa *et al.*, 2000). However, this concept, at least the part related to the sorbitol pathway, is not supported by experimental studies demonstrating the absence of any suppression of diabetes-induced sorbitol pathway hyperactivity activity in the peripheral nerve (Love *et al.*, 1996; Kishi *et al.*, 1999; Obrosova *et al.* 2000c; Stevens *et al.*, 2000), as well as lens (Obrosova *et al.*, 1998; Lee and Chung, 1999; Obrosova and Stevens, 1999) and retina (Obrosova *et al.*, 2001) by antioxidants, e.a., those neutralizing superoxide anion radicals [DL-α-lipoic acid (Packer *et al.*, 1995), taurine(Kilic *et al.*, 1999) and probucol (Ito *et al.*, 1998)]. Furthermore, the aforementioned premise does not explain why interventions with *totally different* pharmacological agents, i.e., inhibitors of AR, glycation, and PKC, effectively prevent or reverse diabetic complications, including diabetic peripheral neuropathy. Based on evidence of contribution of increased AR activity (Lou *et al.*, 1988; Lowitt *et al.*, 1995; Hohman *et al.*, 1997; Lee and Chung, 1999; Obrosova and Fathallah, 2000; Obrosova *et al.*, 2000c; Gupta *et al.*, 2002), the Maillard reaction (Yim *et al.*, 1995), the interaction of advanced glycation end products (AGE) with their receptors (Yan *et al.*, 1994), and, recently, PKC activation (Inoguchi *et al.*, 2000) to hyperglycemia-induced oxidative injury, it would be more logical to assume that the common component for the pathways leading to diabetic complications, i.e., oxidative stress, is localized *downstream* from the primary hyperglycemia-initiated mechanism(s). The input of the aforementioned mechanism(s) to free radical damage varies for different tissue sites and stages of diabetic complications. The absence of AGE accumulation in the peripheral nerve (Ryle *et al.*, 1995) and the failure of aminoguanidine to counteract *nerve* oxidative stress in short-term diabetes (Kihara *et al.*, 1991) suggest that advanced glycation plays no role in nerve oxidative injury in early DN. The contribution of PKC is unclear considering that the enzyme activity has been reported downregulated (Kim *et al.*, 1991), unchanged (Cameron *et al.*, 1999; Nakamura *et al.*, 1999), and upregulated (Kishi *et al.*, 1999) in the peripheral nerve in early diabetes, and no measurements of oxidative stress other than GSH (Cameron *et al.*, 1999) has been performed in PKC inhibitor-treated rats. Based on endothelial cell culture studies (Giardino *et al.*, 1996; Inoguchi *et al.*, 2000), however, one could assume that inhibitors of both nonenzymatic glycation and PKC counteract diabetes-associated oxidative stress in *vasa nervorum*. A number of findings (Lowitt *et al.*, 1995; Berti-Mattera *et al.*, 1996; Hohman *et al.*, 1997; Ishii *et al.*, 1998; Kuruvilla and Eichberg, 1998; Zatechka *et al.*, 2000; Gupta *et al.*, 2002), including those from our group (Obrosova *et al..*, 2000c; 2002b), suggest that increased AR activity has a key primary role in all signal transduction and metabolic changes that lead to diabetic complications, including oxidative stress.

V. Role for Aldose Reductase (AR) in Diabetes-Induced Oxidative Stress in Peripheral Nerve and Endothelium

A. AR Inhibitor Treatment: Beneficial or Detrimental?

Numerous studies of leading experimental groups have demonstrated that diabetes-induced peripheral nerve conduction deficits (Tomlinson *et al.*, 1982; Hotta *et al.*, 1986; Calcutt *et al.*, 1994; Cameron *et al.*, 1994, 1997; Raccah *et al.*, 1998), metabolic imbalances (Berti-Mattera *et al.*, 1996; Kuruvilla and Eichberg, 1998; Sima and Sugimoto, 1999), neurotrophic changes (Diemel *et al.*, 1992; Mizisin *et al.*, 1997; Ohi *et al.*, 1998), and morphological abnormalities (Yasuda *et al.*, 1989; Kamijo *et al.*, 1994; Kato *et al.*, 2000) of diabetic peripheral neuropathy are prevented or reversed by structurally diverse aldose reductase inhibitors. This implicates increased activity of the sorbitol pathway of glucose metabolism in the pathogenesis of diabetic neuropathy (DN). The role for AR in the pathogenesis of peripheral DN is supported by at least four other lines of evidence, i.e., (1) similarity of a number of functional, metabolic, and morphological abnormalities in animal models of diabetes and galactose feeding (Cameron *et al.*, 1992; Kamijo *et al.*, 1994; Kalichman *et al.*, 1998); (2) potentiation of galactose-induced neuropathy in transgenic mice expressing human AR (Yagihashi *et al.*, 1996) and the absence of functional deficits of DN in AR knockout (AR $-/-$) mice (Ho *et al.*, 2000); (3) identification of a high AR protein level as an independent risk factor for DN in patients with both type 1 (insulin-dependent) and type 2 (noninsulin-dependent) diabetes mellitus (Ito *et al.*, 1997); and (4) finding of a 30.2% increase in the frequency of the Z-2 allele of the AR gene, known to be associated with a two- to three fold AR expression in patients with DN compared with neuropathy-free diabetics (Heesom *et al.*, 1998). Despite this compelling evidence, two groups (Rittner *et al.*, 1999; and M. Brownlee, presentation at the Intracellular Oxidative Stress Symposium, the Annual Meeting of the American Diabetes Association, Philadelphia, PA, 2001) have hypothesized that the *key physiological role* of aldose reductase is the detoxification of lipid peroxidation products, and, therefore, the effects of aldose reductase inhibitors on diabetic complications should be detrimental rather than beneficial. However, this premise is not supported by a study on aldose reductase inhibitor-treated nondiabetic animals that did not reveal any appearance of oxidative stress (Hohman *et al.*, 1997). Furthermore, the pattern of neuronal dysfunction induced in rats by diabetes closely resembles that induced by sustained oxidative stress, but not AR inhibition, in nondiabetic rats (Tomlinson *et al.*, 1994; Cameron *et al.*, 1994, 1997, 1998; Nakamura *et al.*, 1999; Hounsom *et al.*, 2001). Indeed, the effects of aldose reductase inhibitors and antioxidants in diabetic animal models are unidirectional in that both classes of agents prevent or delay the development of diabetic complications, including diabetic peripheral neuropathy (Tomlinson, 1994).

The ability of AR to metabolize 4-hydroxynonenal has been demonstrated in a number of *in vitro* studies (Srivastava *et al.*, 1995; Vander Jagt *et al.*, 1995; Del Corso *et al.*, 1998; He *et al.*, 1998). Furthermore, the products of 4-hydroxynonenal metabolism by AR, i.e., glutathione-4-hydroxynonenal conjugate, 1,4-dihydroxy-2-nonene, and 4-hydroxy-2-nonenoic acid, have also been identified in organs perfused *in situ* (Siems *et al.*, 1995; Grune *et al.*, 1997; Srivastava *et al.*, 1998) or in tissues (Srivastava *et al.*, 1998) and red blood cells (Srivastava *et al.*, 2000) incubated *in vitro*. All the aforementioned studies employed additions of relatively high concentrations of 4-hydroxynonenal, ($100-200\ \mu M$). However, despite a continuing 6-year debate about the potential involvement of AR in the metabolism of α,β-unsaturated aldehydes, e.a., in diabetes (Vander Jagt *et al.*, 1995; Ansari *et al.*, 1996; Rittner *et al.*, 1999), no evidence of elevated concentrations of the products of 4-hydroxynonenal metabolism by AR in tissue sites for diabetic complications and, furthermore of the presence of those products *in vivo* has been generated.

At the same time, it is well known that AR inhibition counteracts, rather than exacerbates, hyperglycemia-induced oxidative stress in the peripheral nerve (Lowitt *et al.*, 1995; Hohman *et al.*, 1997; Obrosova *et al.*, 2000c), lens (Lee and Chung *et al.*, 1999; Obrosova and Fathallah, 2000), endothelial (Gupta *et al.*, 2002), and smooth muscle cells (Nakamura *et al.*, 2001). In particular, it has been reported that aldose reductase inhibitors prevent or correct lipid peroxidation product accumulation (Lowitt *et al.*, 1995), GSH depletion (Hohman *et al.*, 1997; Obrosova *et al.*, 2000c), increase in GSSG/GSH ratio (Hohman *et al.*, 1997), downregulation of superoxide dismutase and quinone reductase (Obrosova *et al.*, 2000c) in the peripheral nerve of diabetic rats, as well as the production of superoxide anion radicals by high glucose-exposed vascular endothelium (Gupta *et al.*, 2002). Studies in our laboratory revealed that a short (2-week) treatment with an adequate dose of aldose reductase inhibitor is sufficient to reverse the depletion of key nonenzymatic antioxidants, GSH and ascorbate, and to bring about normal lipid peroxidation in early DN (Obrosova *et al.*, 2002). Furthermore, in contrast to diabetic wild-type mice (AR+/+), diabetic AR knockout (AR−/−) mice do not develop GSH depletion in the peripheral nerve and diabetic peripheral neuropathy in general (Ho *et al.*, 2001).

B. How Does Increased AR Activity Lead to Depletion of the Key
    Nonenzymatic Antioxidant, Glutathione, in the Peripheral Nerve:
    NADPH Deficiency, Osmotic Stress, or Nonenzymatic Glycation?

Evidence from at least six groups suggests that sorbitol pathway
hyperactivity has a key role in hyperglycemia-induced GSH depletion
in the peripheral nerve (Hohman *et al.*, 1997; Obrosova *et al.*, 2000c),
lens (Lou *et al.*, 1988; Lee and Chung, 1999; Obrosova and Fathallah,
2000), and smooth muscle cells (Nakamura *et al.*, 2001). Several concepts
have been proposed regarding the biochemical mechanism(s) linking
increased sorbitol pathway activity and GSH depletion in target tissues for
diabetic complications, but no consensus has been achieved. At least two
groups (Hohman *et al.*, 1997; Lee and Chung, 1999) have suggested that
hyperglycemia-induced GSH deficiency in peripheral nerve results from
depletion of NADPH, a cofactor shared by AR and glutathione reductase,
and resulting slowing of the glutathione redox cycle. However, this concept
is not supported by the absent or minor reciprocal increase of GSSG
concentration in concert with the decrease in GSH in both diabetic nerve
(Nagamatsu *et al.*, 1995; Hohman *et al.*, 1997; Stevens *et al.*, 2000) and
lens (Mitton and Trevithick, 1994; Obrosova *et al.*, 1998; Obrosova and
Stevens, 1999). In addition, NADPH deficiency in the diabetic peripheral
nerve has never been documented. Studies in other tissues, e.g., lens,
revealed that diabetes-induced NADPH deficiency is minor ($\sim$15%) or
absent (Lee *et al.*, 1985; Lou *et al.*, 1988). Apparently, GSH depletion
in a number of target tissues for diabetic complications, i.e., peripheral
nerve (Cameron *et al.*, 1999; Obrosova *et al.*, 2000c), lens (Lou *et al.*, 1988;
Saito, 1995; Obrosova *et al.*, 1998; Lee and Chung, 1999), renal cortex
(Aragno *et al.*, 1999; Obrosova *et al.*, 2000b), is due to a decrease in total
glutathione rather than impairment of the NADPH-dependent reduction
of GSSG to GSH. The concept of "NADPH deficiency" is not supported
by studies with a sorbitol dehydrogenase inhibitor (SDI). Administration
of doses of 50–250 mg/kg/day to diabetic rats associated with sorbitol
accumulation above the "diabetic threshold" in the peripheral nerve
(Cameron *et al.*, 1997; Obrosova *et al.*, 1999b), as well as the lens (Geisen
et al., 1994; Obrosova *et al.*, 1999a), exacerbated lipid peroxidation product
accumulation (Obrosova *et al.*, 1999a,b) and GSH depletion (Geisen
et al., 1994; Obrosova *et al.*, 1999a,b) in both tissues. Therefore, sorbitol
accumulation-linked osmotic stress rather than NADPH deficiency with
resulting slowing of the glutathione redox cycle is responsible for diabetes-
induced GSH depletion in the lens and peripheral nerve. This conclusion
does not contradict the concept of "decreased glutathione biosynthesis"
because osmotic stress can disrupt GSH biosynthesis by affecting the uptake
of the amino acid cysteine (Mitton *et al.*, 1997, 1999), the rate-limiting step

in GSH biosynthesis (*Tachi et al.*, 1998). However, the "osmotic concept" is not supported by studies in diabetic AR-overexpressing SDH knockout (AR+/SDH−) mice that developed higher lens sorbitol accumulation but less manifested GSH depletion than diabetic AR-overexpressing mice with a normal SDH content [(AR+/SDH+), Lee and Chung, 1999].

Some investigators suggest that hyperglycemia-induced GSH depletion occurs due to glycation or decreased expression of the key enzyme of glutathione biosynthesis, γ-glutamyl cysteine synthetase (Murakami, 1991; Urata *et al.*, 1996) and glycation of glutathione reductase (Blakytny and Harding, 1997). Increased sorbitol pathway activity contributes to nonenzymatic glycation/glycoxidation by providing the important glycation agents, i.e., fructose and fructose 3-phosphate (Grandhee and Monnier, 1991; Dills, 1993; Lal *et al.*, 1995), methylglyoxal (Phillips *et al.*, 1993), and 3-deoxyglucosone (Hamada *et al.*, 1996, 2000; Niwa, 1999; Tsukushi *et al.*, 1999), and promoting the accumulation of advanced glycation end products, i.e., pentosidine (Nagaraj *et al.*, 1994) and N^{ε}-(carboxymethyl)lysine (Tsukushi *et al.*, 1999; Hamada *et al.*, 2000), that generate oxidative stress by interacting with their receptors (Yan *et al.*, 1994; Schmidt *et al.*, 1999; Wautier *et al.*, 2001). However, the role of glycation/glycoxidation in diabetes-associated inhibition of GSH biosynthesis is not supported by studies (1) demonstrating the lack of any antioxidant activity of the inhibitor of nonenzymatic glycation, aminoguanidine, in the peripheral nerve of diabetic rats (Kihara *et al.*, 1991); (2) the lack of any correction of GSH depletion or decrease of GSH/cysteine ratio, the index of the rate of glutathione biosynthesis, in the retina of aminoguanidine-treated diabetic rats vs untreated diabetic group (Agardh *et al.*, 2000); and (3) exacerbation rather than correction of diabetes-induced GSH depletion in lens and peripheral nerve by SDI treatment (Geisent *et al.*, 1994; Obrosova *et al.*, 1999a,b) that did not affect intracellular glucose and reduced intracellular fructose concentrations markedly (Obrosova *et al.*, 1999a,b). Of interest, Ou and Wolff (1993) found that aminoguanidine inhibits catalase and generates hydrogen peroxide *in vitro*. The latter is in contrast with the study of Giardino *et al.* (1998), who described antioxidant effects of aminoguanidine in retinal Müller cells exposed to 10 μM hydrogen peroxide. Antioxidant properties of aminoguanidine in the diabetic retina have been described by Kowluru *et al.* (2000); however, the retinal GSH concentrations in the aforementioned study are at least threefold higher than those reported by *four* other groups (Winkler and Giblin, 1983; Organisciak *et al.*, 1984; Agardh *et al.*, 1998; Obrosova *et al.*, 2000a), and no specific controls have been performed to exclude interference of the thiobarbituric acid-reactive substance (TBARS) assay of lipid peroxidation products with glucose (Gutteridge, 1981). No antioxidant properties of aminoguanidine have been revealed in another study that, in addition to TBARS, used the most specific and sensitive marker of lipid peroxidation, i.e., F_2-isoprostane (Reckelhoff *et al.*, 1999).

C. Role for AR in Hyperglycemia-Induced Ascorbate Depletion

Studies of our group (Obrosova *et al.*, 2002a; Table I) revealed that total and free AA concentrations are decreased and the DHAA/AA ratio is increased in the peripheral nerve of diabetic rats. AA has an important role in antioxidative defense, particularly in phenoxyl radical neutralization and α-tocopherol recycling (Stoyanovsky *et al.*, 1995). Diabetic subjects, particularly patients with poor glycemic control, have decreased plasma AA concentrations and increased AA oxidation to DHAA (Yue *et al.*, 1990). The mechanisms of AA depletion in tissue sites for diabetic complications, including nerve (Obrosova *et al.*, 2002a), lens (Mitton and Trevithick, 1994; Saito, 1995; Mitton *et al.*, 1997, 1999; Lindsay *et al.*, 1998; Obrosova *et al.*, 1999a), and kidneys (Lindsay *et al.*, 1998; Obrosova *et al.*, 2000d) have not been studied in detail. However, reports indicate that AA concentrations are related inversely to sorbitol pathway activity (Saito, 1995; Lindsay *et al.*, 1998). The study of our group suggests that AA depletion in the diabetic precataractous lens is mediated by sorbitol accumulation and intralenticular osmotic stress but not by nonenzymatic glycation (Obrosova *et al.*, 1999a). Lens AA concentrations are reduced by L-buthionine (S, R)-sulfoximine, an inhibitor of glutathione biosynthesis (Maitra *et al.*, 1995; Packer *et al.*, 1995), because GSH and other cellular thiols play an important role in vitamin C homeostasis by regenerating AA from DHAA and semiascorbyl radicals (Packer *et al.*, 1995). Therefore, it is not surprising that changes in the glutathione and ascorbate systems of antioxidative defense occur in parallel.

So far, it is unclear whether AA depletion in tissues of diabetic animals is a primary response or a secondary phenomenon occurring due to the depletion of GSH because GSH concentrations are also decreased in major sites for diabetic complications, including lens (Lou *et al.*, 1988; Mitton and Trevithick, 1994; Saito, 1995; Mitton *et al.*, 1997; Obrosova *et al.*, 1998, 1999a,b; Obrosova and Fathallah, 2000) peripheral nerve (Nagamatsu *et al.*, 1995; Cameron *et al.*, 1999; Obrosova *et al.*, 1999c, 2000c,d; Stevens *et al.*, 2000), and kidney (Aragno *et al.*, 1999; Obrosova *et al.*, 2000b). Both diabetes-induced GSH and ascorbate depletion in peripheral nerve and lens are reversed by an aldose reductase inhibitor treatment (Saito, 1995; Obrosova *et al.*, 2002b). However, the glutathione and ascorbate systems of antioxidative defense could respond differently to some experimental conditions, e.g., to treatment with another antioxidant, taurine. In our study

TABLE II

DIABETES-INDUCED CHANGES IN GLUTATHIONE AND ASCORBATE SYSTEMS OF
ANTIOXIDATIVE DEFENSE IN PERIPHERAL NERVE AND EFFECTS OF DIETARY 1% TARINE
SUPPLEMENTATION (MEAN ± SEM, $n = 7–10$)

	Control	Diabetic	Diabetic + taurine
GSH	0.719 ± 0.029	0.435 ± 0.028**	0.447 ± 0.027**
GSSG	0.038 ± 0.004	0.030 ± 0.003	0.024 ± 0.004*
GSSG/GSH	0.053 ± 0.006	0.067 ± 0.007	0.049 ± 0.006
Total AA	0.267 ± 0.021	0.178 ± 0.012**	0.250 ± 0.026[#]
AA	0.219 ± 0.023	0.133 ± 0.015*	0.209 ± 0.025[#]
DHAA	0.044 ± 0.003	0.043 ± 0.005	0.030 ± 0.004*[#]
DHAA/AA	0.211 ± 0.026	0.370 ± 0.077*	0.152 ± 0.017[##]

[a]From Obrosova *et al.* (2002a). The duration of diabetes was 6 weeks. Concentrations of GSH, GSSG, total AA, AA, and DHAA are expressed in μmol/g wet weight. Significantly different compared with controls ($p < 0.05$* and <0.01**). Significantly different compared with untreated diabetic rats ($p < 0.05$[#] and <0.01[##]).

(Table II; Obrosova *et al.*, 2002a), normal total and free AA concentrations and the ascorbate redox state, but not GSH concentrations, were preserved in the peripheral nerve of diabetic rats fed a taurine-supplemented diet. Apparently, this effect of exogenous taurine is not mediated by ascorbate regeneration by GSH, or an osmotic mechanism, because nerve sorbitol concentrations were similar in diabetic rats fed taurine-supplemented and regular diets (Pop-Busui *et al.*, 2001).

D. ANTIOXIDANT PROPERTIES OF TAURINE AND ROLE OF AR IN
HYPERGLYCEMIA-INDUCED TAURINE DEPLETION

The abundant amino acid taurine is a most interesting compound that acts as (1) an osmolyte (Stevens *et al.*, 1993; Nagelhus *et al.*, 1994; Burg, 1997; Schaffer *et al.*, 2000); (2) a neurotransmitter (Rokkas *et al.*, 1995; Benton *et al.*, 2001); (3) a membrane stabilizer (Qi *et al.*, 1995; Chahine *et al.*, 1998); (4) a modulator of Ca^{2+} homeostasis (Michalk *et al.*, 1996; El Idrissi *et al.*, 1999); (5) a regulator of protein kinase C-dependent phosphorylation (Lima and Cubillos, 1998; Azuma *et al.*, 2000); and (6) an endogenous antioxidant (Eppler and Dawson, 2001; Benton *et al.*, 2001). The mechanism(s) of the antioxidant activity of taurine remains poorly understood, although the ability of taurine to decrease diabetes-induced lipid peroxidation has been demonstrated in peripheral nerve (Obrosova *et al.*, 2002a), kidney (Trachtman *et al.*, 1995), lens (Obrosova and Stevens, 1999), retina (Obrosova *et al.*, 2001), liver, and pancreas (Lim *et al.*, 1998). Several groups (Simmonds *et al.*, 1992; Raschke *et al.*, 1995; Gumuslu *et al.*, 1996) reported that taurine decreases luminol-dependent chemiluminescence elicited by chemically generated hydroxyl radicals and *t*–butyl hydroperoxide, whereas others (Aruoma *et al.*, 1988) indicated that hypotaurine, a taurine precursor, rather than taurine itself, has antioxidative properties against the aforementioned ROS. Both taurine and hypotaurine scavenge hypochlorite (Brestel, 1985; Aruoma *et al.*, 1988; Frenkel *et al.*, 1986; Dekigai *et al.*, 1995), which is known to form from hydrogen peroxide in the presence of copper ions Cu^{2+} (Frenkel *et al.*, 1986), i.e., under conditions of diabetes-associated transition metal imbalance, and is involved in inflammation (Daumer *et al.*, 2000; Wu *et al.*, 2000). It has been demonstrated (Engelmann *et al.*, 2000; Herdener *et al.*, 2000) that hypochlorite also reacts with superoxide anions to yield the highly reactive hydroxyl radicals. Thus, taurine, the hypochlorite scavenger, counteracts hydroxyl radical formation. Spin-trapping experiments (Kilic *et al.*, 1999) have revealed that taurine scavenges superoxide, and Erdem *et al.* (2000) have demonstrated the ability of taurine to activate superoxide dismutase. In addition, taurine accelerates the catabolism of norepinephrine (Chabine *et al.*, 1994), which can autoxidize and thus contribute to ROS generation. We have demonstrated that the antioxidant effects of taurine in the diabetic peripheral nerve, but not retina (Obrosova *et al.*, 2002a), are, at least in part, mediated through the ascorbate system of antioxidative defense, i.e., increase in free AA concentrations and AA/DHAA ratio. The studies of two groups implicate increased AR activity and resulting osmotic stress in taurine depletion in high glucose-exposed cells (Stevens *et al.*, 1999) and galactose-fed rat lens (Malone *et al.*, 1993). In both cases, taurine deficiency was effectively prevented by aldose reductase inhibitor treatment.

E. Role for AR in Diabetes-Induced Lipid Peroxidation in Peripheral Nerve

Two studies, including one from our laboratory (Lowitt *et al.*, 1995; Obrosova *et al.*, 2002), indicate that diabetes-induced lipid peroxidation in the peripheral nerve is prevented or reversed by an aldose reductase inhibitor treatment. The arrest of diabetes-induced lens lipid peroxidation by structurally diverse aldose reductase inhibitors has also been reported (Yeh and Ashton, 1990; Saito, 1995; Obrosova and Fathallah, 2000).

F. Role for AR in Hyperglycemia-Induced Reactive Oxygen Species Generation

To evaluate the role of AR in hyperglycemia-induced ROS generation, we have performed experiments with human RPE-47 cells transfected stably with the human AR gene. The cells have been cultured for 2 days in either 5 or 30 mM glucose with or without two structurally diverse aldose reductase inhibitors (100 μM sorbinil or 1 μM fidarestat) added for 3 h before the end of experiment. Ten microliters of 10 μM 5-(and-6-)-chloromethyl-2',7'-dichlorodihydrofluorescein diacetate (CM-H$_2$DCFDA), the dichlorofluorescein derivative with the best retention properties among all the studied analogs (Xie $et\ al.$, 1999), was added 30 min before the end of experiments. The cells were washed and trypsinized, and CM-DCF fluorescence, the index of ROS generation, was measured by flow cytometry (λ excitation: 480 nm; λ emission: 520 nm). The cell concentration for flow cytometry was not less than 10^5 cells in 1 ml buffer. Hyperglycemia-induced ROS production was higher in cells cultured in 30 mM glucose than in 5 mM glucose, and this increase was arrested by 100 μM sorbinil and by 1 μM fidarestat (Fig. 5). None of the aldose reductase inhibitors counteracted oxidative stress produced by buthionine (S, R)-sulfoximine, the inhibitor of glutathione biosynthesis (Maitra $et\ al.$, 1995), in human nontransfected RPE-47 cells under normoglycemic conditions. We conclude, therefore, that neither sorbinil nor fidarestat acts as a direct antioxidant independently of its inhibitory effect on the sorbitol pathway activity.

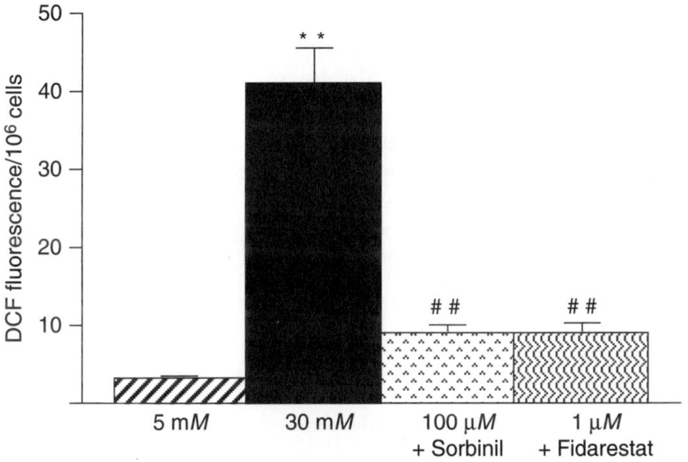

FIG. 5. Effect of two ARIs, sorbinil and fidarestat, on ROS production by human RPE-47 cells transfected stably with the human AR gene (mean ± SEM, $n = 12$). ROS production is expressed as DCF fluorescence/10^6 cells.

G. Role for AR in Hyperglycemia-Associated Increase in Superoxide Anion Radical Abundance

Contrary to the report (Nishikawa *et al.*, 2000) implying that increased AR activity and sorbitol accumulation develop as consequences of hyperglycemia-induced superoxide production two groups have found that increased superoxide anion radical abundance in the plasma of diabetic patients (Fondelli *et al.*, 1993) and in the high glucose-exposed rabbit vascular endothelium (Gupta *et al.*, 2002) is blunted by the aldose reductase inhibitor tolrestat and abrogated by the aldose reductase inhibitor zopolrestat, respectively, thus indicating that increased AR activity is, at least in part, *responsible* for hyperglycemia-associated superoxide production.

VI. Conclusion

The existing observations provide clear evidence of the important role of oxidative stress in diabetes-induced nerve blood flow and conduction deficits, metabolic imbalances, impaired neurotrophic support, and morphological abnormalities characteristic for diabetic peripheral neuropathy. At this point, of particular interest is the identification of diabetes-induced changes in signal transduction mechanisms developing *consequent* to oxidative stress and evaluation of their roles in the pathogenesis of diabetic peripheral neuropathy.

The studies of leading laboratories, *without exception,* indicate that the effects of aldose reductase inhibitors and antioxidants are unidirectional — both aldose reductase inhibitors and antioxidants prevent and correct manifestations of diabetic peripheral neuropathy. AR has a key role in oxidative stress in the diabetic peripheral nerve. The availability of molecular fluorescent probes for the detection of specific free radicals, hydroxyl and superoxide anion radicals and singlet oxygen, by flow cytometry makes it possible to clarify the role of AR vs other factors in the hyperglycemia-induced generation of specific ROS in Schwann cells, sensory neurons, dorsal root, and sympathetic ganglia of the PNS, as well as endothelial cells.

References

Acworth, I. N., Bogdanov, M. B., McCabe, D. R., and Beal, M. F. (1999). Estimation of hydroxyl free radical levels in vivo based on liquid chromatography with electrochemical detection. *Methods Enzymol.* **300**, 297–313.

Adler, V., Yin, Z., Fuchs, S. Y., Benezra, M., Rosario, L., Tew, K. D., Pincus, M. R., Sardana, M., Henderson, C. J., Wolf, C. R., Davis, R. J., and Ronai, Z. (1999). Regulation of JNK signaling by GSTp. *EMBO J.* **18**, 1321–1334.

Agardh, C. D., Agardh, E., Qian, Y., and Hultberg, B. (1998). Glutathione levels are reduced in diabetic rat retina but are not influenced by ischemia followed by recirculation. *Metabolism* **47**, 269–272.

Ansari, N. H., Wang, L., and Srivastava, S. K. (1996). Role of lipid aldehydes in cataractogenesis: 4-Hydroxynonenal-induced cataract. *Biochem. Mol. Med.* **58**, 25–30.

Aragno, M., Tamagno, E., Gatto, V., Brignardello, E., Parola, S., Danni, O., and Boccuzzi, G. (1999). Dehydroepiandrosterone protects tissues of streptozotocin-treated rats against oxidative stress. *Free Radic. Biol. Med.* **26**, 1467–1474.

Aruoma, O. I., Halliwell, B., Hoey, B. M., and Butler, J. (1988). The antioxidant action of taurine, hypotaurine and their metabolic precursors. *Biochem. J.* **256**, 251–255.

Azuma, M., Takahashi, K., Fukuda, T., Ohyabu, Y., Yamamoto, I., Kim, S., Iwao, H., Schaffer, S. W., and Azuma, J. (2000). Taurine attenuates hypertrophy induced by angiotensin II in cultured neonatal rat cardiac myocytes. *Eur. J. Pharmacol.* **403**, 181–188.

Benton, R. L., Ross, C. D., and Miller, K. E. (2001). Spinal taurine levels are increased 7 and 30 days following methylprednisolone treatment of spinal cord injury in rats. *Brain Res.* **893**, 292–300.

Berciano, M. T., Fernandez, R., Pena, E., Calle, E., Villagra, N. T., and Lafarga, M. (1999). Necrosis of Schwann cells during tellurium-induced primary demyelination: DNA fragmentation, reorganization of splicing machinery, and formation of intranuclear rods of actin. *J. Neuropathol. Exp. Neurol.* **58**, 1234–1243.

Berti-Mattera, L., Day, N., Peterson, R. G., and Eichberg, J. (1996). An aldose reductase inhibitor but not myo-inositol blocks enhanced polyphosphoinositide turnover in peripheral nerve from diabetic rats. *Metabolism* **45**, 320–327.

Blakytny, R., and Harding, J. J. (1997). Bovine and human alpha-crystallins as molecular chaperones: Prevention of the inactivation of glutathione reductase by fructation. *Exp. Eye Res.* **64**, 1051–1058.

Brand, P. W. (1982). The diabetic foot. *In* "Diabetes Mellitus" (M. Ellenberg and H. Rifkin, eds.), pp. 829–849. Medical Examination Publishing Co., New York.

Bravenboer, B., Kappelle, A. C., Hamers, F. P., van Buren, T., Erkelens, D. W., and Gispen, W. H. (1992). Potential use of glutathione for the prevention and treatment of diabetic neuropathy in the streptozotocin-induced diabetic rat. *Diabetologia* **35**, 813–814.

Brestel, E. P. (1985). Co-oxidation of luminol by hypochlorite and hydrogen peroxide implications for neutrophil chemiluminescence. *Biochem. Biophys. Res. Commun.* **126**, 482–488.

Burg, M. B. (1997). Renal osmoregulatory transport of compatible organic osmolytes. *Curr. Opin. Nephrol. Hypertens.* **6**, 430–433.

Calcutt, N. A., Freshwater, J. D., and Campana, W. M. (2000). Hyperalgesia mediated by spinal prostaglandin E_2 and signalling through the p38 pathway in diabetic rats. *J. Periph. Nerv. Syst.* **5**, 183–184.

Cameron, N. E., and Cotter, M. A. (1994). Effects of evening primrose oil treatment on sciatic nerve blood flow and endoneurial oxygen tension in streptozotocin-diabetic rats. *Acta Diabetol.* **31**, 220–225.

Cameron, N. E., and Cotter, M. A. (1995). Neurovascular dysfunction in diabetic rats: Potential contribution of autoxidation and free radical examined using transition metal chelating agents. *J. Clin. Invest.* **96**, 1159–1163.

Cameron, N. E., and Cotter, M. A. (1996). Rapid reversal by aminoguanidine of the neurovascular effects of diabetes in rats: Modulation by nitric oxide synthase inhibition. *Metabolism* **45**, 1147–1152.

Cameron, N. E., and Cotter, M. A. (2001). Effects of an extracellular metal chelator on neurovascular function in diabetic rats. *Diabetologia* **44**, 621–628.

Cameron, N. E., Cotter, M. A., Archibald, V., Dines, K. C., and Maxfield, E. K. (1994a). Antioxidant and pro-oxidant effects on nerve conduction velocity, endoneurial blood flow and oxygen tension in non-diabetic and streptozotocin-diabetic rats. *Diabetologia* **37**, 449–459.

Cameron, N. E., Cotter, M. A., Basso, M., and Hohman, T. C. (1997). Comparison of the effects of inhibitors of aldose reductase and sorbitol dehydrogenase on neurovascular function, nerve conduction and tissue polyol pathway metabolites in streptozotocin-diabetic rats. *Diabetologia* **40**, 271–281.

Cameron, N. E., Cotter, M. A., Dines, K. C., Maxfield, E. K., Carey, F. and Mirrlees, D. J. (1994b) Aldose reductase inhibition, nerve perfusion, oxygenation and function in streptozotocin-diabetic rats: Dose-response considerations and independence from a myo-inositol mechanism. *Diabetologia* **37**, 651–663.

Cameron, N. E., Cotter, M. A., Ferguson, K., Robertson, S. and Radcliffe, M. A. (1991). Effects of chronic α-adrenergic receptor blockade on peripheral nerve conduction, hypoxic resistance, polyols, Na^+-K^+-ATPase activity, and vascular supply in STZ-D rats. *Diabetes* **40**, 1652–1658.

Cameron, N. E., Cotter, M. A., Horrobin, D. H., and Tritschler, H. J. (1998). Effects of alpha-lipoic acid on neurovascular function in diabetic rats: Interaction with essential fatty acids. *Diabetologia* **41**, 390–399.

Cameron, N. E., Cotter, M. A., Jack, A. M., Basso, M. D., and Hohman, T. C. (1999). Protein kinase C effects on nerve function, perfusion, Na(+), K(+)-ATPase activity and glutathione content in diabetic rats. *Diabetologia* **42**, 1120–1130.

Cameron, N. E., Cotter, M. A., and Maxfield, E. K. (1993). Anti-oxidant treatment prevents the development of peripheral nerve dysfunction in streptozotocin-diabetic rats. *Diabetologia* **36**, 299–304.

Cameron, N. E., Cotter, M. A., Robertson, S., and Cox, D. (1992). Muscle and nerve dysfunction in rats with experimental galactosaemia. *Exp. Physiol.* **77**, 89–108.

Cameron, N. E., Jack, A. M., and Cotter, M. A. (2001). Effect of alpha-lipoic acid on vascular responses and nociception in diabetic rats. *Free Radic. Biol. Med.* **31**, 125–135.

Chabine, R., Hanna, J., Abou Khalil, K., Cheav, S. L., Hatala, R., Bouchi, N., and Mounayar, A. (1994). Taurine and myocardial noradrenaline. *Arz.-Forschung* **44**, 126–128.

Chahine, R., and Feng, J. (1998). Protective effects of taurine against reperfusion-induced arrhythmias in isolated ischemic rat heart. *Arz.-Forschung.* **48**, 360–364.

Clerk, A., Michael, A., and Sugden, P. H. (1998). Stimulation of multiple mitogen-activated protein kinase sub-families by oxidative stress and phosphorylation of the small heat shock protein, HSP25/27, in neonatal ventricular myocytes. *Biochem. J.* **333**, 581–589.

Coppey, L. J., Gellett, J. S., Davidson, E. P., Dunlap, J. A., Lund, D. D., and Yorek, M. A. (2001). Effect of antioxidant treatment of streptozotocin-induced diabetic rats on endoneurial blood flow, motor nerve conduction velocity, and vascular reactivity of epineurial arterioles of the sciatic nerve. *Diabetes* **50**, 1927–1937.

Cotter, M. A., Cameron, N. E., and Hohman, T. C. (1998). Correction of nerve conduction and endoneurial blood flow deficits by the aldose reductase inhibitor, tolrestat, in diabetic rats. *J. Periph. Nerv. Syst.* **3**, 217–223.

Cuzzocrea, S., McDonald, M. C., Filipe, H. M., Costantino, G., Mazzon, E., Santagati, S., Caputi, A. P., and Thiemermann, C. (2000). Effects of tempol, a membrane-permeable radical scavenger, in a rodent model of carrageenan-induced pleurisy. *Eur. J. Pharmacol.* **390**, 209–222.

Daumer, K. M., Khan, A. U., and Steinbeck, M. J. (2000). Chlorination of pyridinium compounds: Possible role of hypochlorite, N-chloramines, and chlorine in the oxidation of pyridinoline cross-links of articular cartilage collagen type II during acute inflammation. *J. Biol. Chem.* **275**, 34681–34692.

Dekigai, H., Murakami, M., and Kita, T. (1995). Mechanism of *Helicobacter pylori*-associated gastric mucosal injury. *Digest. Dis. Sci.* **40**, 1332–1339.

Del Corso, A., Dal Monte, M., Vilardo, P. G., Cecconi, I., Moschini, R., Banditelli, S., Cappiello, M., Tsai, L., and Mura, U. (1998). Site-specific inactivation of aldose reductase by 4-hydroxynonenal. *Arch. Biochem. Biophys.* **350**, 245–248.

Dewhurst, M., Omawari, N., and Tomlinson, D. R. (1997). Aminoguanidine effects on endoneurial vasoactive nitric oxide and on motor nerve conduction velocity in control and streptozotocin-diabetic rats. *Br. J. Pharmacol.* **120**, 593–598.

Diabetes Control and Complications Trial Research Group. (1993). The effect of intensive treatment of diabetes on the development and progression of long-term complications in insulin-dependent diabetes mellitus. *N. Engl. J. Med.* **329**, 977–986.

Diabetes in America. (1995). 2nd Edition. NIH Publication No. 95-1468. National Institutes of Diabetes and Digestive and Kidney Diseases, Bethesda, MD.

Diemel, L. T., Stevens, E. J., Willars, G. B., and Tomlinson, D. R. (1992). Depletion of substance P and calcitonin gene-related peptide in sciatic nerve of rats with experimental diabetes: Effects of insulin and aldose reductase inhibition. *Neurosci. Lett.* **137**, 253–256.

Dills, W. L., Jr. (1993). Protein fructosylation: Fructose and the Maillard reaction. *Am. J. Clin. Nutr.* **58**(5 Suppl.), 779S–787S.

El Idrissi, A., and Trenkner, E. (1999). Growth factors and taurine protect against excitotoxicity by stabilizing calcium homeostasis and energy metabolism. *J. Neurosci.* **19**, 9459–9468.

Eppler, B., and Dawson, R., Jr. (2001). Dietary taurine manipulations in aged male Fischer 344 rat tissue: Taurine concentration, taurine biosynthesis, and oxidative markers. *Biochem. Pharmacol.* **62**, 29–39.

Fernyhough, P., Gallagher, A., Averill, S. A., Priestley, J. V., Hounsom, L., Patel, J., and Tomlinson, D. R. (1999). Aberrant neurofilament phosphorylation in sensory neurons of rats with diabetic neuropathy. *Diabetes* **48**, 881–889.

Fondelli, C., Signorini, A. M., Borgogni, L., and Gragnoli, G. (1993). Tolrestat and superoxide anion production in type-2 diabetic patients. *Diabetologia* **36**(Suppl. 1), A204.

Frenkel, K., Blum, F., and Troll, W. (1986). Copper ions and hydrogen peroxide form hypochlorite from NaCl thereby mimicking myeloperoxidase. *J. Cell. Biochem.* **30**, 181–193.

Garcia Soriano F., Virag, L., Jagtap, P., Szabo, E., Mabley, J. G., Liaudet, L., Marton, A., Hoyt, D. G., Murthy, K. G., Salzman, A., Southan, G. J., and Szabo, C. (2001). Diabetic endothelial dysfunction: The role of poly(ADP-ribose) polymerase activation. *Nature Med.* **7**, 108–113.

Garrett, N. E., Malcangio, M., Dewhurst, M., and Tomlinson, D. R. (1997). Alpha-lipoic acid corrects neuropeptide deficits in diabetic rats via induction of trophic support. *Neurosci. Lett.* **222**, 191–194.

Giardino, I., Edelstein, D., and Brownlee, M. (1996). BCL-2 expression or antioxidants prevent hyperglycemia-induced formation of intracellular advanced glycation endproducts in bovine endothelial cells. *J. Clin. Invest.* **97**, 1422–1428.

Giardino, I., Fard, A. K., Hatchell, D. L., and Brownlee, M. (1998). Aminoguanidine inhibits reactive oxygen species formation, lipid peroxidation, and oxidant-induced apoptosis. *Diabetes* **47**, 1114–1120.

Goldhaber, J. I., and Qayyum, M. S. (2000). Oxygen free radicals and excitation-contraction coupling. *Antioxid. Redox Signa.* **2**, 55–64.

Grandhee, S. K., and Monnier, V. M. (1991). Mechanism of formation of the Maillard protein cross-link pentosidine: Glucose, fructose, and ascorbate as pentosidine precursors. *J. Biol. Chem.* **266**, 11649–11653.

Grune, T., Siems, W. G., and Petras, T. (1997). Identification of metabolic pathways of the lipid peroxidation product 4-hydroxynonenal in in situ perfused rat kidney. *J. Lipid Res.* **38**, 1660–1665.

Gumuslu, S., Yucel, G., Aydin, M., Yesilkaya, A., Demir, A. Y., and Aksu, T. A. (1996). The role of antioxidants in the prevention of t-butyl hydroperoxide-induced chemiluminescence. *Int. J. Clin. Lab. Res.* **26**, 119–123.

Gupta, S., Chough, E., Daley, J., Oates, P., Tornheim, K., Ruderman, N. B., and Keaney, J. F. (2002). Hyperglycemia increases superoxide anion production in rabbit endothelium leading to decreased Na^+,K^+-ATPase activity. *Am. J. Physiol.* **282**, C560–C566.

Gutteridge, J. M. (1981). Thiobarbituric acid-reactivity following iron-dependent free radical damage to amino acids and carbohydrates. *FEBS Lett.* **128**, 343–346.

Halstead, J. R., Roefs, M., Ellson, C. D., D'Andrea, S., Chen, C., D'Santos, C. S., and Divecha, N. (2001). A novel pathway of cellular phosphatidylinositol(3,4,5)-trisphosphate synthesis is regulated by oxidative stress. *Curr. Biol.* **11**, 386–395.

Hamada, Y., Araki, N., Horiuchi, S., and Hotta, N. (1996). Role of polyol pathway in nonenzymatic glycation. *Nephrol. Dialys. Transplant.* **11**(Suppl. 5), 95–98.

Hamada, Y., Nakamura, J., Naruse, K., Komori, T., Kato, K., Kasuya, Y., Nagai, R., Horiuchi, S., and Hotta, N. (2000). Epalrestat, an aldose reductase inhibitor, reduces the levels of Nepsilon-(carboxymethyl)lysine protein adducts and their precursors in erythrocytes from diabetic patients. *Diab. Care.* **23**, 1539–1544.

He, Q., Khanna, P., Srivastava, S., van Kuijk, F. J., and Ansari, N. H. (1998). Reduction of 4-hydroxynonenal and 4-hydroxyhexenal by retinal aldose reductase. *Biochem. Biophys. Res. Commun.* **247**, 719–722.

Heesom, A. E., Millward, A., and Demaine, A. G. (1998). Susceptibility to diabetic neuropathy in patients with insulin dependent diabetes mellitus is associated with a polymorphism at the 5′ end of the aldose reductase gene. *J. Neur. Neurosurg. Phych.* **64**, 213–216.

Ho, E. C. M., Lam, K. S. L., and Chung, S. S. M. (2000). Aldose reductase-deficient mice are protected from reduction of nerve conduction associated with diabetes. *J. Periph. Nerv. Syst.* **5**, 169–188.

Ho, E. C. M., Lam, K. S. L., Chung, S. S. M., and Chung, S. K. (2001). Aldose-reductase-deficient mice are alleviated from the depletion of GSH in the peripheral nerve and MNCV deficit associated with diabetes. *Diabetes* **50**(Suppl. 2), A59.

Hohman, T. C., Banas, D., Basso, M., Cotter, M. A., and Cameron, N. E. (1997). Increased oxidative stress in experimental diabetic neuropathy. *Diabetologia* **40** (Suppl. 1), A549.

Hohman, T. C., Cotter, M. A., and Cameron, N. E. (2000). ATP-sensitive K(+) channel effects on nerve function, Na(+), K(+) ATPase, and glutathione in diabetic rats. *Eur. J. Pharmacol.* **397**, 335–341.

Hotta, N., Koh, N., Sakakibara, F., Nakamura, J., Hamada, Y., Wakao, T., Hara, T., Mori, K., Naruse, K., Nakashima, E., and Sakamoto, N. (1996). Effect of propionyl-L-carnitine on motor nerve conduction, autonomic cardiac function, and nerve blood flow in rats with streptozotocin-induced diabetes: Comparison with an aldose reductase inhibitor. *J. Pharmacol. Exp. Ther.* **276**, 49–55.

Hounsom, L., Horrobin, D. F., Tritschler, H., Corder, R., and Tomlinson, D. R. (1998). A lipoic acid-gamma linolenic acid conjugate is effective against multiple indices of experimental diabetic neuropathy. *Diabetologia* **41**, 839–843.

Hounsom, L, Corder, R., Patel, J., and Tomlinson, D. R. (2001). Oxidative stress participates in the breakdown of neuronal phenotype in experimental diabetic neuropathy. *Diabetologia* **44**, 424–428.

Inoguchi, T., Li, P., Umeda, F., Yu, H. Y., Kakimoto, M., Imamura, M., Aoki, T., Etoh, T., Hashimoto, T., Naruse, M., Sano, H., Utsumi, H., and Nawata, H. (2000). High glucose level and free fatty acid stimulate reactive oxygen species production through protein kinase C-dependent activation of NAD(P)H oxidase in cultured vascular cells. *Diabetes* **49**, 1939–1945.

Ishii, H., Tada, H., and Isogai, S. (1998). An aldose reductase inhibitor prevents glucose-induced increase in transforming growth factor-beta and protein kinase C activity in cultured mesangial cells. *Diabetologia* **41**, 362–364.

Ito, M., Suzuki, Y., Ishihara, M., and Suzuki, Y. (1998) Anti-ulcer effects of antioxidants: Effect of probucol. *Eur. J. Pharmacol.* **354**, 189–196.

Ito, T., Nishimura, C., Takahashi, Y., Saito, T., and Omori, Y. (1997). The level of erythrocyte aldose reductase: A risk factor for diabetic neuropathy? *Diab. Res. Clin. Pract.* **36**, 161–167.

Jennings, P. E. (2000). Vascular benefits of gliclazide beyond glycemic control. *Metabolism* **49**(10 Suppl. 2), 17–20.

Kalichman, M. W., Powell, H. C., and Mizisin, A. P. (1998). Reactive, degenerative, and proliferative Schwann cell responses in experimental galactose and human diabetic neuropathy. *Acta Neuropathol.* **95**, 47–56.

Kamijo, M., Basso, M., Cherian, P. V., Hohman, T. C., and Sima, A. A. (1994). Galactosemia produces aldose reductase inhibitor-preventable nodal changes similar to those of diabetic neuropathy. *Diab. Res. Clin. Pract.* **25**, 117–129.

Karasu, C., Dewhurst, M., Stevens, E. J., and Tomlinson, D. R. (1995). Effects of antioxidant treatment on sciatic nerve dysfunction in streptozotocin-diabetic rats; comparison with essential fatty acids. *Diabetologia* **38**, 129–134.

Kato, N., Mizuno, K., Makino, M., Suzuki, T., and Yagihashi, S. (2000). Effects of 15-month aldose reductase inhibition with fidarestat on the experimental diabetic neuropathy in rats. *Diab. Res. Clin. Pract.* **50**, 77–85.

Kihara, M., Schmelzer, J. D., Poduslo, J. F., Curran, G. L., Nickander, K. K., and Low, P. A. (1991). Aminoguanidine effects on nerve blood flow, vascular permeability, electrophysiology, and oxygen free radicals. *Proc. Natl. Acad. Sci. USA* **88**, 6107–6111.

Kilic, F., Bhardwaj, R., Caufeild, J., and Trevithick, J. R. (1999). Modelling cortical cataractogenesis 22: Is in vitro reduction of damage in model diabetic rat cataract by taurine due to its antioxidant activity? *Exp. Eye Res.* **69**, 291–300.

Kim, J., Rushovich, E. H., Thomas, T. P., Ueda, T., Agranoff, B. W., and Greene, D. A. (1991). Diminished specific activity of cytosolic protein kinase C in sciatic nerve of streptozocin-induced diabetic rats and its correction by dietary myo-inositol. *Diabetes* **40**, 1545–1554.

Kishi, Y., Schmelzer, J. D., Yao, J. K., Zollman, P. J., Nickander, K. K., Tritschler, H. J., and Low, P. A. (1999) Alpha-lipoic acid: Effect on glucose uptake, sorbitol pathway, and energy metabolism in experimental diabetic neuropathy. *Diabetes* **48**, 2045–2051.

Kowluru, R. A., Engerman, R. L., and Kern, T. S. (2000). Abnormalities of retinal metabolism in diabetes or experimental galactosemia. VIII. Prevention by aminoguanidine. *Curr. Eye Res.* **21**, 814–819.

Kuruvilla, R., and Eichberg, J. (1998). Depletion of phospholipid arachidonoyl-containing molecular species in a human Schwann cell line grown in elevated glucose and their restoration by an aldose reductase inhibitor. *J. Neurochem.* **71**, 775–783.

Lal, S., Szwergold, B. S., Taylor, A. H., Randall, W. C., Kappler, F., and Brown, T. R. (1995). Production of fructose and fructose-3-phosphate in maturing rat lenses. *Invest. Ophthalmol. Vis. Sci.* **36**, 969–973.

Langeveld, C. H., Schepens, E., Stoof, J. C., Bast, A., and Drukarch, B. (1995). Differential sensitivity to hydrogen peroxide of dopaminergic and noradrenergic neurotransmission in rat brain slices. *Free Radic. Biol. Med.* **19**, 209–217.

Lee, A. Y. W., and Chung, S. S. M. (1999). Contribution of polyol pathway to oxidative stress in diabetic cataract. *FASEB J.* **13**, 23–30.

Lee, S. M., Schade, S. Z., and Doughty, C. C. (1985). Aldose reductase, NADPH and NADP+ in normal, galactose-fed and diabetic rat lens. *Biochim. Biophys. Acta* **841**, 247–253.

Li, X., Song, L., and Jope, R. S. (1998). Glutathione depletion exacerbates impairment by oxidative stress of phosphoinositide hydrolysis, AP-1, and NF-kappaB activation by cholinergic stimulation. *Brain Res.* **53**, 196–205.

Li, X. M., Juorio, A. V., Qi, J., and Boulton, A. A. (1998). L-deprenyl potentiates NGF-induced changes in superoxide dismutase mRNA in PC12 cells. *J. Neurosci. Res.* **53**, 235–238.

Lim, E., Park, S., and Kim, H. (1998). Effect of taurine supplementation on the lipid peroxide formation and the activities of glutathione-related enzymes in the liver and islet of type I and II diabetic model mice. *Adv. Exp. Med. Biol.* **442**, 99–103.

Lima, L., and Cubillos, S. (1998). Taurine might be acting as a trophic factor in the retina by modulating phosphorylation of cellular proteins. *J. Neurosci. Res.* **53**, 377–384.

Lindsay, R. M., Walker, S. A., McGuigan, C. C., Smith, W., and Baird, J. D. (1998). Tissue ascorbic acid and polyol pathway metabolism in experimental diabetes. *Diabetologia* **41**, 516–523.

Love, A., Cotter, M. A., and Cameron, N. E. (1996). Effects of the sulphydryl donor N-acetyl-L-cysteine on nerve conduction: Perfusion, maturation and regeneration following freeze damage in diabetic rats. *Eur. J. Clin. Invest.* **26**, 698–706.

Love, A., Cotter, M. A., and Cameron, N. E. (1996). Nerve function and regeneration in diabetic and galactosemic rats: Antioxidant and metal chelator effects. *Eur. J. Pharmacol.* **31**, 433–439.

Lou, M. F., Dickerson, J. E., Garadi, R., and York, B. M., Jr. (1988). Glutathione depletion in the lens of galactosemic and diabetic rats. *Exp. Eye Res.* **46**, 517–530.

Lounsbury, K. M., Hu, Q., and Ziegelstein, R. C. (2000). Calcium signaling and oxidant stress in the vasculature. *Free Radic. Biol. Med.* **28**, 1362–1369.

Low, P. A., and Nickander, K. K. (1991). Oxygen free radical effects in sciatic nerve in experimental diabetes. *Diabetes* **40**, 873–877.

Low, P. A., Nickander, K. K., and Tritschler, H. J. (1997). The roles of oxidative stress and antioxidant treatment in experimental diabetic neuropathy. *Diabetes* **46**(Suppl. 2), S38–S42.

Lowitt, S., Malone, J. I., Salem, A. F., Korthals, J., and Benford, S. (1995). Acetyl-L-carnitine corrects the altered peripheral nerve function of experimental diabetes. *Metabolism* **44**, 677–680.

Luo, X., and Lehotay, D. C. (1997). Determination of hydroxyl radicals using salicylate as a trapping agent by gas chromatography-mass spectrometry. *Clin. Biochem.* **30**, 41–46.

Mackensen, G. B., Patel, M., Sheng, H., Calvi, C. L., Batinic-Haberle, I., Day, B. J., Liang, L. P., Fridovich, I., Crapo, J. D., Pearlstein, R. D., and Warner, D. S. (2001). Neuroprotection from delayed postischemic administration of a metalloporphyrin catalytic antioxidant. *J. Neurosci.* **21**, 4582–4592.

Maitra, I., Serbinova, E., Tritschler, H., and Packer, L. (1995). α–Lipoic acid prevents buthionine-sulfoximine-induced cataract formation in newborn rats. *Free Radic. Biol. Med.* **18**, 823–829.

Malone, J. I., Benford, S. A., and Malone, J., Jr. (1993). Taurine prevents galactose-induced cataracts. *J. Diab. Its Complicat.* **7**, 44–48.

Mattson, M. P., Lovell, M. A., Furukawa, K., and Markesbery, W. R. (1995). Neurotrophic factors attenuate glutamate-induced accumulation of peroxides, elevation of intracellular Ca^{2+} concentration, and neurotoxicity and increase antioxidant enzyme activities in hippocampal neurons. *J. Neurochem.* **65**, 1740–1751.

Maulik, N., Sato, M., Price, B. D., and Das, D. K. (1998). An essential role of NFkappaB in tyrosine kinase signaling of p38 MAP kinase regulation of myocardial adaptation to ischemia. *FEBS Lett.* **429**, 365–369.

McManis, P. G., Low, P. A., and Yao, J. K. (1986). Relationship between nerve blood flow and intercapillary distance in peripheral nerve edema. *Am. J. Physiol.* **251**, E92–E97.

Michalk, D. V., Wingenfeld, P., Licht, C., Ugur, T., and Siar, L. F. (1996). The mechanisms of taurine mediated protection against cell damage induced by hypoxia and reoxygenation. *Adv. Exp. Med. Biol.* **403**, 223–232.

Mitton, K. P., Dzialoszynski, T., Sanford, S. E., and Trevithick, J. R. (1997). Cysteine and ascorbate loss in the diabetic rat lens prior to hydration changes. *Curr. Eye Res.* **16**, 564–571.

Mitton, K. P., Linklater, H. A., Dzialoszynski, T., Sanford, S. E., Starkey, K., and Trevithick, J. R. (1999). Modelling cortical cataractogenesis 21: In diabetic rat lenses taurine supplementation partially reduces damage resulting from osmotic compensation leading to osmolyte loss and antioxidant depletion. *Exp. Eye Res.* **69**, 279–289.

Mitton, K. P., and Trevithick, J. R. (1994). High-performance liquid chromatography — electrochemical detection of antioxidants in vertebrate lens: Glutathione, tocopherol, and ascorbate. *Methods Enzymol.* **233**, 523–539.

Mizisin, A. P., Calcutt, N. A., DiStefano, P. S., Acheson, A., and Longo, F. M. (1997). Aldose reductase inhibitors increases CNTF-like bioactivity and protein in sciatic nerves from galactose-fed and normal rats. *Diabetes* **46**, 647–652.

Mizisin, A. P., Shelton, G. D., Wagner, S., Rusbridge, C., and Powell, H. C. (1998). Myelin splitting, Schwann cell injury and demyelination in feline diabetic neuropathy. *Acta Neuropathol.* **95**, 171–174.

Mizuno, K., Kato, N., Makino, M., Suzuki, T., and Shindo, M. (1999). Continuous inhibition of excessive polyol pathway flux in peripheral nerves by aldose reductase inhibitor fidarestat leads to improvement of diabetic neuropathy. *J. Diabet. Its Complicat.* **13**, 141–150.

Murakami, K. (1991). Glutathione metabolism in erythrocytes from patients with diabetes mellitus. *Hokkaido Igaku Zasshi — Hokkaido J. Med. Sci.* **66**, 29–40.

Myhre, O., and Fonnum, F. (2001). The effect of aliphatic, naphthenic, and aromatic hydrocarbons on production of reactive oxygen species and reactive nitrogen species in rat brain synaptosome fraction: The involvement of calcium, nitric oxide synthase, mitochondria, and phospholipase A. *Biochem. Pharmacol.* **62**, 119–128.

Nagamatsu, M., Nickander, K. K., Schmelzer, J. D., Raya, A., Wittrock, D. A., Tritschler H., and Low, P. A. (1995). Lipoic acid improves nerve blood flow, reduces oxidative stress, and improves distal nerve conduction in experimental diabetic neuropathy. *Diabe. Care* **18**, 1160–1167.

Nagamatsu, M., Schmelzer, J. D., Zollman, P. J., Smithson, I. L., Nickander, K. K., and Low, P. A. (1996). Ischemic reperfusion causes lipid peroxidation and fiber degeneration. *Muscle Nerve* **19**, 37–47.

Nagaraj, R. H., Prabhakaram, M., Ortwerth, B. J., and Monnier, V. M. (1994). Suppression of pentosidine formation in galactosemic rat lens by an inhibitor of aldose reductase. *Diabetes* **43**, 580–586.

Nagelhus, E. A., Amiry-Moghaddam, M., Lehmann, A., and Ottersen, O. P. (1994). Taurine as an organic osmolyte in the intact brain: Immunocytochemical and biochemical studies. *Adv. Exp. Med. Biol.* **359**, 325–334.

Nair, P., Tammariello, S. P., and Estus, S. (2000). Ceramide selectively inhibits apoptosis-associated events in NGF-deprived sympathetic neurons. *Cell Death Differ.* **7**, 207–214.

Nakamura, J., Kasuya, Y., Hamada, E., Nakashima, K., Naruse, Y., Yasuda, K., Kato, N., and Hotta, N. (2001). Glucose-induced hyperproliferation of cultured rat aortic smooth muscle cells through polyol pathway hyperactivity. *Diabetologia* **44**, 480–487.

Nakamura, J., Kato, K., Hamada, Y., Nakayama, M., Chaya, S., Nakashima, E., Naruse, K., Kasuya, Y., Mizubayashi, R., Miwa, K., *et al.* (1999). A protein kinase C-beta-selective inhibitor ameliorates neural dysfunction in streptozotocin-induced diabetic rats. *Diabetes* **48**, 2090–2095.

Nickander, K. K., McPhee, B. R., Low, P. A., and Trischler, H. (1996). Alpha-lipoic acid: Antioxidant potency against lipid peroxidation of neural tissues in vitro and implications for diabetic neuropathy. *Free Radic. Biol. Med.* **21**, 631–639.

Nickander, K. K., Schmelzer, J. D., Rohwer, D. A., and Low, P. A. (1994). Effect of α-tocopherol deficiency on indices of oxidative stress in normal and diabetic peripheral nerve. *J. Neurol. Sci.* **126**, 6–14.

Nishikawa, T., Edelstein, D., and Brownlee, M. (2000). The missing link: A single unifying mechanism for diabetic complications. *Kidney Int.* **58**(Suppl. 77), S26–S30.

Niwa, T. (1999). 3-Deoxyglucosone: Metabolism, analysis, biological activity, and clinical implication. *J. Chromatogr. B. Biomed. Sci. Appl.* **731**, 23–36.

Oates, P. J., and Mylari, B. L. (1999). Aldose reductase inhibitors: Therapeutic implications for diabetic complications. *Exp. Opin. Invest. Drugs* **8**, 2095–2119.

Obrosova, I. G., Cao, X., Greene, D. A., and Stevens, M. J. (1998). Diabetes-induced changes in lens antioxidant status, glucose utilization and energy metabolism: Effect of DL-α-lipoic acid. *Diabetologia* **41**, 1442–1450.

Obrosova, I. G., and Fathallah, L. (2000). Evaluation of an aldose reductase inhibitor on lens metabolism, ATPases and antioxidative defense in streptozotocin-diabetic rats: An intervention study. *Diabetologia* **43**, 1048–1055.

Obrosova, I. G., Fathallah, L., and Greene, D. A. (2000a). Early changes in lipid peroxidation and antioxidative defense in diabetic rat retina: Effect of DL-alpha-lipoic acid. *Eur. J. Pharmacol.* **398**, 139–146.

Obrosova, I. G., Fathallah, L., and Lang, H. J. (1999a). Interaction between osmotic and oxidative stress in diabetic precataractous lens: Studies with a sorbitol dehydrogenase inhibitor. *Biochem. Pharmacol.* **58**, 1945–1954.

Obrosova I. G., Fathallah L., Lang H. J., and Greene D. A. (1999b). Evaluation of a sorbitol dehydrogenase inhibitor on diabetic peripheral nerve metabolism: A prevention study. *Diabetologia* **42**, 1187–1194.

Obrosova, I. G., Fathallah, L., Liu, E., and Nourooz-Zadeh, J. (2000b). Early changes in lipid peroxidation and antioxidative defense in diabetic rat renal cortex: Effect of DL-α-lipoic acid. *Diabetes* **49**(Suppl. 1), A56.

Obrosova, I. G., Fathallah, L., Stevens, M. J. (2002a). Taurine counteracts oxidative stress and nerve growth factor deficit in early experimental diabetic neuropathy. *Exp. Neurol.* **172**, 211–219.

Obrosova, I. G., Greene, D. A., and Lang, H. J. (2000c). Antioxidative defense in diabetic peripheral nerve: Effect of DL-α-lipoic acid, aldose reductase inhibitor and sorbitol dehydrogenase inhibitor. *In* "Antioxidants in Diabetes Management" (L. Packer, P. Roesen, H. Tritschler, G. King, and A. Azzi, eds.), pp. 93–110. Dekker, New York.

Obrosova, I. G., Minchenko, A. G., Marinescu, V., Fathallah, L., Kennedy, A., Stockert, C. M., Frank, R. N., and Stevens, M. J. (2001). Antioxidants attenuate early up regulation of retinal vascular endothelial growth factor in streptozotocin-diabetic rats. *Diabetologia* **44**, 1102–1110.

Obrosova, I. G., and Stevens, M. J. (1999). Effect of dietary taurine supplementation on GSH and NAD(P)-redox status, lipid peroxidation, and energy metabolism in diabetic precataractous lens. *Invest. Ophthalmol. Vis. Sci.* **40**, 680–688.

Obrosova, I. G., Van Huysen, C., Fathallah, L., Cao, X., Greene, D. A., and Stevens, M. J. (2002b). An aldose reductase inhibitor reverses early diabetes-induced changes in peripheral nerve function, metabolism and antioxidative defense. *FASEB J.* **16**, 123–125.

Obrosova, I. G., Van Huysen, C., Fathallah, L., Cao, X., Stevens, M. J., and Greene, D. A. (2000d). Evaluation of α_1-adrenoceptor antagonist on nerve function, metabolism and antioxidative defense. *FASEB J.* **14**, 1548–1558.

Ohi, T., Saita, K., Furukawa, S., Ohta, M., Hayashi, K., and Matsukura, S. (1998). Therapeutic effects of aldose reductase inhibitor on experimental diabetic neuropathy through synthesis/secretion of nerve growth factor. *Exp. Neurol.* **151**, 215–220.

Okabe, E., Tsujimoto, Y., and Kobayashi, Y. (2000). Calmodulin and cyclic ADP-ribose interaction in Ca^{2+} signaling related to cardiac sarcoplasmic reticulum: Superoxide anion radical-triggered Ca^{2+} release. *Antioxid. Redox Signal.* **2**, 47–54.

Organisciak, D. T., Wang, H. M., and Kou, A. L. (1984). Ascorbate and glutathione levels in the developing normal and dystrophic rat retina: Effect of intense light exposure. *Curr. Eye Res.* **3**, 257–267.

Ou, P., and Wolff, S. P. (1993). Aminoguanidine: A drug proposed for prophylaxis in diabetes inhibits catalase and generates hydrogen peroxide in vitro. *Biochem. Pharmacol.* **46**, 1139–1144.

Packer, L., Witt, E. H., and Tritschler, H. J. (1995). α-Lipoic acid as a biological antioxidant. *Free Radic. Biol. Med.* **19**, 227–250.

Paget, C., Lecomte, M., Ruggiero, D., Wiernsperger, N., and Lagarde, M. (1998). Modification of enzymatic antioxidants in retinal microvascular cells by glucose or advanced glycation end products. *Free Radic. Biol. Med.* **25**, 121–129.

Phillips, S. A., Mirrlees, D., and Thornalley, P. J. (1993). Modification of the glyoxalase system in streptozotocin-induced diabetic rats: Effect of the aldose reductase inhibitor Statil. *Biochem. Pharmacol.* **46**, 805–811.

Plaschke, K., Kopitz, J., Weigand, M. A., Martin, E., and Bardenheuer, H. J. (2000). The neuroprotective effect of cerebral poly(ADP-ribose)polymerase inhibition in a rat model of global ischemia. *Neurosci. Lett.* **284**, 109–112.

Pop-Busui, R., Sullivan, K. A., Van Huysen, C., Bayer, L., Cao, X., Towns, R., and Stevens, M. J. (2001). Depletion of taurine in experimental diabetic neuropathy: Implications for nerve metabolic, vascular, and functional deficits. *Exp. Neurol.* **168**, 259–272.

Qian, M., and Eaton, J. W. (2000). Glycochelates and the etiology of diabetic peripheral neuropathy. *Free Radic. Biol. Med.* **28**, 652–656.

Qian, M., Liu, M., and Eaton, J. W. (1998). Transition metals bind to glycated proteins forming redox active "glycochelates": Implications for the pathogenesis of certain diabetic complications. *Biochem. Biophys. Res. Commun.* **250**, 385–389.

Reckelhoff, J. F., Hennington, B. S., Kanji, V., Racusen, L. C., Schmidt, A. M., Yan, S. D., Morrow, J., Roberts, L. J., II, and Salahudeen, A. K. (1999). Chronic aminoguanidine attenuates renal dysfunction and injury in aging rats. *Am. J. Hyperten.* **12**, 492–498.

Romero, F. J., Martinez-Blasco, A., Bosch-Morell, F., Marin, N., Trenor, C., and Roma, J. (1999). Glycemic control and not protein kinase C inhibition prevents the early decrease of glutathione peroxidase activity in peripheral nerve of diabetic mice. *J. Periph. Nerv. Syst.* **4**, 265–269.

Ryle, C., Leo, C. K., and Dinghy, M. (1997). Non-enzymatic glycation of peripheral and central nervous system proteins in experimental diabetes mellitus. *Muscle Nerve* **20**, 577–584.

Qi, B., Yamagami, T., Naruse, Y., Sokejima, S., and Kagamimori, S. (1995). Effects of taurine on depletion of erythrocyte membrane Na-K ATPase activity due to ozone exposure or cholesterol enrichment. *J. Nutr. Sci. Vitaminol.* **41**, 627–634.

Raccah, D., Coste, T., Cameron, N. E., Dufayet, D., Vague, P., and Hohman, T. C. (1998). Effect of the aldose reductase inhibitor tolrestat on nerve conduction velocity, Na/K ATPase activity, and polyols in red blood cells, sciatic nerve, kidney cortex, and kidney medulla of diabetic rats. *J. Diabet. Its Complicat.* **12**, 154–162.

Raschke, P., Massoudy, P., and Becker, B. F. (1995). Taurine protects the heart from neutrophil-induced reperfusion injury. *Free Radic. Biol. Med.* **19**, 461–471.

Rittner, H. L., Hafner, V., Klimiuk, P. A., Szweda, L. I., Goronzy, J. J., and Weyand, C. M. (1999). Aldose reductase functions as a detoxification system for lipid peroxidation products in vasculitis. *J. Clin. Invest.* **103**, 1007–1013.

Rokkas, C. K., Cronin, C. S., Nitta, T., Helfrich, L. R., Jr., Lobner, D. C., Choi, D. W., and Kouchoukos, N. T. (1995). Profound systemic hypothermia inhibits the release of neuro-transmitter amino acids in spinal cord ischemia. *J. Thor. Cardiovasc. Surg.* **110**, 27–35.

Sagara, M., Satoh, J., Wada, R., Yagihashi, S., Takahashi, K., Fukuzawa, M., Muto, G., Muto, Y., and Toyota, T. (1996). Inhibition of development of peripheral neuropathy in streptozotocin-induced diabetic rats with N-acetylcysteine. *Diabetologia* **39**, 263–269.

Saito, H. (1995). Biochemical changes in lens, aqueous humor and vitreous body and effects of aldose reductase inhibitor (TAT) on rats with experimental diabetes. *Nippon Ika Daigaku Zasshi — J. Nippon Med. School* **62**, 339–350.

Sasaki, H., Schmelzer, J. D., Zollman, P. J., and Low, P. A. (1997). Neuropathology and blood flow of nerve, spinal roots and dorsal root ganglia in longstanding diabetic rats. *Acta Neuropathol.* **93**, 118–128.

Schaffer, S., Takahashi, K., and Azuma, J. (2000). Role of osmoregulation in the actions of taurine. *Amino Acids* **19**, 527–546.

Schmidt, A. M., Yan, S. D., Wautier, J. L., and Stern, D. (1999). Activation of receptor for advanced glycation end products: A mechanism for chronic vascular dysfunction in diabetic vasculopathy and atherosclerosis. *Circ. Res.* **84**, 489–497.

Schraufstatter, I. U., Hinshaw, D. B., Hyslop, P. A., Spragg, R. G., and Cochrane, C. G. (1986). Oxidant injury of cells: DNA strand-breaks activate polyadenosine diphosphate-ribose polymerase and lead to depletion of nicotinamide adenine dinucleotide. *J. Clin. Invest.* **77**, 1312–1320.

Servitja, J. M., Masgrau, R., Pardo, R., Sarri, E., and Picatoste, F. (2000). Effects of oxidative stress on phospholipid signaling in rat cultured astrocytes and brain slices. *J. Neurochem.* **75**, 788–794.

Shojaee, N., Patton, W. F., Hechtman, H. B., and Shepro, D. (1999). Myosin translocation in retinal pericytes during free-radical induced apoptosis. *J. Cell Biochem.* **75**, 118–129.

Sima, A. A., and Sugimoto, K. (1999). Experimental diabetic neuropathy: An update. *Diabetologia* **42**, 773–788.

Siems, W. G., Grune, T., and Esterbauer, H. (1995). 4-Hydroxynonenal formation during ischemia and reperfusion of rat small intestine. *Life Sci.* **57**, 785–789.

Simbulan-Rosenthal, C. M., Rosenthal, D. S., Iyer, S., Boulares, A. H., and Smulson, M. E. (1998). Transient poly(ADP- ribosyl)ation of nuclear proteins and role of poly(ADP-ribose) polymerase in the early stages of apoptosis. *J. Biol. Chem.* **273**, 13703–13712.

Simmonds, N. J., Allen, R. E., Stevens, T. R., Van Someren, R. N., Blake, D. R., and Rampton, D. S. (1992). Chemiluminescence assay of mucosal reactive oxygen metabolites in inflammatory bowel disease. *Gastroenterology* **103**, 186–196.

Soldatenkov, V. A., and Smulson, M. (2000). Poly(ADP-ribose)polymerase in DNA damage-response pathway: Implications for radiation oncology. *Int. J. Cancer (Radiat. Oncol. Invest.)* **90**, 59–67.

Srivastava, S., Chandra, A., Ansari, N. H., Srivastava, S. K., and Bhatnagar, A. (1998). Identification of cardiac oxidoreductase(s) involved in the metabolism of the lipid peroxidation-derived aldehyde-4-hydroxynonenal. *Biochem. J.* **329**, 469–475.

Srivastava, S., Chandra, A., Bhatnagar, A., Srivastava, S. K., and Ansari, N. H. (1995). Lipid peroxidation product, 4-hydroxynonenal and its conjugate with GSH are excellent substrates of bovine lens aldose reductase. *Biochem. Biophys. Res. Commun.* **217**, 741–746.

Srivastava, S., Chandra, A., Wang, L. F., Seifert, W. E., Jr., DaGue, B. B., Ansari, N. H., Srivastava, S. K., and Bhatnagar, A. (1998). Metabolism of the lipid peroxidation product, 4-hydroxy-trans-2-nonenal, in isolated perfused rat heart. *J. Biol. Chem.* **273**, 10893–10900.

Srivastava, S., Dixit, B. L., Cai, J., Sharma, S., Hurst, H. E., Bhatnagar, A., and Srivastava, S. K. (2000). Metabolism of lipid peroxidation product, 4-hydroxynonenal (HNE) in rat erythrocytes: Role of aldose reductase. *Free Radic. Biol. Med.* **29**, 642–651.

Stevens, M. J., Lattimer, S. A., Kamijo, M., Van Huysen, C., Sima, A. A., and Greene, D. A. (1993). Osmotically-induced nerve taurine depletion and the compatible osmolyte hypothesis in experimental diabetic neuropathy in the rat. *Diabetologia* **36**, 608–614.

Stevens, M. J., Hosaka, Y., Masterson, J. A., Jones, S. M., Thomas, T. P., and Larkin D. D. (1999). Downregulation of the human taurine transporter by glucose in cultured retinal pigment epithelial cells. *Am. J. Physiol.* **277**(4 Pt 1), E760–E771.

Stevens, M. J., Obrosova, I. G., Cao, X., Van Huysen, C., and Greene, D. A. (2000). Effect of DL-α-lipoic acid on peripheral nerve conduction, blood flow, energy metabolism and oxidative stress in experimental diabetic neuropathy. *Diabetes* **49**, 1006–1016.

Stoyanovsky, D. A., Goldman, R., Darrow, R. M., Organisciak, D. T., and Kagan, V. E. (1995). Endogenous ascorbate regenerates vitamin E in the retina directly and in combination with exogenous dihydrolipoic acid. *Curr. Eye Res.* **14**, 181–189.

Szabo C. (1998). Role for poly(ADP-ribose)synthetase in inflammation. *Eur. J. Pharmacol.* **350**, 1–19.

Tachi, Y., Okuda, Y., Bannai, C., Okamura, N., and Bannai, S. (1998). High concentration of glucose causes impairment of the function of the glutathione redox cycle in human vascular smooth muscle cells. *FEBS Lett.* **42**, 19–22.

The DCCT Research Group. (1995). The effect of intensive diabetes therapy on the development and progression of neuropathy. *Ann. Int. Med.* **122**, 561–568.

Thies, R. L., and Autor, A. P. (1991). Reactive oxygen injury to cultured pulmonary artery endothelial cells: Mediation by poly(ADP-ribose) polymerase activation causing NAD depletion and altered energy balance. *Arch. Biochem. Biophys.* **286**, 353–363.

Tomlinson, D. R. (1993). Aldose reductase inhibitors and the complications of diabetes mellitus. *Diab. Med.* **10**, 214–230.

Tomlinson, D. R. (1998). Future prevention and treatment of diabetic neuropathy. *Diabet. Metab.* **24**(Suppl. 3), 79–83.

Tomlinson, D. R. (1999). Mitogen-activated protein kinases as glucose transducers for diabetic complications. *Diabetologia* **42**, 1271–1281.

Tomlinson, D. R., Dewhurst, M., Stevens, E. J., Omawari, N., Carrington, A. L., and Vo, P. A. (1998). Reduced nerve blood flow in diabetic rats: Relationship to nitric oxide production and inhibition of aldose reductase. *Diabet. Med.* **15**, 579–585.

Tomlinson, D. R., Fernyhough, P., and Diemel, L. T. (1996). Neurotrophins and peripheral neuropathy. *Phil. Trans. R. Soc. Lond. B Biol. Sci.* **351**, 455–462.

Tomlinson, D. R., Fernyhough, P., and Diemel, L. T. (1997). Role of neurotrophins in diabetic neuropathy and treatment with nerve growth factors. *Diabetes* **46** (Suppl. 2), S43–S49.

Tomlinson, D. R., Stevens, E. J., and Diemel, L. T. (1994). Aldose reductase inhibitors and their potential for the treatment of diabetic complications. *Trends Pharmacol. Sci.* **15**, 293–297.

Trachtman, H., Futterweit, S., Maesaka, J., Ma, C., Valderrama, E., Fuchs, A., Tarectecan, A. A., Rao, P. S., Sturman, J. A., Boles, T. H., and Baynes, J. (1995). Taurine ameliorates chronic streptozocin-induced diabetic nephropathy in rats. *Am. J. Physiol.* **269**, F429–F438.

Tsukushi, S., Katsuzaki, T., Aoyama, I., Takayama, F., Miyazaki, T., Shimokata, K., and Niwa, T. (1999). Increased erythrocyte 3-DG and AGEs in diabetic hemodialysis patients: Role of the polyol pathway. *Kidney Int.* **55**, 1970–1976.

Tuck, R. R., Schmelzer, J. D., and Low, P. A. (1984). Endoneurial blood flow and oxygen tension in the sciatic nerves of rats with experimental diabetic neuropathy. *Brain* **107**, 935–950.

UK Prospective Diabetes Study Group. (1998). Intensive blood glucose control with sulphonylureas or insulin compared with conventional treatment and risk of complications in patients with type II diabetes (UK PDS 33). *Lancet* **352**, 837–853.

Urata, Y., Yamamoto, H., Goto, S., Tsushima, H., Akazawa, S., Yamashita, S., Nagataki, S., and Kondo T. (1996). Long exposure to high glucose concentration impairs the responsive expression of gamma-glutamylcysteine synthetase by interleukin-1beta and tumor necrosis factor-alpha in mouse endothelial cells. *J. Biol. Chem.* **271**, 15146–15152.

Vander Jagt, D. L., Kolb, N. S., Vander Jagt, T. J., Chino, J., Martinez, F. J., Hunsaker, L. A., and Royer, R. E. (1995). Substrate specificity of human aldose reductase: Identification of 4-hydroxynonenal as an endogenous substrate. *Biochim. Biophys. Acta* **1249**, 117–126.

Walisser, J. A., and Thies, R. L. (1999). Poly(ADP-ribose) polymerase inhibition in oxidant-stressed endothelial cells prevents oncosis and permits caspase activation and apoptosis. *Exp. Cell Res.* **251**, 401–413.

Whisler, R. L., Newhouse, Y. G., Beiqing, L., Karanfilov, B. K., Goyette, M. A., and Hackshaw, K. V. (1994). Regulation of protein kinase enzymatic activity in Jurkat T cells during oxidative stress uncoupled from protein tyrosine kinases: Role of oxidative changes in protein kinase activation requirements and generation of second messengers. *Lymphokine Cytokine Res.* **13**, 399–410.

Winkler, B. S., and Giblin, F. J. (1983). Glutathione oxidation in retina: Effects on biochemical and electrical activities. *Exp. Eye Res.* **36**, 287–297.

Wu, S. M., and Pizzo, S. V. (2001). Alpha(2)-Macroglobulin from rheumatoid arthritis synovial fluid: Functional analysis defines a role for oxidation in inflammation. *Arch.Biochem. Biophys.* **391**, 119–126.

Xie, Z., Kometiani, P., Liu, J., Li, J., Shapiro, J. I., and Askari, A. (1999). Intracellular reactive oxygen species mediate the linkage of Na+/K+-ATPase to hypertrophy and its marker genes in cardiac myocytes. *J. Biol. Chem.* **274**, 19323–19328.

Yagihashi, S., Yamagishi, S., Wada, R., Sugimoto, K., Baba, M., Wong, H. G., Fujimoto, J., Nishimura, C., and Kokai, Y. (1996) Galactosemic neuropathy in transgenic mice for human aldose reductase. *Diabetes* **45**, 56–59.

Yan, S. D., Schmidt, A. M., Anderson, G. M., Zhang, J., Brett, J., Zou, Y. S., Pinsky, D., and Stern, D. (1994) Enhanced cellular oxidant stress by the interaction of advanced glycation end products with their receptors/binding proteins. *J. Biol. Chem.* **269**, 9889–9897.

Yasuda, H., Sonobe, M., Yamashita, M., Terada, M., Hatanaka, I., Huitian, Z., and Shigeta, Y. (1989) Effect of prostaglandin E1 analogue TFC 612 on diabetic neuropathy in streptozocin-induced diabetic rats. Comparison with aldose reductase inhibitor ONO 2235. *Diabetes* **38**, 832–838.

Yeh, L. A., and Ashton, M. A. (1990). The increase in lipid peroxidation in diabetic rat lens can be reversed by oral sorbinil. *Metabolism* **39**, 619–622.

Yim, H. S., Kang, S. O., Hah, Y. C., Chock, P. B., and Yim, M. B. (1995). Free radicals generated during the glycation reaction of amino acids by methylglyoxal: A model study of protein-cross-linked free radicals. *J. Biol. Chem.* **270**, 28228–28233.

Yoshida, T., Maulik, N., Engelman, R. M., Ho, Y. S., and Das, D. K. (2000). Targeted disruption of the mouse Sod I gene makes the hearts vulnerable to ischemic reperfusion injury. *Circ.Res.* **86**, 264–269.

Yue, D. K., McLennan, S., McGill, M., Fisher, E., Heffernan, S., Capogreco, C., and Turtle, J. R. (1990). Abnormalities of ascorbic acid metabolism and diabetic control: Differences between diabetic patients and diabetic rats. *Diabet. Res. Clin. Pract.* **9**, 239–244.

Zatechka, D. S., Kador, P. F., and Lou, M. F. (2000). Effect of aldose reductase inhibitor on the lens signal transduction cascades under diabetic conditions. *Exp. Eye Res.* **71** (Suppl.1), S42.
Zingarelli, B., Szabo, C., and Salzman, A. L. (1999). Blockade of poly(ADP-ribose)synthetase inhibits neutrophil recruitment, oxidant generation, and mucosal injury in murine colitis. *Gastroenterology* **116**, 335–345.

GLYCATION IN DIABETIC NEUROPATHY: CHARACTERISTICS, CONSEQUENCES, CAUSES, AND THERAPEUTIC OPTIONS

Paul J. Thornalley

Department of Biological Sciences, University of Essex, Colchester
Essex CO4 3SQ, United Kingdom

Glycation is the nonenzymatic reaction of glucose, α-oxoaldehydes, and other saccharide derivatives with proteins, nucleotides, and lipids. Early glycation adducts (fructosamines) and advanced glycation adducts (AGEs) are formed. "Glycoxidation" is a term used for glycation processes involving oxidation. Sural, peroneal, and saphenous nerves of human diabetic subjects contained AGEs in the perineurium, endothelial cells, and pericytes of endoneurial microvessels and in myelinated and unmyelinated fibres localized to irregular aggregates in the cytoplasm and interstitial collagen and basement membranes. Pentosidine content was increased in cytoskeletal and myelin protein extracts of the sural nerve of human subjects and cytoskeletal proteins of the sciatic nerve of streptozotocin-induced diabetic rats. AGEs in the sciatic nerve of diabetic rats were decreased by islet transplantation. Improved glycemic control of diabetic patients may be expected to decrease protein glycation in the nerve. Protein glycation may decrease cytoskeletal assembly, induce protein aggregation, and provide ligands for cell surface receptors. The receptor for advanced glycation and products (RAGE) was expressed in peripheral neurons. It is probable that high intracellular glucose concentration is an important trigger for increased glycation, leading to increased formation of methylglyoxal, glyoxal, and

3-deoxyglucosone that glycate proteins to form AGEs intracellularly and extracellularly. Oxidative stress enhances these processes and is, in turn, enhanced by AGE/RAGE interactions. An established therapeutic strategy to prevent glycation is the use of α-oxoaldehyde scavengers. Available therapeutic options for trial are high-dose nicotinamide and thiamine therapies to prevent methylglyoxal formation. Future possible therapeutic strategies are RAGE antagonists and inducers of the enzymatic antiglycation defense. More research is required to understand the role of glycation in the development of diabetic neuropathy.   © 2002, Elsevier Science (USA).

I. Glycation: A Definition

Glycation is the nonenzymatic reaction of glucose, α-oxoaldehydes, and other saccharide derivatives with proteins, nucleotides, and lipids (Thornalley, 1999). It involves a complex series of parallel and sequential reactions, often termed collectively "the Maillard reaction," that form many different adducts, some of which are fluorescent and colored "browning pigments." Studies of glycation in diabetic neuropathy have investigated glycation of proteins only, but glycation of guanyl nucleotides and basic phospholipids also occurs.

II. Nomenclature: Early and Advanced Glycation

Historically, glycation of lysyl side chain and N-terminal amino groups by glucose was the major glycation process studied. Glucose reacts with amino groups initially to form a Schiff's base, which undergoes an Amadori rearrangement to form N-(1-deoxy-D-fructos-1-yl)amino acid or fructosamine (Fischer and Winterhalter, 1981; Neglia *et al.*, 1983; Thornalley, 1999) (see Fig. 1a). Reactive α-oxoaldehydes (glyoxal, methylglyoxal, and 3-deoxyglucosone) are also important precursors of advanced glycation adducts (AGEs) *in vivo* (Fig. 1b). Early glycation adducts are fructosamine derivatives of N-terminal amino groups and lysyl side chains of proteins (Fig. 1c). AGEs are other stable, glycation adducts — hydroimidazolones of glyoxal, methylglyoxal, and 3-deoxyglucosone (G-H, MG-H, and 3DG-H, respectively), N^{ϵ}-carboxymethyl-lysine (CML), N^{ϵ}-(1-carboxyethyl)lysine CEL, pyrraline, bis(lysyl) imidazolium cross-links glyoxal-, methylglyoxal-, and 3-deoxyglucosone-derived lysine dimers (GOLD, MOLD, and DOLD respectively), fluorophores pentosidine and argpyrimidine, tetrahydropyrimidine, and others (Thornalley, 1999) (Fig. 1d). "Glycoxidation" is a

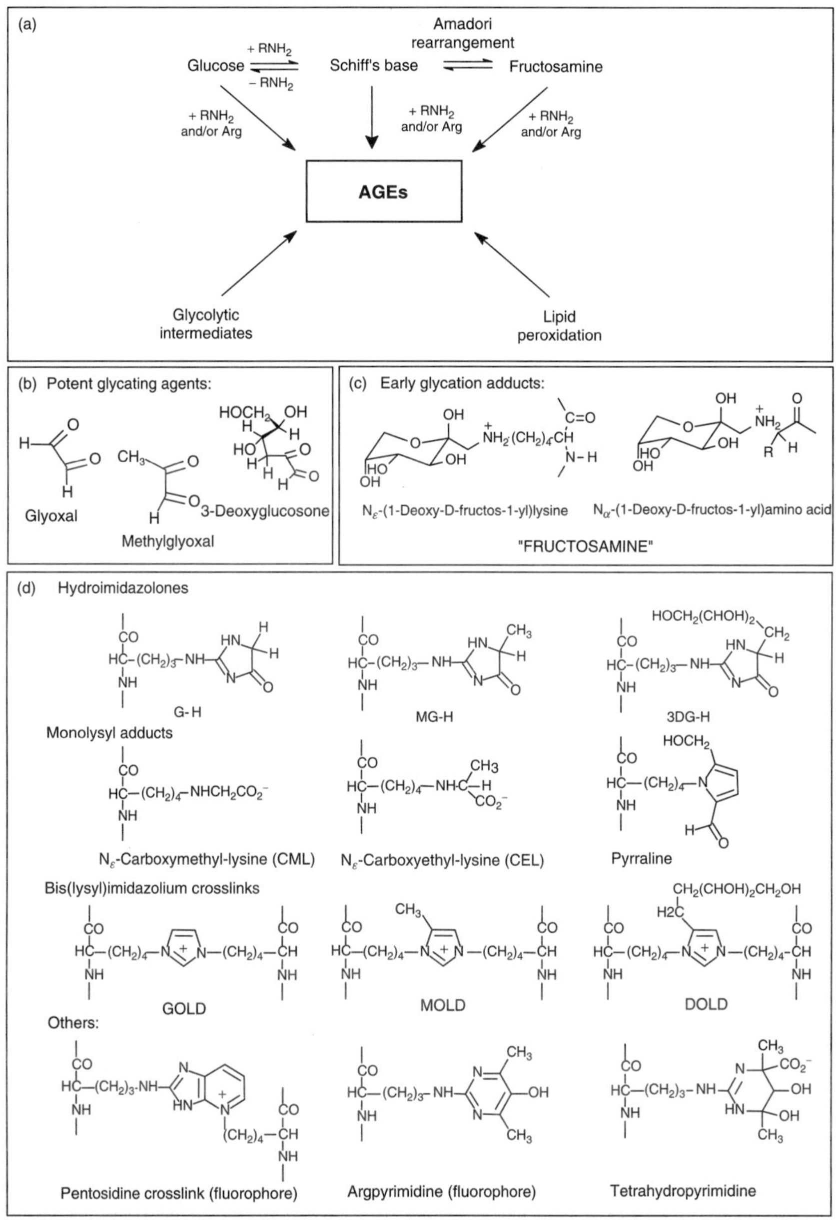

FIG. 1. Formation of early glycation adducts and advanced glycation end products. (a) Formation of advanced glycation end products (AGEs), (b) reactive α-oxoaldehydes in physiological glycation process, (c) early glycation adducts (fructosamines), and (d) molecular structures of AGEs.

term used for glycation processes involving oxidation. AGEs thereby formed are called "glycoxidation products" (Baynes and Thorpe, 1999). AGEs pentosidine and CML are glycoxidation products.

At least four processes are involved in the formation of AGEs in physiological systems: (i) monosaccharide autoxidation (autoxidative glycosylation) — the degradation of saccharide unattached to a protein, (ii) Schiff's base fragmentation, (iii) fructosamine degradation, and (iv) direct reaction of α,β-dicarbonyl compounds formed from the degradation of glycolytic intermediates and lipid peroxidation with proteins (Thornalley et al., 1999) (Fig. 1a). For brevity, these processes are not discussed further herein but have been covered in detail elsewhere (Thornalley, 1999; Baynes and Thorpe, 1999; Thornalley et al., 1999).

Glycation adducts are assayed by chromatographic and immunochemical techniques. Glycation is usually (i.e., in the nondiabetic state) a minor posttranslational modification. In proteins, the extent of glycation depends on concentration, reactivity and half-life of the protein substrate, and the concentration and reactivity of the glycating agent. The range of extent of glycation is typically 0.1–1% of lysine and arginine residues in proteins. For serum albumin, the concentration ranges (mol/mol albumin) of Schiff's base adduct, fructosamine, and AGEs are ca. 1–5, 6–15, and 0.01–7% (Day et al., 1979; Degenhardt et al., 1998). Only fructosyl-lysine, CML/CEL, pentosidine, and AGEs of unknown structure have been studied in diabetic neuropathy to date.

III. Evidence for Increased Glycation in Diabetic Neuropathy

Early glycation of peripheral and central nervous system proteins of alloxan-induced (AXN) diabetic rats and normal controls was studied by Vlassara and co-workers in 1981. Using sodium [^{3}H]borohydride, which reduces fructosyl-lysine to N^{ϵ}-2-[^{3}H]glucitol-lysine, they found a two- to fivefold increased incorporation of radioactivity into protein extracts of sciatic nerve and brain of AXN rats after 3 months of diabetes (Vlassara et al., 1983). This was also studied in streptozotocin-induced (STZ) diabetic rats, where glycation of tubulin in brain tissue was found. Tubulin from STZ diabetic rats had a 20% decreased assembly into microtubules (Williams et al., 2001). Cullum et al. (1991) did not find increased glycation of brain tubulin in STZ diabetic rats, however, but they did find a fourfold increase in the incorporation of radioactivity into tubulin extracted and purified from sciatic nerve after 2 weeks of diabetes. They also found increased glycation of actin in microfilaments in the brain and spinal cord of STZ diabetic rats

(Mclean *et al.*, 1992). In cytoskeletal and myelin protein extracts of sural nerve of human subjects, markers of fructosyl-lysine content — furosine in acid hydrolysates and [³H]borohydride reduction — were assayed in the high molecular mass protein fraction. No significant increase of either of these markers was found in diabetes mellitus with respect to normal controls (Ryle and Donaghy, 1995).

The presence of AGEs in peripheral nerve tissue of human diabetic and nondiabetic subjects was investigated by Sugimoto *et al.* (2001) using immunohistochemical techniques with the murine anti-AGE monoclonal antibody 6D12. This antibody was initially claimed to recognize only the AGE epitope CML (Horiuchi *et al.*, 1991), but this has been reappraised recently and it also recognizes CEL (Nagai *et al.* 1999). CML is an AGE compound that has been studied widely. It is formed from the oxidative degradation of fructosyl-lysine (Ahmed *et al.*, 1986) and by the reactions of glyoxal and ascorbic acid with lysine residues in proteins. CEL is formed by the reaction of methylglyoxal, a potent glycating agent formed from the degradation of triosephosphates (Phillips and Thornalley, 1993), with lysine residues in proteins (Ahmed *et al.*, 1997). CML is a glycoxidation product but CEL is not. Sural, peroneal, and saphenous nerve tissue were obtained from donors. AGE immunoreactivity was localized in the perineurium, endothelial cells, and pericytes of endoneurial microvessels and in myelinated and unmyelinated fibers. AGE immunoreactivity was associated with irregular aggregates in the cytoplasm of endothelial cells, pericytes, axoplasm, and Schwann cells of both myelinated and unmyelinated fibers. Interstitial collagen and basement membranes of the perineurium also contained AGEs. AGE immunoreactivity was found in normal and diabetic nerve but was more intense in the latter. This increased immunoreactivity correlated with a decreased density of myelinated fibers but not directly with the degeneration of nerve fibers (Sugimoto *et al.*, 2001). The pentosidine content of cytoskeletal and myelin protein extracts of the sural nerve of human subjects was increased significantly (Ryle and Donaghy, 1995). Pentosidine was also increased in cytoskeletal proteins of the sciatic nerve of STZ diabetic rats after 8 weeks of diabetes (Ryle *et al.*, 1997).

AGEs were detected in brain, spinal cord, and sciatic nerve of STZ diabetic rats by immunoassay with polyclonal anti-AGE antibodies of unidentified AGE epitope specificity. AGEs were increased at all sites after 4, 8, and 12 months. Islet transplantation after 4 and 8 months returned blood glucose and nervous tissue AGEs to normal levels. This illustrates that there is a turnover of AGE-modified proteins in the peripheral and central nervous system (Sensi *et al.*, 1998). The accumulation of AGEs in the peripheral nerve in diabetic neuropathy, therefore, may occur as a result of increased formation and decreased clearance of AGE-modified proteins. Improvement of glycemic control in diabetic patients may be expected to decrease protein glycation in the nerve whenever improved glycemic control is initiated. It is not known if there would be an accompanying clinical benefit. There is a recognized effect of "glycemic memory" where established poor glycemic control followed by a period of improved glycemic control was associated with the continued development of microvascular complications without significant benefit from the improved glycemic control (Engerman and Kern, 1987).

Glycated rat serum albumin had increased penetration of the sciatic nerve of control and STZ diabetic rats than unmodified albumin. The entry of glycated serum albumin into the sciatic nerve was greater in control than in STZ diabetic rats (Patel *et al.*, 1991). An increased albumin content has been found in the endoneurium of diabetic patients (Ohi *et al.*, 1985; Poduslo *et al.*, 1988).

To date, therefore, there have been very few studies following the changes of glycation adducts, particularly AGEs, in peripheral nerve during the development of diabetic neuropathy. This is an area of diabetes research that deserves further attention.

IV. Indirect Evidence for Involvement of Increased Glycation in Diabetic Neuropathy: Effects of Antiglycation Agents

Aminoguanidine (Pimagedine) has been a pioneering therapeutic agent in studies of glycation in the development of diabetic complications. It intervenes in glycation reactions to scavenge α-oxoaldehydes, particularly methylglyoxal, glyoxal, 3-deoxyglucosone, and α,β-dicarbonyls in glycated proteins (Thornalley *et al.*, 2000). It is also an inhibitor of inducible nitric oxide synthase (Tilton *et al.*, 1993) and amine oxidase (Yu and Zou, 1997) and has antioxidant activity (Giardino *et al.*, 1998). Aminoguanidine (50 mg/kg) inhibited the decrease (57%) in nerve blood flow in STZ diabetic rats after 8 weeks of diabetes and improved the conduction velocity of sciatic-tibial and caudal nerves over 16 to 24 weeks, respectively (Kihara *et al.*, 1991). The improvement of peripheral nerve conduction velocity in STZ diabetic rats by aminoguanidine was corroborated in two later studies (Cameron *et al.*, 1992; Wada *et al.*, 1999). Aminoguanidine (25 mg/kg, sc) for 16 weeks decreased the expansion of endoneurial microvessels in STZ diabetic rats (Sugimoto and Yagihashi, 1997). Aminoguanidine did not affect the incidence or ultrastructural appearance of neuroaxonal dystrophy, a feature of autonomic neuropathy, in STZ diabetic rats (Schmidt *et al.*, 1996). A study in STZ diabetic baboons showed no evidence of improvement by aminoguanidine of declining nerve conduction velocity and autonomic dysfunction (heart rate response to positional change) (Birrell *et al.*, 2000).

None of the studies just given determined AGEs or α-oxoaldehydes in diabetic animal models. Aminoguanidine adducts of glyoxal, methylglyoxal, and 3-deoxyglucosone were formed in aminoguanidine-treated STZ diabetic rats (Araki *et al.*, 1998), although the residual concentrations of these α-oxoaldehydes were not measured. It also decreased aortal pentosidine and cross-linking, but it did not prevent increased albumin permeation of the sciatic nerve (Nyengaard *et al.*, 1997; Brownlee *et al.*, 1986). Aminoguanidine failed to prevent the diabetes-induced increase in CML and pentosidine in rat skin collagen (Degenhardt *et al.*, 1999) and collagen-associated fluorescence of skin, kidney, and aorta (Nyengaard *et al.*, 1997). Thus, where aminoguanidine has shown functional beneficial effects in experimental neuropathy, the observations have not been accompanied by demonstrations of decreased formation of AGEs. It is not clear, therefore, if the beneficial effects of aminoguanidine on neuropathy can be attributed to inhibition of AGE accumulation.

V. Glycation-Related Correlates in Risk Analysis of Clinical Diabetic Neuropathy

The outcome of long-term clinical trials of the effect of intensive therapy of hyperglycemia in clinical diabetes mellitus type 1 (DCCT) and type 2 (UKPDS) has linked hyperglycemia with the incidence of microvascular complications, including diabetic neuropathy (The Diabetes Control and Complications Trial Research Group, 1993; UK Prospective Diabetes Study Group, 1998). In the DCCT trial, slower median and peroneal motor nerve conduction, lower Valsalva ratio (indicating autonomic neuropathy) was associated with higher glycated hemoglobin HbA_{1c}. Generally, increased HbA_{1c} was associated with electrophysiological impairment (The DCCT Research Group, 1998). Variables related to the glycation of skin (furosine, CML, and pentosidine) were related positively to the incidence of neuropathy and decreased nerve condition velocity in a subset of the DCCT cohort (Monnier *et al.*, 1999). In this analysis, fructosamine and AGEs of proteins are risk markers for the development of diabetic neuropathy because fructosamine is an indicator of medium term glycemic control (Goldstein *et al.*, 1995) and AGEs are indicators of medium or long-term glycemic control, depending on the AGE and protein substrate studied (Beisswenger *et al.*, 1993; Salmela *et al.*, 1995). [Note that some AGEs are relatively short-lived — they have half-lives from *ca.* 1 day–2 weeks, whereas others are long-lived and their concentrations reflect glycemic status and half-life of the protein substrate (Verzijl *et al.*, 2000)].

VI. Functional Effects of Increased Glycation in Diabetic Nerve

The modification and cross-linking of proteins by AGEs may result in significant alterations in their structural and functional characteristics. Unlike Amadori products, AGEs are formed irreversibly. Therefore, they may accumulate on long-lived proteins *in vivo* such as nerve myelin. The chemical stability of the AGE and the turnover of protein substrate are determinants of the AGE content of proteins. The minimal modification of proteins found *in vivo* has three major effects on protein structure and function: (i) decreased functional polymerization of tubulin and actin to form microtubules and microfilaments of the cytoskeleton, (ii) decreased protein solubility and susceptibility to enzymatic digestion of AGE-modified proteins, and (iii) binding to cell surface receptors.

It is not clear if the glycation of tubulin and actin in diabetes *in vivo* leads to a significant deficit of the microtubular and microfilament cytoskeleton (Williams *et al.*, 2001; Cullum *et al.*, 1991; Mclean *et al.*, 1992). Sugimoto and co-workers (2001) found AGEs localized to irregular aggregates in peripheral neurons and endoneurial cells. This may be an effect of AGE-mediated protein cross-linking and a change in protein surface charge and hydrophobicity by AGE modification (Westwood and Thornalley, 1995).

Proteins modified with AGEs bind to cell surface receptors and other proteins. Several AGE receptors and AGE-binding proteins have been proposed: the receptor for advanced glycation end products (RAGE) (Neeper *et al.*, 1992), oligosaccharyl transferase complex protein 48 (OST-48) (Yang *et al.*, 1991), 80K-H protein (Yang *et al.*, 1991), galectin-3 (Vlassara *et al.*, 1995), and others (Li *et al.*, 1995; Araki *et al.*, 1995; Ohgami *et al.*, 2001). RAGE expression was found in peripheral neurons (Brett *et al.*, 1993). OST-48 is a component of the enzymatic glycosylation complex of the endoplasmic reticulum (Fu and Kreibich, 2000) and hence is expected to be present in neuronal cell bodies. 80K-H is an adaptor protein linked to fibroblast growth factor receptor-3 (FGFR3), which is expressed at low levels in the peripheral nerve (Grothe *et al.*, 2001); hence, 80K-H is probably expressed in the peripheral nerve too. In the nervous system, galectin-3 is only expressed in microglia (Pesheva *et al.*, 1998). There is doubt concerning whether some of these proteins are indeed AGE receptors: OST-48 and 80K-H are not on the cell surface but have an endoplasmic reticulum and cytoplasmic location, respectively (reviewed in

Thornalley, 1998). For the RAGE receptor, however, it has been established that the receptor has a cell surface location, it binds proteins glycated by AGEs *in vivo,* and the AGE epitope recognition characteristics have been investigated.

VII. Receptor for Advanced Glycation End Products

The receptor for advanced glycosylation end products is a member of the immunoglobulin superfamily of proteins (Neeper *et al.,* 1992; Schmidt *et al.,* 1994). Human RAGE has an extracellular V-shaped domain of 320 amino acids with two N-linked oligosaccharides, a membrane-spanning domain of 21 amino acids, and a short cytoplasmic domain of 41 amino acids. Two forms of RAGE (35 and 45 kDa) were detected by immunoblotting in most tissues. The proteolytic 35-kDa fragment of the extracellular domain was termed soluble RAGE (sRAGE), which has been used as an affinity ligand for AGE-modified proteins (Yan *et al.,* 1994) and as a pharmacological agent to prevent the vascular effects of advanced glycation in experimental diabetes *in vivo* (Renard *et al.,* 1997). An alternative gene splicing has been found to express a form of RAGE that lacks the 19 amino acids of the membrane-spanning domain. This form of RAGE, termed hRAGEsec, was composed of 321 amino acids, had a molecular mass of 36 kDa, and was secreted (Malherbe *et al.,* 1999).

RAGE is thought to exist in a complex with lactoferrin-like protein (LF-L) (Fig. 2). The binding site for RAGE in LF-L was in the C-terminal half of the protein (amino acids 357–708). The dissociation constant K_D for LF-L binding to RAGE was 100 pM. The binding affinities of the AGE-modified protein to RAGE, LF-L, and RAGE-LF-L complex were remarkably similar (Schmidt *et al.,* 1994). AGE-BSA binding to RAGE had a K_D value of *ca.* 100 nM (Neeper *et al.,* 1992).

In the central nervous system, RAGE also bound amphoterin. Amphoterin is a DNA-binding protein that is strongly expressed in developing neurons and mediates neurite outgrowth. Amphoterin bound to RAGE with high affinity: the K_D value was 6.4 nM. The high expression of RAGE in cultured rat embryonic neurons and its binding to amphoterin suggest that RAGE may promote neurite formation during normal brain development (Hori *et al.,* 1995). Indeed, amphoterin increased the transcription and translation of RAGE in neuronal cells, a response linked to the Sp-1-binding element in the RAGE gene promoter (Li and Schmidt, 1997). RAGE also binds S100/calgranulin proteins. S100 proteins are a closely related family of acidic calcium-binding proteins with a molecular mass of

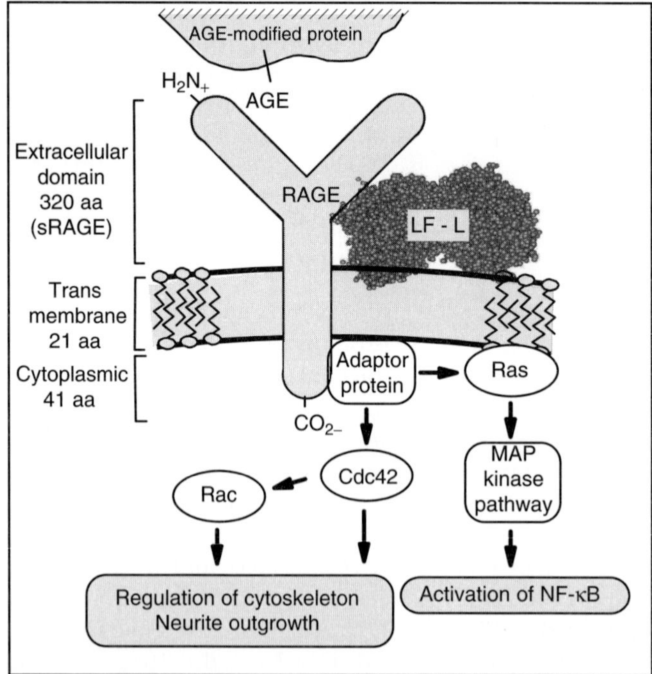

FIG. 2. Receptor for advanced glycation end products (RAGE).

10–14 kDa (Schafer and Heizmann, 1996). The binding of both ampho-terin and S100/calgranulins to RAGE in neuronal cells stimulated neurite outgrowth and activation of NF-κB. This was associated with increased cell survival, related to the RAGE-mediated increased expression of the anti-apoptotic protein Bcl-2. This response was promoted by low concentrations of S100/calgranulins (nM), whereas high concentrations (μM) induced apoptosis by induction of oxidative stress. Both trophic and cytotoxic effects were mediated by RAGE (Huttunen *et al.*, 2001).

Signal transduction studies with neuronal cells suggested that Cdc42 and Rac of the Rho family of small GTPases were involved in signal transduction for amphoterin-stimulated, RAGE-mediated neurite outgrowth, whereas Ras was involved in the RAGE-mediated activation of NF-κB (Huttunen *et al.*, 1999) (Fig. 2). Binding of AGE-BSA to RAGE induced oxidative stress, as indicated by increased thiobarbituric acid reactive substances (TBARS) and the expression of hemoxygenase. This was linked to the Ras pathway, leading to the activation of NF-κB (Yan *et al.*, 1994; Huttunen *et al.*, 1999).

sRAGE linked to an affinity matrix bound AGE-modified proteins from physiological samples. This demonstrated that RAGE binds proteins modified minimally by AGEs found *in vivo* (Yan *et al.*, 1994). CML was an AGE epitope recognized by RAGE. Albumin modified by CML was bound specifically by RAGE: bovine serum albumin (BSA) with 18 CML residues per mole had a K_D value of 76 nM. sRAGE prevented the binding. Ovalbumin modified minimally with CML (0.66 mol CML/mol) activated NF-κB in endothelial cells, and activation was blocked by anti-RAGE IgG. Hence, CML is a probable AGE epitope for RAGE linked to cell activation *in vivo* (Kislinger *et al.*, 1999). Evidence suggests that methylglyoxal-derived hydroimidazolone may also be a RAGE recognition factor (Ng *et al.*, 2001).

VIII. Raised Intracellular Glucose: Trigger for Increased Glycation by α-Oxoaldehydes

A key feature of the locus of development of diabetic complications is that intracellular glucose rises with hyperglycemia in cells that take up glucose by the insulin-independent glucose transporter, GLUT1. Cells forming the perineurium and endoneurial capillaries, plus Schwann cells, all express GLUT1 (Muona *et al.*, 2001). In aged animals, the neuronal glucose transporter GLUT3 was also detected in perineurium and endoneurial vessel walls, paranodal Schwann cells, and axonal membranes (Magnani *et al.*, 1996). During hyperglycemia, endoneurial capillary endothelial cells, pericytes, Schwann cells, and neurons will internalize increased amounts of glucose, causing supranormal intracellular glucose concentrations. Increased metabolic flux through glycolysis leads to an accumulation of triosephosphates and an increased formation of methylglyoxal (Thornalley, 1988; Phillips and Thornalley, 1993). Fructose-1,6-bisphosphate, the major component of the triosephosphate pool, was increased 60–80% in the sciatic nerve of STZ diabetic rats (Thurston *et al.*, 1995). Increased flux through the polyol pathway leads to an increased formation of fructose in the sciatic nerve in STZ diabetic rats (Poulsom *et al.*, 1983), and increasing the formation of fructose-3-phosphate by 3-phosphokinase catalyzed phosphorylation, which then degrades to 3-deoxyglucosone (Petersen *et al.*, 1990). Increased electron flux into complex II of respiration increases the formation of hydrogen peroxide by mitochondria and induction of oxidative stress (Nishikawa *et al.*, 2000), leading to the formation of glyoxal from lipid peroxidation (Fu *et al.*, 1996). Oxidative stress may further exacerbate triosephosphate accumulation and methylglyoxal formation by inactivating glyceraldehyde-3-phosphate dehydrogenase and depleting cells of the NAD^+ cofactor by the activation of poly(ADP-ribose) polymerase (PARP) (Hyslop *et al.*, 1988; Soriano *et al.*, 2001). Glycation of nucleotides may also activate PARP. Oxidative stress also enhances the concentrations of glyoxal, methylglyoxal, and 3-deoxyglucosone by *in situ* depletion of cellular cofactors required for α-oxoaldehyde detoxification: GSH for glyoxalase I-catalyzed detoxification of glyoxal and methylglyoxal, and NADPH for the detoxification of 3-deoxyglucosone (Abordo *et al.*, 1999). These potent glycating agents may be formed inside cells and glycate proteins therein, but they also cross cell membranes to glycate proteins extracellularly. Extracellular glycation forms CML, methylglyoxal-derived hydroimidazolone, and other AGEs that may activate RAGE (Kislinger *et al.*, 1999; Ng *et al.*, 2001). Activation of RAGE is linked to activation of an NADPH oxidase in some cell types (Wautier *et al.*, 2001), further exacerbating oxidative stress.

Activation of NADPH oxidase by AGEs is a potentially important source of oxidative stress in diabetes. NADPH oxidase forms hydrogen peroxide, via superoxide dismutation, at a very high rate — equivalent to a flux of 10–500 m*M* per day, depending on the cell type (Rossi, 1986). The only other source of hydrogen peroxide that has a flux similar to this is mitochondrial respiration — equivalent to a flux of 10–20 m*M* per day (Chance *et al.*, 1979). Compared to this, fluxes of peroxide generated by glycation directly are very low: monosaccharide autoxidation generates hydrogen peroxide at a rate of *ca.* 20 n*M* per day and fructosamine degradation at a rate of *ca.* 2 μ*M* per day (Thornalley *et al.*, 1999; Smith and Thornalley, 1992). If substantiated further, therefore, RAGE-mediated activation of an NADPH oxidase could be an important source of oxidative stress in diabetes.

Increased intracellular glucose concentration is now perceived to play an important role in diabetic neuropathy. The increased formation of AGEs and oxidative stress (Sugimoto *et al.*, 2001; Hounsom *et al.*, 2001) are consequences of this.

IX. Therapeutic Options

It is not yet clear if early or advanced glycation adducts are risk factors for diabetic neuropathy, hence it is not known if decreasing these adducts will prevent the development of diabetic neuropathy. Assuming that glycation adducts, particularly AGEs, are risk factors for diabetic neuropathy, there are several strategies to prevent glycation and its consequences.

A. Suppressing Formation of Methylglyoxal

This may be achieved by suppressing the accumulation of triosephos-phates in hyperglycemia. Nicotinamide supplementation may counter the depletion of cellular NAD$^+$ in hyperglycemia (Wahlberg *et al.*, 2000). Increased NAD$^+$ levels will feed through to NADP$^+$ and NADPH, and the decline in these metabolites in diabetes may also be prevented; consti-tutive nitric oxide synthase activity should also be preserved. Normalization of NAD$^+$ will also diminish the ability of the polyol pathway to limit the availability of NAD$^+$ (Williamson *et al.*, 1993). The pharmacokinetics and safety profile of nicotinamide have been well studied (Kuip *et al.*, 2000). Hence, the design of prospective therapeutic dosing schedules is relatively straightforward.

An alternative approach is to stimulate the consumption of glyceraldehyde-3-phosphate (GA3P) by the nonoxidative pentosephosphate pathway. GA3P and fructose-6-phosphate (F-6-P) are converted to ribose-5-phosphate (R-5-P): $2GA3P + 4F\text{-}6\text{-}P \rightarrow 6R\text{-}5\text{-}P$ (Fig. 3). This may be achieved by high-dose thiamine supplementation, which maximizes the activity of transketolase with a saturating concentration of thiaminepyrophosphate (TPP) cofactor. Increased R-5-P may stimulate *de novo* purine synthesis by increasing the availability of ribosylpyrophosphate.

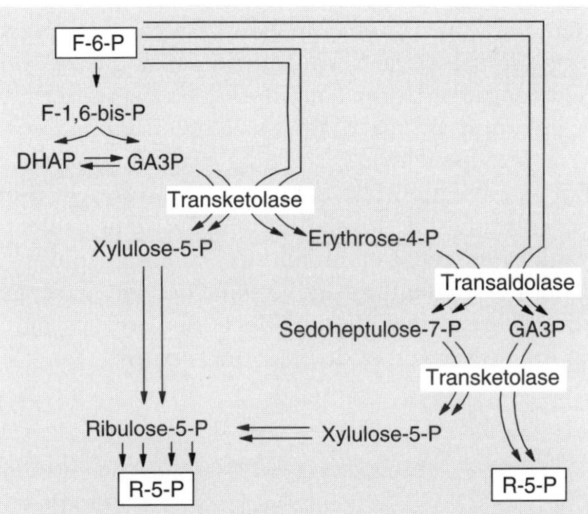

Fig. 3. The nonoxidative pentosephosphate pathway. F-6-P, fructose-6-phosphate; F-1,6-bis-P, fructose-1,6-bisphosphate; DHAP, dihydroxyacetonephosphate; GA3P, glyceral-dehyde-3-phosphate; R-5-P, ribose-5- phosphate.

Triosephosphate accumulation and increased formation of methylglyoxal induced by hyperglycemia in red blood cells *in vitro* were suppressed by high-dose thiamine (Thornalley *et al.*, 2001). Streptozotocin-induced diabetic rats given 70 mg/kg thiamine or 100 mg/kg *S*-benzoylthiaminephosphate (Benfotiamine) daily for 6 months had decreased development of diabetic neuropathy and advanced glycation (Hammes *et al.*, 1998). Clinical trials of Benfotiamine decreased the formation of methylglyoxal-derived AGEs (Lin *et al.*, 2000) and suppressed the development of diabetic neuropathy (Winkler *et al.*, 1999). Clinical diabetes mellitus was associated with a mild thiamine deficiency in some instances. In a study of diabetic subjects (type not specified), 76% of subjects had a thiamine concentration lower than the lower limit of the normal range (Saito *et al.*, 1987). In a study of type 2 diabetic subjects, 18% of subjects had a thiamine concentration lower than the lower limit of the normal range (Havivi *et al.*, 1991). When subjects were given 20–50 mg thiamine per day (20- to 50-fold the normal daily requirement), 14 of 24 subjects had maximum, TPP-saturated transketolase activity (Saito *et al.*, 1987). This work and related studies now provide a biochemical basis for the prevention of diabetic complications by high-dose thiamine therapy. The prevention of diabetic neuropathy by thiamine now deserves attention.

B. α-Oxoaldehyde Scavengers

Aminoguanidine was a prototype agent for the preventive therapy of diabetic complications. It reacts with physiological glyoxal, methylglyoxal, and 3-deoxyglucosone to form 3-amino-1,2,4-triazine derivatives (Fig. 4) and prevent glycation by these agents *in vitro* and *in vivo*. Studies of the kinetics of α-oxoaldehyde-scavenging reactions of aminoguanidine suggested that it was kinetically competent to scavenge α-oxoaldehydes and decrease related AGE formation *in vivo*. This effect was limited, however, by the rapid renal elimination of aminoguanidine (Thornalley *et al.*, 2000). An aminoguanidine-like α-oxoaldehyde scavenger that has a longer pharmacokinetic half-life and decreased toxicity may find future clinical use in the prevention of diabetic neuropathy

C. Other Strategies

Other strategies to prevent glycation effects in diabetic neuropathy are the development of AGE receptor antagonists to block the activation of RAGE during periods of excessive AGE accumulation (Ng *et al.*, 2001). Alternatively, pharmacological agents that stimulate the activities of

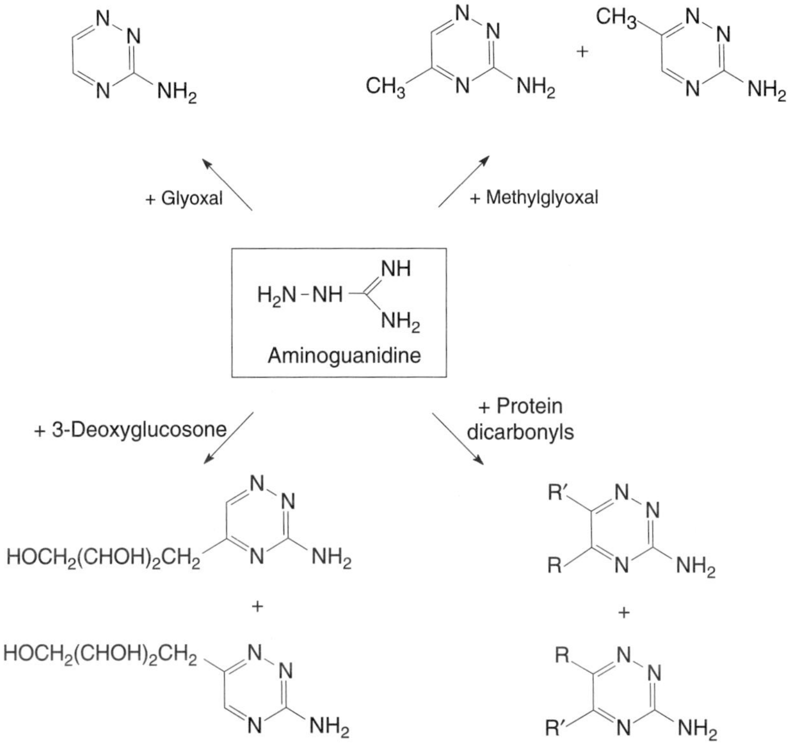

FIG. 4. Scavenging of α,β-dicarbonyl compounds by aminoguanidine.

enzymes that detoxify α-oxoaldehydes (glyoxalase I and 3-deoxyglucosone reductase) may be developed.

X. Conclusion

The role of glycation in diabetic neuropathy is often implicated but has been little studied and is poorly understood. More research is required to study the formation, turnover, and functional events of glycation adducts in the peripheral nerve in diabetes. Because glycation and oxidative stress are interlinked processes, a full description of the pathological events in the development of diabetic neuropathy cannot be achieved without it. It will be interesting to see in the years to come if nicotinamide, thiamine, and antioxidants can prevent diabetic neuropathy.

References

Abordo, E. A., Minhas, H. S., and Thornalley, P. J. (1999). Accumulation of α-oxoaldehydes during oxidative stress: A role in cytotoxicity. *Biochem. Pharmacol.* **58**, 641–648.

Ahmed, M. U., Brinkmann Frye, E., Degenhardt, T. P., Thorpe, S. R., and Baynes, J. W. (1997). N^ϵ-(Carboxyethyl)lysine, a product of chemical modification of proteins by methylglyoxal, increases with age in human lens proteins. *Biochem. J.* **324**, 565–570.

Ahmed, M. U., Thorpe, S. R., and Baynes, J. W. (1986). Identification of N^ϵ-carboxymethyl-lysine as a degradation product of fructoselysine in glycated protein. *J. Biol. Chem.* **261**, 4889–4894.

Araki, A., Glomb, M. A., Takahashi, M., and Monnier, V. M. (1998). Determination of dicarbonyl compounds as aminotriazines during the Maillard reaction and *in vivo* detection in aminoguanidine-treated rats. *In* "The Maillard Reaction in Foods and Medicine" (J. O'Brien, H. E. Nursten, M. J. C. Crabbe, and J. M. Ames eds.), p. 400. RSC Publ., Cambridge, UK.

Araki, N., Higashi, T., Mori, T., Shibayama, R., Kawabe, Y., Kodama, T., Takahashi, K., Shichiri, M., and Horiuchi, S. (1995). Macrophage scavenger receptor mediates the endocytic uptake and degradation of advanced glycation end-products of the Maillard reaction. *Eur. J. Biochem.* **230**, 408–415.

Baynes, J. W., and Thorpe, S. R. (1999). Role of oxidative stress in diabetic complications: A new perspective on an old paradigm. *Diabetes* **48**, 1–9.

Beisswenger, P. J., Moore, L. L., and Curphey, T. J. (1993). Relationship between glycemic control and collagen-linked advanced glycosylation end products in type I diabetes. *Diabetes Care* **16**, 689–694.

Birrell, A. M., Hefferman, S. J., Ansselin, A. D., McLennan, S., Church, D. K., Gillin, A. G., and Yue, D. K. (2000). Functional and structural abnormalities in the nerves of type 1 diabetic baboons: Aminoguanidine treatment does not improve nerve function. *Diabetologia* **43**, 110–116.

Brett, J., Schmidt, A.-M., Yan, S. D., Zou, S., Weidman, E., Pinsky, D., Nowygrod, R., Neeper, M., Przysiecki, C., Shaw, A., Migheli, A., and Stern, D. (1993). Survey of the distribution of a newly characterized receptor for advanced glycation end-products in tissues. *Am. J. Pathol.* **143**, 1699–1712.

Brownlee, M., Vlassara, H., Kooney, A., Ulrich, P., and Cerami, A. (1986). Aminoguanidine prevents diabetes-induced arterial wall protein cross-linking. *Science* **232**, 1629–1632.

Cameron, N. E., Cotter, M. A., Dines, K., and Love, A. (1992). Effects of aminoguanidine on peripheral nerve function and polyol pathway metabolites in streptozotocin-diabetic rats. *Diabetologia* **35**, 946–950.

Chance, B., Sies, H., and Boveris, A. (1979). Hydroperoxide metabolism in mammalian organs. *Physiol. Rev.* **59**, 527–605.

Cullum, N. A., Mahon, J., Stringer, K., and Mclean, W. G. (1991). Glycation of rat sciatic nerve tubulin in experimental diabetes mellitus. *Diabetologia* **34**, 387–398.

Day, J. F., Thorpe, S. R., and Baynes, J. W. (1979). Nonenzymatically glycosylated albumin (*in vitro* preparation and isolation from normal human serum). *J. Biol. Chem.* **254**, 595–597.

Degenhardt, T. P., Fu, M.-X., Voss, E., Reiff, K., Neidlein, R., Strein, K., Thorpe, S. R., Baynes, J. W., and Reiter, R. (1999). Aminoguanidine inhibits albuminuria but not the formation of advanced glycation end-products in skin collagen of diabetic rats. *Diabetes Res. Clin. Pract.* **43**, 81–89.

Degenhardt, T. P., Thorpe, S. R., and Baynes, J. W. (1998). Chemical modification of proteins by methylglyoxal. *Cell. Mol. Biol.* **44**, 1139–1145.

Engerman, R. L., and Kern, T. S. (1987). Progression of incipient diabetic retinopathy during good glycemic control. *Diabetes* **36**, 808–812.

Fischer, R. W., and Winterhalter, K. H. (1981). The carbohydrate moiety in hemoglobin A1c is present in the ring form. *FEBS Lett.* **135**, 145–147.

Fu, J., and Kreibich, G. (2000). Retention of subunits of the oligosaccharyltransferase complex in the endoplasmic reticulum. *J. Biol. Chem.* **275**, 3984–3990.

Fu, M.-X., Requena, J. R., Jenkins, A. J., Lyons, T. J., Baynes, J. W., and Thorpe, S. R. (1996). The advanced glycation end product, N^ϵ-(carboxymethyl)lysine, is a product of both lipid peroxidation and glycoxidation reactions. *J. Biol. Chem.* **271**, 9982–9986.

Giardino, I., Fard, A. K., Hatchell, D. L., and Brownlee, M. (1998). Aminoguanidine inhibits reactive oxygen species formation, lipid peroxidation, and oxidant-induced apoptosis. *Diabetes* **47**, 1114–1120.

Goldstein, D. E., Malone, J. I., Little, R. R., Nathan, D., Lorenz, R. A., and Petersen, C. M. (1995). Tests for glycaemia in diabetes. *Diabetes Care* **18**, 896–909.

Grothe, C., Meisenger, C., and Claus, P. (2001). *In vivo* expression and localization of the fibroblast growth factor system in the intact and lesioned rat peripheral nerve and spinal ganglia. *J. Comp. Neurol.* **434**, 342–357.

Hammes, H.-P., Bretzel, R. G., Federlin, K., Horiuchi, S., Niwa, T., and Stracke, H. (1998). Benfotiamin inhibits the formation of advanced glycation end products in diabetic rats. *Diabetologia* **41**, 1164–1164. [Abstract]

Havivi, E., Bar On, H., Reshef, A., and Raz, I. (1991). Vitamins and trace metals status in non insulin dependent diabetes mellitus. *Int. J. Vit. Nutr. Res.* **61**, 328–333.

Hori, O., Brett, J., Slattery, T., Cao, R., Zhang, J., Chen, J. X., Nagashima, M., Lundh, E. R., Vijay, S., Nitecki, D., Morser, J., Stern, D., and Schmidt, A.-M. (1995). The receptor for advanced glycation endproducts (RAGE) is a cellular binding site for amphoterin. *J. Biol. Chem.* **270**, 25752–25761.

Horiuchi, S., Araki, N., and Morino, Y. (1991). Immunochemical approach to characterize advanced glycation end products of the Maillard reaction. *J. Biol. Chem.* **266**, 7329–7332.

Hounsom, L., Corder, R., Patel, J., and Tomlinson, D. R. (2001). Oxidative stress participates in the breakdown of neuronal phenotype in experimental diabetic neuropathy. *Diabetologia* **44**, 424–428.

Huttunen, H. J., Fages, C., and Rauvala, H. (1999). Receptor for advanced glycation end products (RAGE)-mediated neurite outgrowth and activation of NF-κB require the cytoplasmic domain of the receptor but different downstream signaling pathways. *J. Biol. Chem.* **274**, 19919–19924.

Huttunen, H. J., Kuja-Panula, J., Sorci, G., Agneletti, A. L., Donato, R., and Rauvala, H. (2001). Coregulation of neurite outgrowth and cell survival by amphoterin and S100 proteins through receptor for advanced glycation end products (RAGE) activation. *J. Biol. Chem.* **275**, 40096–40105.

Hyslop, P. A., Hinshaw, D. B., Halsey, W. A., and Shraufstatter, I. U. (1988). Mechanisms of oxidant-mediated cell injury: The glycolytic and mitochondrial pathways of ADP phosphorylation are major intracellular targets inactivated by hydrogen peroxide. *J. Biol. Chem.* **263**, 1665–1675.

Kihara, M., Schmelzer, J. D., Poduslo, J. F., Curran, G. L., Nickander, K. K., and Low, P. A. (1991). Aminoguanidine effects on nerve blood flow, vascular permeability, electrophysiology, and oxygen free radicals. *Proc. Natl. Acad. Sci. USA* **88**, 6107–6111.

Kislinger, T., Fu, C., Huber, B., Qu, W., Taguchi, A., Yan, S. D., Hofmann, M., Yan, S. F., Pischetsrieder, M., Stern, D., and Schmidt, A.-M. (1999). N^ϵ-(Carboxymethyl)lysine adducts of proteins are ligands for receptor for advanced glycation end products that activate cell signaling pathways and modulate gene expression. *J. Biol. Chem.* **274**, 31740–31749.

Kuip, M., Douek, I. F., Moore, W. P. T., Gillmor, H. A., McLean, A. E. M., Bingley, P. J., and Gale, E. A. M. (2000). Safety of high-dose nicotinamide: A review. *Diabetologia* **43**, 1345.

Li, J., and Schmidt, A.-M. (1997). Characterization and functional analysis of the promoter of RAGE, the receptor for advanced glycation end products. *J. Biol. Chem.* **272**, 16498–16506.

Li, M. Y., Tan, A. X., and Vlassara, H. (1995). Antibacterial activity of lysozyme and lactoferrin is inhibited by binding of advanced glycation-modified proteins to a conserved motif. *Nature Med.* **1**, 1057–1061.

Lin, J., Alt, A., Liersch, J., Bretzel, R. G., Brownlee, M., and Hammes, H.-P. (2000). Benfotiamine inhibits intracellular formation of advanced glycation endproducts in vivo. *Diabetes* **49**, A143–A143. [Abstract]

Magnani, P., Cherian, P. V., Gould, G. W., Greene, D. A., Sima, A. A. F., and Brosius, F. C., III, (1996). Glucose transporters in rat peripheral nerve: Paradonal expression of GLUT1 and GLUT3. *Metabolism* **45**, 1466–1473.

Malherbe, P., Richards, J. G., Gaillard, H., Thompson, A., Diener, C., Schuler, A., and Huber, G. (1999). cDNA cloning of a novel secreted isoform of the human receptor for advanced glycation end products and characterization of cells co-expressing cell-surface scavenger receptors and Swedish mutant amyloid precursor protein. *Mol. Brain Res.* **71**, 159–170.

Mclean, W. G., Pekiner, C., Cullum, N. A., and Casson, I. F. (1992). Posttranslational modifications of nerve cytoskeletal proteins in experimental diabetes. *Mol. Neurobiol.* **6**, 225–237.

Monnier, V. M., Bautista, O., Kenny, D., Sell, D. R., Fogarty, J., Dahms, W., Cleary, P. A., Lachin, J., Genuth, S., and DCCT Skin Collagen Ancillary Study Group. (1999) Skin collagen glycation, glycoxidation, and crosslinking are lower in subjects with long-term intensive versus conventional therapy of type 1 diabetes. *Diabetes* **48**, 870–880.

Muona, P., Sollberg, S., Peltonen, J., and Uitto, J. (2001). Glucose transporters of rat peripheral nerve. Differential expression of GLUT1 gene by Schwann cells and perineurial cells *in vivo* and *in vitro*. *Diabetes* **41**, 1587–1596.

Nagai, R., Thornalley, P. J., Jono, T., and Horiuchi, S. (1999). Detection of N^{ϵ}-(carboxyethyl)lysine in glucose-modified AGE-proteins: Implication of an important role of methylglyoxal as an active intermediate in AGE formation. *Diabetes* **48**, A32. [Abstract]

Neeper, M., Schmidt, A.-M., Brett, J., Yan, S. D., Wang, F., Pan, Y.-C. E., Elliston, K., Stern, D., and Shaw, A. (1992). Cloning and expression of a cell surface receptor for advanced glycosylation endproducts of proteins. *J. Biol. Chem.* **267**, 14998–15004.

Neglia, C. I., Cohen, H. J., Garber, A. R., Ellis, P. D., Thorpe, S. R., and Baynes, J. W. (1983). ^{13}C NMR investigation of nonenzymatic glucosylation of protein. *J. Biol. Chem.* **258**, 14279–14283.

Ng, R., Argirov, O. K., Ahmed, N., Weigle, B., and Thornalley, P. J. (2002). Human serum albumin minimally modified by methylglyoxal binds to human mononuclear leukocytes via the RAGE receptor and is displaced by N-carboxymethyl-lysine and hydroimidazolone AGE epitopes. *Proc. Int. Maillard Symp. Kumamto Japan*, in press.

Nishikawa, T., Edelstein, D., Liang Du, X., Yamagishi, S., Matsumura, T., Kaneda, Y., Yorek, M. A., Beede, D., Oates, P. J., Hammes, H.-P., Giardino, I., and Brownlee, M. (2000). Normalizing mitochondrial superoxide production blocks three pathways of hyperglycaemia damage. *Nature* **404**, 787–790.

Nyengaard, J. R., Chang, K., Berhorst, S., Reiser, K. M., Williamson, J. R., and Tilton, R. G. (1997). Discordant effect of guanidines on renal structure and function and on regional vascular dysfunction and collagen changes in diabetic rats. *Diabetes* **46**, 94–106.

Ohgami, N., Nagai, R., Ikemoto, M., Arai, H., Kuniyasu, A., Horiuchi, S., and Nakayama, H. (2001). CD36, a member of the class B scavenger receptor family, as a receptor for advanced glycation end products. *J. Biol. Chem.* **276**, 3195–3202.

Ohi, T., Poduslo, J. F., and Dyck, P. J. (1985). Increased endoneurial albumin in diabetic polyneuropathy. *Neurology* **35**, 1790–1791.

Patel, V. J., Misra, V. P., Dandona, P., and Thomas, P. K. (1991). The effect of non-enzymatic glycation of serum proteins on their permeation into peripheral nerve in normal and streptozotocin-diabetic rats. *Diabetologia* **34**, 78–80.

Pesheva, P., Urschel, S., Frei, K., and Probstmeier, R. (1998). Murine microglial cells express functionally active galectin-3 *in vitro. J. Neurosci. Res.* **51**, 49–57.

Petersen, A., Szwergold, B. S., Kappler, F., Weingarten, M., and Brow, T. W. (1990). Identification of sorbitol 3-phosphate and fructose 3-phosphate in normal and diabetic human erythrocytes. *J. Biol. Chem.* **265**, 17424–17427.

Phillips, S. A., and Thornalley, P. J. (1993). The formation of methylglyoxal from triose phosphates: Investigation using a specific assay for methylglyoxal. *Eur. J. Biochem.* **212**, 101–105.

Poduslo, J. F., Curran, G. L., and Dyck, P. J. (1988). Increase in albumin, IgG, and IgM blood–nerve barrier indices in human diabetic neuropathy. *Proc. Natl. Acad. Sci. USA* **85**, 4879–4883.

Poulsom, R., Mirrlees, D., Earl, D. C. N., and Heath, H. (1983). The effects of an aldose reductase inhibitor upon the sorbitol pathway, fructose-1-phosphate and lactate in the retina and nerve of streptozotocin-diabetic rats. *Exp. Eye Res.* **36**, 754–760.

Renard, C., Chappey, O., Wautier, M. P., Nagashima, M., Lundh, E. R., Morser, J., Zhao, L., Schmidt, A.-M., Scherrmann, J.-M., and Wautier, J. L. (1997). Recombinant advanced glycation endproduct receptor pharmacokinetics in normal and diabetic rats. *Mol. Pharmacol.* **52**, 54–62.

Rossi, F. (1986). The O_2^--forming NADPH oxidase of the phagocytes: Nature, mechanism of activation and function. *Biochim. Biophys. Acta* **853**, 65–89.

Ryle, C., and Donaghy, M. (1995). Non-enzymatic glycation of peripheral nerve proteins in human diabetics. *J. Neurol. Sci.* **129**, 62–68.

Ryle, C., Leow, C. K., and Donaghy, M. (1997). Nonenzymatic glycation of peripheral and central nervous system proteins in experimental diabetes. *Muscle Nerve* **20**, 577–584.

Saito, N., Kimura, M., Kuchiba, A., and Itokawa, Y. (1987). Blood thiamine levels in outpatients with diabetes mellitus. *J. Nutr. Sci. Vitaminol.* **33**, 421–430.

Salmela, P. I., Oikarinen, A. I., Ukkola, O., Karajalainen, A., Linnaluoto, M., Puukka, R., and Ryhanen, L. (1995). Improved metabolic control in patients with non-insulin-dependent diabetes mellitus is associated with a slower accumulation of glycation products in collagen. *Eur. J. Clin. Invest.* **25**, 494–500.

Schafer, B. W., and Heizmann, C. W. (1996). The S100 family of EF-hand calcium-binding proteins: Functions and pathology. *T.I.B.S.* **21**, 134–140.

Schmidt, A.-M., Mora, R., Cao, R., Yan, S.-D., Brett, J., Ramakrishnan, R., Tsang, T. E., Simionescu, M., and Stern, D. (1994). The endothelial cell binding site for advanced glycation endproducts consists of a complex: An integral membrane protein and a lactoferrin-like polypeptide. *J. Biol. Chem.* **269**, 9882–9888.

Schmidt, R. E., Dorsey, D. A., Beaudet, L. N., Resier, K. M., Williamson, J. R., and Tilton, R. G. (1996). Effect of aminoguanidine on the frequency of neuroaxonal dystrophy in the superior mesenteric sympathetic autonomic ganglia of rats with streptozotocin-induced diabetes. *Diabetes* **45**, 284–290.

Sensi, M., Morano, S., Morelli, S., Castaldo, P., Sagratella, E., Grazia de Rossi, M., Andreani, D., Caltabiano, V., Vetri, M., Purrello, F., and Di Mario, U. (1998). Reduction of advanced glycation end-products (AGE) levels in nervous tissue proteins of diabetic Lewis rats following islet transplants is related to different durations of poor metabolic control. *Eur. J. Neurosci.* **10**, 2768–2775.

Smith, P. R., and Thornalley, P. J. (1992). Mechanism of the degradation of non-enzymatically glycated proteins under physiological conditions. *Eur. J. Biochem.* **210**, 729–739.

Soriano, F. G., Virag, L., Jagtap, P., Szabo, E., Mabley, J. G., Liaudet, L., Marton, M., Hoyt, D. G., Murthy, K. G. K., Salzman, A. L., Southan, G. J., and Szabo, C. (2001). Diabetic endothelial dysfunction: The role of poly(ADP-ribose) polymerase activation. *Nature Med.* **7**, 108–113.

Sugimoto, K., Nishizawa, Y., Horiuchi, S., and Yagihashi, S. (2001). Localization in human diabetic peripheral nerve of N^ϵ-carboxymethyl-lysine-protein adducts, an advanced glycation endproduct. *Diabetologia* **40**, 1380–1387.

Sugimoto, K., and Yagihashi, S. (1997). Effects of aminoguanidine on structural alterations of microvessels in peripheral nerve of streptozotocin diabetic rats. *Microvasc. Res.* **53**, 105–112.

The DCCT Research Group. (1998). Factors in the development of diabetic neuropathy: Baseline analysis of neuropathy in feasibility phase of diabetes control and complications trial (DCCT). *Diabetes* **37**, 486–471.

The Diabetes Control and Complications Trial Research Group. (1993). The effect of intensive treatment of diabetes on the development and progression of long-term complications in insulin-dependent diabetes mellitus. *N. Engl. J. Med.* **327**, 977–986.

Thornalley, P. J. (1988). Modification of the glyoxalase system in human red blood cells by glucose *in vitro*. *Biochem. J.* **254**, 751–755.

Thornalley, P. J. (1998). Cell activation by glycated proteins: AGE receptors, receptor recognition factors and functional classification of AGEs. *Cell. Mol. Biol.* **44**, 1013–1023.

Thornalley, P. J. (1999). Clinical significance of glycation. *Clin. Lab.* **45**, 263–273.

Thornalley, P. J., Jahan, I., and Ng, R. (2001). Suppression of the accumulation of triosephosphates and increased formation of methylglyoxal in human red blood cells during hyperglycaemia by thiamine *in vitro*. *Japan. J. Biochem.* **129**, 543–549.

Thornalley, P. J., Langborg, A., and Minhas, H. S. (1999). Formation of glyoxal, methylglyoxal and 3-deoxyglucosone in the glycation of proteins by glucose. *Biochem. J.* **344**, 109–116.

Thornalley, P. J., Yurek-George, A., and Argirov, O. K. (2000). Kinetics and mechanism of the reaction of aminoguanidine with the α-oxoaldehydes, glyoxal, methylglyoxal and 3-deoxyglucosone under physiological conditions. *Biochem. Pharmacol.* **60**, 55–65.

Thurston, J. H., McDougal, D. R., Jr, Hauhart, R. E., and Schulz, D. W. (1995). Effects of acute, subacute and chromic diabetes on carbohydrate and energy metabolism in rat sciatic nerve. *Diabetes* **44**, 190–195.

Tilton, R. G., Chang, K., Hasan, K. S., Smith, S. R., Petrash, J. M., Misko, T. P., Moore, W. M., Currie, M. G., Corbett, J. A., McDaniel, M. L., and Williamson, J. R. (1993). Prevention of diabetic vascular dysfunction by guanidines: Inhibition of nitric oxide synthase versus advanced glycation end-product formation. *Diabetes* **42**, 221–232.

UK Prospective Diabetes Study Group. (1998). Intensive blood-glucose control with sulphonylureas or insulin compared with conventional treatment and risk of complications in patients with type 2 diabetes (UKPDS 33). *Lancet* **352**, 837–853.

Verzijl, N., DeGroot, J., Thorpe, S. R., Bank, R. A., Shaw, J. N., Lyons, T. J., Bijlsma, J. W. J., Lafeberi, F. P. J. G., Baynes, J. W., and TeKoppele, J. M. (2000). Effect of collagen turnover on the accumulation of advanced glycation end products. *J. Biol. Chem.* **275**, 39027–39031.

Vlassara, H., Brownlee, M., and Cerami, A. (1981). Non-enzymatic glycosylation of peripheral nerve protein in diabetes mellitus. *Proc. Natl. Acad. Sci. USA* **78**, 5190–5192.

Vlassara, H., Brownlee, M., and Cerami, A. (1983). Excessive glycosylation of peripheral and central nervous system myelin components in diabetic rats. *Diabetes* **32**, 670–674.

Vlassara, H., Li, Y. M., Imani, F., Wojciechowicz, D., Yang, Z., Liu, F.-T., and Cerami, A. (1995). Identification of galectin-3 as a high-affinity binding protein for advanced glycation end products (AGE): A new member of the AGE-receptor family. *Mol. Med.* **1**, 634–646.

Wada, R., Sugo, M., Nakano, M., and Yagihashi, S. (1999). Only limited effects of aminoguanidine treatment on peripheral nerve function, [Na$^+$,K$^+$]-ATPase activity and thrombomodulin expression in streptozotocin-induced diabetic rats. *Diabetologia* **42**, 743–747.

Wahlberg, G., Adamson, U., and Svensson, J. (2000). Pyridine nucleotides in glucose metabolism and diabetes: A review. *Diabetes Metab. Rev.* **16**, 33–42.

Wautier, M. P., Chappey, O., Corda, S., Stern, D. M., Schmidt, A.-M., and Wautier, J. L. (2001). Activation of NADPH oxidase by AGE links oxidant stress to altered gene expression via RAGE. *Am. J. Physiol.* **280**, E685–E694.

Westwood, M. E., and Thornalley, P. J. (1995). Molecular characteristics of methylglyoxal-modified bovine and human serum albumins: Comparison with glucose-derived advanced glycation end product-modified serum albumins. *J. Prot. Chem.* **14**, 359–372.

Williams, S. K., Howarth, N. L., Devenny, J. J., and Bitensky, M. W. (2001). Structural and functional consequences of increased tubulin glycosylation in diabetes mellitus. *Proc. Natl. Acad. Sci. USA* **79**, 6546–6550.

Williamson, J. R., Chang, K., Frangos, M., Hasan, K. S., Ido, T., Kawamura, T., Nyengaard, J. R., van den Enden, M., Kilo, C., and Tilton, R. G. (1993). Hyperglycaemic pseudohypoxia and diabetic complications. *Diabetes* **42**, 801–813.

Winkler, G., Pal, B., Nagybeganyi, E., Ory, I., Porochnavec, M., and Kempler, P. (1999). Effectiveness of different benfotiamine dosage regimens in the treatment of painful diabetic neuropathy. *Arzneimittel-forschung Drug Res.* **49**, 220–224.

Yan, S. D., Schmidt, A.-M., Anderson, G. M., Zhang, J., Brett, J., Zou, Y. S., Pinsky, D., and Stern, D. (1994). Enhanced cellular oxidant stress by the interaction of the advanced glycation end products with their receptors/binding proteins. *J. Biol. Chem.* **269**, 9889–9897.

Yang, Z., Makita, Z., Horii, Y., Brunelle, S., Cerami, A., Sepajpal, P., Suthanthiran, M., and Vlassara, H. (1991). Two novel rat liver membrane proteins that bind advanced glycosylation end products: Relationship to macrophage receptor for glucose-modified proteins. *J. Exp. Med.* **174**, 515–524.

Yu, H. P., and Zou, D. M. (1997). Aminoguanidine inhibits semicarbazide-sensitive amino oxidase activity: Implication for advanced diabetic complications. *Diabetologia* **40**, 1243–1250.

PART II
SECONDARY CHANGES

PROTEIN KINASE C CHANGES IN DIABETES: IS THE CONCEPT RELEVANT TO NEUROPATHY?

Joseph Eichberg

Department of Biology and Biochemistry, University of Houston, Houston, Texas 77204

Protein kinase C (PKC) comprises a superfamily of isoenzymes, many of which are activated by 1,2-diacylglycerol (DAG) in the presence of phosphatidylserine. In order to be capable of DAG activation, PKC must first undergo a series of phosphorylations at three conserved sites. PKC isoforms phosphorylate a wide variety of intracellular target proteins and have multiple functions in signal transduction-mediated cellular regulation. An elevation in DAG levels and an increase in composite PKC activity and/or certain isoforms occurs in several nonneural tissues from diabetic animals, including the vasculature. The ability of isoform-specific PKC inhibitors to antagonize diabetes-induced abnormalities has implicated altered PKCβ activity in the onset of several diabetic complications. In contrast to many other tissues, DAG levels fall in diabetic nerve and a consistent pattern of change in PKC activity has not been observed. Treatments that alter PKC activity affect nerve Na$^+$,K$^+$-ATPase activity, but the mechanism involved is not well understood. Inhibition of PKCβ in diabetic rats appears to correct reduced nerve blood flow and decreased nerve conduction velocity. These and other findings indicate that changes in the neurovasculature exert adverse effects during the pathogenesis of diabetic neuropathy. Still unresolved is a clear-cut role for PKC in the development of abnormalities in neural cell metabolism. Further progress will depend on a more complete understanding of the functions of individual PKC isoforms in nerve. Future investigation could focus profitably on biochemical processes in nerve cells that modulate PKC activity and that are altered in diabetes, such as vascular endothelial growth factor levels and production of reactive oxygen species arising from oxidative stress. © 2002, Elsevier Science (USA).

I. Introduction

On the basis of abundant evidence, the onset of diabetic neuropathy is clearly the consequence of a multifactorial and complex sequence of events. The outcome of the Diabetes Control and Complications Trial (1993) made it apparent that diabetic complications, including those affecting peripheral nerve, arise because of sustained hyperglycemia and can be ameliorated by maintaining serum glucose levels as nearly euglycemic as possible. A multitude of biochemical alterations is triggered by sustained elevated glucose in susceptible tissues, including nerve. Those perturbations that are most widely considered to play a role in the pathogenesis of diabetic neuropathy include enhanced polyol pathway activity, glycation of cell constituents, especially proteins, and the onset of oxidative stress due to a heightened production of reactive oxygen and nitrogen species combined with weakened antioxidant defenses. In addition, diabetic nerve is characterized by reduced Na^+,K^+-ATPase activity, which can cause difficulties in maintaining normal nerve membrane resting potential, and deficits in neurotrophic support.

The actions and interactions involved in this panoply of abnormal processes are generally considered to lead to diminished nerve conduction velocity and reduced nerve blood flow, which are widely regarded as the physiological hallmarks of diabetic neuropathy, particularly in experimental animals. However, in explaining progressive nerve dysfunction, there is as yet no consensus regarding either the relative contribution or sequence of events that result in emerging pathology within the nerve tissue and the development of neurovascular defects.

Virtually all of the biochemical changes associated with diabetic neuropathy are either modulated by or impact on the ubiquitous protein kinase C (PKC) superfamily of enzymes. As will be become apparent, altered PKC function has been strongly implicated in several diabetic complications. This chapter reviews the evidence that PKC is likewise important in the onset of diabetic neuropathy.

II. Molecular Features of Protein Kinase C (PKC)

Soon after the first description of PKC in the late 1970s, this enzyme activity was recognized to be a serine/threonine kinase, which is dependent on phosphatidylserine and Ca^{2+} and is activated by 1,2-diacylglycerol (DAG) and phorbol esters (Nishizuka, 1984). Over the ensuing years, biochemical and molecular biological approaches have revealed that PKC consists of a superfamily of molecules comprising at least 12 isoenzymes (Mellor and Parker, 1998). These isoforms can be divided into several distinct subfamilies. The classical PKC (cPKC) family has been studied most extensively in normal and diabetic tissues and includes the α, the alternative splicing products, βI and βII, and γ isoforms. The requirements for optimal activity of these enzymes conform to the original description of PKC. Novel PKCs (nPKCs) consist of δ, ε, η, μ, and θ isoforms, which are nonresponsive to Ca^{2+}, but are activated by DAG in the presence of phosphatidylserine. Members of the atypical PKC (aPKC) group include ι (and its species homologue, λ) and ζ, which do not require Ca^{2+} and are not stimulated by DAG or phorbol ester. In addition, a group of PKC-related kinases, exists, identified through sequence homology, which, like aPKCs, respond to neither Ca^{2+} nor DAG.

The structural design of all PKC isoforms is similar and consists of conserved (C1–C4) and variable (V1–V5) sequences (Fig. 1). Functionally, the enzyme structure is defined by C1 and C2 or C2-like domains, a kinase domain, and the V5 region (Mellor and Parker, 1998). cPKCs and nPKCs contain two C1 domains, C1a and C1b, whereas aPKCs possess one, C1a-like, motif. These domains are characterized by the presence of a zinc finger motif that coordinates two Zn^{2+} ions. Evidence suggests that the C1b domain provides the binding site for DAG or phorbol ester. It should be noted that other mammalian proteins, notably the chimerin family, bind phorbol esters, and therefore studies of PKC function that utilize these substances must be interpreted with caution.

Near the N terminus of the C1 domain is the pseudosubstrate site, so called because it has the characteristics of the consensus sequence motifs that are phosphorylated by PKC, but has an alanine in place of the predicted serine/threonine target. In nonactivated PKC, this domain interacts and masks the catalytic site, thereby suppressing autophosphorylation.

The C2 domain, which is immediately C terminal to the C1 domain, has been considered the binding site for Ca^{2+} and phosphatidylserine. The βII isoform binds approximately eight phosphatidylserine molecules (Mosior and Newton, 1998). Johnson *et al.*, (2000) have demonstrated that while the C2 domain binds phosphatidylserine in a Ca^{2+}-dependent manner, the C1B domain also binds this phospholipid in a Ca^{2+}-independent fashion and confers the specificity observed for the phosphatidylserine head group. The C2 domain was initially believed to occur only in cPKCs, but it is now accepted that Ca^{2+}-independent nPKCs and aPKCs contain a region with homology to the C2 domain, but which is missing critical aspartyl residues

                           JOSEPH EICHBERG

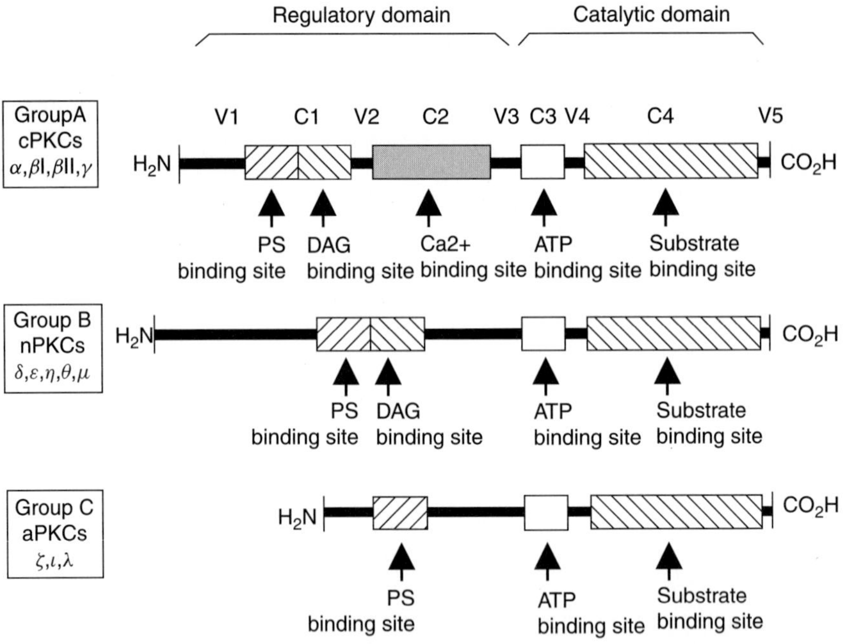

FIG. 1. Catalytic and regulatory domain structures of conventional, novel, and atypical PKC isoforms showing conserved regions (C1–C4), variable regions (V1–V5), and binding sites for Ca^{2+}, phosphatidylserine (PS), DAG, and ATP. Adapted with permission from Idris *et al.* (2001).

needed for Ca^{2+} binding. This domain also interacts with an anchoring group of proteins known as receptors for activated PKC (RACKs), which are believed to participate in directing active PKCs to specific membrane sites (Mochly-Rosen and Gordon, 1998).

The kinase domain contains the catalytic site, including ATP-binding and substrate-binding sites, and is well conserved among PKC isoforms. The V5 region is a relatively short sequence at the C terminal end of PKC, which has an important regulatory role in providing specificity for protein–protein interactions and hence the localization of isoforms. This has been shown most clearly for the βI and βII isoforms, which differ in sequence only in the V5 region, and in U937 monocytic cells, which are mainly found in microtubules and secretory granules, respectively (Kiley and Parker, 1995). It has been shown that PKC βII in the nuclear membrane is specifically activated by phosphatidylglycerol via a mechanism that involves binding of the lipid to a site in the V5 region unique to this isoform (Murray and Fields, 1998; Gokman-Polar and Fields, 1998). The V5 region also contains sites that undergo phosphorylation during the activation of PKC (see next section).

III. Activation and Regulation of PKC

Activation of PKCs requires two distinct events that occur sequentially and regulate the subcellular localization and function of the enzyme (Dempsey *et al.*, 2000). PKC first undergoes several ordered phosphorylations, which are initiated by a 3-phosphoinositide-dependent kinase (PDK-1), followed by binding of the phosphorylated molecule to the lipid second messenger, DAG (Fig. 2). Utilizing the PKC βII isoform, Newton and collaborators (Newton, 2001) have elucidated the sequence of phosphorylation events in a series of elegant studies. Newly synthesized PKC molecules are phosphorylated at three conserved sites. PDK-1 catalyzes the first and rate-limiting phosphorylation in all PKC isoforms at a threonine in the activation loop, a region near the active site. For cPKCs and nPKCs, this phosphorylation triggers rapid autophosphorylation at two sites in the

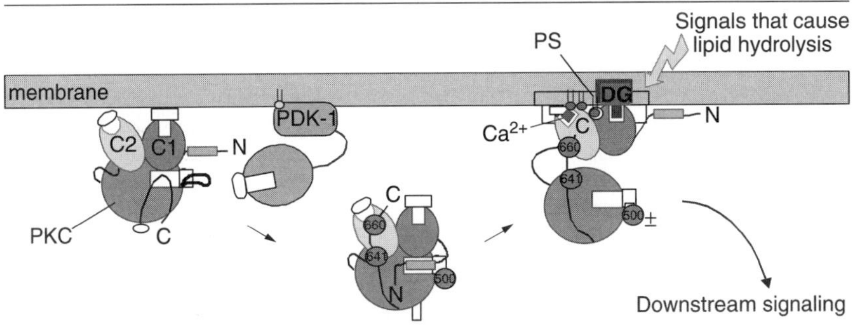

FIG. 2. Schematic representation of the two modes of regulation of PKC: (1) phosphorylation triggered by 3-phosphoinositide-dependent protein kinase 1 (PDK-1) and (2) allosteric control mediated by the lipid second messenger diacylglycerol (DAG). Unphosphorylated PKC is associated with the membrane where it is phosphorylated by PDK-1 at a site on its activation loop. The immediate consequence of this phosphorylation event is autophosphorylation at two positions in the carboxy terminus. The phosphorylated enzyme is then released into the cytosol where it is maintained in an inactive conformation by binding of an autoinhibitory sequence, the pseudosubstrate, to the substrate-binding cavity. Membrane recruitment is achieved when PKC binds to DAG and phosphatidylserine. This provides the energy to expel the pseudosubstrate sequence from the substrate-binding cavity, thus activating PKC for downstream signaling. C, carboxy terminus; N, amino terminus; circled numbers, phosphorylation sites. Adapted with permission from Dempsey *et al.* (2000).

carboxy terminus, one within a turn motif conserved among all classes of PKCs and the other within a hydrophobic motif. Other findings have refined this model and suggested that PDK-1 initially associates with the carboxy terminus of PKC and then carries out the first phosphorylation. PDK-1 then dissociates from PKC, thereby unmasking the carboxy terminus and allowing autophosphorylation to proceed (Gao *et al.*, 2001). Moreover, in unphosphorylated PKC, the pseudosubstrate sequence is displaced from the active site, thus rendering the activation loop accessible, and only after phosphorylation does the pseudosubstrate region block the active site (Dutil and Newton, 2000).

Binding of phosphorylated PKC to DAG is preceded by generation of this plasma membrane lipid, mainly through G-protein-coupled receptor-mediated hydrolysis of phosphatidylinositol-4,5-bisphosphate to yield DAG and inositol 1,4,5-trisphosphate. The latter hydrolysis product triggers liberation of Ca^{2+} into the cytosol from intracellular stores. The released DAG and Ca^{2+} then bring about binding of PKC to the membrane via the C1 and C2 domains, and the energy of these interactions helps dissociate the pseudosubstrate sequence from the active site, thus fully activating the enzyme. Unsaturated free fatty acids can stimulate most PKC isoforms, albeit at high concentrations ($>20-100 \ \mu M$), and can act synergistically with DAG (Shimomura *et al.*, 1991). Arachidonic acid has been shown to promote translocation of PKCs from cytosol to the plasma membrane of neutrophils at concentrations up to several orders of magnitude below that required for enzyme activation (O'Flaherty *et al.*, 2001). Prolonged activation of PKC is believed to involve association of individual isoforms with specific membrane-associated RACKs (Dempsey *et al.*, 2000).

IV. PKC and Nonneural Diabetic Complications

Since the early 1990s, steadily accumulating evidence has strongly linked retinal, renal, and cardiovascular complications of diabetes with elevations in DAG levels and increases in both PKC expression and activity. A complete evaluation of these findings is beyond the scope of this chapter and the subject has been well reviewed elsewhere (Ways and Sheetz, 2000; Idris *et al.*, 2001). Here only the highlights are summarized, especially as they may bear on a possible role of PKC in diabetic neuropathy. In a variety of nonneural diabetic tissues and derived cultured cell lines exposed to elevated glucose, measurement of DAG mass revealed an increase (Craven and DeRubertis, 1989; Craven *et al.*, 1990; Inoguchi *et al.*, 1992; Xia *et al.*, 1994). Radiolabeling studies suggested that the mechanism involves increased *de novo* synthesis of DAG. Concomitant with this increase, a rise in PKCßII isoform expression was observed in vascular tissues and cells (Inoguchi *et al.*, 1992). Enhanced PKC activity was associated with localization of an increased proportion of total PKC in the membrane vs the cytosol fraction, and this altered enzyme distribution occurred as a result of hypergalactosemia as well as hyperglycemia (Xia *et al.*, 1994). Other investigations have shown that intracellular changes in expression and redistribution of PKC isoforms in diabetic tissues are varied and complex (Tang *et al.*, 1993; Haller *et al.*, 1995; Kang *et al.*, 1999). For the most part, the precise functional significance of these changes is obscure. However, evidence has been presented that activation of both PKCβ and PKCδ accompanies the induction of endothelin-1 expression in endothelial cells exposed to high glucose (Park *et al.*, 2000).

There are indications that nonesterified fatty acids, which are elevated in diabetic tissues, including nerve (Chattopadhyay *et al.*, 1992), may influence PKC activity in several ways. In smooth muscle cells, saturated fatty acids stimulated DAG synthesis and PKC activity. Fatty acids also appear to regulate PKC activation in a manner that is independent of increased *de novo* DAG formation (Lu *et al.*, 2000). Moreover, fatty acids inhibit the activation of DAG kinase α in smooth muscle cells, although not the basal activity of the enzyme (Du *et al.*, 2001). Reduced DAG kinase activity has been reported in tissues from diabetic rats (Nobe *et al.*, 1998), an alteration that will tend to sustain DAG concentrations. In contrast, α-tocopherol was shown to prevent the increase in membrane-associated PKC activity and DAG levels in vascular tissues and cultured smooth muscle cells and also to antagonize enhanced expression of the PKCßII isoform (Kunisaki *et al.*, 1994; Ganz and Seftel, 2000). The mechanism of this effect can be attributed in part to the ability of this natural antioxidant to stimulate DAG kinase (Tran *et al.*, 1994; Lee *et al.*, 1999). In these experiments, α-tocopherol also antagonized the excessive production of reactive oxygen species (Ganz and Seftel, 2000), but whether stimulation of DAG kinase is wholly due to its antioxidant properties is not clear.

The value of a pharmacological approach in which inhibitors are employed to study the role of PKC in diabetic complications has, until recently, been limited because most of the compounds used, such as staurosporine and related substances, show little or no specificity for PKC isoforms. An exciting new vista has opened with the development of bisindolylmaleimide-based inhibitors, which exhibit an over 60-fold greater selectivity for PKC ßI and ßII compared to other cPKCs, as well as most other PKCs tested. They are much less potent against other classes of protein kinases (Jirousek *et al.*, 1996; Ways *et al.*, 2000). These compounds, LY333531 and its analog, LY379196, compete with ATP at its binding site and inhibit the β isoforms of PKC with a K_i of ca. 5 nM. The ability of a sufficient dose of orally administered LY333531 to ameliorate diabetes-induced dysfunctions in retina, kidney, and vasculature of rats and *db/db* mice has been demonstrated by several investigators (Ishii *et al.*, 1996; Kowluru *et al.*, 1998; Koya *et al.*, 2000). The inhibitor has also been found to be effective against neural deficits (Nakamura *et al.*, 1999), which is discussed further later. Finally, it should be mentioned that epalrestat, an aldose reductase inhibitor, like LY33351, antagonized enhanced protein kinase C activity in smooth muscle cells grown in elevated glucose (Nakamura *et al.*, 2001). However, only epalrestat normalized the increased level of the PKCßII isoform, suggesting that polyol pathway activity may modulate the activity and expression of this isoform.

V. PKC and 1,2-Diacylglycerol in Normal and Diabetic Nerve

PKC activity was first described in peripheral nerve as Ca^{2+}- and phosphatidylserine-dependent protein kinase activity that was stimulated by phorbol esters and phosphorylated myelin basic protein (for a review, see Eichberg and Iyer, 1996). P_0, the principal protein of peripheral nerve myelin, was subsequently identified as a PKC substrate (Brunden and Poduslo, 1987), and P_0 phosphorylation is increased in endoneurial preparations from streptozotocin-induced diabetic rats (Schrama *et al.*, 1987; Rowe-Rendleman and Eichberg, 1994). PKC activity measurements in homogenates or cytosol and membrane fractions of diabetic rat nerve have shown a decrease (Kim *et al.*, 1991; Mathew *et al.*, 1997; Kowluru *et al.*, 1998), no change (Simpson and Hawthorne, 1988; Borghini *et al.*, 1994; Nakamura *et al.*, 1999), or an increase (Kishi *et al.*, 1999). Wide discrepancies in the level and proportion of activities in these fractions have been reported, perhaps in part due to technical differences in tissue fractionation, assay conditions, duration and severity of diabetes, and animal model employed. Moreover, in diabetic vs normal animals, the distribution of PKC activity between cytosolic and particulate fractions from nerve was unaltered (Table I). The interpretation of these findings is difficult because they reflect composite PKC activities in a tissue containing heterogeneous cell types so that alterations in the activity of one or more specific isoforms might be masked. Of considerable interest is a report that total PKC activity in endoneurial membranes in diabetic mice is reduced and the decrease is more pronounced in transgenic animals overexpressing aldose reductase. In contrast, activity in the perineurium, where the bulk of microvessels is located, is increased (Yamagishi *et al.*, 2001). These findings suggest

TABLE I

COMPOSITE PROTEIN KINASE C ACTIVITY IN RAT NERVE[a]

Nerve from normal rats			Nerve from diabetic rats			
PKC activity (pmol/mg protein/min)						
Homogenate	Cytosol	Membranes	Homogenate	Cytosol	Membranes	Reference
1114 ± 127	1660 ± 216	835 ± 93	736 ± 50	832 ± 138^{b}	964 ± 120	Kim *et al.* (1991)
~1000[c]	~630[c]	~370[c]	~870[c]	~562[c]	~308[c]	Borghini *et al.* (1994)
	1170 ± 73	1020 ± 87		1250 ± 40	900 ± 38	Mathew *et al.* (1997)
34 ± 4			22 ± 4^{d}			Kowluru *et al.* (1998)
	21 ± 6	214 ± 52		14 ± 3	189 ± 33	Nakamura *et al.* (1999)
	15789 ± 1144	3864 ± 648		24331 ± 2333^{e}	5876 ± 956	Kishi *et al.* (1999)

[a]All animals had streptozotocin-induced diabetes except for Mathew *et al.* (1997), who used the Zucker fatty diabetic rat (ZDF/GMI *fa*). Duration of diabetes varied from 1 to 3 months.
[b]Different from normal cytosol, $p < 0.05$.
[c]Values based on authors' estimate of total PKC activity in normal animals.
[d]Different from normal homogenate, $p < 0.01$.
[e]Different from normal cytosol, $p < 0.01$.

an influence of the polyol pathway on endoneurial PKC activity that is separable from alterations in microvessel PKC activity.

Initial examination of PKC isoforms in nerve revealed that PKCα is abundant and that relatively little of the PKCβ isoforms are present (Kikkawa *et al.*, 1989). Subsequent immunochemical analysis for individual PKC family members in nerve has revealed the presence of α, βI, βII, γ, δ, and ε isoforms (Borghini *et al.*, 1994; Rowe-Rendleman and Eichberg, 1994; Roberts and Mclean, 1997; Matthew *et al.*, 1997). cPKCs were also detected in spinal cord and dorsal root ganglia (Roberts and McLean, 1997). Among cell types within nerve, Schwann cells in culture contained α, βII, δ, and ε isoforms. Variable results have been obtained as to the impact of diabetes on the level and distribution of nerve PKC isoforms. Borghini *et al.* (1994) reported that streptozotocin-induced diabetes of up to 12 weeks duration had no effect on the abundance or distribution of isoforms. Roberts and McLean (1997) reported there is no change in expression of this isoform in streptozotocin-induced diabetes, but found that translocation occurred from cytosol to membrane structures. There was also a fall in immunoreactivity against the βII isoform in the cytosol, but no alterations in βI and γ isoforms. In diabetic animals, the mRNA levels of PKCα in dorsal root ganglia, which contain only sensory neuron cell bodies, was decreased markedly. In contrast, PKCα mRNA and immunoreactivity were doubled in spinal cord, the location of motor neuron cell bodies. Again, no changes in PKCβII expression were seen. Purified peripheral myelin from diabetic animals exhibited increased PKCα and PKCε expression, a loss of PKCβ isoforms, and an increase in total PKC activity (Setton-Avruj *et al.*, 2001). A somewhat different pattern was evident in nerve from the ZDF rat, an obese spontaneously diabetic strain, in which PKCα underwent translocation from membrane to cytosol, as compared to lean controls (Mathew *et al.*, 1997). The existing state of knowledge leaves much to be clarified regarding the distribution of PKC isoforms in nerve structures. For example, it is unknown whether PKCs are localized differentially within nerve fiber classes or whether certain isoforms are concentrated differentially in neuronal cell bodies, the axonal membrane, at the nodes of Ranvier, or in Schwann cells *in vivo*.

The level of DAG in diabetic nerve dissected largely free of vascular elements has been reported to be decreased 30–40% as compared to normal (Zhu and Eichberg, 1990; Ido *et al.*, 1994), in contrast to its elevation in other tissues from diabetic animals. A partial explanation may be that the metabolic fate of glucose in nerve is different from many nonneural tissues in that enhanced *de novo* DAG synthesis could occur to a much lesser extent under hyperglycemic conditions. DAG can also arise as a result of phospholipase C-mediated hydrolysis of phosphoinositide-4,5-bisphosphate or via phosphatidic acid phosphohydrolase-catalyzed breakdown of phosphatidic acid, which in turn can be generated either by *de novo* formation or through phospholipase D action, mainly on phosphatidylcholine. From the molecular species profile of rat nerve DAG, it appears that both *de novo* DAG synthesis and hydrolysis of preexisting phospholipids participate in generation of this lipid (Zhu and Eichberg, 1990). The sites of DAG production by these various pathways within nerve compartments are unknown.

VI. PKC and Na$^+$,K$^+$-ATPase

Initial evidence suggested that PKC is involved in the mechanism leading to reduced Na$^+$,K$^+$-ATPase activity and diminished nerve conduction velocity, which are hallmarks of experimental diabetic neuropathy. Thus Greene and co-workers (Lattimer *et al.*, 1989) found that exposure of nerve to phorbol myristoyl acetate corrected this deficit. This group proposed that the reduction of *myo*-inositol, in diabetic nerve, which most likely occurs to maintain osmotic balance in compensation for the accumulation of sorbitol, impairs phosphoinositide metabolism and consequently DAG production, resulting in diminished PKC activity (Greene *et al.*, 1989).

A number of other investigators have conducted studies, utilizing PKC inhibitors, that have created controversy as to whether PKC activation or inhibition is associated with decreased nerve Na^+,K^+-ATPase activity. Injection of alloxan diabetic mice with the nonselective PKC inhibitors H-7 and calphostin C was shown to restore or maintain Na^+,K^+-ATPase activity and simultaneously to reduce membrane-associated PKC activity (Hermenegildo *et al.*, 1992; Hermenegildo *et al.*, 1993). Somewhat similar results were obtained by Cameron *et al.* (1999), who upon treating diabetic rats with the PKC inhibitors WAY 151003 or chelerythrine found a partial correction of neural Na^+,K^+-ATPase activity. Tomlinson *et al.* (1993), using endoneurial preparations from diabetic animals, found that diminished $^{86}Rb^+$ uptake, a measure of Na^+,K^+ activity, was improved by the PKC inhibitor H-7 and by a more selective PKC blocker, Ro 31-8220, but also by phorbol dibutyrate. These seemingly paradoxical data could lend support to the hypothesis that PKC activity in nerve that modulates the Na^+,K^+-ATPase is enhanced rather than reduced in diabetic nerve, but only if the stimulatory action of phorbol esters operates via a PKC-independent mechanism.

The modulatory effect of PKC on Na^+,K^+-ATPase has most frequently been considered to be the result of direct phosphorylation. In several types of intact cells, endogenous phosphorylation occurs on ser^{18} of the α subunit in a Ca^{2+}-dependent manner, but the modification has no effect on kinetic properties of Na^+,K^+-ATPase (Feschenko and Sweadnor, 1997). In some nonneural cell types, as well as in normal and diabetic nerve, forskolin, an agent that raises cAMP levels, reduced PKC-mediated Na^+,K^+-ATPase phosphorylation (Borghini *et al.*, 1994; Feschenko *et al.*, 2000). Thus existing data suggest there are as yet undefined interactions between PKC- and cAMP-dependent pathways.

Evidence has also been presented that supports an alternative means of PKC involvement in Na^+ pump inhibition. In this mechanism, PKC phosphorylates and thereby activates cytosolic phospholipase A_2 to release arachidonic acid from phospholipids. Both arachidonic acid and its metabolite, prostaglandin E_2, then inhibit the pump, as measured by ouabain-sensitive $^{86}Rb^+$ uptake (Xia *et al.*, 1995). In diabetes, a greater sustained activation of PKC is considered to enhance arachidonic acid and prostaglandin production, leading to diminished Na^+ pump activity. This mechanism was shown to operate in smooth muscle cells exposed to elevated glucose levels, but whether it does so in nerve has not been investigated. It should be noted that an unexplained finding by Xia *et al.* (1995) was that a PKC inhibitor also restored activity of the pump, despite reducing the extent of phospholipase A_2 phosphorylation. This observation again points to mediation of Na^+,K^+-ATPase activity by PKC via multiple pathways.

VII. PKC Actions: Neural versus Neurovascular?

The increasingly recognized importance of the DAG/PKC pathway in the pathogenesis of vascular and other nonneural diabetic complications has prompted several investigators to ask whether activation of PKC in the nerve vasculature might play a role in the onset of diabetic neuropathy. In streptozotocin-induced diabetic rats, the nonselective PKC inhibitors WAY151 003 and chelerythrine improved the reduced nerve blood flow and motor nerve conduction velocity deficit (Cameron $et\ al.$, 1999). This effect of WAY151 003 was prevented if a nitric oxide synthase inhibitor was given simultaneously. Treatment of diabetic rats with cremophor, a drug capable of complexing with DAG, also restored both reduced nerve conduction velocity and decreased blood flow toward normal, and this corrective effect was likewise antagonized by the blockade of nitric oxide synthesis (Jack $et\ al.$, 1999). Most interestingly, administration of the PKCβ-selective inhibitor LY 333351 to diabetic animals prevented development of a motor nerve conduction drop and diminished blood flow, although the agent was without detectable effect on composite PKC activity in the nerves $in\ vitro$ (Kowluru $et\ al.$, 1998, Nakamura $et\ al.$, 1999). An altered activity of PKCβ might have been masked because, as noted previously, it is a relatively minor component of the PKC isoform population in nerve. Alternatively, LY 333351 might exert other, nonspecific actions on the tissue, unrelated to its inhibition of PKCβ. Nonetheless, available evidence tends to support a role for PKCβ activation, whether localized in vascular elements or in neural cells, in bringing about decreased blood flow and conduction velocity in diabetic nerve. Still remaining to be accomplished is identification of those proteins phosphorylated by PKC that participate in this process. Such downstream targets are likely to include MAP kinases (Tomlinson, 1999; Igarashi $et\ al.$, 1999).

Some investigators have championed the idea that endoneurial ischemia brought about by decreased blood flow is the sole or principal cause of diabetic neuropathy (Cameron and Cotter, 1994; Koya and King, 1998). However, there are difficulties in accepting this conclusion unequivocally. For one thing, in human diabetic neuropathy, functional sensory nerve fiber deficits can occur without evident dysfunction of neighboring motor nerve fibers (Thomas and Tomlinson, 1993). Because both fiber types would presumably be subject to a similar degree of ischemia, an explanation is needed for the greater susceptibility of sensory fibers to damage. In this regard, isolated rat sensory nerve fibers exhibit a greater sensitivity to hyperglycemic hypoxia as compared to motor nerve fibers (Schneider $et\ al.$, 1992). For another, the presence of reduced blood flow, although widely reported in experimental diabetic neuropathy, is not observed universally (Zochodne and Ho, 1992; Chang *et al.*, 1997) and may in some cases reflect experimental artifacts (Ido *et al.*, 1997; Dines *et al.*, 1999). Moreover, Tomlinson *et al.* (1996) have demonstrated that the extent of skeletal muscle wasting during diabetes contributes significantly to reduced sciatic nerve blood flow. The degree of diminished nerve blood flow in human diabetes is uncertain, at least during early stages of polyneuropathy [reviewed by Zochodne (1999) and in this volume]. Finally, patients with Mendenhall's syndrome, in which a mutation in the insulin receptor gene leads to persistent uncontrolled hyperglycemia, exhibit severe neuropathy, but only mild microangiopathy in comparison with vascular abnormalities in diabetic patients (Malik *et al.*, 1995). Overall, our present state of knowledge strongly suggests that neurovascular deficits contribute to the pathogenesis of diabetic neuropathy, particularly in the latter stages, but that neural metabolic changes are likely to be important as well.

VIII. Emerging and Potential Roles for PKC in Nerve

Another promising development that may have therapeutic application is the finding that intramuscular gene transfer and overexpression of vascular endothelial growth factor (VEGF-1 or VEGF-2) elicits beneficial effects in diabetic neuropathy. In both diabetic rats and rabbits, this treatment normalizes nerve blood flow and improves vascularity, as well as motor and sensory sciatic nerve conduction velocity (Schratzberger *et al.*, 2001). As an endothelial cell mitogen, VEGF promotes angiogenesis, in part via receptor-mediated phospholipase $C\gamma$, to generate DAG and inositol-1,4,5-trisphosphate. In response, the proportion of membrane-associated Ca^{2+}-dependent PKC isoforms, α and ßII, increases. The PKCβ-selective inhibitor LY333531 blocks the mitogenic effects of VEGF on endothelial cell proliferation (Xia *et al.*, 1996). Immunochemical analysis has indicated that VEGF is also expressed abundantly in neuronal cell bodies in dorsal root ganglia and nerve fibers of the streptozotocin-induced diabetic rat, whereas normal animals show little or no expression (Samii *et al.*, 1999). Therefore, it may be that diabetes-induced upregulation of VEGF in neural cells activates PKC-mediated responses in the diabetic state that are independent of the vasculature.

Growing evidence supports an important role for oxidative stress in the pathogenesis of diabetic neuropathy and other diabetic complications, and the participation of PKC in this process seems probable (Cameron and Cotter, 1999; Rosen *et al.*, 2001). While the pathways involved are complex, formation of advanced glycation end products (AGE products) is one mechanism that can lead to the production of reactive oxygen species (Yan *et al.*, 1994). In cultured mesangial cells, AGE products, acting through specific receptors (RAGE), increase intracellular oxidative stress and cause translocation of PKCßII (but not PKCα) from cytosol to membranes over similar time courses (Scivittaro *et al.*, 2000). Such a process might profitably be investigated in neural cells. Superoxide generated in the mitochondrial electron transport chain may also modulate PKC activity. Nishikawa *et al.* (2000) presented evidence that reactive oxygen species are elevated and PKC activity is enhanced in cultured endothelial cells grown in medium containing an elevated glucose level. When the cells are treated either with an inhibitor of electron transport or with Mn^{2+}-dependent superoxide dismutase so as to reduce superoxide concentrations, PKC activity drops to levels found in cells grown in 5 mM glucose. It is of interest that superoxide is capable of stimulating the activity of several PKC isoforms in hippocampal preparations, including PKCα and PKCβII, by oxidation of thiols and release of zinc from the enzyme (Knapp and Klann, 2000). This increase of enzyme activity occurred in the absence of Ca^{2+}, phosphatidylserine, and DAG, but not in the presence of these cofactors. Thus, PKC activation of this kind would not be detected under conventional assay conditions. Further, other studies have implicated PKC, especially PKCβ, in the activation of NADPH oxidase in endothelial cells and neutrophils (Dekker *et al.*, 2000; Korchak and Kilpatrick, 2001; Hink *et al.*, 2001). These findings lead to the speculation that elevated levels of glucose in the nerve could trigger a mutually interdependent activation of PKC and NADPH oxidase, which could enhance oxidative stress and have adverse consequences for nerve function.

Cytosolic NADPH oxidase, which has long been known to be a potent generator of superoxide in cells of hematopoietic origin, was identified in sympathetic neurons by Tammariello *et al.* (2000). These authors showed that activity of this enzyme contributes both to oxidative stress and to apoptosis under conditions of nerve growth factor deprivation. Because diabetic nerve is characterized by a reduced availability of nerve growth factor, as well as by diminished nerve growth factor receptor function (Fernyhough *et al.*, 1995; Delacroix *et al.*, 1998), the possibility arises that superoxide-mediated PKC activation could affect neuronal survival adversely. Indeed, neurons in dorsal root and superior cervical ganglia have been demonstrated to undergo apoptosis upon exposure to elevated glucose levels (Russell *et al.*, 1999; Russell and Feldman, 1999).

A little investigated area is the possible relationship between PKC activity and the onset of pain, which frequently accompanies diabetic neuropathy. Experimentally diabetic rats display hyperalgesia, and it has been suggested that this response is associated with hyperexcitability, which is exhibited by

C-fiber afferents in these animals (Ahlgren and Levine, 1992). Injection of nonisoform-selective PKC inhibitors into streptozotocin-induced diabetic rats reduced C-fiber hyperexcitability, but had no effect on control animals (Ahlgren and Levine, 1994). The ε isoform of PKC has been implicated in the regulation of nociceptor function, and both nonselective PKC and selective PKCε inhibitors, when injected intradermally, ameliorated hyperalgesia in painful alcoholic neuropathy in the rat (Khasar *et al.*, 1999; Dina *et al.*, 2000). Whether PKCε is involved in the development of painful diabetic neuropathy remains to be investigated.

IX. Conclusions and Future Directions

Existing information provides strong evidence that alterations in PKC activity and expression are involved in mechanisms that bring about vascular disturbances in diabetes. Specific isoforms of this enzyme family are being recognized as integral to these perturbations, and their identification is already forming the basis of possible therapeutic intervention. While there is thus little doubt about the importance of neurovascular dysfunction in diabetic neuropathy, it is not yet clear how early in the course of pathogenesis these defects are manifested. Moreover, we lack sufficient understanding of the downstream events that are influenced by PKC activation or altered expression. Such interactions likely include modifications in MAP kinase and phosphatidylinositol-3-kinase-mediated phosphorylation cascades, but how these signaling pathways regulate physiological functions that go awry in diabetic complications is still largely obscure.

Our knowledge of the distribution and functions of individual PKC isoforms in neuronal cell bodies and axons, as well as in Schwann cells, is still in a rudimentary stage. The failure to observe PKC changes in diabetic nerve that resemble those in other affected tissues is puzzling, but may reflect the heterogeneous nature of peripheral nerve structures, which could mask responses of specific PKC isoforms when whole nerve is studied. The use of homogeneous nerve-derived cell cultures to avoid this problem, while often valuable in unraveling basic biochemical mechanisms, is unsatisfying in that such preparations cannot reproduce the complex environment that exists in the nerve *in situ*.

Despite these difficulties, it is likely that novel roles for PKC relevant to the onset of diabetic neuropathy await discovery. Much data suggest that the enzyme is involved in the regulation of Na^+,K^+-ATPase, possibly via multiple mechanisms. Interactions remain to be explored between PKC and other molecules and processes that are disturbed in diabetic nerve, including the polyol pathway, nitric oxide formation, formation of AGE products, growth factors, and neurotrophins. Some of these abnormalities are discussed in other chapters of this volume. Particularly promising areas would seem to be the participation of PKC in VEGF-mediated events and the impact of oxidative stress on PKC activity. New insights into how specific isoforms of PKC are involved in nerve metabolism will probably be necessary prior to formulation of a novel testable hypothesis that implicates the enzyme in neural metabolic alterations occurring in diabetic nerve. Such identification of new roles for PKC could expand the potential for therapeutic applications of selective PKC inhibitors (Parker, 1999). Consequently, investigations of individual PKC isoforms in the progression of changes leading to diabetic neuropathy should be pursued well into the future.

References

Ahlgren, S. C., and Levine, J. D. (1994). Protein kinase C inhibitors decrease hyperalgesia and C-fiber hyperexcitability in the streptozotocin-diabetic rat. *J. Neurophysiol.* **72**, 684–692.

Ahlgren, S. C., White, D. M., and Levine, J. D. (1992). Increased responsiveness of sensory neurons in the saphenous nerve of the streptozotocin-diabetic rat. *J. Neurophysiol.* **68**, 2077–2085.

Borghini, I., Ania-Laherta, A., Regazzi, R., Ferrari, G., Gjinovci, A., Wollheim, C. B., and Pralong, W. F. (1994). Alpha, beta I, beta II, delta and epsilon protein kinase C isoforms and compound activity in the sciatic nerve of normal and diabetic rats. *J. Neurochem.* **62**, 686–696.

Borghini, I., Geering, K., Gjinovci, A., Wollhein, C.B., and Pralong, W. F. (1994). In vivo phosphorylation of the Na, K-ATPase a subunit in sciatic nerves of control and diabetic rats: Effects of protein kinase modulators. *Proc. Natl. Acad. Sci. USA* **91**, 6211–6215.

Brunden, K. R., and Poduslo, J. F. (1987). A phorbol ester-sensitive kinase catalyzes the phosphorylation of P_0 glycoprotein in myelin. *J. Neurochem.* **46**, 1863–1872.

Cameron, N. E., and Cotter, M. A. (1994). The relationship of vascular changes to metabolic factors in diabetes mellitus and their role in the development of peripheral nerve complications. *Diabet. Metab. Rev.* **10**, 189–224.

Cameron, N. E., and Cotter, M. A. (1999). Effect of antioxidants on nerve and vascular dysfunction in experimental diabetes. *Diabet. Res. Clin. Prac.* **45**, 137–146.

Cameron, N. C., Cotter, M. A., Jack, A. M., Basso, M. D., and Hohman, T. C. (1999). Protein kinase C effects on nerve function, perfusion, Na(+),K(+)-ATPase activity and glutathione content in diabetic rats. *Diabetologia* **42**, 1120–1130.

Chang, K., Ido, Y., Lejeune, W., Williamson, J. R., and Tilton, R. G. (1997). Increased sciatic nerve blood flow in diabetic rats: Assessment by "molecular" vs. particulate microspheres. *Am. J. Physiol.* **273**, 164–173.

Chattopadhyay, J., Thompson, E. W., and Schmid, H. H. (1992). Nonesterified fatty acids in normal and diabetic rat sciatic nerve. *Lipids* **27**, 513–517.

Craven, P. A., Davidson, C. M., and DeRubertis, F. R. (1990). Increase in diacylglycerol mass in isolated glomeruli by glucose from de novo synthesis of glycerolipids. *Diabetes* **39**, 667–674.

Craven, P. A., and De Rubertis, F. R. (1989). Protein kinase C is activated in glomeruli from streptozotocin diabetic rats: Possible mediation by glucose. *J. Clin. Invest.* **83**, 1667–1675.

Dekker, L. V., Leitges, M., Altschuler, G., Mistry, N., McDermott, A., Roes, J., and Segal, A. W. (2000). Protein kinase C-β contributes to NADPH oxidase activation in neutrophils. *Biochem. J.* **347**, 285–289.

Delacroix, J. D., Michael, G. J., Priestley, J. V., Tomlinson, D. R., and Fernyhough, P. (1998). Effect of nerve growth factor treatment on p75NTR gene expression in lumbar dorsal root ganglia of streptozotocin-induced diabetic rats. *Diabetes* **47**, 1779–1785.

Dempsey, E. C., Newton, A. C., Mochly-Rosen, D., Fields, A. P., Reyland, M. E., Insel, P. A., and Messing, R. O. (2000). Protein kinase C isozymes and the regulation of diverse cell responses. *Am. J. Physiol. Lung Cell Mol. Physiol.* **279**, L429–L438.

Diabetes Control and Complications Trial Research Group. (1993). The effect of intensive treatment of diabetes on the development and progression of long-term complications in insulin-dependent diabetes mellitus. *N. Engl. J. Med.* **329**, 977–986.

Dina, O. A., Barletta, J. A., Chen, X., Mutero, A., Martin, A., Messing, R. O., and Levine, J. D. (2000). Key role for the epsilon isoform of protein kinase C in painful alcoholic neuropathy in the rat. *J. Neurosci.* **20**, 8614–8619.

Dines, K. C., Calcutt, N. A., Nunag, K. D., Mizisin, A. P., and Kalichman, M. W. (1999). Effects of hindlimb temperature on sciatic nerve laser Doppler vascular conductance in control and streptozotocin-diabetic rats. *J. Neurol. Sci.* **163**, 17–24.

Du, X., Jiang, Y., Qian, W., Lu, X., and Walsh, J. P. (2001). Fatty acids inhibit growth factor-induced diacylglycerol kinase α activation in vascular smooth muscle cells. *Biochem. J.* **357**, 275–282.

Dutil, E. M., and Newton, A. C. (2000). Dual role of pseudosubstrate in the coordinated regulation of protein kinase C by phosphorylation and diacylglycerol. *J. Biol. Chem.* **275**, 10697–10701.

Eichberg, J., and Iyer, S. (1996). Phosphorylation of myelin proteins: Recent advances. *Neurochem. Res.* **21**, 527–535.

Fernyhough, P., Diemel, L. T., Hardy, Brewster, W. J., Mohuiddin, L., and Tomlinson, D. R. (1995). Human recombinant nerve growth factor replaces deficient neurotrophic support in the diabetic rat. *Eur. J. Neurosci.* **7**, 1107–1110.

Feschenko, M. S., Stevenson, E., and Sweadnor, K. J. (2000). Interaction of protein kinase C and cAMP-dependent pathways in the phosphorylation of the Na,K-ATPase. *J. Biol. Chem.* **275**, 34693–34700.

Feschenko, M. S., and Sweadnor, K. J. (1997). Phosphorylation of Na,K-ATPase by protein kinase C at ser18 occurs in intact cells but does not result in direct inhibition of ATP hydrolysis. *J. Biol. Chem.* **272**, 17726–17733.

Ganz, M. B., and Seftel, A. (2000). Glucose-induced changes in protein kinase C and nitric oxide are prevented by vitamin E. *Am. J. Physiol. Endocrinol. Metab.* **278**, E146–E152.

Gao, T., Toker, A., and Newton, A. C. (2001). The carboxyl terminus of protein kinase C provides a switch to regulate its interaction with the phosphoinositide-dependent kinase, PDK-1. *J. Biol. Chem.* **276**, 19588–19596.

Gokmen-Polar, Y., and Fields, A. P. (1998). Mapping of a molecular determinant for protein kinase C betaII isozyme function. *J. Biol. Chem.* **273**, 20261–20266.

Greene, D. A., Lattimer-Greene, S. A., and Sima, A. A. F. (1989). Pathogenesis of diabetic neuropathy: Role of altered phosphoinositide metabolism. *Crit. Rev. Neurobiol.* **5**, 143–219.

Haller, H., Baur, E., Quass, P., Behrend, M., Lindschau, C., Distler, A., and Luft, F. C. (1995). High glucose concentrations and protein kinase C isoforms in vascular smooth muscle cells. *Kidney Int.* **47**, 1057–1067.

Hermenegildo, C., Felipo, V., Minana, M.-D., and Grisolia, S. (1992). Inhibition of protein kinase C restores Na$^+$,K$^+$-ATPase activity in sciatic nerve of diabetic mice. *J. Neurochem.* **58**, 1246–1249.

Hermenegildo, C., Felipo, V., Minana, M.-D., Romero, F. J., and Grisolia, S. (1993). Sustained recovery of Na$^+$-K$^+$-ATPase activity in sciatic nerve of diabetic mice by administration of H7 or calphostin C, inhibitors of PKC. *Diabetes* **42**, 257–262.

Hink, U., Li, H., Mollnau, H., Oelze, M., Matheis, E., Hartmann, M., Skatchkov, M., Thaiss, F., Stahl, R. A., Warnholtz, A., Meinertz, A., Griendling, K., Harrison, D. G., Forstermann, U., and Munzel, T. (2001). Mechanisms underlying endothelial cell dysfunction in diabetes mellitus. *Circ. Res.* **88**, E14–E22.

Ido, Y., Chang, K., Lejeune, W., Tilton, R. G., Monafo, W. W., and Williamson, J. R. (1997). Diabetes impairs sciatic nerve hyperemia induced by surgical trauma: Implications for diabetic neuropathy. *Am. J. Physiol.* **273**, E174–184.

Ido, Y., McHowat, J., Chang, K. C., Arrigoni-Martelli, E., Orfalian, Z., Kilo, C., Corr, P. B., and Williamson, J. R. (1994). Neural dysfunction and metabolic imbalances in diabetic rats: Prevention by acetyl-L-carnitine. *Diabetes* **43**, 1269–1277.

Idris, L., Gray, S., and Donnelly, R. (2001). Protein kinase C activation: Isozyme-specific effects on metabolism and cardiovascular complications in diabetes. *Diabetologia* **44**, 659–673.

Igarashi, M., Wakasaki, H., Takahara, N., Ishii, H., Jiang, Z. Y., Yamauchi, T., Kuboki, K., Meier, M., Rhodes, C. J., and King, G. L. (1999) Glucose or diabetes activates p38 mitogen-activated protein kinase via different pathways. *J. Clin. Invest.* **103**, 183–195.

Inoguchi, T., Battan, R., Handler, E., Sportsman, J. R., Heath, W., and King, G. L. (1992). Preferential elevation of protein kinase C isoform beta II and diacylglycerol levels in the aorta and heart of diabetic rats: Differential reversibility to glycemic control by islet cell transplantation. *Proc. Natl. Acad. Sci. USA* **89**, 11059–11063.

Inoguchi, T., Li, P., Umeda, F., Yu, H. Y., Kakimoto, M., Imamura, M., Aoki, T., Etoh, T., Hashimoto, T., Naruse, M., Sano, H., Utsumi, H., and Nawata, H. (2000). High glucose level and free fatty acid stimulate reactive oxygen species production through protein kinase C-dependent activation of NAD(P)H oxidase in cultured vascular cells. *Diabetes* **49**, 1939–1945.

Ishii, H., Jirousek, M. R., D. Koya., Takagi, C, Xia, P., Clermont, A., Bursell, S. E., Kern, T. S., Ballas, W. F., Heath, W. R., Stramm, L., Feener, E., and King, G. L. (1996). Amelioration of vascular dysfunction in diabetic rats by an oral PKC β inhibitor. *Science* **272**, 728–731.

Jack, A. M., Cameron, N. E., and Cotter, M. A. (1999). Effects of the diacylglycerol complexing agent, cremophor, on nerve conduction velocity and perfusion in diabetic rats. *J. Diabet. Complicat.* **13**, 2–9.

Jirousek, M. R., Gillig, J. R., Gonzalez, C. M., Health, W. F., McDonald, J. H. 3rd, Neel, D. A., Rito, C. J., Singh, U., Stramm, L. E., Melikian-Badalian, A., Baevsky, M., Ballas, L. M., Hall, S. E., Winneroski, L. L., and Faul, M. M. (1996). (S)-13-[(dimethylamino)methyl]-10,11,14,15-tetrahydro-4,9:16,21-dimetheno-1H,13H-dibenzo[e,k]pyrrolo[3,4-h][1,4,13]oxadiazacyclohexadecene-1,3(2H)-d ione (LY333531) and related analogues: Isozyme selective inhibitors of protein kinase C beta. *J. Med. Chem.* **39**, 2664–2671.

Johnson, J. E., Giorgione, J., and Newton, A. C. (2000). The C1 and C2 domains of protein kinase C are independent membrane targeting modules, with specificity for phosphatidylserine conferred by the C1 domain. *Biochemistry* **39**, 11360–11369.

Kang, N., Alexender, G., Park, J. K., Maasch, C., Buchwalow, I., Luft, F. C., and Haller, H. (1999). Differential expression of protein kinase C isoforms in streptozotocin-induced diabetic rats. *Kidney Int.* **56**, 1737–1750.

Khasar, S. G., Lin, Y. H., Martin, A., Dadgar, J., McMahon, T., Wang, D., Hundle, B., Aley, K. O., Isenberg, W., McCarter, G., Green, P. G., Hodge, C. W., Levine, J. D., and

Messing, R. O. (1999). A novel nociceptor signaling pathway revealed in protein kinase C epsilon mutant mice. *Neuron* **24**, 253–260.

Kikkawa, U., Kishimoto, A., and Nishizuka, Y. (1989). The protein kinase C family: Heterogeneity and its implications. *Annu. Rev. Biochem.* **58**, 31–44.

Kiley, S., and Parker, P. J. (1995). Differential localization of protein kinase C isozymes in U937 cells: Evidence for distinct isozyme functions during monocyte differentiation. *J. Cell Sci.* **108**, 1003–1016.

Kim, J., Kyriazi, H., and Greene, D. A. (1991). Normalization of Na$^+$-K$^+$-ATPase activity in isolated membrane fraction from sciatic nerves of streptozotocin-induced diabetic rats by dietary *myo*-inositol supplementation *in vivo* or protein kinase C agonists *in vitro*. *Diabetes* **40**, 558–567.

Kishi, Y., Schmelzer, J. D., Yao, J. K., Zollman, P. J., Nickander, K. K., Tritschler, H. J., and Low, P. A. (1999). α-Lipoic acid: Effect on glucose uptake, sorbitol pathway and energy metabolism in experimental diabetic neuropathy. *Diabetes* **48**, 2045–2051.

Knapp, L. T., and Klann, E. (2000). Superoxide-induced stimulation of protein kinase C via thiol modification and modulation of zinc content. *J. Biol. Chem.* **275**, 24136–24145.

Korchak, H. M., and Kilpatrick, L. E. (2001). Roles for βII-protein kinase C and RACK1 in positive and negative signaling for superoxide anion generation in differentiated HL-60 cells. *J. Biol. Chem.* **276**, 8910–8917.

Kowluru, R. A., Jirousek, M. R., Stramm, L., Farid, N., Engerman, R. L., and Kern., T. S. (1998). Abnormalities of retinal metabolism in diabetes or experimental galactosemia: Relationship between protein kinase C and ATPases. *Diabetes* **47**, 464–469.

Koya, D., Haneda, H., and Nakagawa, H. (2000). Amelioration of accelerated diabetes mesangial expansion by treatment with a PKC beta inhibitor in diabetic *db/db* mice, a rodent model of type 2 diabetes. *FASEB J.* **14**, 439–447.

Koya, D., and King, G. L. (1998). Protein kinase C activation and the development of diabetic complications. *Diabetes* **47**, 859–866.

Kunisaki, M., Bursell, S. E., Umeda, F., Nawata, H., and King, G. L. (1994). Normalization of diacylglycerol-protein kinase C activation by vitamin E in aorta of diabetic rats and cultured rat smooth muscle cells exposed to elevated glucose levels. *Diabetes* **43**, 1372–1377.

Lattimer, S. A., Sima, A. A. F., and Greene, D. A. (1989). *In vivo* correction of Na$^+$,K$^+$-ATPase activity in diabetic nerve by protein kinase C agonists. *Am. J. Physiol.* **256**, E264–E269.

Lee, I. K., Koya, D., Ishi, H., Kanoh, H., and King, G. L. (1999). d-alpha tocopherol prevents the hyperglycemia induced activation of diacylglycerol (DAG)–protein kinase C pathway in vascular smooth muscle cells by an increase of DAG kinase activity. *Diabet. Res. Clin. Pract.* **45**, 183–190.

Lee, T.-S., Saltsman, K. A., Ohashi, H., and King, G. L. (1989). Activation of protein kinase C activity by elevation of glucose concentration: Proposal for a mechanism in the development of diabetic vascular complications. *Proc. Natl. Acad. Sci. USA* **86**, 5141–5145.

Lu, X., Yang, X. Y., Howard, R. L., and Walsh, J. P. (2000). Fatty acids modulate protein kinase C activation in porcine vascular smooth muscle cells independently of their effect on de novo diacylglycerol synthesis. *Diabetologia* **43**, 1136–1144.

Malik, R. A., Kumar, S., and Boulton, A. J. (1995). Mendenhall's syndrome: Clues to the aetiology of human diabetic neuropathy. *J. Neurol. Neurosurg. Psychiatry* **58**, 493–495.

Mathew, J., Bianchi, R., McLean, W. G., and Eichberg, J. (1997). Phosphoinositide metabolism, Na, K-ATPase and protein kinase C are altered in peripheral nerve from Zucker diabetic fatty rats. *Neurosci. Res. Commun.* **20**, 21–30.

Mellor, H., and Parker, P. J. (1998). The extended protein kinase C superfamily. *Biochem. J.* **332**, 281–292.

Mochly-Rosen, D., and Gordon, A. S. (1998). Anchoring proteins for protein kinase C: A means for isozyme selectivity. *FASEB J.* **12**, 35–42.

Mosior, H., and Newton, A. C. (1998). Mechanism of the apparent cooperativity in the interaction of protein kinase C with phosphatidylserine. *Biochemistry* **37**, 17271–17279.

Murray, N. R., and Fields, A. P. (1998). Phosphatidylglycerol is a physiologic activator of nuclear protein kinase C. *J. Biol. Chem.* **273**, 11514–11520.

Nakamura, J., Kasuya, V., Hamada, Y., Nakashima, E., Naruse, K., Yasuda, Y., Kato, K., and Hotta, N. (2001). Glucose-induced hyperproliferation of cultured rat aortic smooth muscle cells through polyol pathway hyperactivity. *Diabetologia* **44**, 480–487.

Nakamura, J., Kato, K., Hamada, Y., Nakayama, M., Chaya, S., Nakashima, E., Naruse, K., Kasuya, Y., Mizubayashi, R., Miwa, K., Yasuda, Y., Kamiya, H., Ienaga, K., Sakakibara, F., Koh, N., and Hotta, N. (1999). A protein kinase C-β-selective inhibitor ameliorates neural dysfunction in streptozotocin-induced diabetic rats. *Diabetes* **48**, 290–2095.

Newton, A. C. (2001). Protein kinase C: Structural and spatial regulation by phosphorylation, cofactors and macromolecular interactions. *Chem. Rev.* **101**, 2353–2364.

Nishikawa, T., Edelstein, D., Du, X. L., Yamagishi, S.-I., Matsumara, T., Kaneda, Y., Yorek, M. A., Beebe, D., Oates, P. J., Hammes, H.-P., Giardino, I., and Brownlee, M. (2000). Normalizing mitochondrial superoxide production blocks three pathways of hyperglycaemic damage. *Nature* **404**, 787–790.

Nishizuka, Y. (1984). The role of protein kinase C in cell surface signal transduction and tumour promotion. *Nature* **308**, 693–698.

Nobe, K., Sakai, Y., and Momose, K. (1998). Alterations of diacylglycerol kinase in streptozotocin-induced diabetic rats. *Cell Signal.* **10**, 465–471.

O'Flaherty, J. T., Chadwell, B. A., Kearns, M. W., Sergeant, S., and Daniel, L. (2001). Protein kinase C translocation responses to low concentrations of arachidonic acid. *J. Biol. Chem.* **276**, 24743–24750.

Park, J. Y., Takahara, N., Gabriele, A., Chou, E., Naruse, K., Suzuma, K., Yamauchi, T., Ha, S. W., Meier, M., Rhodes, C. J., and King, G. L. (2000). Induction of endothelin-1 expression by glucose: An effect of protein kinase C activation. *Diabetes* **49**, 1239–1248.

Parker, P. J. (1999). Inhibition of protein kinase C-do we, can we, and should we? *Pharmacol. Ther.* **82**, 263–267.

Roberts, R. E., and McLean, W. G. (1997). Protein kinase C isozyme expression in sciatic nerve and spinal cords of experimentally diabetic rats. *Brain Res.* **754**, 147–156.

Rosen, P., Nawoth, P. P., King, G., Moller, W., Tritshchler, H. J., and Packer, L. (2001). The role of oxidative stress in the onset and progression of diabetes and its complications: A summary of a Congress series sponsored by UNESCO-MCBN, the American Diabetes Association and the German Diabetic Society. *Diabet. Metab. Res. Rev.* **17**, 189–212.

Rowe-Rendleman, C. R., and Eichberg, J. (1994). P_0 phosphorylation in nerves from normal and diabetic rats: Role of protein kinase C and turnover of phosphate groups. *Neurochem. Res.* **19**, 1023–1031.

Russell, J. W., and Feldman, E. L. (1999). Insulin-like growth factor-I prevents apoptosis in sympathetic neurons exposed to high glucose. *Horm. Metab. Res.* **31**, 90–96.

Russell, J. W., Sullivan, K. A., Windebank, A. J., Herrmann, D. N., and Feldman, E. L. (1999). Neurons undergo apoptosis in animal and cell culture models of diabetes. *Neurobiol. Dis.* **6**, 347–363.

Samii, A., Unger, J., and Lange, W. (1999). Vascular endothelial growth factor expression in peripheral nerves and dorsal root ganglia in diabetic neuropathy in rats. *Neurosci. Lett.* **262**, 159–162.

Schneider, U., Jund, R., Nees, S., and Grafe, P. (1992). Differences in sensitivity to hyperglycemic hypoxia of isolated rat sensory and motor nerve fibers. *Ann. Neurol.* **31**, 605–610.

Schrama, L. H., Berti-Mattera, L. N., and Eichberg, J. (1987). Altered protein phosphorylation in sciatic nerve from rats with streptozotocin-induced diabetes. *Diabetes* **36**, 1254–1260.

Schratzberger, P., Walter, D. H., Rittig, K., Bahlmann, F. H., Pola, R. Curry, C., Silver, M., Krainin, J. G., Weinberg, D. H., Ropper, A. H., and Isner, J. M. (2001). Reversal of experimental diabetic neuropathy by VEGF gene transfer. *J. Clin. Invest.* **107**, 1083–1092.

Scivittaro, V., Ganz, M. B., and Weiss, M. F. (2000). AGEs induce oxidative stress and activate protein kinase βII in neonatal mesangial cells. *Am. J. Physiol. Renal Physiol.* **278**, F676–F683.

Setton-Avruj, C., Goedelmann, C. J., Soto, E. F., and Pasquini, J. M. (2001). Protein kinase C in peripheral myelin from control and streptozotocin-induced diabetic rats: Regulation by phosphatidic acid phospohydrolase. *J. Neurochem.* **78** (Suppl.) 1, 143.

Shimomura, T., Asoka, Y., Oka, M., Yoshida, K., and Nishizuka, Y. (1991). Synergistic action of diacylglycerol and unsaturated fatty acids for PKC activation: Its possible implication. *Proc. Natl. Acad. Sci. USA* **88**, 5149–5153.

Simpson, C. M., and Hawthorne, J. N. (1988). Reduced Na^{++} K^+-ATPase activity in peripheral nerve of streptozotocin-diabetic rats: A role for protein kinase C? *Diabetologia* **31**, 297–303.

Tammariello, S. P., Quinn, M. T., and Estus, S. (2000). NADPH oxidase contributes directly to oxidative stress and apoptosis in nerve growth factor-deprived sympathetic neurons. *J. Neurosci.* **20**, RC53 (1–5).

Tang, E. Y., Parker, P. J., Beattie, J., and Houslay, M. D. (1993). Diabetes induces selective alterations in the expression of protein kinase C isoforms in hepatocytes. *FEBS Lett.* **326**, 117–123.

The Diabetes Control and Complications Trial Research Group (1993). The effect of intensive treatment of diabetes on the development and progression of long-term complications in insulin-dependent diabetes mellitus. *N. Engl. J. Med.* **329**, 977–986.

Thomas, P. K., and Tomlinson, D. R. (1993). Diabetic and hypoglycemic neuropathy. *In* "Peripheral Neuropathy" (P. J. Dyck, P. K. Thomas, J. W. Griffin, P. A. Low, and J. F. Poduslo, eds.), pp. 1219–1250, Saunders, Philadelphia.

Tomlinson, D. R. (1999). Mitogen-activated protein kinases as glucose transducers in diabetic complications. *Diabetologia* **42**, 1271–1281.

Tomlinson, D. R., Ettlinger, C. B., and Lockett, M. J. (1993). Modulation of the sodium-potassium pump of peripheral nerve-pharmacological manipulation of protein kinase C. *Mol. Neuropharmacol.* **3**, 133–138.

Tomlinson, D. R., Riaz, S., and Stevens, E. J. (1996). Dependence of sciatic nerve blood flow on hind-limb muscle mass: An explanation for the deficits in nerve perfusion in experimental diabetes. *Diabet. Nutr. Metab.* **9**, 258–266.

Tran, K., Proulx, P. R., and Chan, A. C. (1994). Vitamin E suppresses diacylglycerol (DAG) level in thrombin-stimulated endothelial cells through an increase of DAG kinase activity. *Biochim. Biophys. Acta* **1212**, 193–202.

Way, K. J., Chou, E., and King, G. L. (2000). Identification of PKC-isoform-specific biological actions using pharmacological approaches. *Trends Pharmacol. Sci.* **21**, 181–186.

Ways, D. K., and Sheetz, M. J. (2000). The role of protein kinase C in the development of the complications of diabetes. *Vit. Horm.* **60**, 149–193.

Xia, P., Aiello, L. P., Ishii, H., Jiang, Z. Y., Park, D. J., Robinson, G. S., Takagi, H., Newsome, W. P., Jirousek, M. R., and King, G. L. (1996). Characterization of vascular endothelial growth factor's effect on the activation of protein kinase C, its isoforms, and endothelial cell growth. *J. Clin. Invest.* **98**, 2018–2026.

Xia, P., Inoguchi, T., Kern, T. S., Engerman, R. L., Oates, P. J., and King, G. L. (1994). Characterization of the mechanism for the chronic activation of diacylglycerol-protein kinase C pathway in diabetes and hypergalactosemia. *Diabetes* **43**, 1122–1129.

Xia, P., Kramer, R., and King, G. L. (1995). Identification of the mechanism for the inhibition of Na^+,K^+-adenosine triphosphatase by hyperglycemia involving activation of protein kinase C and cytosolic phospholipase A_2. *J. Clin. Invest.* **96**, 733–740.

Yamagishi, S.-I., Masuta, N., Okamoto, K., and Yagihashi, S. (2001). Alterations of protein kinase C activity in peripheral nerve of STZ-induced diabetic mice overexpressing human aldose reductase. *Diabetes* **50** (Suppl. 1), A190.

Yan, S. D., Schmidt, A. M., Anderson, G. M., Zhang, J., Brett, J., Zou, Y. S., Pinsky, D., and Stern, D. (1994). Enhanced oxidant stress by the interaction of advanced glycation end products with their receptors/binding proteins. *J. Biol. Chem.* **269**, 9889–9897.

Yu, H. V., Inoguchi, T., Kakimoto, M., Nakashima, N., ImamuDra, M., Hashimoto, T., Umeda, F., and Nawata, H. (2001). Saturated non-esterified fatty acids stimulate de novo diacylglycerol synthesis and protein kinase C activity in cultured aortic smooth muscle cells. *Diabetologia* **44**, 614–620.

Zhu, X., and Eichberg, J. (1990). 1,2-Diacylglycerol content and its arachidonyl-containing molecular species are reduced in sciatic nerve from streptozotocin-induced diabetic rats. *J. Neurochem.* **55**, 1087–1090.

Zochodne, D. W. (1999). Diabetic neuropathies: Features and mechanisms. *Brain Pathol.* **9**, 369–391.

Zochodne, D. W., and Ho, L. T. (1992). Normal blood flow but lower oxygen tension in diabetes of young rats: Microenvironment and the influence of sympathectomy. *Can. J. Physiol. Pharmacol.* **70**, 651–659.

ARE MITOGEN-ACTIVATED PROTEIN KINASES GLUCOSE TRANSDUCERS FOR DIABETIC NEUROPATHIES?

Tertia D. Purves and David R. Tomlinson

Neuroscience Division, University of Manchester, School of Biological Sciences
Manchester M13 9PT, United Kingdom

I. Introduction

The Incidence and severity of diabetic neuropathy are increased by poor glycemic control (Diabetes Control and Complications Trial Research Group, 1993), indicating that excess glucose may be the biochemical trigger in the pathogenesis of diabetic neuropathy. Distal symmetrical polyneuropathy is the most commonly recognized form of diabetic neuropathy and is composed primarily of sensory deficits and symptoms, which generally predominate over motor involvement (Brown and Greene, 1984; Thomas and Tomlinson, 1992). The symptoms of sensory neuropathy vary depending on the class of nerve fiber involved. The loss of large sensory fibers diminishes light touch, proprioception, and mechanoreception. Loss of small fibers diminishes pain and temperature perception, predisposing to injuries that can lead to diabetic foot ulcers. Slowing of nerve conduction velocity is attributed to several physiological and anatomical abnormalities (Greene *et al.*, 1988). The reversible stage of conduction velocity deficits precedes structural nerve damage. However, as diabetes progresses, the reversible component of nerve conduction slowing diminishes as nerves degenerate (Pfeifer and Schumer, 1995). The pathology of diabetic neuropathy is multifocal with changes in axons, Schwann cells, microvascular elements in the endoneurium, and extracellular matrix (Thomas and Tomlinson, 1992; Dyck and Giannini, 1996; Zochodne, 1999).

The changes in cellular phenotype brought about by high glucose, which in sensory neurons *in vivo*, translates to a degeneration of axons, require an initial change in gene expression. Thus a critical feature of the etiology of neuropathies is the mechanisms by which raised extracellular glucose alters the pattern of gene expression, which constitutes cell phenotype.

II. Mitogen-Activated Protein (MAP) Kinases

Mitogen-activated protein kinases transduce signals from the cell surface to the nucleus where they are responsible for activating specific transcription factors, gene expression, and eventually initiating a cellular response (Fig. 1). Three distinct groups of MAP kinases have been identified in mammalian cells: extracellular signal-regulated kinase (ERK), c-Jun N-terminal kinase (JNK), and p38. They play a central role in diverse cellular

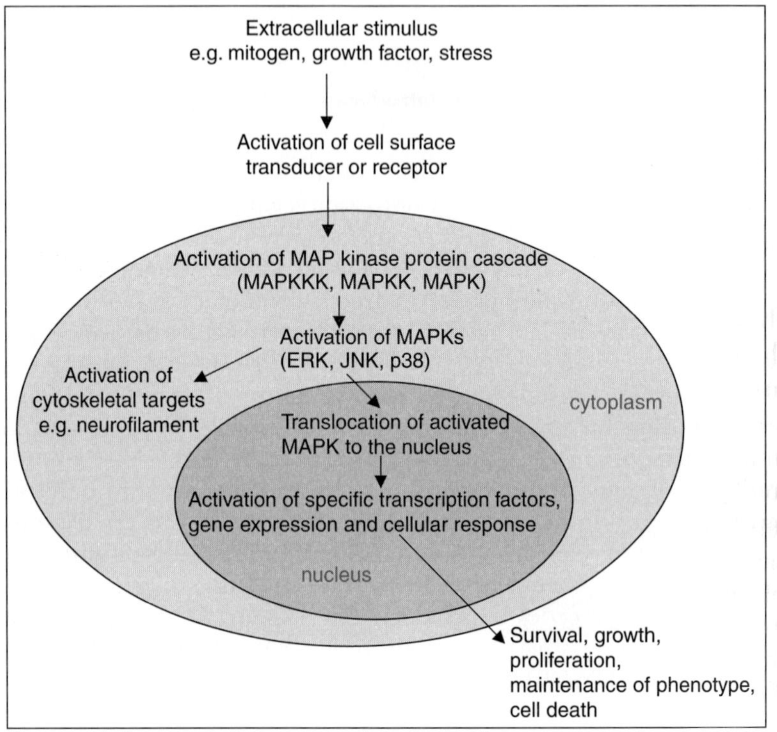

Fig. 1. The MAP kinase cascade.

responses, including cell survival, growth, differentiation, maintenance of phenotype, and cell death (Geilen *et al.*, 1996). Historically, ERK has been classified as responding to mitogens and growth factors, implicated in growth and survival and referred to as a MAP kinase. JNK and p38 have been described as stress-activated protein (SAP) kinases as they respond predominantly to different stress stimuli [e.g., ultra violet (UV) radiation, oxidants, cytokines] (Ono and Han, 2000; Davis, 2000). However, ERK has also been implicated in stress responses, including apoptosis. (Derkinderen *et al.*, 1999), and it is becoming clear that stimuli that activate these kinases overlap and all three MAP kinases can be considered to be (at least potential) SAP kinases. An important factor determining cell fate is integration of the different MAP kinase pathways and the balance between their activities. (Xia *et al.*, 1995).

The ERK pathway has many roles in neurons, including synaptic plasticity, long-term potentiation, survival, and growth (Grewal *et al.*, 1999; Kaplan and Miller, 2000). There are currently five mammalian ERK isoforms identified, referred to as ERK1 through ERK5 (Boulton *et al.*, 1991; Lee *et al.*, 1995; Zhou *et al.*, 1995; Turgeon *et al.*, 2000). Little is known about the regulation and targets of ERK3, ERK4, and ERK5. ERK has been implicated in neuronal survival, but it is becoming evident that ERK plays a more important role in neurite outgrowth, as inhibition studies of the ERK pathway failed to result in cell death (Xia *et al.*, 1995; Klesse *et al.*, 1999). In these studies, ERK activity contributed to, but was not essential for, cell survival.

The JNK p46 and p54 isoforms were the first to be isolated and defined as SAP kinases that regulate the activity of the transcription factor, c-Jun (Lin *et al.*, 1993; Woodgett *et al.*, 1996). JNK is activated by treatment of cells with cytokines such as tumor necrosis factor-α (TNFα) and interleukin-1 (IL-1) and by exposure of cells to many forms of environmental stress, including osmotic stress, UV radiation, redox stress, and mechanical stress (Ip and Davis, 1998). It is now known that JNK is required for embryonic morphogenesis and that this pathway contributes to the regulation of cell proliferation and apoptosis (Davis, 2000). JNK also contributes to the function of some differentiated cells (Davis, 2000).

p38 was discovered in response to lipopolysaccharide (LPS), endotoxin, and cytokine suppressive anti-inflammatory drugs (CSAID) (Han *et al.*, 1994; Lee *et al.*, 1994). p38 shares the phosphorylation motif Thr-Gly-Tyr with HOG1, a yeast MAP kinase required for cellular osmoregulation (Han *et al.*, 1994; Raingeaud *et al.*, 1995). p38 is activated by cytokines (e.g., IL-1, TNFα) and plays a role in the inflammatory response (Ono and Han, 2000). p38 is also activated by heat shock, cell stretch, oxidants, and ischemia/reperfusion (Ono and Han, 2000). More recently, p38 has been shown to be activated by mitogens and growth factors, including nerve growth factor (NGF), fibroblast growth factor (FGF), and platelet-derived growth factor (PDGF), in certain cell types and is linked to growth, differentiation, and survival (Morooka and Nishida, 1998; Ono and Han, 2000).

It is apparent that the relationship among the three MAP kinase pathways and whether they serve as survival or death signals is cell type and stimuli specific. In PC12 cells, activation of both p38 and ERK MAP kinase pathways by NGF was required for neurite outgrowth (Morooka and Nishida, 1998). ERK is known to play a role in neuronal survival and differentiation and in neurite outgrowth (Derkinderen *et al.*, 1999; English *et al.*, 1999; Grewal *et al.*, 1999). ERK has been proposed to play a role in neuronal excitotoxic cell death and oxidative stress-induced cell death (Derkinderen *et al.*, 1999; Bhat and Zhang, 1999), but the precise role of ERK activation is unknown. JNK has, in some situations, been associated with apoptosis, as well as with cell survival, differentiation, and neuroregeneration (Ip and Davis, 1998; Tibbles and Woodgett, 1999; Davis, 2000).

MAP kinases are components of a three kinase regulatory cascade (Treisman, 1996; Garrington and Johnson, 1999; Graves and Krebs, 1999). MAP kinase activation requires dual phosphorylation on Thr and Tyr within the motif Thr-X-Tyr, where X represents Glu in ERK, Pro in JNK, and Gly in p38 (Kyriakis and Avruch, 1996). MAP kinase (MAPK) kinases form a highly conserved group that is activated through the phosphorylation of conserved serine and threonine residues by a more diverse group of MAPK (MAPKK) kinases. Within the three MAP kinases pathways, MAP kinases, MAPK kinases, and MAPKK kinases are usually conserved in their catalytic domains, but can differ considerably in their regulation domains and substrate specificities. MAP kinases differ in their regulation and substrate specificities, but are highly conserved in their primary structure and mode of activation. Experimentally, phosphorylation and activation of MAP kinases are examined using immunoblotting techniques with antibodies raised against phosphorylated epitopes of the kinases (P) and antibodies raised against the kinases irrespective of phosphorylation, hence referred to as total (T) kinase and reflecting changes in protein expression. Therefore, levels of phosphorylated kinase and changes in kinase expression can be determined, and the ratio of phosphorylated to total kinase (P/T) reflects activation of the specific kinase.

III. Glucose, Diabetes, and MAP Kinases

Several hypotheses for the mechanism of action of high glucose in diabetes have been proposed, including hyperosmolarity, production of advanced glycation end products (AGE), increased polyol pathway flux, protein kinase C (PKC) activation, and glucose-induced oxidative stress or direct effects of the glucose molecule itself (Zochodne, 1999; Fernyhough and Tomlinson, 1999; Carrington and Litchfield, 1999). It is likely that glucose and its metabolites mediate some or all of the adverse effects by altering various signal transduction pathways, which are used by cells to perform their functions and to maintain cellular integrity. The potential for a pivotal role for oxidative stress in the etiology of diabetic neuropathy (Mohamed *et al.*, 1999; Laight *et al.*, 2000; Obrosova *et al.*, 2000; Stevens *et al.*, 2000) has prompted one unifying hypothesis involving mitochondrial dysfunction to explain the complications of diabetes (Nishikawa *et al.*, 2000).

MAP kinases have been implicated in the etiology of diabetic complications. All three groups of MAP kinases (JNK, p38, and ERK) can be activated by osmotic perturbations derived from glucose itself or from the polyol pathway (Cohen, 1997; Kultz and Burg, 1998), by oxidative stress.(Wang *et al.*, 1998), and by AGE via the receptors of AGE (RAGE) (Thornalley, 1998; Tomlinson, 1999). High glucose levels phosphorylate ERKs in rat glomeruli and mesangial cells. (Haneda *et al.*, 1997; Awazu *et al.*, 1999), implicating MAP kinases in the etiology of diabetic nephropathy. Different mechanisms of MAP kinase activation were identified (Awazu *et al.*, 1999). The total protein content of ERK was increased, as was the phosphorylation of ERK. MAP kinase phosphatase-1 (MKP-1), a phosphatase that inactivates ERK, was 60% of control in diabetic kidney and in cultured mesangial cells treated with high glucose. The decreased activity of MKP-1 was dependent on elevated PKC activity (Awazu *et al.*, 1999). Enhanced levels of PKC activity in mesangial cells have been shown to be dependent on high glucose and polyol pathway activity (Kapor-Drezgic *et al.*, 1999). Igarashi *et al.* (1999) have characterized the mechanisms by which the elevation of glucose levels activated p38 MAP kinase in cultured vascular cells and aorta from diabetic rats (Igarashi *et al.*, 1999). A time- and dose-dependent activation of p38 with elevated glucose in aortic smooth muscle cells was observed. They confirmed that in aorta from diabetic rats, p38 levels (both total and phosphorylated) were elevated compared to aorta from control rats. It was shown that moderate hyperglycemia activated p38 by a PKC-δ isoform-dependent pathway, but glucose at extremely elevated levels activated p38 kinase by hyperosmolarity via a PKC-independent pathway. PKC can feed into the ERK pathway via activation of MAPK/ERK kinase (MEK) (Kaplan and Miller, 2000).

Oxidative stress is a potent activator of MAP kinases in various cell types (Liu *et al.*, 1996; Cantoni *et al.*, 1996; Yu *et al.*, 1997; Elbirt *et al.*, 1998; Wang *et al.*, 1998; Iordanov and Magun, 1999). Evidence for the involvement of oxidative stress in the pathogenesis of diabetic complications has arisen from studies showing the attenuation or prevention of functional and structural abnormalities in nerves of diabetic animals by antioxidants such as DL-α-lipoic acid (Laight *et al.*, 2000; Stevens *et al.*, 2000). Cells are susceptible to damage by reactive oxygen species (ROS). ROS, which are derived naturally in many metabolic processes, are dealt with by intracellular antioxidant mechanisms such as the glutathione redox recycling pathway (Sies, 1985). Extracellular-derived ROS deplete cellular defense mechanisms, thereby causing oxidative stress by reducing the capacity of the cell to defend against oxidants. Oxidative stress can lead to cellular damage, apoptosis, or necrosis depending on the severity.

Molecular oxygen is the energy source that supports all mammalian tissues, serving as the electron acceptor in the oxidative process that provides this energy. Molecular oxygen is an important component of many pathological processes that result in oxidative injury. As defined by Halliwell (1992), "*oxidative stress is an increase in the generation of oxygen derived species beyond the ability of antioxidant defences to cope with them.*" The major source of oxidative stress in cells arises from the production of ROS and reactive nitrogen species (RNS), which are derived from enzymatic and nonenzymatic processes. Naturally occurring free radicals typically have an oxygen- or nitrogen-based unpaired electron and these include superoxide anion ($O_2 \cdot ^-$), hydroxyl radical ($^\bullet OH$), nitric oxide ($\cdot NO$), and peroxynitrite ($ONOO^-$). The concentrations of individual ROS and RNS formed by cells and tissues are tightly controlled by specific metabolizing and scavenging systems. It is becoming widely accepted that both ROS and RNS play important roles in intra- and intercellular signalling (Rhee, 1999). Oxidative stress has been implicated in the etiology of many neurodegenerative disease processes (Halliwell, 1992), including Parkinson's disease (Cassarino *et al.*, 2000), Alzheimer's disease (Lezoualc'h *et al.*, 1998), amyotrophic lateral sclerosis (Coyle and Puttfarcken, 1993; Migheli *et al.*, 1997), in the general process of aging (Berlett & Stadtman, 1997), and diabetes (Mohamed *et al.*, 1999).

A major source of ROS in the cell arises via the generation of H_2O_2 during normal cellular metabolism. The process, known as redox cycling, occurs during autooxidation, which yields superoxide, and occurs when electronegative compounds intercept electrons from normal cellular electron transport systems in mitochondria and endoplasmic reticulum and reduce O_2. The ROS, superoxide, is converted to hydrogen peroxide by superoxide dismutase (SOD). Other cellular sources of H_2O_2 involve enzymatic two-electron transfer to O_2. Enzymes such as urate oxidase, acyl CoA oxidase, and monoamine oxidase are some of the enzymes that yield H_2O_2. Monamine oxidase is important for the oxidative deamination of biogenic amines to their corresponding aldehydes and H_2O_2, and this enzyme represents the largest source of intracellular H_2O_2. Intracellular H_2O_2 is metabolized by glutathione peroxidase, myeloperoxidase, and catalase, which maintain intracellular H_2O_2 levels at about 1 nM. Nonenzymatic decomposition of H_2O_2 occurs via a Fenton reaction whereby reduced transition metal complexes, usually involving Cu^+ or Fe^{+2}, facilitate the homolytic decomposition of H_2O_2, resulting in the release of the highly reactive hydroxyl radical ($^\bullet$OH). Damaged cells release metal ions into the surrounding tissues, creating a source of metal catalysts that results in the oxidative damage of surrounding tissue.

$^\bullet$OH is a potent electrophile that diffuses easily from the site of production before acting randomly with any molecular target, including DNA, protein, and membrane lipids. Oxidation of thiol groups of proteins results in the loss of enzyme function, transmembrane metabolite and ion transport, and contractile function (Sies, 1985). Peroxidation of polyunsaturated fatty acids, cholesterol, and other lipids result in altered fluidity and permeability characteristics of cell membranes (Sies, 1985). Oxidative scission and chemical alterations of deoxyribose and of the purine and pyrimidine bases of nucleic acids result in genomic damage (Sies, 1985; Halliwell, 1992). Oxidative destruction of polysaccharides also occurs and can result in altered function, such as the loss of viscosity of hyaluronic acid (Sies, 1985). An additional mechanism of DNA damage is the increase of intracellular-free Ca^{+2} concentrations by oxidative stress, leading to the activation of nuclease enzymes within the nucleus (Halliwell, 1992). Increased intracellular Ca^{+2} can also activate Ca^{+2}-stimulated proteases that can cause phenomena, such as the blebbing of the plasma membrane (Halliwell, 1992).

During oxidative stress, cellular defense mechanisms become overwhelmed and the redox balance of the cells is shifted to a more oxidized state. Redox cycles in which glutathione (GSH) is involved are important in maintaining the redox balance of the cells and hence the antioxidant capacity of the cells. The GSH redox cycle involves the oxidation of GSH to glutathione disulfide (GSSG) by glutathione peroxidase (GSH Px) and the concomitant conversion of H_2O_2 to H_2O. The reduction of GSSG to GSH catalyzed by glutathione reductase (GR) is dependent on NADPH as a cofactor. As oxidative stress in cells progresses and the antioxidant defense mechanisms of the cells are challenged, a depletion of cofactors such as NADPH occurs, causing further redox imbalances within the cell. Experimental interruption of the glutathione redox cycle provides evidence for the consequences of elimination of NADPH oxidation by this cycle and increased levels of H_2O_2, and hence oxidative stress. Overactivity of metabolic processes that require NADPH as a cofactor, such as the polyol pathway, deplete intracellular cofactors available for the GSH cycle and thereby generate oxidative stress.

Hydrogen peroxide activates all three MAP kinases; however, the combination of kinases and the extent to which they are activated is apparently cell type specific. In smooth muscle cells, H_2O_2 induced ERK activation that resulted in cell death (Cantoni *et al.*, 1996). In hepatoma cells (Elbirt *et al.*, 1998), Rat1 fibroblasts, and PC12 cells (Liu *et al.*, 1996; Iordanov and Magun, 1999) different combinations of MAP kinases were activated. In an oligodendrocyte cell line (CG-4), activation of all three MAP kinases by H_2O_2 was linked to stress-induced cell death, which was prevented by blocking ERK signaling (Bhat and Zhang, 1999). In Hela or HepG2 cells, butylated hydroxyanisole (BHA), which generates intracellular H_2O_2, activated ERK2, followed by a slower activation of JNK1 (Yu *et al.*, 1997; Wang *et al.*, 1998). Pretreatment with antioxidants such as N-acetyl-L-cysteine or vitamin E attenuated ERK2 activation but not JNK activation (Yu *et al.*, 1997). There is, however, little work aimed at testing the involvement of MAP kinases in the damage induced by oxidative stress in diabetes. RAGE activation by AGE binding generated reactive oxygen species (Yan *et al.*, 1994) and activated MAP kinases in PC12 and rat pulmonary artery smooth muscle cells (Simm *et al.*, 1997; Lander *et al.*, 1997; Thornalley, 1998). In cultured endothelial cells, high glucose treatment resulted in the production of ROS via increased oxidative phosphorylation in mitochondria (Nishikawa *et al.*, 2000) by the depletion of NADPH and the subsequent compromising of intracellular antioxidant defense mechanisms (Wüllner *et al.*, 1999). In human endothelial cells (HUVECs), high glucose generated an increase in ROS, an activation of the transcription factor, nuclear factor-κB (NFκB), and the induction of apoptosis (Du *et al.*, 1999). High glucose has been shown to trigger apoptosis in vascular endothelial cells (Baumgartner-Parzer *et al.*, 1995). It has been reported that human SH-SY5Y neuroblastoma cells undergo apoptosis when exposed to high levels of glucose (Russell and Feldman, 1999). In the same cells, glucose treatment resulted in the activation of both ERK and JNK MAP kinases (Cheng and Feldman, 1998). Russell *et al.* (1999) have shown an induction of apoptosis in DRG from diabetic rats, rats made acutely hyperglycemic by glucose infusion, and cultured embryonic DRG neurons treated with elevated glucose (Russell *et al.*, 1999).

IV. MAP Kinase Activation in Primary Sensory Neurons

It is proposed that MAP kinases provide a focus for the synergy of the different biochemical and metabolic changes brought about by high glucose. Cultures of adult rat DRG neurons have been used as a tool for examining neuronal MAP kinase signaling pathways in response to different diabetes-related cellular stressors, including high glucose and oxidative stress. As expressed by Lindsay *et al.* (1991), *"the heterogeneity of the neuronal and glial components of the nervous system pose a barrier when trying to elucidate the physiological and biochemical processes of subclasses of neuronal and glial cell populations and tissue culture is used to simplify such studies by isolating or enriching for specific cell populations allowing for the analysis of specific components of the nervous system."* The cell bodies of peripheral sensory neurons are located in the spinal ganglia and are readily dissectable. Dissected neurons can be dissociated into single cell suspensions and maintained in culture. Separation of neurons from nonneuronal cells, myelin, nonviable cells, and other cellular debris utilizes a simple filtration procedure based on size and two separation procedures based on cell density. Dissociation and culturing of neurons involve axotomy close to the cell body and disruption of the extracellular environment; therefore, the predominant phenotype of cultured primary cells is one of regeneration. Evidence that DRG neurons in culture are in an axotomized state is provided by the fact that DRG cultures express NPY to about the same extent as axotomized DRG *in vivo*; in contrast, *in vivo* uninjured neurons do not contain detectable levels of NPY-like immunoreactivity (Kerekes *et al.*, 2000). Peripheral nerve regeneration comprises the formation of axonal sprouts, their outgrowth as regenerating axons, and the reinnervation of the original targets (Ide, 1996). Primary sensory neuronal cultures provide a model system encompassing the first two steps of regeneration. Importantly, it has been established that there is no significant difference between distributions of cell sizes or neurochemically defined subpopulations of cells from L4/L5 ganglia, *in vivo* and in culture (McMahon *et al.*, 1994; Gavazzi *et al.*, 1999). L4 and L5 DRG neurons project peripherally into the sciatic nerve.

In the study reported by Purves *et al.* (2001), sensory neuron cultures were treated with diabetes-related stressors and the activation of MAP kinases examined. Sensory neuron cultures were treated with glucose concentrations reflecting physiological plasma glucose levels in diabetic rats. H_2O_2 was used as a direct extracellular oxidative stress and DEM as an indirect oxidative stress. DEM was used to induce an oxidative stress by binding to and thereby depleting intracellular glutathione levels, compromising the antioxidant defense mechanism of the neurons. (Ito *et al.*, 1998). By binding to GSH, DEM also prevents the neutralization of lipid peroxidation by conjugation. This was an attempt to mimic the *in vivo* situation whereby the increased polyol pathway flux has been proposed to induce an oxidative stress in diabetes by depleting the cofactor NADPH that is required for the effective functioning of the glutathione redox recycling pathway. H_2O_2 was used as an extracellular oxidative stress, whereby H_2O_2 breaks down rapidly to produce hydroxyl radicals via the Fenton reaction. These oxidative stressors serve as partial surrogates for the oxidative stress proposed to occur in diabetic nerve.

As reported by Purves *et al.* (2001), glucose treatment of sensory neurons resulted in a dose-dependent (10, 25, 50, 100, 150, and 200 mM) increase in p38 and JNK activation at 16 hr of treatment but did not increase ERK activation at this time point (Fig. 2A). Oxidative stresses activated a different combination of MAPKs. H_2O_2 or DEM activated p38 and ERK MAP kinase and not JNK MAP kinase (Fig. 2B). An additive effect of p38 activation in response to cotreatment with DEM/high glucose or H_2O_2/high glucose was reported, suggesting alternative mechanisms of high glucose versus oxidative stress-dependent MAP kinase activation in the relatively short-term (16 hr) cell culture treatment regime (Purves *et al.*, 2001). In addition, a low concentration of H_2O_2 (2 μM) did not activate p38, but when used in conjunction with high glucose, p38 was activated to a greater extent than when cells were treated with high glucose alone (Purves *et al.*, 2001). It was proposed that cells exposed to low levels of oxidative stress were more susceptible to glucose-induced stress.

Cytotoxicity assays, lactate dehydrogenase (LDH), and MTT have been used to examine the effect of high glucose and oxidative stress on sensory neuron cell viability (Purves *et al.*, 2001). High glucose did not result in decreased cell viability using either the LDH or the MTT cytotoxicity assay, indicating that the combination of p38 and JNK MAP kinase activation does not result in cell death. H_2O_2treatment resulted in a dose-dependent decrease in cell survival as measured by the MTT assay. At control glucose (10 mM), H_2O_2 started decreasing cell survival significantly at 500 μM, whereas in cells treated with 50 mM glucose, H_2O_2 started decreasing cell survival at 200 μM. This indicated an increased susceptibility of cells treated with high glucose to oxidative stress. In the same study, H_2O_2 (0, 75, 200, and 500 μM) resulted in a dose-dependent activation of ERK and p38 MAP kinases. It was proposed that there is a minimum level of ERK and p38 activation that results in cellular damage, and this minimum level occurs at lower H_2O_2 concentrations if cells are treated concomitantly with high glucose. The decreases in cell survival induced by H_2O_2 were attenuated significantly by the addition of either an ERK pathway-specific inhibitor, U0126 (Favata *et al.*, 1998), or a p38 pathway-specific inhibitor, SB202190 (Lee *et al.*, 1999). These results implicate the p38 and ERK MAP kinase pathways in the neuronal damage arising from H_2O_2-induced oxidative stress (Purves *et al.*, 2001). DEM treatment also resulted in cellular damage, which was partially prevented by the addition of either U0126 or SB202190.

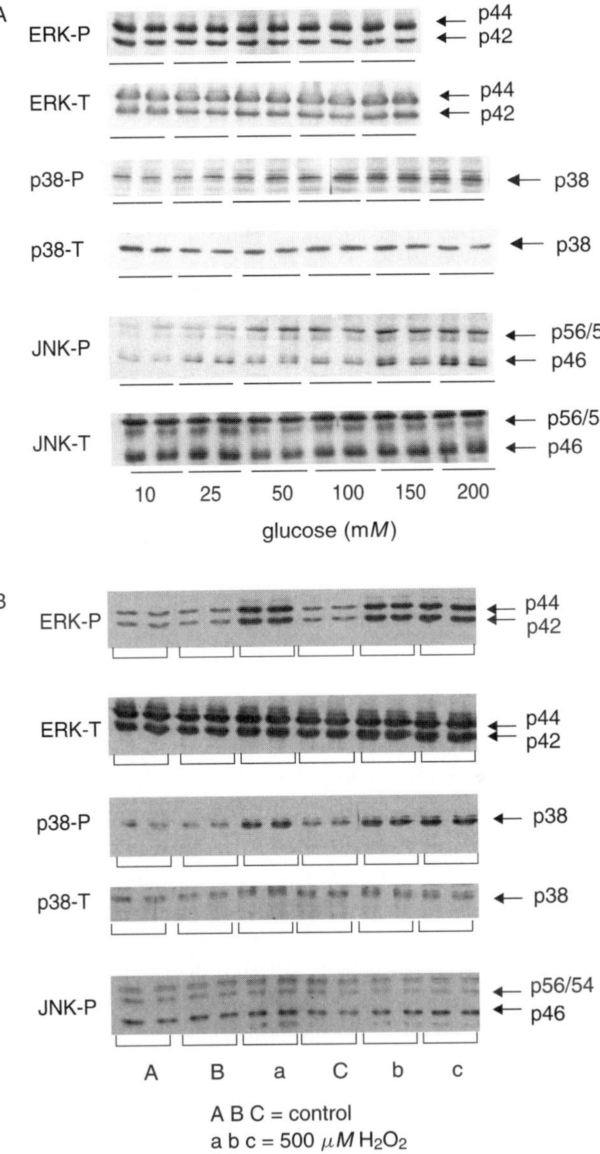

FIG. 2. Activation of MAP kinases in Western blots from primary cultures of adult rat dorsal root ganglia by raised glucose concentration and by oxidative stress. Immunoreactivity for total protein (T), via a nonphosphorylated epitope, and for activated enzyme (P), via a phospho-specific epitope, is shown. Neither raised glucose (A) nor hydrogen peroxide (B) affected total protein, but both increased activation of p38 and JNK. Hydrogen peroxide also activated ERK, but raised glucose did not.

Addition of the inhibitors together did not have an additive effect on the prevention of cell death.

The fact that the addition of the two inhibitors together in primary sensory neurons did not prevent the decrease in cell viability to a greater extent than that seen with each inhibitor separately provides a possible clue as to the mechanism of action of cell death. It provides evidence that both ERK and p38 activation are required for the induction of cell death in sensory neuron cultures and that activation of either pathway alone is not sufficient. This is supported by the fact that cells treated with high glucose showed no decrease in cell viability and this treatment activated only JNK and p38, not ERK MAP kinase. It has been reported previously that combinations of MAP kinase activation and suppression are required to elicit cellular responses. In PC12 cells, NGF treatment activated ERK and resulted in differentiation (Xia *et al.*, 1995). Removal of NGF resulted in activation of JNK and p38 with the concomitant suppression of ERK activation and culminated in apoptosis. The hypothesis was proposed that the cellular response depends on the integration of multiple signals; a greater amount of ERK activation relative to that of JNK or p38 promoted survival, whereas a greater amount of JNK or p38 activity relative to ERK triggered apoptosis (Xia *et al.*, 1995). Thus it is this opposing balance of MAP kinase activities that determines cellular outcome. In the study by Purves *et al.* (2001), it was hypothesized that both p38 and ERK must be activated in sensory neurons, in conjunction, above a minimum level to induce cell death.

High glucose exposure in primary DRG neurons did not result in cell death at 16 hr (Purves *et al.*, 2001). At this early time point in culture it would appear that the metabolism of high glucose concentrations, e.g., via the polyol pathway or possibly through increased rates of pyruvate-driven oxidative phosphorylation in the mitochondria, is not linked to overproduction of ROS or other factors that may result in oxidative stress. In endothelial cells at longer periods in cell culture (1 week), high glucose treatment resulted in the production of ROS (Du *et al.*, 1999) via increased oxidative phosphorylation in mitochondria (Nishikawa *et al.*, 2000), by the depletion of NADPH, and the subsequent compromising of intracellular antioxidant defense mechanisms (Wüllner *et al.*, 1999). Therefore, it would appear that although glucose has the capacity to induce an oxidative stress, this has not occurred in the culture paradigm presented by Purves *et al.* (2001). Interestingly, in the study the decrease in cell viability seen when cells were treated with H_2O_2 occurred at a lower H_2O_2 concentration when cells were cotreated with high glucose, indicating that cells treated with high glucose are more susceptible to oxidative stress. This may be a result of the activation of p38 by glucose such that lower concentrations of oxidative stress are then required to activate p38 (and ERK) to levels high enough to induce death. Other studies have shown cell death in response to glucose using embryonic sensory neurons and doses of glucose up to 300 mM for 24 hr (Russell *et al.*, 1999). These differences may be due to the susceptibility of developing neurons to programmed cell death and the high glucose concentrations translating more rapidly from primary effects of glucose to oxidative stress effects.

V. MAP Kinases and Neuropathy in the Streptozotocin Rat Model of Diabetes

Evidence from streptozotocin (STZ)-induced diabetic rats, a model of type I diabetes, is accumulating, implicating MAP kinase activation in the etiology of diabetic neuropathy. Although complete homology of neuropathic manifestations has not been obtained in animal models, they show an adequate resemblance with respect to functional, biochemical, and structural deficits seen in diabetic patients. STZ rats are the most commonly used animal model of type I diabetes (Rakieten *et al.*, 1963). STZ rats consistently show delayed motor nerve conduction velocities (Jakobsen and Lundbæk, 1976; Sharma and Thomas, 1987), reversible with insulin treatment and therefore attributable to hyperglycemia (Greene *et al.*, 1975; Jakobsen, 1979). STZ-induced hyperglycemia elicits polyol pathway activation in rats (Greene and Lattimer, 1983, 1986). Morphological changes are mild in the peripheral nerve of STZ rats, lacking apparent fiber loss or marked fiber degeneration seen in human diabetes (Sharma and Thomas, 1987). However, minor but significant and reproducible structural changes have been demonstrated in the peripheral nerve of STZ rats (Jakobsen, 1976; Jakobsen and Lundbæk, 1976). Systematic morphometric studies have shown that myelinated fiber atrophy occurs (Yagihashi *et al.*, 1990) and is most severe in distal nerve, whereas proximal nerve fibers were expanded (Medori *et al.*, 1985, 1988). Loss of axonal caliber has been detected at 12 weeks in STZ rats (Mizisin *et al.*, 1999). Although care has to be taken when relating work carried out in diabetic rats to human diabetes, there is sufficient evidence to suggest similarities between the two, and animal models of diabetes are therefore a useful tool to investigate the etiology of this disease.

A general increase in MAP kinase signaling has been observed in L4/L5 DRG, sciatic nerve, and sural nerve of STZ rats, although statistical significance varied (Fernyhough *et al.*, 1999; Purves *et al.*, 2001) (T. Purves and D. R. Tomlinson, unpublished data). Studies using L4/L5 DRG from rats with diabetes of 8 and 12 week duration indicated a significant increase in p38 and ERK MAP kinase activation in 8-week STZ rats compared to age-matched control rats and a significant increase in p56/54 JNK MAP kinase activation in 12-week STZ rats compared to age-matched controls (Purves *et al.*, 2001). Similar findings have been reported previously (Fernyhough *et al.*, 1999). In L4/L5 DRG from 12-week STZ rats, elevated levels of phosphorylated and total ERK, JNK, and p38 MAP kinases were detected compared to age-matched control animals (Fig. 3A; T. Purves and D. R. Tomlinson, unpublished data). In sciatic nerve, a statistically significant increase in phosphorylation and activation of p44 and p42 ERK MAP kinase and p38 MAP kinase in 12-week STZ rats compared to age-matched controls has been observed (data not shown; T. Purves and D. R. Tomlinson, unpublished data). A significant increase in p44 ERK phosphorylation in the sural nerve of 12-week STZ rats compared to age-matched controls has been observed, and levels of total and phosphorylated JNK and p38 were elevated in the sural nerve of 12-week STZ rats compared to age-matched controls (Fig. 3B; T. Purves and D. R. Tomlinson, unpublished data). Previous reports have indicated an increase in JNK but not ERK activation in 12-week STZ rat sural nerve (Fernyhough *et al.*, 1999).

It is apparent from aforementioned data that the severity of diabetes, reflected by plasma glucose concentration, determines the extent of MAP kinase activation. In our study, significant increases of p38 and ERK activation at 8 weeks diabetes was reported, and plasma glucose levels were 6.17 ± 0.55 g/liter (Purves *et al.*, 2001). In a separate study (in the same laboratory, thereby eliminating interlaboratory differences), MAP kinases tended to be increased in DRG, sciatic, and sural nerve of 12-week diabetic rats, but not all the increases reached significance (Fig. 3). The plasma glucose concentration of the diabetic group was 4.51 ± 0.25 g/liter. This adds weight to the hypothesis that it is the glucose and the biochemical changes that it initiates that are responsible for MAP kinase activation.

PGP9.5 has been used to determine whether the ratio of neuronal protein to nonneuronal protein remained constant between tissue samples within the same group and between control and diabetic groups. Levels of PGP9.5, a neuron-specific marker (Doran *et al.*, 1983), remained essentially unchanged in DRG, sural nerve, and sciatic nerve from age-matched control rats compared to 12-week STZ rats, indicating that changes in the ratio of protein contributed by the neuronal component and protein contributed by the nonneuronal component of the tissue were not responsible for the changes in MAP kinases (Purves *et al.*, 2001).

Immunohistochemical analyses confirm the expression of MAP kinases in the neuronal component of sciatic nerve and DRG. Studies have shown immunoreactivity against phosphorylated and total JNK present in the cytoplasm and nuclei of sensory neuron perikarya of control and STZ rats (Fernyhough *et al.*, 1999). In the same study, JNK was localized to the axons of sensory neurons (Fernyhough *et al.*, 1999). Colocalization studies of ligated (single ligature) sciatic nerve of normal rats have been carried out with antibodies against calcitonin gene-related protein (CGRP), a marker of small nociceptive fibers, and antibodies against phosphorylated p38 (Fig. 4C, see also color insert; A. Middlemas and D. R. Tomlinson, unpublished data). Damage to small nociceptive fibers in diabetes as a result of high glucose can predispose to diabetic foot ulcers due to a loss of pain sensation. ERK-T and ERK-P have been located in DRG and axons and colocalized with CGRP (Averill *et al.*, 2001). Immunocytochemical

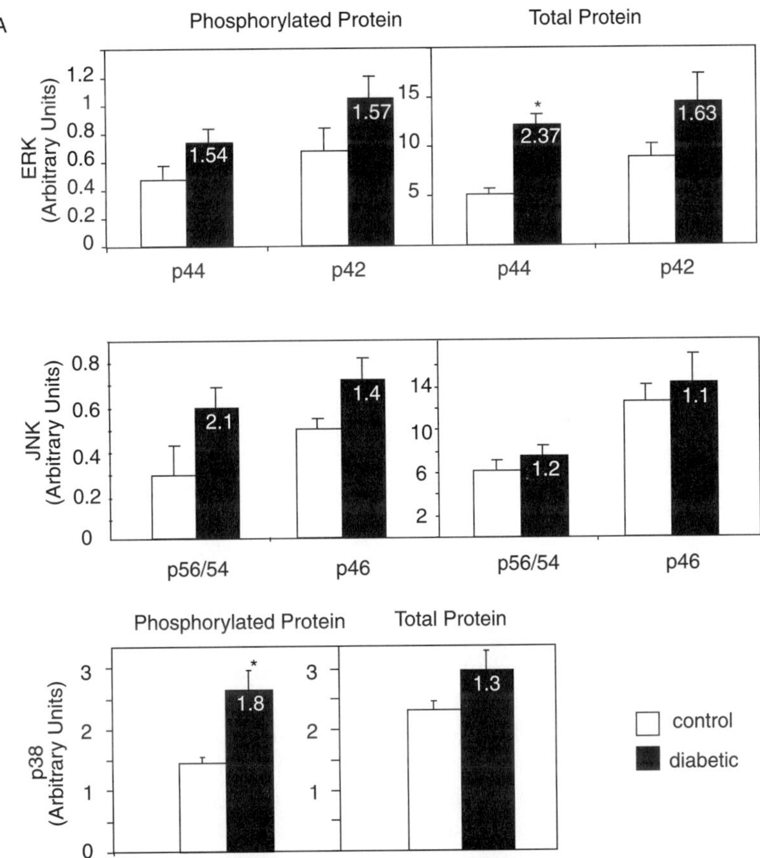

FIG. 3. Activation of MAP kinases by streptozotocin-induced diabetes in dorsal root ganglia (DRG; A) and sural nerves (B) from rats. Data are group means ±1 SEN showing total and phosphorylated protein (see Fig. 2 legend) for each sample.

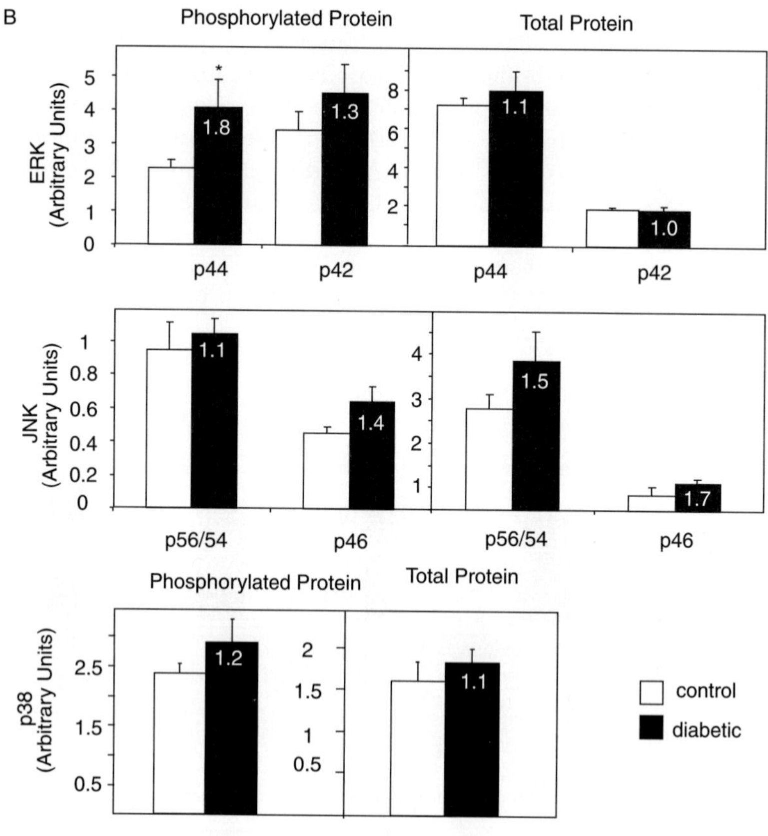

FIG. 3. (*continued*)

analyses of cultured DRG show MAP kinase expression in cell bodies of sensory neurones (Fig. 4A, see also color insert). This evidence confirms that MAP kinases are present in axons, nuclei, and perikarya of sensory neurons. It does not, however, eliminate the possibility that changes in MAP kinases also occur in satellite cells of the axons and DRG. There is evidence of MAP kinase (JNK, p38, and ERK) expression in satellite cells of the DRG *in vivo*; however, the specific non neuronal cell types have not been identified (Averill *et al.*, 2001) (Tomlinson, D. R. and Middlemas A. unpublished data). The increased p38 activation by high glucose was confirmed, by immunocytochemistry, to be in the neuronal component of the sensory neuron cultures (Fig. 4B, see also color insert) (Purves *et al.*, 2001).

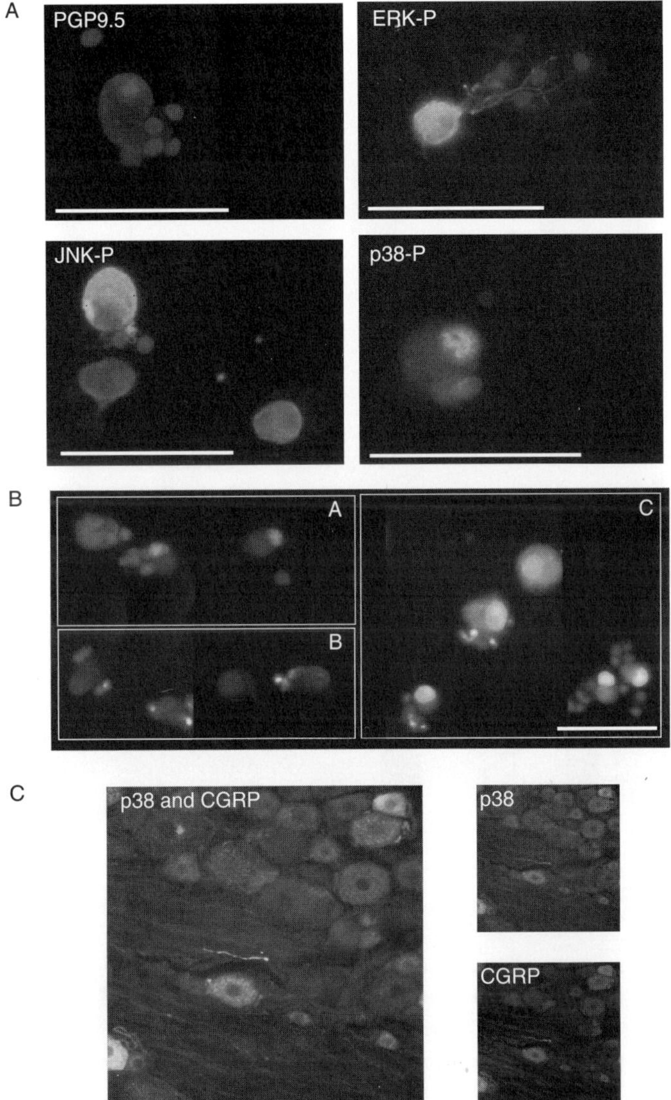

FIG. 4. Activation of MAP kinases in cultures of dorsal root ganglia (DRG) from adult rats shown by immunoreactivity (A and B) and colocalization of MAP kinase p38 with CGRP in sectioned intact DRG from rat (C). (A) Neuron-specific immunoreactivity to PGP 9.5 (red), with nonneurons PGP negative, but with DAPI-positive nuclei stained blue. Staining for phospho-MAP kinases, ERK, JNK, and p38, was confined to the neurons. (B) p38 activation by 50 mM glucose (subfigure C) as compared to 10 mM glucose (subfigure A). A negative control, without primary antibody, is shown in subfigure B. (C) Specific staining for p38 and CGRP in the small micrographs, with the combined picture in the larger micrograph. (See also color insert.)

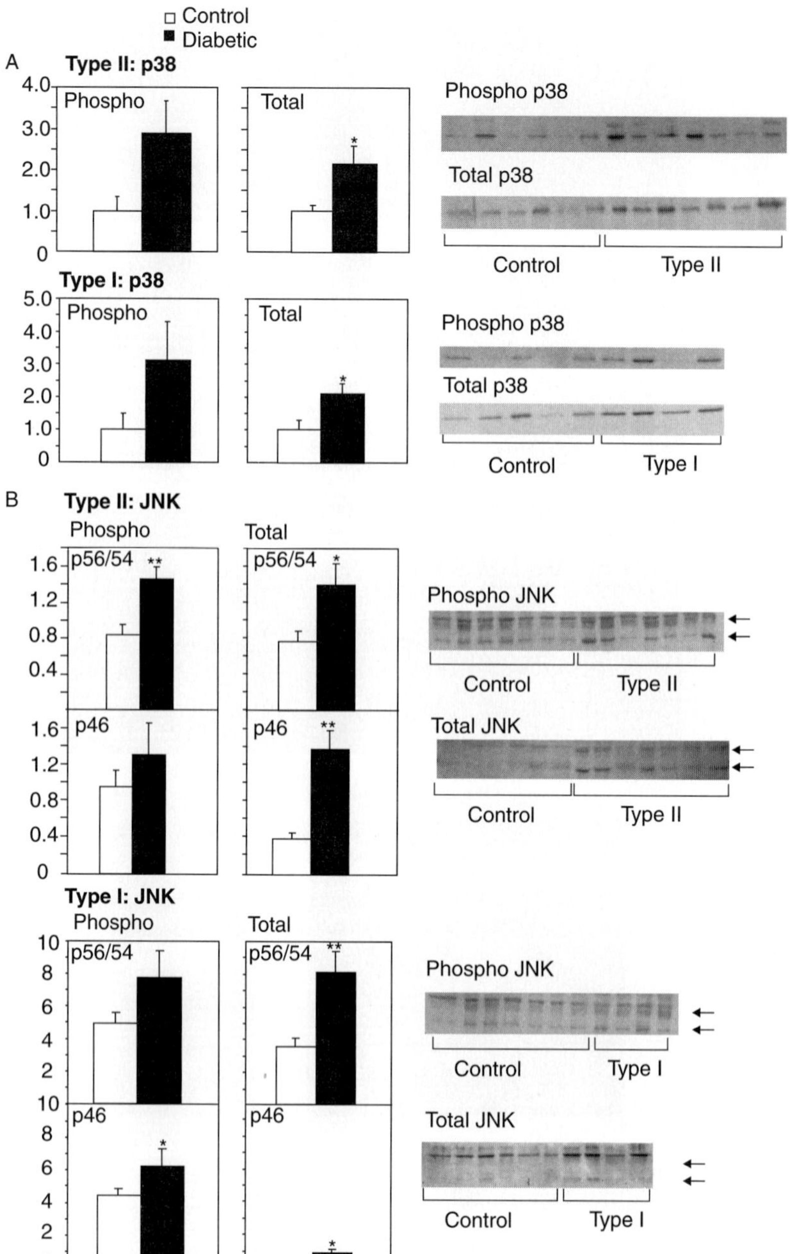

FIG. 5. Activation of MAP kinases p38 and JNK in sural nerve biopsy specimens from patients with either type I or type II diabetes and from nondiabetic subjects. Organization is similar to Fig. 3.

Further evidence for a role of MAP kinase activation in sensory nerve has been provided by STZ rats (8-week duration) subjected to double midsciatic nerve ligature (12 hr) compared to age-matched control rats. Accumulation of protein distal to a ligature is indicative of retrogradely transported protein. In order to ensure that accumulation was a result of diabetes and not local injury due to the ligature, a double ligature paradigm was used (ligatures 1 cm apart). The 1-cm portion distal to the ligatures represents retrogradely transported MAP kinase. This is corrected for the MAP kinase in the 0.5 cm of nerve between the two ligatures, furthest away from the distal portion, which accounts for local synthesis and retrograde transport as a result of injury due to the ligatures. MAP kinases were detected using immunoblotting, and the results indicated retrograde transport of total and phosphorylated JNK MAPK, and phosphorylated p38 MAPK in 8-week STZ rats was increased significantly compared to age-matched control rats (Delcroix *et al.*, 2000).

VI. MAP Kinase Activation in Sural Nerve from Type I and Type II Diabetic Patients

MAP kinase activation in sural nerve from type I and II diabetic patients has been reported (Fig. 5) (Purves *et al.*, 2001). Sural nerve from type I and type II diabetic patients and control sural nerve from nondiabetic amputees were subjected to immunoblotting. There was an increase in p38 activation in type I and type II diabetic patients, which was not significant; however, total p38 was increased significantly in both type I and type II patients. Results indicated a significant increase of total p56/54 JNK in type I and type II diabetic patients and a significant increase in the phosphorylation of p56/54 JNK in type II and p46 JNK in type I patients. Data for ERK fluctuated between patients and no significant changes were seen. These results from human sural nerve specimens indicate that changes in MAP kinase activation associated with diabetes are remarkably similar to those seen in diabetic rats.

VII. Role of MAP Kinase Activation

This evidence indicates a role for MAP kinases in the etiology of diabetic neuropathy (Fig. 6), and the question arises: *Does the activation of MAP kinases in the peripheral nervous system in diabetes play a neuroprotective of neurodegenerative role?* As described earlier, MAP kinases have been implicated in the

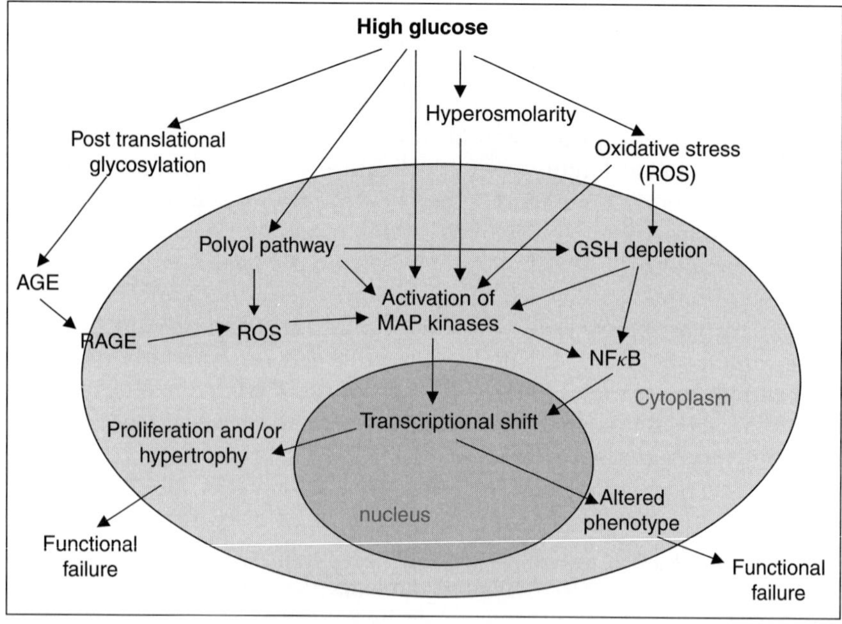

FIG. 6. Synopsis of pathways by which MAP kinases synergize different inputs to diabetes complications.

etiology of other diabetic complications. Accumulating evidence in the literature indicates that MAP kinase activation, particularly ERK and p38, is damaging in the diabetic milieu. Data from diabetic rats and diabetic patients, in combination with the implication of ERK and p38 activation in sensory cell damage *in vitro,* suggest a role for MAP kinases in the etiology of diabetic neuropathy (Purves *et al.,* 2001). Further evidence implicating p38 in a neurodegenerative role arises from an *in vivo* study in which the treatment of STZ rats with a p38 inhibitor for the final 5 weeks of a 12-week diabetic protocol attenuated the decrease in nerve conduction velocity seen in untreated diabetics compared to controls (S. Agthong and D. R. Tomlinson, unpublished data).

The role of JNK in diabetes and in sensory neurons in response to diabetes-related stressors is uncertain. The duality of JNK is apparent in the literature with evidence for both cell death and cell survival functions. JNK activation has been associated with the induction of apoptosis in cultured neonatal cells and neuronal cell lines after neurotrophic factor deprivation (Xia *et al.,* 1995; Eilers *et al.,* 1998; Watson *et al.,* 1998; Maroney *et al.,* 1998; Mielke and Herdegen, 2000). Inhibition of JNK supports survival of chick embryonic neurons and rat embryonic motor neurons (Borasio *et al.*, 1998; Maroney *et al.*, 1998). JNK activation in response to UV light has been linked to cytochrome *c* release from the mitochondria and apoptosis. (Tournier *et al.*, 2000). However, activation of JNK does not always result in cell death (Herdegen *et al.*, 1997). Rat sympathetic neuronal apoptosis is not wholly dependent on JNK activation (Virdee *et al.*, 1997), and other groups have suggested that JNK has roles other than apoptosis in adult mice central neurons (Xu *et al.*, 1997). In HeLa or HepG2 cells treated with butylated hydroxyanisole (BHA), which generates intracellular H_2O_2, there was rapid activation of ERK2 and a slower activation of JNK1. Pretreatment with antioxidants such as vitamin E and *N*-acetylcysteine attenuated the ERK response but not the JNK response (Yu *et al.*, 1997), implicating ERK activation as an early component of the damaging effects, with the possibility that JNK activation formed part of a protective response. Studies in NIH/3T3 cells indicated that JNK-induced activation of AP-1 resulted in the induction of GSH synthesis in response to alkylating agents, and a role for JNK in the regulation of a cellular defense mechanism was suggested (Wilhelm *et al.*, 1997). After sciatic nerve injury, rapid, long-term upregulation of JNK activity occurs as part of the neuronal stress response, but no apoptosis ensues and the upregulation of JNK and its downstream target, c-Jun, may be required for regeneration (Kenney and Kocsis, 1998; Herdegen *et al.*, 1998). Furthermore, JNK has been shown to be involved in neurite outgrowth in PC12 cells (Yao *et al.*, 1997). In support of this, a further study in differentiated PC12 cells, triturated to shear neurites and replated, reported that a JNK inhibitor (CEP-1347) suppressed neurite outgrowth (Maroney *et al.*, 1999). In the same study, inhibition of JNK protected against cell death induced by trophic factor withdrawal and UV irradiation (Maroney *et al.*, 1999). In sensory neuron cultures, high glucose resulted in the activation of a combination of MAP kinases (p38, JNK) that did not result in cell damage; oxidative stress led to the activation of a combination of MAP kinases (ERK, p38) that culminated in cell death (Purves *et al.*, 2001). In light of this evidence and the evidence described earlier, it is speculated that JNK activation may be involved in a neuronal protective response.

Although many studies have implicated ERK activation as protective, largely in clonal cell lines (Xia *et al.*, 1995; Guyton *et al.*, 1996), other studies have implicated it in cell death (Bhat and Zhang, 1999; Satoh *et al.*, 2000; Stanciu *et al.*, 2000). There are many more published examples of these opposing effects of ERK activation and only a few are referred to. Studies using U0126 have implicated ERK MAP kinase in the neuronal damage induced by glutamate or hypoxia in a neuronal cell line (HT22) and in primary immature cortical neurons (Satoh *et al.*, 2000; Stanciu *et al.*, 2000).

U0126 protected both cell types against these oxidative stresses. Similar observations have been made in other systems. In an oligodendrocyte cell line (CG4), H_2O_2 activated both ERK and p38 and resulted in apoptosis, and addition of the MEK inhibitor, PD98059, blocked cell death (Bhat and Zhang, 1999). In a smooth muscle line, H_2O_2-induced phosphorylation of ERK resulted in cytotoxicity (Cantoni *et al.*, 1996). In rat pleural mesothelial cells, H_2O_2-induced apoptosis was abrogated by the MEK inhibitor PD98059 (Jimenez *et al.*, 1997).

An important question to address is how can activation of these kinases exhibit such diverse and apparently opposing effects depending on cell type and/or type of stimulation? There are a number of possible mechanisms that enable signaling pathways to switch roles. The combination of the MAP kinases activated plays an important role in biological outcome (Xia *et al.*, 1995). In the study reported by Purves *et al.* (2001), the activation of both p38 and ERK was required to induce cell death in response to oxidative stress. If either of the pathways were blocked, cell death was decreased; if the pathways were blocked simultaneously, there was no additive effect on the inhibition of cell death, which would be expected if the pathways were acting independently. Thus, ERK activation alone may be involved in neurite outgrowth and survival (Sjögreen *et al.*, 2000), whereas ERK activation in conjunction with p38 appears to be damaging.

Temporal organization of MAP kinase activities plays an important role in the generation of specific biological responses (Schaeffer and Weber, 1999; Fanger, 1999). For example, in PC12 cells, ERK initiates two opposing cellular responses: cell proliferation and cell differentiation. Sustained activation of ERK and nuclear translocation lead to differentiation and transient activation leads to cell growth (Marshall, 1995). Upstream regulators of MAP kinases can modulate the persistence of activation. MAP kinase in response to NGF is controlled by Ras and Rap1 (York *et al.*, 1998). Rap1 was shown to mediate persistent ERK activation, and Ras was shown to be responsible for the early activation of ERK (York *et al.*, 1998).

The subcellular or spatial localization of MAP kinases plays an important role in determining the outcome of kinase activation. Architectural organization or the multimeric organization of signaling proteins into complexes, which are localized to distinct subcellular regions, is recognized as an important mechanism that influences the regulation of MAP kinase pathways (Schaeffer and Weber, 1999; Fanger, 1999). Spatial localization is of particular importance in neuronal cells due to their bipolar structure. In addition, extracellular stimuli can induce relocation of specific signal transduction proteins. The model proposed by Coffey *et al.* (2000) (in cerebellar granule neurons), whereby distinct pools of JNK serve different functions, provides a basis for understanding the apparently conflicting roles of JNK.

In this model, the JNK scaffold protein, JIP, is proposed to compete with c-Jun for JNK binding and therefore to channel JNK signaling away from the c-Jun pathway toward the expression of other genes. It is thus perceivable that such a mechanism could switch JNK activation from inducing apoptosis to a protective role. Cellular location of regulators of MAP kinase pathways also determine the downstream responses. Again in Coffey's model, JNK is associated with MKK4 in the cytoplasm and is proposed to have cytosolic targets (such as cytoskeletal elements); JNK is associated with MKK7 in the nucleus where it is assumed to have transcription factor targets. In support of this hypothesis, studies in embryonic DRG cultures showed differing responses of JNK isoforms in neurites and cell bodies to a JNK inhibitor, suggesting the presence of different signaling pathways in the two neuronal compartments (O'Ferrall *et al.*, 2000).

It is speculated that the glucose-dependent activation of JNK and p38 in primary sensory neurons (Purves *et al.*, 2001) is not the result of a glucose-induced oxidative stress, but is the result of a direct effect of the glucose molecule itself, elevated polyol pathway intermediates, glycation of proteins, or transcription-dependent induction of activators of JNK and p38, such as gene 33. In culture, high glucose resulted in a sustained activation of p38 and JNK at 16 hr (Purves *et al.*, 2001). The sustained activation of JNK can be explained by the function of an immediate early gene, gene 33, that is induced transcriptionally by a diverse array of extracellular stimuli. Induction of gene 33 requires activation of JNK and, in a feedback loop, transient expression of gene 33 results in the selective activation of JNK (Makkinje *et al.*, 2000). Previous studies have reported increased gene 33 expression levels in kidney, soon after onset of diabetes, that are sustained throughout the progression of diabetic nephropathy (Makkinje *et al.*, 2000) and thus it is possible that such a mechanism is responsible for the sustained activation of JNK *in vitro* in sensory neurons.

A second hypothesis used to explain the sustained activation of JNK in primary sensory neurons reported by Purves *et al.* (2001) involves a model whereby glucose, via pyruvate, results in the production and release of H_2O_2 from mitochondria and this H_2O_2 activates cytosolic JNK (Nemoto *et al.*, 2000). In a feedback loop, activated JNK increases glycogen synthase via decreasing glycogen synthase kinase β (GSKβ) activity (phosphorylated GSKβ is inactive). If GSKβ is inactive, this allows activity of glycogen synthase, which converts glucose to glycogen. This results in decreased glucose available to the mitochondria and therefore a fall in the production of mitochondrial oxidants. Results reported by Purves *et al.* (2001) support this model in that high glucose did not result in oxidative stress, but cells with high glucose were more susceptible to oxidative stress. This hypothesis also provides a model for the suggested protective role of JNK in primary sensory neurons in response to high glucose.

As indicated earlier in this account, evidence exists for reduced MAPK phosphatase activity in the kidney in diabetes. This represents an area that is ripe for study. The half-life of activated MAPKs has not been studied systematically, but it is clear that ERK activation, for example, is a relatively short-duration phenomenon in neurons and in other cells (Marshall, 1995). Neuron cell bodies from diabetic rats appear to show sustained steady-state activation of ERK as well as the other MAPKs. Taken together, these phenomena indicate that defects in the inactivation of MAPKs in diabetes may also contribute to events downstream of their activation.

VIII. What Is the Molecular Basis for the Development of Nerve Conduction Deficits and Structural Abnormalities in Diabetes?

There is no clear answer to this question. It is known that MAP kinase signaling cascades phosphorylate downstream targets other than transcription factors. As is described in the chapter by Fernyhough and Schmidt, hyperphosphorylation of cytoskeletal elements such as neurofilaments has been proposed as part of the neurodegenerative process and there is evidence that this is mediated by MAP kinases, including JNK (Giasson and Mushynski, 1996; Fernyhough *et al.*, 1999) and ERK (Veeranna *et al.*, 1998). Hyperphosphorylation of neurofilaments has been linked to the etiology of diabetic neuropathy (Fernyhough *et al.*, 1999). It is proposed that hyperphosphorylation of neurofilaments in the perikarya of neurons results in the decreased delivery of neurofilaments to the distal nerve, contributing to distal axonopathy (Schmidt *et al.*, 1997a,b; Fernyhough *et al.*, 1999). In addition, axon size is increased proximally and decreased distally, which is likely to be a result of impaired transport of hyperphosphorylated cytoskeletal elements (Medori *et al.*, 1985, 1988).

Further molecular mechanisms initiated by MAP kinase activation are not yet known, including and especially the range of gene targets for these groups of enzymes. Many ion channel proteins are involved in the generation of action potentials and in the regulation of threshold and latency of firing. The expression and posttranslational regulation of these in diabetes have not been studied, but it is certainly conceivable that reduced expression of one or more of these could form a downstream target of MAP kinases.

At end stage neuropathy, nerves become demyelinated as a result of Schwann cell malfunction and degeneration (Thomas and Lascelles, 1965;

Powell *et al.*, 1977; Mizisin *et al.*, 1998). As described elsewhere in this volume (see chapters by Eckersley and Tolkovsky), activation of JNK as a prelude to alteration in Schwann cell phenotype and even apoptosis is a possibility in diabetes. Thus, elucidation of the downstream targets of MAP kinases may well provide a map for the molecular origins of many aspects of diabetic neuropathies.

References

Averill, S., Delcroix, J.-D., Tomlinson, D. R., Fernyhough, P., and Priestley, J. V. (2001). Nerve growth factor modulates the activation status and fast axonal transport of Erk 1/2 in adult nociceptive neurones. *Mol. Cell. Neurosci.* **18**, 183–196.

Awazu, M., Ishikura, K., Hida, M., and Hoshiya, M. (1999). Mechanisms of mitogen-activated protein kinase activation in experimental diabetes. *J. Am. Soc. Nephrol.* **10**, 738–745.

Baumgartner-Parzer, S. M., Wagner, L., Pettermann, M., Grillari, J., Gessl, A., and Waldhausl, W. (1995). High-glucose-triggered apoptosis in cultured endothelial cells. *Diabetes* **44**, 1323–1327.

Berlett, B. S., and Stadtman, E. R. (1997). Protein oxidation in aging, disease, and oxidative stress. *J. Biol. Chem.* **272**, 20313–20316.

Bhat, N. R., and Zhang, P. (1999). Hydrogen peroxide activation of multiple mitogen-activated protein kinases in an oligodendrocyte cell line: Role of extracellular signal-regulated kinase in hydrogen peroxide-induced cell death. *J. Neurochem.* **72**, 112–119.

Borasio, G. D., Horstmann, S., Anneser, J. M. H., Neff, N. T., and Glicksman, M. A. (1998). CEP-1347/KT7515, a JNK pathway inhibitor, supports the *in vitro* survival of chick embryonic neurons. *NeuroReport* **9**, 1435–1439.

Boulton, T. G., Nye, S. H., Robbins, D. J., Ip, N. Y., Radziejewska, E., Morgenbesser, S. D., DePinho, R. A., Panayotatos, N., Cobb, M. H., and Yancopoulos, G. D. (1991). ERKs: A family of protein-serine/threonine kinases that are activated and tyrosine phosphorylated in response to insulin and NGF. *Cell* **65**, 663–675.

Brown, M. J., and Greene, D. A. (1984). Diabetic neuropathy: Pathophysiology and management. *In* "Peripheral Nerve Disorders" (A. K. Asbury and R. G. Gilliatt, eds.), p. 126. Butterworths, London.

Cantoni, O., Boscoboinik, D., Fiorani, M., Stauble, B., and Azzi, A. (1996). The phosphorylation state of MAP-kinases modulates the cytotoxic response of smooth muscle cells to hydrogen peroxide. *FEBS Lett.* **389**, 285–288.

Carrington, A. L., and Litchfield, J. E. (1999). The aldose reductase pathway and nonenzymatic glycation in the pathogenesis of diabetic neuropathy: A critical review for the end of the 20th century. *Diabet. Rev.* **7**, 275–299.

Cassarino, D. S., Halvorsen, E. M., Swerdlow, R. H., Abramova, N. N., Parker, W. D., Sturgill, T. W., and Bennett, J. P. (2000). Interaction among mitochondria, mitogen-activated protein kinases, and nuclear factor-kappaB in cellular models of Parkinson's disease. *J. Neurochem.* **74**, 1384–1392.

Cheng, H.-L., and Feldman, E. L. (1998). Bidirectional regulation of p38 kinase and c-Jun N-terminal protein kinase by insulin-like growth factor-I. *J. Biol. Chem.* **273**, 14560–14565.

Cohen, D. M. (1997). Mitogen-activated protein kinase cascades and the signaling of hyperosmotic stress to immediate early genes. *Comp. Biochem. Physiol. A. Physiol.* **117**, 291–299.

Coyle, J. T., and Puttfarcken, P. (1993). Oxidative stress, glutamate, and neurodegenerative disorders. *Science* **262**, 689–695.

Davis, R. J. (2000). Signal transduction by the JNK group of MAP kinases. *Cell* **103**, 239–252.

Delcroix, J.-D., Fernyhough, P., and Tomlinson, D. R. (2000). A retrograde stress signal in diabetic neuropathy? *Diabet. Med.* **17**(Suppl. 1), 29. [Abstract]

Derkinderen, P., Enslen, H., and Girault, J. A. (1999). The ERK/MAP-kinases cascade in the nervous system. *NeuroReport* **10**, R24–R34.

Diabetes Control and Complications Trial Research Group. (1993). The effect of intensive treatment of diabetes on the development and progression of long-term complications in insulin-dependent diabetes mellitus. *N. Engl. J. Med.* **329**, 977–986.

Doran, J. F., Jackson, P., Kynoch, P. A. M., and Thompson, R. J. (1983). Isolation of PGP 9.5, a new human neurone-specific protein detected by high-resolution two-dimensional electrophoresis. *J. Neurochem.* **40**, 1542–1547.

Du, X., Stockklauser-Färber, K., and Rösen, P. (1999). Generation of reactive oxygen inter-mediates, activation of NF-KB, and induction of apoptosis in human endothelial cells by glucose: Role of nitric oxide synthase? *Free Radic. Biol. Med.* **27**, 752–763.

Dyck, P. J., and Giannini, C. (1996). Pathologic alterations in the diabetic neuropathies of humans: A review. *J. Neuropathol. Exp. Neurol.* **55**, 1181–1193.

Eilers, A., Whitfield, J., Babij, C., Rubin, L. L., and Ham, J. (1998). Role of the Jun kinase pathway in the regulation of c-Jun expression and apoptosis in sympathetic neurons. *J. Neurosci.* **18**, 1713–1724.

Elbirt, K. K., Whitmarsh, A. J., Davis, R. J., and Bonkovsky, H. L. (1998). Mechanism of sodium arsenite-mediated induction of heme oxygenase-1 in hepatoma cells: Role of mitogen-activated protein kinases. *J. Biol. Chem.* **273**, 8922–8931.

English, J., Pearson, G., Wilsbacher, J., Swantek, J., Karandikar, M., Xu, S., and Cobb, M. H. (1999). New insights into the control of MAP kinase pathways. *Exp. Cell Res.* **253**, 255–270.

Fanger, G. R. (1999). Regulation of the MAPK family members: Role of subcellular localization and architectural organization. *Histol. Histopathol.* **14**, 887–894.

Favata, M. F., Horiuchi, K. Y., Manos, E. J., Daulerio, A. J., Stradley, D. A., Feeser, W. S., Van-Dyk, D. E., Pitts, W. J., Earl, R. A., Hobbs, F., Copeland, R. A., Magolda, R. L., Scherle, P. A., and Trzaskos, J. M. (1998). Identification of a novel inhibitor of mitogen-activated protein kinase kinase. *J. Biol. Chem.* **273**, 18623–18632.

Fernyhough, P., Gallagher, A., Averill, S., Priestley, J. V., Hounsom, L., and Tomlinson, D. R. (1999). Aberrant neurofilament phosphorylation in sensory neurons of rats with diabetic neuropathy. *Diabetes* **48**, 881–889.

Fernyhough, P., and Tomlinson, D. R. (1999). The therapeutic potential of neurotrophins for the treatment of diabetic neuropathy. *Diabet. Rev.*, **7**, 300–311.

Garrington, T. P., and Johnson, G. L. (1999). Organization and regulation of mitogen-activated protein kinase signaling pathways. *Curr. Opin. Cell Biol.* **11**, 211–218.

Gavazzi, I., Kumar, R. D. C., McMahon, S. B., and Cohen, J. (1999). Growth responses of different subpopulations of adult sensory neurons to neurotrophic factors *in vitro. Eur. J. Neurosci.* **11**, 3405–3414.

Geilen, C. C., Wieprecht, M., and Orfanos, C. E. (1996). The mitogen-activated protein kinases system (MAP kinase cascade): Its role in skin signal transduction. A review. *J. Dermatol. Sci.* **12**, 255–262.

Giasson, B. I., and Mushynski, W. E. (1996). Aberrant stress-induced phosphorylation of perikaryal neurofilaments. *J. Biol. Chem.* **271**, 30404–30409.

Graves, J. D., and Krebs, E. G. (1999). Protein phosphorylation and signal transduction. *Pharmacol. Ther.* **82**, 111–121.

Greene, D. A., De Jesus, P. V., Jr., and Winegrad, A. I. (1975). Effects of insulin and dietary myoinositol on impaired peripheral motor nerve conduction velocity in acute streptozotocin diabetes. *J. Clin. Invest.* **55**, 1326–1336.

Greene, D. A., and Lattimer, S. A. (1983). Impaired rat sciatic nerve sodium-potassium adenosine triphosphatase in acute streptozocin diabetes and its correction by dietary *myo*-inositol supplementation. *J. Clin. Invest.* **72**, 1058–1063.

Greene, D. A., and Lattimer, S. A. (1986). Protein kinase C agonists acutely normalize decreased ouabain-inhibitable respiration in diabetic rabbit nerve: Implications for (Na,K)-ATPase regulation and diabetic complications. *Diabetes* **35**, 242–245.

Greene, D. A., Lattimer, S. A., and Sima, A. A. F. (1988). Pathogenesis and prevention of diabetic neuropathy. *Diabet. Metab. Rev.* **4**, 201–221.

Grewal, S. S., York, R. D., and Stork, P. J. (1999). Extracellular-signal-regulated kinase signalling in neurons. *Curr. Opin. Neurobiol.* **9**, 544–553.

Guyton, K. Z., Liu, Y., Gorospe, M., Xu, Q., and Holbrook, N. J. (1996). Activation of mitogen-activated protein kinase by H_2O_2: Role in cell survival following oxidant injury. *J. Biol. Chem.* **271**, 4138–4142.

Halliwell, B. (1992). Reactive oxygen species and the central nervous system. *J. Neurochem.* **59**, 1609–1623.

Han, J., Lee, J. D., Bibbs, L., and Ulevitch, R. J. (1994). A MAP kinase targeted by endotoxin and hyperosmolarity in mammalian cells. *Science* **265**, 808–811.

Haneda, M., Araki, S., Togawa, M., Sugimoto, T., Isono, M., and Kikkawa, R. (1997). Mitogen-activated protein kinase cascade is activated in glomeruli of diabetic rats and glomerular mesangial cells cultured under high glucose conditions. *Diabetes* **46**, 847–853.

Herdegen, T., Claret, F. X., Kallunki, T., Martin-Villalba, A., Winter, C., Hunter, T., and Karin, M. (1998). Lasting N-terminal phosphorylation of c-Jun and activation of c-Jun N-terminal kinases after neuronal injury. *J. Neurosci.* **18**, 5124–5135.

Herdegen, T., Skene, P., and Bähr, M. (1997). The c-Jun transcription factor: Bipotential mediator of neuronal death, survival and regeneration. *Trends Neurosci.* **20**, 227–231.

Ide, C. (1996). Peripheral nerve regeneration. *Neurosci. Res.* **25**, 101–121.

Igarashi, M., Wakasaki, H., Takahara, N., Ishii, H., Jiang, Z. Y., Yamauchi, T., Kuboki, K., Meier, M., Rhodes, C. J., and King, G. L. (1999). Glucose or diabetes activates p38 mitogen-activated protein kinase via different pathways. *J. Clin. Invest.* **103**, 185–195.

Iordanov, M. S., and Magun, B. E. (1999). Different mechanisms of c-Jun NH_2-terminal kinase-1 (JNK1) activation by ultraviolet-B radiation and by oxidative stressors. *J. Biol. Chem.* **274**, 25801–25806.

Ip, Y. T., and Davis, R. J. (1998). Signal transduction by the c-Jun N-terminal kinase (JNK)—from inflammation to development. *Curr. Opin. Cell Biol.* **10**, 205–219.

Ito, H., Okamoto, K., and Kato, K. (1998). Enhancement of expression of stress proteins by agents that lower the levels of glutathione in cells. *Biochim. Biophys. Acta* **1397**, 223–230.

Jakobsen, J. (1976). Axonal dwindling in early experimental diabetes. II. A study of isolated nerve fibres. *Diabetologia* **12**, 547–553.

Jakobsen, J. (1979). Early and preventable changes of peripheral nerve structure and function in insulin-deficient diabetic rats. *J. Neurol. Neurosurg. Psychiat.* **42**, 509–518.

Jakobsen, J., and Lundbæk, K. (1976). Neuropathy in experimental diabetes: An animal model. *Br. Med. J.* **2**, 278–279.

Jimenez, L. A., Zanella, C., Fung, H., Janssen, Y. M., Vacek, P., Charland, C., Goldberg, J., and Mossman, B. T. (1997). Role of extracellular signal-regulated protein kinases in apoptosis by asbestos and H_2O_2. *Am. J. Physiol.* **273**, L1029–L1035.

Kaplan, D. R., and Miller, F. D. (2000). Neurotrophin signal transduction in the nervous system. *Curr. Opin. Neurobiol.* **10**, 381–391.

Kapor-Drezgic, J., Zhou, X. P., Babazono, T., Dlugosz, J. A., Hohman, T., and Whiteside, C. (1999). Effect of high glucose on mesangial cell protein kinase C-δ and -ϵ is polyol pathway-dependent. *J. Am. Soc. Nephrol.* **10**, 1193–1203.

Kenney, A. M., and Kocsis, J. D. (1998). Peripheral axotomy induces long-term c-Jun amino-terminal-kinase-1 activation and activator protein-1 binding activity by c-Jun and junD in adult rat dorsal root ganglia *in vivo. J. Neurosci.* **18**, 1318–1328.

Kerekes, N., Landry, M., Lundmark, K., and Hökfelt, T. (2000). Effect of NGF, BDNF, bFGF, aFGF and cell density on NPY expression in cultured rat dorsal root ganglion neurones. *J. Auton. Nerv. Syst.* **81**, 128–138.

Klesse, L. J., Meyers, K. A., Marshall, C. J., and Parada, L. F. (1999). Nerve growth factor induces survival and differentiation through two distinct signaling cascades in PC12 cells. *Oncogene* **18**, 2055–2068.

Kultz, D., and Burg, M. (1998). Evolution of osmotic stress signaling via map kinase cascades. *J. Exp. Biol.* **201**, 3015–3021.

Kyriakis, J. M., and Avruch, J. (1996). Protein kinase cascades activated by stress and inflammatory cytokines. *BioEssays* **18**, 567–577.

Laight, D. W., Carrier, M. J., and Änggård, E. E. (2000). Antioxidants, diabetes and endothelial dysfunction. *Cardiovasc. Res.* **47**, 457–464.

Lander, H. M., Tauras, J. M., Ogiste, J. S., Hori, O., Moss, R. A., and Schmidt, A. M. (1997). Activation of the receptor for advanced glycation end products triggers a $p21^{ras}$-dependent mitogen-activated protein kinase pathway regulated by oxidant stress. *J. Biol. Chem.* **272**, 17810–17814.

Lee, J. C., Kassis, S., Kumar, S., Badger, A., and Adams, J. L. (1999). p38 mitogen-activated protein kinase inhibitors: Mechanisms and therapeutic potentials. *Pharmacol. Ther.* **82**, 389–397.

Lee, J. C., Laydon, J. T., McDonnell, P. C., Gallagher, T. F., Kumar, S., Green, D., McNulty, D., Blumenthal, M. J., Heys, J. R., Landvatter, S. W., Strikler, J. E., McLaughlin, M. M., Siemens, I. R., Fisher, S. M., Livi, G. P., White, J. R., Adams, J. L., and Young, P. R. (1994). A protein kinase involved in the regulation of inflammatory cytokine biosynthesis. *Nature* **372**, 739–746.

Lee, J.-D., Ulevitch, R. J., and Han, J. (1995). Primary structure of BMK1: A new mammalian map kinase. *Biochem. Biophys. Res. Commun.* **213**, 715–724.

Lezoualc'h, F., Sagara, Y., Holsboer, F., and Behl, C. (1998). High constitutive NF-κB activity mediates resistance to oxidative stress in neuronal cells. *J. Neurosci.* **18**, 3224–3232.

Lin, L.-L., Wartmann, M., Lin, A. Y., Knopf, J. L., Seth, A., and Davis, R. J. (1993). $cPLA_2$ is phosphorylated and activated by MAP kinase. *Cell* **72**, 269–278.

Lindsay, R. M., Evison, C. J., and Winter, J. (1991). Culture of adult mammalian peripheral neurons. *In* "Cellular and Molecular Neurobiology: A Practical Approach" (J. Chad and H. Wheal, eds.), pp. 3–17. Oxford Univ. Press, Oxford.

Liu, Y., Guyton, K. Z., Gorospe, M., Xu, Q., Lee, J. C., and Holbrook, N. J. (1996). Differential activation of ERK, JNK/SAPK and P38/CSBP/RK map kinase family members during the cellular response to arsenite. *Free Radic. Biol. Med.* **21**, 771–781.

Makkinje, A., Quinn, D. A., Chen, A., Cadilla, C. L., Force, T., Bonventre, J. V., and Kyriakis, J. M. (2000). Gene 33/Mig-6, a transcriptionally inducible adapter protein that binds GTP-Cdc42 and activates SAPK/JNK: A potential marker transcript for chronic pathologic conditions, such as diabetic nephropathy. Possible role in the response to persistent stress. *J. Biol. Chem.* **275**, 17838–17847.

Maroney, A. C., Finn, J. P., Bozyczko-Coyne, D., O'Kane, T. M., Neff, N. T., Tolkovsky, A. M., Park, D. S., Yan, C. Y. I., Troy, C. M., and Greene, L. A. (1999). CEP-1347 (KT7515), an inhibitor of JNK activation, rescues sympathetic neurons and neuronally differentiated PC12 cells from death evoked by three distinct insults. *J. Neurochem.* **73**, 1901–1912.

Maroney, A. C., Glicksman, M. A., Basma, A. N., Walton, K. M., Knight, E., Murphy, Bartlett, B. A., Finn, J. P., Angeles, T., Matsuda, Y., Neff, N. T., and Dionne, C. A. (1998). Motoneuron apoptosis is blocked by CEP-1347 (KT 7515), a novel inhibitor of the JNK signaling pathway. *J. Neurosci.* **18**, 104–111.

Marshall, C. J. (1995). Specificity of receptor tyrosine kinase signaling: Transient versus sustained extracellular signal-regulated kinase activation. *Cell* **80**, 179–185. [Abstract]

McMahon, S. B., Armanini, M. P., Ling, L. H., and Phillips, H. S. (1994). Expression and coexpression of Trk receptors in subpopulations of adult primary sensory neurons projecting to identified peripheral targets. *Neuron* **12**, 1161–1171.

Medori, R., Autilio-Gambetti, L., Jenich, H., and Gambetti, P. (1988). Changes in axon size and slow axonal transport are related in experimental diabetic neuropathy. *Neurology* **38**, 597–601.

Medori, R., Autilio-Gambetti, L., Monaco, S., and Gambetti, P. (1985). Experimental diabetic neuropathy: Impairment of slow transport with changes in axon cross-sectional area. *Proc. Natl. Acad. Sci. USA* **82**, 7716–7720.

Medori, R., Jenich, H., Autilio-Gambetti, L., and Gambetti, P. (1988). Experimental diabetic neuropathy: Similar changes of slow axonal transport and axonal size in different animal models. *J. Neurosci.* **8**, 1814–1821.

Mielke, K., and Herdegen, T. (2000). JNK and p38 stresskinases: Degenerative effectors of signal-transduction-cascades in the nervous system. *Prog. Neurobiol.* **61**, 45–60.

Migheli, A., Piva, R., Atzori, C., Troost, D., and Schiffer, D. (1997). c-Jun, JNK/SAPK kinases and transcription factor NF-κB are selectively activated in astrocytes, but not motor neurons, in amyotrophic lateral sclerosis. *J. Neuropathol. Exp. Neurol.* **56**, 1314–1322.

Mizisin, A. P., Calcutt, N. A., Tomlinson, D. R., Gallagher, A., and Fernyhough, P. (1999). Neurotrophin-3 reverses nerve conduction velocity deficits in streptozotocin-diabetic rats. *J. Periph. Nerv. Syst.* **4**, 211–221.

Mizisin, A. P., Shelton, G. D., Wagner, S., Rusbridge, C., and Powell, H. C. (1998). Myelin splitting, Schwann cell injury and demyelination in feline diabetic neuropathy. *Acta Neuropathol. (Berl.)* **95**, 171–174.

Mohamed, A. K., Bierhaus, A., Schiekofer, S., Tritschler, H., Ziegler, R., and Nawroth, P. P. (1999). The role of oxidative stress and NF-kappaB activation in late diabetic complications. *BioFactors* **10**, 157–167.

Morooka, T., and Nishida, E. (1998). Requirement of p38 mitogen-activated protein kinase for neuronal differentiation in PC12 cells. *J. Biol. Chem.* **273**, 24285–24288.

Nemoto, S., Takeda, K., Yu, Z. X., Ferrans, V. J., and Finkel, T. (2000). Role for mitochondrial oxidants as regulators of cellular metabolism. *Mol. Cell. Biol.* **20**, 7311–7318.

Nishikawa, T., Edelstein, D., and Brownlee, M. (2000). The missing link: A single unifying mechanism for diabetic complications. *Kidney Int.* **58**, S26–S30.

Nishikawa, T., Edelstein, D., Du, X. L., Yamagishi, S., Matsumura, T., Kaneda, Y., Yorek, M. A., Beebe, D., Oates, P. J., Hammes, H. P., Giardino, I., and Brownlee, M. (2000). Normalizing mitochondrial superoxide production blocks three pathways of hyperglycaemic damage. *Nature* **404**, 787–790.

Obrosova, I. G., Fathallah, L., and Greene, D. A. (2000). Early changes in lipid peroxidation and antioxidative defense in diabetic rat retina: Effect of DL-α-lipoic acid. *Eur. J. Pharmacol.* **398**, 139–146.

O'Ferrall, E. K., Robertson, J., and Mushynski, W. E. (2000). Inhibition of aberrant and constitutive phosphorylation of the high-molecular-mass neurofilament subunit by CEP-1347 (KT7515), an inhibitor of the stress-activated protein kinase signaling pathway. *J. Neurochem.* **75**, 2358–2367.

Ono, K., and Han, J. H. (2000). The p38 signal transduction pathway: Activation and function. *Cell Signal.* **12**, 1–13.

Pfeifer, M. A., and Schumer, M. P. (1995). Clinical trials of diabetic neuropathy: Past, present, and future. *Diabetes* **44**, 1355–1361.

Powell, H., Knox, D., Lee, S., Charters, A. C., Orloff, M. J., Garrett, R. S., and Lampert, P. (1977). Alloxan diabetic neuropathy: Electron microscopic studies. *Neurology* **27**, 60–66.

Purves, T., Middlemas, A., Agthong, S., Jude, E. B., Boulton, A., Fernyhough, P., and Tomlinson, D. R. (2001). A role for mitogen-activated protein kinases in the aetiology of diabetic neuropathy. *FASEB J.* **15**, 2508–2514.

Raingeaud, J., Gupta, S., Rogers, J. S., Dickens, M., Han, J., Ulevitch, R. J., and Davis, R. J. (1995). Pro-inflammatory cytokines and environmental stress cause p38 mitogen-activated protein kinase activation by dual phosphorylation on tyrosine and threonine. *J. Biol. Chem.* **270**, 7420–7426.

Rakieten, N., Rakieten, M. L., and Moreshwar, V. N. (1963). Studies on the diabetogenic action of streptozotocin (NSC-37917). *Cancer Chemother. Rep.* **29**, 91–98.

Rhee, S. G. (1999). Redox signaling: Hydrogen peroxide as intracellular messenger. *Exp. Mol. Med.* **31**, 53–59.

Russell, J. W., and Feldman, E. L. (1999). Insulin-like growth factor-I prevents apoptosis in sympathetic neurons exposed to high glucose. *Horm. Metab. Res.* **31**, 90–96.

Russell, J. W., Sullivan, K. A., Windebank, A. J., Herrman, D. N., and Feldman, E. L. (1999). Neurons undergo apoptosis in animal and cell culture models of diabetes. *Neurobiol. Dis.* **6**, 347–363.

Satoh, T., Nakatsuka, D., Watanabe, Y., Nagata, I., Kikuchi, H., and Namura, S. (2000). Neuroprotection by MAPK/ERK kinase inhibition with U0126 against oxidative stress in a mouse neuronal cell line and rat primary cultured cortical neurons. *Neurosci. Lett.* **288**, 163–166.

Schaeffer, H. J., and Weber, M. J. (1999). Mitogen-activated protein kinases: Specific messages from ubiquitous messengers. *Mol. Cell. Biol.* **19**, 2435–2444.

Schmidt, R. E., Beaudet, L. N., Plurad, S. B., and Dorsey, D. A. (1997a). Axonal cytoskeletal pathology in aged and diabetic human sympathetic autonomic ganglia. *Brain Res.* **769**, 375–383.

Schmidt, R. E., Dorsey, D., Parvin, C. A., Beaudet, L. N., Plurad, S. B., and Roth, K. A. (1997b). Dystrophic axonal swellings develop as a function of age and diabetes in human dorsal root ganglia. *J. Neuropathol. Exp. Neurol.* **56**, 1028–1043.

Sharma, A. K., and Thomas, P. K. (1987). Animal models: Pathology and pathophysiology. *In* "Diabetic Neuropathy" (P. J. Dyck, P. K. Thomas, A. K. Asbury, A. I. Winegrad, and D. Porte, eds.), pp. 237–252. Saunders, Philadelphia.

Sies, H. (1985). Oxidative Stress. Academic Press, London.

Simm, A., Munch, G., Seif, F., Schenk, O., Heidland, A., Richter, H., Vamvakas, S., and Schinzel, R. (1997). Advanced glycation endproducts stimulate the MAP-kinase pathway in tubulus cell line LLC-PK1. *FEBS Lett.* **410**, 481–484.

Sjögreen, B., Wiklund, P., and Ekström, P. A. R. (2000). Mitogen activated protein kinase inhibition by PD98059 blocks nerve growth factor stimulated axonal outgrowth from adult mouse dorsal root ganglia *in vitro*. *Neuroscience.* **100**, 407–416.

Stanciu, M., Wang, Y., Kentor, R., Burke, N., Watkins, S., Kress, G., Reynolds, I., Klann, E., Angiolieri, M. R., Johnson, J. W., and DeFranco, D. B. (2000). Persistent activation of ERK contributes to glutamate-induced oxidative toxicity in a neuronal cell line and primary cortical neuron cultures. *J. Biol. Chem.* **275**, 12200–12206.

Stevens, M. J., Obrosova, I., Cao, X. H., Van Huysen, C., and Greene, D. A. (2000). Effects of DL-α-lipoic acid on peripheral nerve conduction, blood flow, energy metabolism, and oxidative stress in experimental diabetic neuropathy. *Diabetes* **49**, 1006–1015.

Thomas, P. K., and Lascelles, R. G. (1965). Schwann-cell abnormalities in diabetic neuropathy. *Lancet* **1**, 1355–1357.

Thomas, P. K., and Tomlinson, D. R. (1992). Diabetic and hypoglycaemic neuropathy. *In* "Peripheral Neuropathy" (P. J. Dyck, P. K. Thomas, J. W. Griffin, P. A. Low, and J. F. Poduslo, eds.), pp. 1219–1250 Saunders, Philadelphia.

Thornalley, P. J. (1998). Cell activation by glycated proteins: AGE receptors, receptor recognition factors and functional classification of AGEs. *Cell. Mol. Neurobiol.* **44**, 1013–1023.

Tibbles, L. A., and Woodgett, J. R. (1999). The stress-activated protein kinase pathways. *Cell. Mol. Life Sci.* **55**, 1230–1254.

Tomlinson, D. R. (1999). Mitogen-activated protein kinases as glucose transducers for diabetic complications. *Diabetologia* **42**, 1271–1281.

Tournier, C., Hess, P., Yang, D. D., Xu, J., Turner, T. K., Nimnual, A., Bar, S. D., Jones, Flavell, R. A., and Davis, R. J. (2000). Requirement of JNK for stress-induced activation of the cytochrome c -mediated death pathway. *Science* **288**, 870–874.

Treisman, R. (1996). Regulation of transcription by MAP kinase cascades. *Curr. Opin. Cell Biol.* **8**, 205–215.

Turgeon, B., Saba-El-Leil, M. K., and Meloche, S. (2000). Cloning and characterization of mouse extracellular-signal-regulated protein kinase 3 as a unique gene product of 100 kDa. *Biochem. J.* **346**(Pt. 1), 169–175.

Veeranna, Amin, N. D., Ahn, N. G., Jaffe, H., Winters, C. A., Grant, P., and Pant, H. C. (1998). Mitogen-activated protein kinases (Erk1,2) phosphorylate Lys-Ser-Pro (KSP) repeats in neurofilament proteins NF-H and NF-M. *J. Neurosci.* **18**, 4008–4021.

Virdee, K., Bannister, A. J., Hunt, S. P., and Tolkovsky, A. M. (1997). Comparison between the timing of JNK activation, c-Jun phosphorylation, and onset of death commitment in sympathetic neurones. *J. Neurochem.* **69**, 550–561.

Wang, X., Martindale, J. L., Liu, Y., and Holbrook, N. J. (1998). The cellular response to oxidative stress: Influences of mitogen-activated protein kinase signalling pathways on cell survival. *Biochem. J.* **333**, 291–300.

Wang, X. T., Martindale, J. L., Liu, Y. S., and Holbrook, N. J. (1998). The cellular response to oxidative stress: Influences of mitogen- activated protein kinase signalling pathways on cell survival. *Biochem. J.* **333**, 291–300.

Watson, A., Eilers, A., Lallemand, D., Kyriakis, J., Rubin, L. L., and Ham, J. (1998). Phosphorylation of c-Jun is necessary for apoptosis induced by survival signal withdrawal in cerebellar granule neurons. *J. Neurosci.* **18**, 751–762.

Wilhelm, D., Bender, K., Knebel, A., and Angel, P. (1997). The level of intracellular glutathione is a key regulator for the induction of stress-activated signal transduction pathways including Jun N-terminal protein kinases and p38 kinase by alkylating agents. *Mol. Cell Biol.* **17**, 4792–4800.

Woodgett, J. R., Avruch, J., and Kyriakis, J. (1996). The stress activated protein kinase pathway. *Cancer Surv.* **27**, 127–138.

Wüllner, U., Seyfried, J., Groscurth, P., Beinroth, S., Winter, S., Gleichmann, M., Heneka, M., Löschmann, P. A., Schulz, J. B., Weller, M., and Klockgether, T. (1999). Glutathione depletion and neuronal cell death: the role of reactive oxygen intermediates and mitochondrial function. *Brain Res.* **826**, 53–62.

Xia, Z., Dickens, M., Raingeaud, J., Davis, R. J., and Greenberg, M. E. (1995). Opposing effects of ERK and JNK-p38 MAP kinases on apoptosis. *Science* **270**, 1326–1331.

Xu, X., Raber, J., Yang, D., Su, B., and Mucke, L. (1997). Dynamic regulation of c-Jun N-terminal kinase activity in mouse brain by environmental stimuli. *Proc. Natl. Acad. Sci. USA* **94**, 12655–12660.

Yagihashi, S., Kamijo, M., and Watanabe, K. (1990). Reduced myelinated fiber size correlates with loss of axonal neurofilaments in peripheral nerve of chronically streptozotocin diabetic rats. *Am. J. Pathol.* **136**, 1365–1373.

Yan, S. D., Schmidt, A. M., Anderson, G. M., Zhang, J., Brett, J., Zou, Y. S., Pinsky, D., and Stern, D. (1994). Enhanced cellular oxidant stress by the interaction of advanced glycation end products with their receptors/binding proteins. *J. Biol. Chem.* **269**, 9889–9897.

Yao, R., Yoshihara, M., and Osada, H. (1997). Specific activation of a c-Jun NH_2-terminal kinase isoform and induction of neurite outgrowth in PC-12 cells by staurosporine. *J. Biol. Chem.* **272**, 18261–18266.

York, R. D., Yao, H., Dillon, T., Ellig, C. L., Eckert, S. P., McCleskey, E. W., and Stork, P. J. S. (1998). Rap1 mediates sustained MAP kinase activation induced by nerve growth factor. *Nature* **392**, 622–626.

Yu, R., Tan, T. H., and Kong, A. T. (1997). Butylated hydroxyanisole and its metabolite tert-butylhydroquinone differentially regulate mitogen-activated protein kinases: The role of oxidative stress in the activation of mitogen-activated protein kinases by phenolic antioxidants. *J. Biol. Chem.* **272**, 28962–28970.

Zhou, G., Bao, Z. Q., and Dixon, J. E. (1995). Components of a new human protein kinase signal transduction pathway. *J. Biol. Chem.* **270**, 12665–12669.

Zochodne, D. W. (1999). Diabetic neuropathies: Features and mechanisms. *Brain Pathol.* **9**, 369–391.

NEUROFILAMENTS IN DIABETIC NEUROPATHY

Paul Fernyhough

School of Biological Sciences, University of Manchester
Manchester M13 9PT, United Kingdom

Robert E. Schmidt

Department of Pathology and Immunology, Washington University School of Medicine
St. Louis, Missouri 63110

I. Introduction
II. Neurofilament Structure
III. Phosphorylation of Neurofilament
    A. Protein Kinases Controlling Neurofilament Phosphorylation
    B. Role of Stress-Activated Protein Kinases in the Nervous System
    C. Cyclin-Dependent Kinase 5 and Glycogen Synthase Kinase-3β in the Nervous System
IV. Axonal Transport of Neurofilament
V. Neurofilaments and Axonal Cytoskeleton: Regulation of Axonal Caliber and Slow Transport
VI. Neurofilament Pathology in Diabetic Sensory and Autonomic Neuropathy
    A. Pathology of Sympathetic Ganglia
    B. Pathology of Dorsal Root Ganglia
    C. Pathology of Peripheral Nerves (Somatic and Autonomic)
VII. Possible Pathogenetic Mechanisms
    A. Accumulation of Neurofilament
    B. Alterations in Neurofilament Synthesis
    C. Synaptic Degradation of Neurofilament
    D. Posttranslational Modification of Neurofilament
    E. Axonal Regeneration and Collateral Sprouting
    F. Effect of Neurotrophic Factors on Neurofilament Properties
VIII. Future Directions
    References

This review discusses the role of abnormal neurofilament (NF) expression, processing, and structure as an etiological factor in diabetic neuropathy. Diabetic sensory and autonomic neuropathy in humans is associated with a spectrum of structural changes in peripheral nerve that includes axonal degeneration, paranodal demyelination, and loss of myelinated fibers — the latter is probably the result of a dying-back of distal axons. NF filaments are composed of three subunit proteins, NFL, NFM, and NFH, and are major constituents of the axonal cylinder. It is clear that any abnormality in synthesis, delivery, or processing of these critical proteins could lead to severe impairments in axon structure and function. This article describes mechanisms of synthesis, phosphorylation, and delivery of NF and discusses how these processes may be abnormal in diabetes. The pathological alterations in the ganglion and peripheral nerve that occur in sensory and autonomic neuropathy will be outlined and related to possible abnormal processing of NF. A major focus is the role of aberrant NF phosphorylation and its possible involvement in the impaired delivery of NF to the distal axon. Identification of stress-activated protein kinases (SAPKs) as NF kinases is discussed in detail and it is proposed that hyperglycemia-induced activation of SAPKs may be a primary etiological event in diabetic neuropathy. © 2002, Elsevier Science (USA).

I. Introduction

Diabetic sensory neuropathy in humans is associated with a spectrum of structural changes in peripheral nerve that includes axonal degeneration, paranodal demyelination, and loss of myelinated fibers — the latter is probably the result of a dying-back of distal axons (Thomas and Tomlinson, 1992; Yagihashi, 1997). In the streptozotocin (STZ)-diabetic rat and Bio-Breeding (BB) rat animal models of type I diabetes, similar structural abnormalities in peripheral nerve have been observed (Yagihashi, 1997). The Diabetes Control and Complications Trial (DCCT) concluded that control of hyperglycemia remains the ideal means of preventing the appearance of complications in diabetes, such as peripheral neuropathy (Diabetes Control and Complications Trial Research Group, 1993). However, such a goal remains an unrealized ideal and research continues to focus on biochemical transducers downstream from hyperglycemia that may directly induce neuropathic sensory nerve damage. This review discusses the role of abnormal neurofilament (NF) expression, processing, and structure as an etiological factor in diabetic neuropathy.

The aim of this review is to provide a brief overview of the neurobiology of NFs in the nervous system and to focus on abnormalities of NF structure/function in diabetes. For detailed information on NF cell biology, the reader should refer to excellent recent reviews (Julien and Mushynski, 1998; Julien, 1999; Grant and Pant, 2000).

II. Neurofilament Structure

NFs are assembled by the copolymerization of the three intermediate filament proteins: NFL (61 kDa), NFM (90 kDa), and NFH (110 kDa)

(Hoffman and Lasek, 1975). NFs have a central domain, of approximately 310 amino acids, in common with other members of the intermediate protein family, which is involved in the formation of coiled-coil dimers. Two coiled-coil dimers align in a staggered manner to form an antiparallel tetramer. NFs are obligate heteropolymers requiring NFL to aggregate with either NFM or NFH (Lee *et al.*, 1993). Neurofilaments have extensive carboxy-terminal tail domains highlighted by a high content of charged amino acids, such as Glu and Lys, as well as phosphoserine in the case of NFM and NFH (Julien and Mushynski, 1998). *In vitro* assemblies of NF show that NFM and NFH tail domains form side-arm projections of about 55 and 63 nm in length, respectively, extending from the NF axis (Nakagawa *et al.*, 1995). The carboxy tail domain regions of NFM and NFH are composed of multiple copies of the sequence motif Lys-Ser-Pro (KSP), and there are 5–12 copies of this motif in NFM of various mammalian species. The number of KSP repeats in NFH is much higher, ranging from 43 to 44 in humans and over 50 in mouse, rat, and rabbit (Julien and Mushynski, 1998) (Fig. 1).

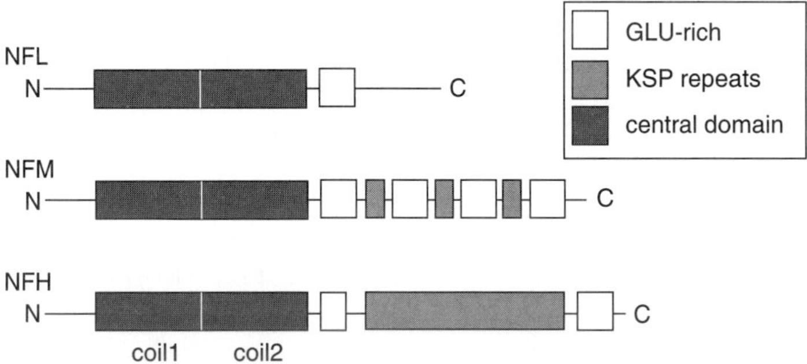

FIG. 1. Structure of neurofilament proteins. Neurofilament polymers are formed by the association of neurofilament subunits at the central domain region of approximately 310 amino acids (black boxes). This highly conserved central domain consists of two coiled regions that form coiled-coil associations on dimer formation (i.e., two NFL molecules), which then form a staggered antiparallel tetramer. Eight tetramers packed together form a single 10-nm neurofilament heteropolymer. The basic unit of any neurofilament polymer can be composed of NFL and NFM or NFL and NFH. KSP regions (gray boxes) are found to be highly enriched in KSP repeats in the carboxy-terminal domains of NFM and NFH. There are 5–12 KSP copies in NFM in mammals, whereas in NFH the number can rise to 50. There are also a limited number of SP sites that do not have an N-terminal lysine. These KSP (or SP) domains are targeted for phosphorylation by cdk5, ERK, GSK-3β, and JNK. Modified from Julien and Mushynski (1998).

III. Phosphorylation of Neurofilament

Phosphorylation of NFM and NFH occurs predominantly at serine residues within KSP motifs of the carboxy-terminal tail domain (Julien and Mushynski, 1998; Grant and Pant, 2000). Newly synthesized NF enters sensory axons in a relatively unphosphorylated state; subsequent slow anterograde axonal transport (see Section IV for a detailed description) delivers the protein into the stable cytoskeleton at more distal segments of the axon (Nixon, 1993). Phosphorylation of NF regulates the movement of NF into this detergent-insoluble fraction (Lewis and Nixon, 1988) and, consequently, a proximo-distal gradient of increasing NF phosphorylation exists in sensory axons and corresponds with decreased rates of slow axonal transport (Archer *et al.*, 1994). In addition to the former, phosphorylation of NFM and NFH carboxy-terminal domains is believed to regulate radial growth during neuronal maturation and, in turn, axonal caliber (see Section V) (Hirokawa and Takeda, 1998). Phosphorylation of the tails of NFM and NFH elevates total negative charge and thus lateral extension of the carboxy-terminal domain, causing increased NF nearest-neighbor spacing and/or increased cross-bridging to other axonal components (Hirokawa and Takeda, 1998). Regions in axons with high NF densities, such as the node of Ranvier and initial axon segments, are characterized by a low level of NF phosphorylation (Hsieh *et al.*, 1994; Nixon *et al.*, 1994). Impaired myelination in the Trembler mouse mutant results in reduced NFH phosphorylation and corresponding increased NF densities, a reduced rate of slow axonal transport, and loss of axon caliber (De Waegh *et al.*, 1992).

A. Protein Kinases Controlling Neurofilament Phosphorylation

The carboxy-terminal KSP domains of NFM and NFH are phosphorylated by several protein kinases, including cyclin-dependent kinase 5 (cdk5), glycogen synthase kinase-3β (GSK-3β), and members of the stress-activated protein kinase (SAPK) family (Julien and Mushynski, 1998; Julien, 1999; Grant and Pant, 2000). The SAPK subgroup of mitogen-activated protein kinases (MAPKs) are activated by a range of cellular stresses — hyperglycemia, hypertonicity, treatment with cytokines such as interleukin-1β (IL-1β) and tumor necrosis factor α (TNFα), oxidative stress, and ultraviolet irradiation being among the most commonly studied (Whitmarsh and Davis, 1996). In PC12 cells, SH-SY5Y cells, and embryonic cortical and peripheral neurons, members of the SAPK family are activated by hyperglycemia (Cheng and Feldman, 1998), hyperosmolarity (Giasson and Mushynski, 1997), glutamate (Brownlees *et al.*, 2000; Ackerley *et al.*, 2000), or changes in neurotrophic support (Deshmukh and Johnson, 1997; Xia *et al.*, 1995). The SAPK family includes extracellular signal-regulated kinase 1/2 (ERK 1/2), c-jun N-terminal protein kinase (JNK; composed of three genes termed JNK1, 2, and 3), and SAPK2 (p38/HOG1) (Whitmarsh and Davis, 1996). Upon activation, this family of kinases binds to many transcription factors, including c-jun (JNK only), ATF2, and ELK-1, and phosphorylates activation domains favoring DNA binding (Whitmarsh and Davis, 1996). Events downstream of transcription factor activation are not well defined in neurons, but in PC12 cells and sympathetic neurons, activation of JNK and SAPK2, triggered by nerve growth factor (NGF) removal, induces apoptosis (Deshmukh and Johnson, 1997; Xia *et al.*, 1995).

B. ROLE OF STRESS-ACTIVATED PROTEIN KINASES IN THE NERVOUS SYSTEM

JNK1 and JNK2 combine to regulate apoptosis during development of the mouse brain (Kuan *et al.*, 1999). In the adult central nervous system (CNS) there is a heterogeneous expression of JNK and activation in response to extracellular stress (Carletti *et al.*, 1995; Xu *et al.*, 1997); e.g., kainate-induced excitotoxicity in the hippocampus is mediated via activation of JNK3 (Yang *et al.*, 1997). In the peripheral nervous system (PNS), axotomy of sensory neurons induces rapid and long-term activation of JNK within the lumbar dorsal root ganglia (DRG) (Kenney and Kocsis, 1998) and this precedes the elevation in phosphorylation of NFM and NFH (Goldstein *et al.*, 1987). The mechanism of JNK activation is unknown but axotomized sensory neurons express, and are exposed to, proinflammatory cytokines, including interleukin-6 (IL-6), IL-1β, and TNFα (Murphy *et al.*, 1995; Rotshenker *et al.*, 1992; Shubayev and Myers, 2001; Copray *et al.*, 2001), and experience a loss of neurotrophic support (Richardson, 1991).

In cultured human SH-SY5Y neuroblastoma cells and in primary hippocampal neurons, phosphorylation of NFM and NFH is mediated, in part, by ERK 1/2, and inhibition of this process retarded neurite outgrowth (Veeranna *et al.*, 1998). Follow-up studies have shown that this pathway is mediated via integrin binding to laminin and fibronectin in primary motoneurons (Li *et al.*, 2001).

C. CYCLIN-DEPENDENT KINASE 5 AND GLYCOGEN SYNTHASE KINASE-3β IN THE NERVOUS SYSTEM

Cdk5 and GSK-3β dependent phosphorylation of the carboxy-terminal domains of NFs has been well characterized in nonneurons transfected with NF constructs; however, the importance of these kinases in the PNS is unclear (Bajaj and Miller, 1997; Guidato *et al.*, 1996; Grant *et al.*, 2001). Cdk5 modulates the rate of fast anterograde axonal transport in squid axoplasm by phosphorylating microtubule/membrane-associated proteins (Ratner *et al.*, 1998). Cdk5 probably plays a minor role in modulating the phosphorylation of NF in sensory neurons. Sensory neurons develop normally in cdk5 null mice, whereas CNS neurons are clearly abnormal (Giasson and Mushynski, 1997). The cofactor, p35, required for the activation of cdk5 in the CNS is not expressed in the PNS and may account for low levels of cdk5 activity in sensory neurons (Giasson and Mushynski, 1997); this low cdk5 activity suggests the presence of alternative activators (Guidato *et al.*, 1996). It is possible, however, that integrin binding to components of the extracellular matrix, such as laminin, may also modulate cdk5 activity and in turn affect NFM and NFH phosphorylation (Li *et al.*, 2000).

The role of GSK-3β in the PNS is relatively unstudied, but presumably modulates glycogen synthesis in an insulin-dependent manner as in nonneuronal cells (Cohen, 1999). Studies in the CNS and PNS demonstrate an important role for GSK-3β in tau phosphorylation and neuronal survival (Mattson, 2001; Cross *et al.*, 2001).

IV. Axonal Transport of Neurofilament

The three NF subunits are transported by anterograde axonal transport down the axon as a part of the slow component of transport (Hoffman and Lasek, 1975). Classically, it was thought that NF subunits moved as a cohesive network; however, this view has been supplanted by a theory based on the dynamic movement of individual cytoskeletal subunits (Hirokawa *et al.*, 1997). The latter hypothesis has been supported by studies showing rapid fast axonal transport of NF subunits (Wang *et al.*, 2000). Real-time video microscopy of neurons expressing green fluorescent protein (GFP)-labeled NF revealed bidirectional movement (strongly favoring the anterograde direction) of NF to be rapid and consistent with rates of fast axonal transport (Roy *et al.*, 2000; Wang *et al.*, 2000). Although this vectorial transport was rapid, it was intermittent and so overall rates of transport of NF could be compared with measured rates of slow anterograde transport. Using a similar methodology, the rate of transport of GFP-NFM was shown to be impaired on glutamate-induced stress in cultured cortical neurons. Under such conditions, NFM phosphorylation was elevated, possibly via activation of JNK (Ackerley *et al.*, 2000).

V. Neurofilaments and Axonal Cytoskeleton: Regulation of Axonal Caliber and Slow Transport

NF proteins are major constituents of the axonal cylinder and form parallel arrays of 10-nm filaments with frequent cross-bridges between NF filaments and microtubules (MTs) or membranous organelles (Hirokawa and Takeda, 1998). NFL forms a 10-nm core filament by itself; upon addition of NFM or NFH, many side arms project from the core (Hirokawa and Takeda, 1998). NFM and NFH have carboxy-terminal tails containing several KSP repeats that are targets for phosphorylation and may regulate inter-NF and inter-MT spacing (Hirokawa and Takeda, 1998). The levels and phosphorylation status of these proteins are believed to determine axonal caliber through mechanisms involving the modulation of inter-NF and inter-MT spacing (Hoffman *et al.*, 1987; De Waegh *et al.*, 1992). Hypophosphorylation of NF in the sciatic nerve of the mutant trembler mouse resulted in increased NF filament density, which in turn was associated with reduced rates of slow axonal transport (De Waegh *et al.*, 1992).

Studies in transgenic mice overexpressing individual NF proteins show that the ratio of NFL:NFM:NFH is crucial for the maintenance of optimal NF densities and axonal caliber in motor axons (Xu *et al.*, 1996). Large myelinated sensory axons exhibit reduced radial growth and, in turn, axonal caliber, in transgenic mice overexpressing mouse NFH (Marszalek *et al.*, 1996) and NFM (Wong *et al.*, 1995). Abnormal levels of NFH and NFM induce an accumulation of NF subunits in sensory perikarya and a reduction in total NF content in sciatic nerve, possibly mediated via inhibition of axonal transport of NF (Wong *et al.*, 1995; Marszalek *et al.*, 1996). While NF levels in sciatic nerve were reduced in both transgenic lines, there was no change in nearest-neighbor NF spacing (Wong *et al.*, 1995; Marszalek *et al.*, 1996). Conversely, in NFH null mice, there is also reduced axial growth in large-caliber sensory axons, but not motor axons (Rao *et al.*, 1998). Increased NFM levels in axons offset the absence of NFH with nearest-neighbor spacing between NFs being unchanged and β-tubulin III protein and MT levels markedly elevated (Rao *et al.*, 1998).

VI. Neurofilament Pathology in Diabetic Sensory and Autonomic Neuropathy

A. PATHOLOGY OF SYMPATHETIC GANGLIA

1. *Pathology of Sympathetic Ganglia in Human Diabetes*

An extensive autopsy study of sympathetic autonomic ganglia of a large number of unselected autopsied adults (Schmidt *et al.*, 1993, 1997; Schroer *et al.*, 1992) identified a reproducible neuropathologic alteration characterized by markedly swollen nerve terminals and distal preterminal axons (neuroaxonal dystrophy) developing in the apparent absence of significant neuron loss (see chapter by Schmidt). Dystrophic axons were more than 10-fold more frequent in prevertebral celiac and superior mesenteric ganglia (SMG) than paravertebral superior cervical ganglia (SCG). Quantitative studies have determined that the frequency of neuroaxonal dystrophy is increased in aged subjects and arises prematurely and at an accelerated rate in diabetic subjects. Both aged and diabetic subjects develop individual dystrophic axons, which are ultrastructurally and immunohistochemically

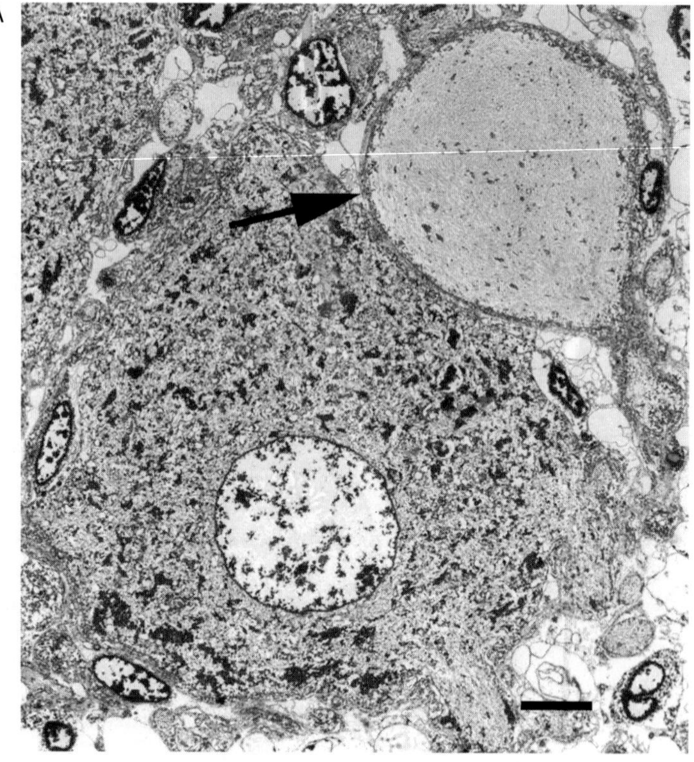

FIG. 2. Typical appearance of dystrophic neurites in diabetic human prevertebral superior mesenteric sympathetic ganglia. (A) A neurofilament-laden dystrophic axon (arrow) is intimately applied to the surface of a principal sympathetic neuronal perikaryon (electron micrograph) Bar: 5 μm. (B) Neurofilaments in dystrophic axons are frequently misaligned, forming twisted skeins, and are surrounded by a thin rim of neurotransmitter containing vesicles (arrows) (electron micrograph). Bar: 4 μm.

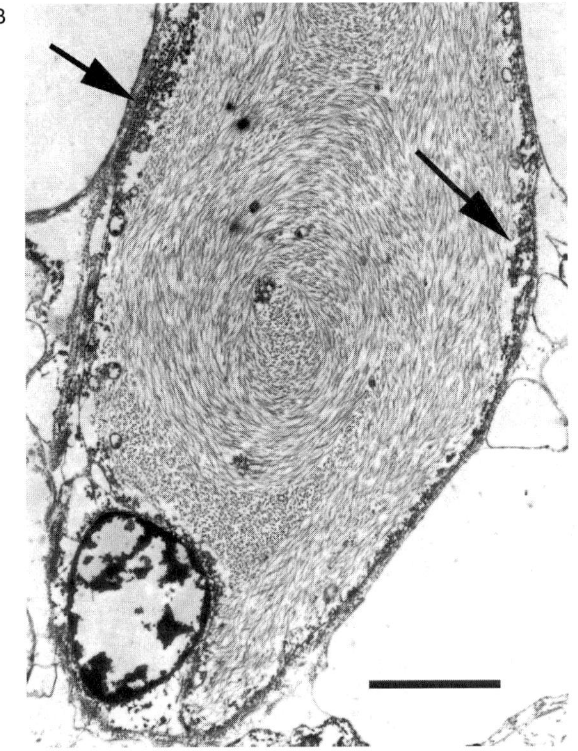

FIG. 2. (*continued*)

indistinguishable. Ultrastructural studies have identified several populations of dystrophic swellings, the most frequent pattern is represented by the accumulation of large numbers of NF (Fig. 2A), which are disorganized, often circumferentially arranged, and with a thin subplasmallemal rim of neurotransmitter containing granules (arrows, Fig. 2B). Using a panel of antibodies directed against NF epitopes of various sizes and phosphorylation states, it was demonstrated that the NF that accumulate in diabetic and aged dystrophic sympathetic nerve terminals consist almost exclusively of extensively phosphorylated NFH epitopes (Fig. 3A, see also color insert). Although individual principal sympathetic neuronal perikarya demonstrate significantly less intense immunoreactivity than dystrophic elements, the intensity of staining is variable from one cell body to another. Antisera directed against NFL, NFM and nonphosphorylated epitopes of NFH preferentially label sympathetic neuronal perikarya and principal dendrites but do not label dystrophic axons, evidence against their origin from principal

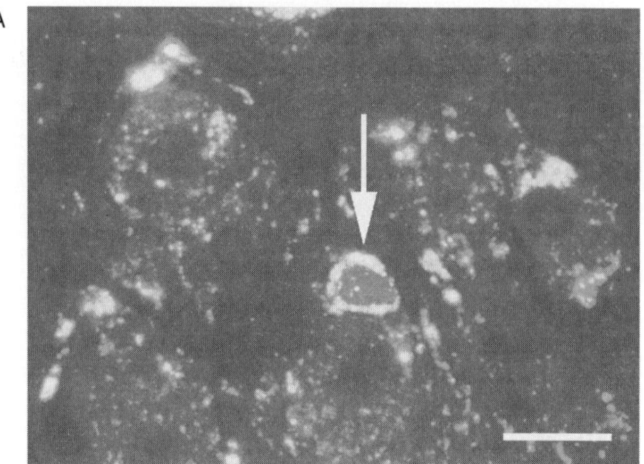

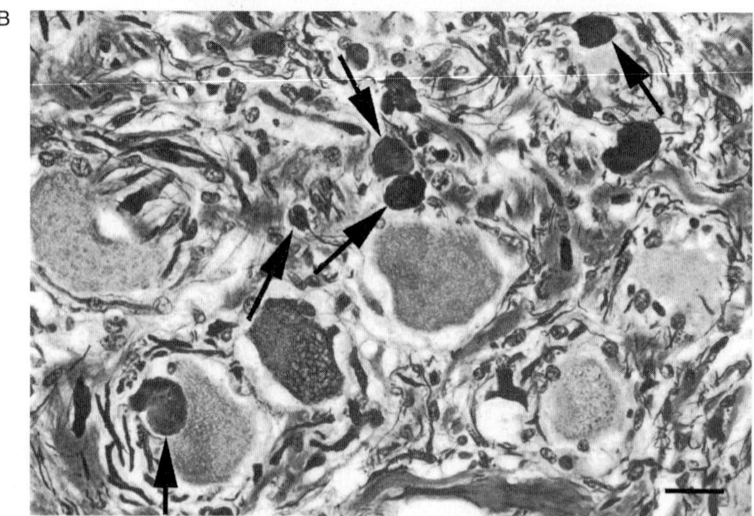

FIG. 3. Neuroaxonal dystrophy in diabetic human prevertebral sympathetic and dorsal root ganglia. (A) Simultaneous immunohistochemical detection of dopamine-β-hydroxylase (red) and highly phosphorylated NF-H (green) in a diabetic sympathetic ganglion shows their typical colocalization in a dystrophic axon (arrow), a pattern reflecting DβH containing neurotransmitter granules surrounding a neurofilamentous core (DβH and SMI-34 immunofluorescence) Bar: 30 μm. (B and C) Markedly enlarged neurofilament-laden dystrophic axons (arrows) are intimately apposed to principal dorsal root ganglion neurons (B: Bielschowsky silver stain; C: 1-μm plastic section) Bars: 20 μm. (See also color insert.)

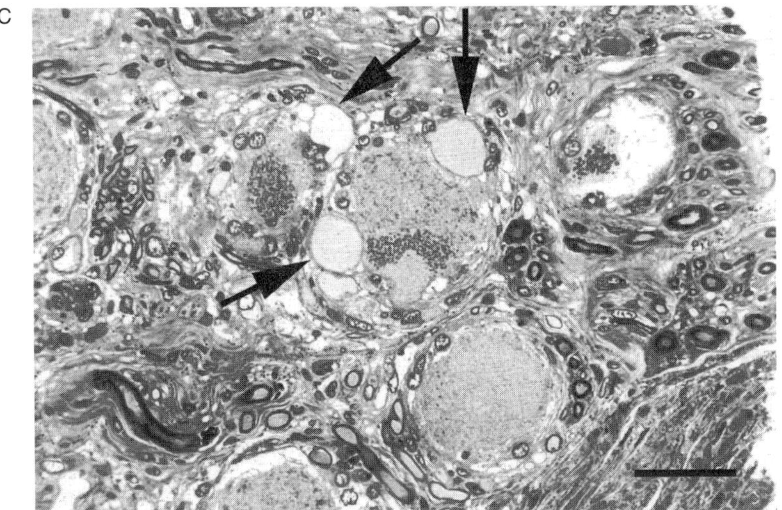

FIG. 3. (*continued*)

dendrites, perikaryal spines, or proximal perisomal axons. Simultaneous immunolocalization of phosphorylated NF proteins and MAP-2 protein (a marker for dendrites and cell bodies) also fails to demonstrate colocalization in dystrophic axons. Peripherin, a 58-kDa cytoskeletal element structurally distinct from any NF subunit, is located in populations of sensory and sympathetic neurons and responds differently than NF proteins in response to injury. Peripherin and NF immunoreactivity colocalize in many dystrophic elements in diabetic sympathetic prevertebral ganglia, a result that may reflect the operation of a shared degradative mechanism, rather than altered synthesis, as a target of diabetes.

2. *Pathology of Sympathetic Ganglia in Experimental Animals*

Changes in diabetic human sympathetic ganglia show significant correspondence with those in the sympathetic autonomic ganglia in STZ- and BB-diabetic rats and genetically diabetic Chinese hamsters. Dystrophic NF containing axons represent only one form of several in STZ- and BB-diabetic rat neuroaxonal dystrophy. NF-laden neurites are found in the ganglionic neuropil (STZ rat) and in proximal axons and occasional neuronal perikarya (BB rat) (Yagihashi and Sima, 1986; Schmidt *et al.*, 1989). Although NF may be admixed with tubulovesicular elements in dystrophic axons, they often occur as relatively pure aggregates. Similar lesions occur in rat sympathetic ganglia as a function of aging (Schmidt *et al.*, 1983). As in human subjects, neuroaxonal dystrophy is significantly more frequent in prevertebral SMG and celiac ganglia than in the SCG. Although small patches of NF can be demonstrated in principal sympathetic neuronal perikarya, they are infrequent and do not clearly differ in diabetic and control animals.

B. Pathology of Dorsal Root Ganglia

1. *Pathology of DRG in Human Diabetes*

In a variety of clinical conditions, including, but not confined to, diabetes and aging, an abnormality of sensory function involving both loss of function and the development of abnormal sensations (i.e., dysesthesias or paresthesias), is thought to reflect abnormal communication between sensory and sympathetic nervous systems (Galer, 1995), which may be sensitive to surgical or chemical sympathectomy. Normal human and rodent DRG represent collections of sensory neuronal perikarya and normally lack intraganglionic synapses. However, in response to peripheral injury of sensory axons in nondiabetic experimental animals, perivascular sympathetic axons sprout into the DRG, establishing new, nondystrophic synapses surrounding the perikarya of DRG neurons (McLachlan *et al.*, 1993). It was anticipated that sympathetic sprouting might occur in diabetic human DRG in response to diabetes-induced distal axonopathy and result in neuroaxonal dystrophy similar to that involving sympathetic terminals in human diabetic sympathetic ganglia.

Examination of thoracic and lumbar DRG from a large series of autopsied human subjects (Schmidt *et al.*, 1997) demonstrated neuroaxonal dystrophy (Figs. 3B and 3C, see also color insert), which developed as a function of diabetes and, to a lesser extent, aging. Dystrophic swellings (5–30 μm in diameter) are typically located within the satellite cell capsule intimately applied to sensory neuronal perikarya. Although most dystrophic axons arise from delicate axons within the ganglionic neuropil, a few originate as swellings arising along the initial sensory axon segment or as recurrent collaterals from the proximal axon. Dystrophic axons terminate on a subpopulation of larger DRG neurons. As in sympathetic ganglia, swollen axons contain either tubulovesicular profiles or NF, which are immunoreactive with antisera to highly phosphorylated NFH epitopes but not poorly phosphorylated NFH, NFM, and NFL epitopes. Surprisingly, dystrophic axons in the DRG are not immunoreactive with antisera to tyrosine hydroxylase or neuropeptide Y, as expected in line with findings in the sympathetic ganglia, although normal noradrenergic sympathetic vascular terminals were identified within the ganglia. Dystrophic swellings in diabetic and aging human DRG are, however, consistently and strongly immunoreactive for calcitonin gene-related peptide (CGRP) and, to a lesser degree, substance P, which may reflect their origin as sprouts from small, potentially nociceptive, DRG neurons. Dystrophic axons in aged and diabetic human DRG are identical in their light microscopic, immuno-histochemical and ultrastructural appearance, suggesting the possibility of shared pathogenetic mechanisms.

2. *Pathology of DRG in Experimental Animals*

Morphometric studies of short-term diabetic DRG have described reduced perikaryal volume (Sidenius and Jakobsen, 1980) and, in chronically diabetic rats, neurons containing large numbers of perikaryal vacuoles (Sasaki *et al.*, 1997). Biochemical and immunocytochemical studies of diabetic rat lumbar DRG demonstrate a two- to threefold increase in NF phosphorylation, which was localized to the perikarya of medium to large neurons (Fernyhough *et al.*, 1999). Small CGRP containing neurons did not show an increase in perikaryal-phosphorylated NFs in STZ-diabetic rats.

3. *Central Projections of Sensory Neurons*

Aggregates of disoriented NF have been demonstrated in the central projection of DRG neurons in the gracile nucleus as an early alteration in BB diabetic rats, which accompanied peripheral axonopathy ("central–peripheral distal axonopathy"; Sima and Yagihashi, 1986).

C. PATHOLOGY OF PERIPHERAL NERVES (SOMATIC AND AUTONOMIC)

1. *Pathology of Human Nerves*

Diabetic symmetrical sensorimotor neuropathy is characterized by endoneurial microvasculopathy and axon loss, thought to reflect a multi-focal ischemic pathogenesis, atrophy, endoneurial microangiopathy, and teased fiber evidence of segmental demyelination. Although swollen axons containing NF are not typically seen in sural nerve biopsies, the surgical sampling site is significantly distant from the nerve termini. A number of studies have begun the analysis of epidermal and dermal sensory and autonomic axons. An increase in the innervation of diabetic skin early in the course of diabetic neuropathy (Properzi *et al.*, 1993) is followed by a loss of epidermal sensory axons (Kennedy and Wendelschafer-Crabb, 1999) in a length-dependent distribution in the skin biopsies of diabetic patients, in some cases associated with abrupt termination at the epidermal border and occasionally scattered swollen axons (Lauria *et al.*, 1998; Vinik *et al.*, 2000; M. Polydefkis, personal communication), in some cases in the absence of losses of significant sural nerve axons.

2. *Pathology of Peripheral Nerves in Experimental Animals*

The density of NF is diminished in the distal sensory axons of chronically diabetic STZ rats (Yagihashi *et al.*, 1990), which may reflect a decrease in levels of transcripts for NF subunits in DRG (Medori *et al.*, 1988a,b; Mohiuddin *et al.*, 1995; Scott *et al.*, 1999). NF containing axonal swellings are reported in the distal ileal mesenteric nerves of STZ-diabetic rats (Schmidt and Scharp, 1982). Decreased expression of all NF genes and α-tubulin mRNA in sensory neurons coincides with decreased NF and microtubule number in large myelinated and unmyelinated axons in 6-month diabetic rats (Scott *et al.*, 1999). Margination and malorientation of microtubules and NF are reproducible alterations of somatic nerves in the BB rat (Sima *et al.*, 1982) and mouse (Sima and Robertson, 1979).

Distal axonal atrophy in sensory axons has been related to a proportional loss of NF coupled with no change in NF nearest-neighbor spacing (or density) in rats with long-term STZ diabetes (Yagihashi *et al.*, 1990) and chronic galactose neuropathy (Nukada *et al.*, 1986). Studies in 12-week STZ-diabetic rats confirm these results with loss of NF in large myelinated fibers of sural (sensory) nerve, but not peroneal (motor) nerve (Mizisin *et al.*, 1999). The NF loss was coincident with loss of axonal caliber in sensory, but not motor, axons and with NF density remaining unchanged. These observations parallel NF changes and sensory axon abnormalities seen in the NFM and NFH transgenic mice described earlier (Wong *et al.*, 1995; Marszalek *et al.*, 1996; Rao *et al.*, 1998) and imply a crucial role for abnormal NF processing in etiology of diabetic sensory neuropathy.

VII. Possible Pathogenetic Mechanisms

A. Accumulation of Neurofilament

The accumulation of axonal NFs is a common alteration in a variety of toxic, genetic, and age-related conditions. Cytoskeletal pathology in the PNS may result from (i) overproduction of filaments, resulting in the accumulation of NF polymers or NF subunits of normal or abnormal composition (Cote *et al.*, 1993; Muma and Cork, 1993); (ii) focal abnormality in the axonal transport of NF (Graham *et al.*, 1984; Griffin *et al.*,

1978; Troncoso *et al.*, 1985); (iii) damage to NF side arms, which may result in NF compaction and dysfunction (Pettus *et al.*, 1994; Okonkwo *et al.*, 1998); (iv) posttranslational modification of NF structure (e.g., excessive phosphorylation, glycosylation, partial proteolysis, advanced glycosylation end product (AGE)-mediated cross-linking) (Vlassara *et al.*, 1994); or (v) abnormality of the terminal degradative apparatus.

B. Alterations in Neurofilament Synthesis

1. *Regulation of Neurofilament Synthesis*

Synthesis of NF is regulated at transcriptional and posttranscriptional levels (Schwartz *et al.*, 1992, 1997; Lindenbaum *et al.*, 1988). NGF can increase protein levels of all the NFs (Lee *et al.*, 1982), mainly through increased stability of the NF message (Lindenbaum *et al.*, 1988). Stability of NFL transcripts is regulated by a 68 nucleotide region at the junction of the coding and 3′-untranslated regions (Canete-Soler, 1998). Binding of the cytoskeleton-associated GTPase exchange factor, p190RhoGEF, to this region protects the NFL transcript from rapid destabilization and degradation (Canete-Soler, 2001). The stability of the other NF transcripts is likely to be regulated in a similar manner. Detailed regulation of the promoter function of NF genes remains to be determined; however, the NF-1-like transcription factor and the brn-3 family of POU domain transcription factors can upregulate the expression of NF transcripts (Schwartz *et al.*, 1997; Smith *et al.*, 1997). NGF can regulate the expression of NF mRNA and the density of NF filaments in sensory neurons *in vivo*. The down-regulation of NF mRNA levels observed in the DRG upon axotomy can be partially prevented by intrathecal NGF treatment (Verge *et al.*, 1990). The axotomy-induced axonal dwindling observed in central projections of sensory neurons as a consequence of NF filament loss could be prevented by NGF treatment (Gold *et al.*, 1991). Furthermore, NF filament loss in distal axons of large myelinated neurons could be induced by treatment with function blocking antibody to NGF (Gold *et al.*, 1991).

2. *Dysfunctional Expression of Neurofilament*

NF triplet proteins must be synthesized in appropriate ratios to avoid producing dysfunctional NF. Transgenic mice selectively overexpressing human NFH develop motor and sensory perikaryal and proximal axonal swellings with distal axonal atrophy (Cote *et al.*, 1993). Alteration in peripherin dynamics may secondarily produce NF dysfunction, as both peripherin and NF proteins have been demonstrated in the same individual NF in cultured PC12 cells (Parysek *et al.*, 1991). NF of aged rats, which show a 60% loss of NFM, develop increased packing density, aggregates of isolated filaments, and higher phosphorylation of NFH in the tail domain. Studies in STZ-diabetic and BB rats show reduced sensory neuron expression of NF mRNA (Mohiuddin *et al.*, 1995; Mohiuddin and Tomlinson, 1997; Liuzzi *et al.*, 1998) and deficient axonal transport of NF in sensory (Macioce *et al.*, 1989) and motor axons (Medori *et al.*, 1988b).

a. Hypothesis 1: Impaired Synthesis of Neurofilaments Results in Suboptimal Delivery of NF to the Distal Axon. Hyperglycemia and/or lack of insulin signaling may result in the direct inhibition of NF synthesis. This could occur at transcriptional and/or posttranscriptional levels. For example, insulin and insulin-like growth factors have been demonstrated to modulate NF mRNA levels in human neuroblastoma SH-SY5Y cells (Wang *et al.*, 1992). Impaired synthesis of NF transcripts and protein will result in reduced delivery to the distal nerve ending (see Section VIII for a discussion of future studies).

C. Synaptic Degradation of Neurofilament

The three proteins that form the NF triplet are preferentially distributed to certain regions of the normal neuron: poorly phosphorylated forms of NFL and NFM, as well as hypophosphorylated forms of NFH, are concentrated in cell bodies and proximal dendrites, whereas hyperphosphorylated forms of NFH are typically axonal constituents. However, neuron populations in sympathetic ganglia may vary significantly in NF content from different chemically coded subpopulations of noradrenergic neurons (Vickers *et al.*, 1990). Although transported anterogradely, NF are not normal constituents of the nerve terminal and are degraded within the terminal axon. Disassembly of slowly transported NF is thought to occur in the preterminal axon in response to calpain activity, which is sensitive to activity-related nerve terminal Ca^{2+} content. In support of this process, NF accumulation has been described in inactive nerve terminals (Bondar and Roots, 1977), and the appearance of phosphorylated NFH epitopes has been reported in the perikarya of quiescent decentralized/deafferented sympathetic neurons (Shaw *et al.*, 1988). Predictably, local administration of protease inhibitors resulted in the accumulation of NF in terminal axons (Roots, 1983). Calpain I (micromolar calpain) or another more selective protease may have a specific target in NF side arms, important for the determination of the density of NF and their proper axonal function. Partial proteolysis of NF side arms may be important for the formation of structural NF aggregates (Pettus and Povlishock, 1996; Okonkwo *et al.*, 1998). Aluminium binds to NF proteins and inhibits calpain-mediated proteolysis, producing protease-resistant high molecular weight complexes in neuronal perikarya (Nixon *et al.*, 1990). Poorly understood processes, such as the role of proteolytic NF fragments on the synthesis of NF in perikarya (Schlaepfer, 1987) and the retrograde transport of some forms of NF proteins (Watson *et al.*, 1993), suggest the possibility that the abnormality of peripheral degradation of NF may influence synthetic processes.

D. POSTTRANSLATIONAL MODIFICATION OF NEUROFILAMENT

1. *Aberrant Phosphorylation of Neurofilament*

Sensitivity of NF to calpain-mediated proteolysis is inhibited by increased carboxyl-terminal phosphorylation (Terada *et al.*, 1998; Pant, 1988). The spinal cord, sciatic nerves, and DRG of diabetic rats (STZ-diabetic and BB rats) contain substantially increased levels of phosphorylated NFM and NFH by Western blot and immunohistochemical analysis in comparison to controls (Pekiner and McLean, 1991; Terada *et al.*, 1998; Fernyhough *et al.*, 1999). The degradation of phosphorylated NF was delayed in Wallerian degeneration following sciatic axotomy in diabetic rats compared to controls, although both groups showed comparable degradation of nonphosphorylated NF protein (Terada *et al.*, 1998). Phosphorylation of NF is also important for the intraaxonal spacing of NF. Disruption of the BPAG1 locus, which normally produces a protein that permits the proper association of NF with the actin cytoskeleton, results in peripheral axonal NF aggregates and eventual degeneration in sensory neurons (Guo *et al.*, 1995). The demonstration of serpins (serine protease inhibitors) within NF aggregates in the motor neurons of patients with amyotrophic lateral sclerosis (ALS) suggests that an imbalance of serine proteases and internalized serpins may have a role in the formation of NF containing axons or cell body inclusions (Chou *et al.*, 1998). Analysis of NF conglomerates, which form in ALS, may exhibit phosphorylation, glycation, nitration, ubiquitination, and cross-linking by Ca^{2+}-dependent transglutaminase (Chou *et al.*, 1998).

Abnormal NF phosphorylation may result in the accumulation of NF in sensory neurons or form axonal swellings containing NF in central and peripheral projections in intoxication with β,β'-iminodipropionitrile (IDPN) (Gold and Austin, 1991), 2,5-hexanedione (King *et al.*, 1993), and acrylamide (Gold *et al.*, 1988), as well as a number of diseases of the CNS, particularly ALS (Brady, 1993). The distribution of NF swellings in proximal or distal axons may reflect the severity of the pathogenetic process, as the toxin IDPN results in proximal or distal NF swellings as a function of its dose; low doses favor a distal localization (Denlinger *et al.*, 1992). In inherited giant axonal neuropathy, there is a relative decrease in side arm size that ultimately results in NF disorientation; however, its appearance differs from 2-hexanedione intoxication in NF size and spacing (King *et al.*, 1993). A transgenic mouse with a modification of NFL amino acid structure that mimics permanent phosphorylation of NFL results in Purkinje cells with somal and proximal axonal NF aggregates (Gibb *et al.*, 1998), suggesting that phosphorylation of NFL represents a mechanism for regulating NF organization *in vivo*. Nitration of tyrosine residues in NF-L inhibits its phosphorylation needed for NF assembly, resulting in NF aggregates in motoneurons (Chou *et al.*, 1996).

2. *SAPKs in Diabetes*

Activation of SAPKs is a component of the osmosensing pathway in mammalian cells (Treisman, 1985). Hyperosmolarity induces activation of JNK and p38 in cultured renal medullary cells (Kultz *et al.*, 1997) and has been demonstrated in the glomeruli of the kidney in STZ-diabetic rats (Awazu *et al.*, 1999; Haneda *et al.*, 1997) and hence a role in nephropathy has been suggested. Advanced glycation end product-mediated activation of ERK1/2 has also been implicated in the etiology of nephropathy (Simm *et al.*, 1997). Vascular dysfunction is associated with PKC activation, which leads to downstream induction of p38 in aortic smooth muscle cells (Igarashi *et al.*, 1999). Studies in STZ-diabetic and BB rats show that JNK and ERK underwent long-term activation in lumbar DRG and sural nerve, which correlated with elevated NFM and NFH phosphorylation (Fernyhough *et al.*, 1999). Additionally, sural nerve biopsies taken from amputated limbs of type 1 diabetic patients revealed a 2.5-fold elevation in activation of JNK (Purves *et al.*, 2001). Treatment of cultures of adult sensory neurons with high concentrations of glucose led to long-term activation of JNK, which correlated with increased NF phosphorylation (Purves *et al.*, 2001). An important feature of this glucose-dependent activation of JNK in cultured sensory neurons was the slow rate of development coupled with a sustained nature but with no evidence of cell death. Such long-term activation of JNK has been observed previously in kidney in STZ-diabetic rats, and in cultured embryonic kidney cells, this process was mediated via gene 33, an inducible adapter protein that regulates the ability of cdc42 to activate JNK (Makkinje *et al.*, 2000).

a. Hypothesis 2: Hyperglycemia-Induced Aberrant NF Phosphorylation Inhibits Axonal Transport of Cytoskeletal Elements and Denudes the Distal Axon. Studies in STZ-diabetic rats show that phosphorylated NF is expressed by sensory neuron cells bodies (Fernyhough *et al.*, 1999). This is abnormal and may suggest either an aggregation or an accumulation of NF complexes in the cell body. High glucose concentrations have been demonstrated to activate SAPKs, which in turn may induce aberrant phosphorylation of NFs within the cell body and proximal axon (Purves *et al.*, 2001). Consequently, transport of newly synthesized NF will be impaired, and hyperphosphorylation of NF at proximal parts of the axon will lead to premature incorporation into the detergent-insoluble immobile fraction of the cytoskeleton. A proximal accumulation of cytoskeletal elements has been suggested previously based on measurement of increased caliber of proximal axons in STZ- and BB-diabetic rats (Medori *et al.*, 1988a,b). Interestingly, large myelinated sensory fibers appear to be most sensitive to functional impairment in diabetes and it is this population of neurons that exhibits the highest level of aberrant NF phosphorylation (Thomas and Tomlinson, 1992; Fernyhough *et al.*, 1999). Negative feedback from accumulation of NF will induce a downregulation of NF transcript production and so a general slowing of NF synthesis and delivery to the distal axon will follow. Furthermore, the proximal accumulation of hyperphosphorylated NF may also impair the transport of other cytoskeletal proteins, i.e., tubulin. The result will be a reduction in bulk delivery of cytoskeletal proteins to the distal axon and axonopathy may follow.

3. *Glycosylation of Neurofilament*

Postsynthetic modification of NF by glycosylation, a process thought to operate in both aging and diabetes, may change the sensitivity of NF to calpains and could result in their excessive accumulation in axonal terminals in aging and diabetes. Glycation of cytoskeletal elements may underlie axonal atrophy and degeneration (Ryle *et al.*, 1997). ALS motoneurons show colocalization of NF-bound AGEs, AGE receptors, and SOD1, suggesting AGE-mediated oxidative stress and protein aggregation (Chou *et al.*, 1999). Hyperglycemia-induced glycosylation of various intracellular and extracellular proteins may be followed by chemical rearrangement into advanced glycosylation end products, which may result in cross-linking of a variety of proteins and subcellular organelles by distinctive linkers such as pentosidine. Treatment of experimental diabetic animals with aminoguanidine, an agent that prevents AGE-dependent protein cross-linking, but not its initial glycosylation step, has resulted in improvement in axonal transport, nerve conduction, and other disturbed functions in diabetic somatic nerves (Brownlee *et al.*, 1988).

E. Axonal Regeneration and Collateral Sprouting

Diabetes and aging are associated with abnormal axonal regeneration and collateral axonal sprouting. Immunohistochemical studies (see chapter by Schmidt) suggest the origin of sympathetic ganglionic dystrophy from noradrenergic sympathetic axons, which most likely reflects local intraganglionic collateral axonal sprouting or cycles of terminal axonal degeneration/regeneration. Proper completion of the regenerative or sprouting response may depend on the initiation of a "stop program" reflecting the switch from axonal elongation to the steady state of the normal nerve terminal. A stop program may require the proteolytic breakdown of NF in the preterminal axon and/or reversal of axonal transport polarity ("turnaround"), the absence of which might result in the delivery of an excessive supply of tubulovesicular elements or NF to the nerve terminal (Liuzzi, 1990). Neuroaxonal dystrophy in the diabetic DRG may arise as part of collateral sprouting and, in the SMG, the normal ongoing process of ganglionic synaptic degeneration and regeneration, turnover which may underlie synaptic plasticity. GAP-43 expression is decreased in diabetic DRG and dystrophic lesions are not routinely GAP-43 immunoreactive in diabetic SMG or DRG (Schmidt *et al.*, 1991, 1997), which may reflect deficiencies in growth-associated processes. Axonal sprouts (but not dystrophic axons) containing CGRP and substance P have been described in the axotomized DRG of nondiabetic rats (McLachlan and Hu, 1998).

F. Effect of Neurotrophic Factors on Neurofilament Properties

1. *Effects of Neurotrophins on Neurofilaments in Diabetes*

In cisplatin-induced sensory neuropathy the accumulation of NF in the DRG and associated aberrant phosphorylation could be ameliorated by treatment with NT-3 (Gao *et al.*, 1995). STZ- and BB-diabetic rats show substantially increased amounts of phosphorylated NFM and NFH in DRG and peripheral somatic nerves (Fernyhough *et al.*, 1999). Treatment of 12-week STZ-diabetic rats with human recombinant NT-3 resulted in a reduced level of NFH phosphorylation in the sural nerve, which coincided with a reversal of the deficit in sensory nerve conduction velocity (Mizisin *et al.*, 1999). However, this treatment regime (three times weekly with sc 1.0 mg/kg NT-3 for the final month of 3) did not succeed in reversing the loss of NF filament number or total NF protein levels in sural nerve. Studies in the galactose-fed model of diabetes show that treatment with NT-3 or BDNF was able to reverse loss of caliber in the central and peripheral (NT-3 only) projections of sensory neurons but had no effect on caliber loss in motor axons (Mizisin *et al.*, 1997, 1998).

2. *NGF and Calpain Activity*

NGF administration is known to decrease calpain activity, perhaps by an increase in the inhibitor calpastatin, resulting in increased phosphorylation (Guroff, 1993), which may have an untoward effect when superimposed on the background of the aged or diabetic ganglionic milieu. Calpain I (micromolar calpain) may have a specific target in NF side arms, which are important for controlling the density of NF and their proper axonal transport function, and have relevance to the formation of structural aggregates in pathologic settings (Pettus and Povlishock, 1996).

3. *Neurotrophic Substances and Collateral Sprouting*

Collateral intraganglionic sprouting of DRG neurons may develop in response to injury of their peripheral processes as part of diabetic distal symmetrical neuropathy. Sympathetic ganglionic neuroaxonal dystrophy could be the result of a similar sprouting stimulus or develop as the result of the effect of diabetes on the normal process of synaptic turnover, a process that has been designated "synaptic dysplasia" (Schmidt, 1996). Upregulation of a variety of neurotrophic substances in Schwann cells or perineuronal satellite cells has been described in the DRG (Heumann *et al.*, 1987; Banner *et al.*, 1994; Ji *et al.*, 1995; Hammarberg *et al.*, 1996), including, but not limited to, leukemia inhibitory factor (LIF), beta fibroblast growth factor (bFGF), glial cell line-derived neurotrophic factor (GDNF), or NGF. Upregulation of neurotrophic substances may result in the local induction of neuronal sprouting in nearby neurons or as recurrent collaterals derived from the same neuron. For example, BDNF is known to be released by neurons as part of an autocrine loop, which may affect adjacent neurons in the ganglionic milieu (Acheson *et al.*, 1995). The concomitant upregulation of the low-affinity neurotrophin receptor p75^{NTR}, which occurs within DRG satellite cells (Zhou *et al.*, 1996) and Schwann cells (Taniuchi *et al.*, 1986) following axonopathy, may bind and locally present neurotrophins to induce ganglionic sprouting. The demonstration that intraganglionic noradrenergic sprouts develop in the uninjured DRG of transgenic mice overexpressing NGF in the skin likely represents such a neurotrophin-induced sprouting response of sympathetic axons (Davis *et al.*, 1998), a mechanism we have considered in the pathogenesis of sympathetic neuroaxonal dystrophy (see chapter by Schmidt). Although regenerative changes may be linked to the development of neuroaxonal dystrophy, it is unclear how this process is involved in the development of NF containing swellings rather than tubulovesicular elements that characterize axonal growth cones.

VIII. Future Directions

The mechanisms of regulation of synthesis of NF and the process of aberrant phosphorylation of NF in diabetes are poorly understood. Studies should be aimed at determining the ability of high glucose concentrations to directly modulate NF synthesis. These studies would have to be performed *in vitro* and should target the transcription factors known to modulate NF transcript production, namely the NF-1-like transcription factor and the POU domain brn-3 family of factors. The complexity of NF transcript regulation also demands that the effects of high glucose on NF mRNA stability be investigated with a focus on the role of the GTPase exchange factor, p190RhoGEF.

SAPK-mediated phosphorylation of NF requires detailed investigation with a focus on how hyperglycemia and/or lack of insulin leads to the activation of kinases such as ERK and JNK. Of particular relevance is the identification of gene 33 as a rapidly inducible regulator of cdc42, which in turn causes sustained activation of JNK (Makkinje *et al.*, 2000). The mechanism of induction of gene 33 may reflect a general process whereby glucose or insulin can regulate the phenotype of cells under conditions of stress.

The availability of transgenic mice with an altered expression or mutant NF proteins should provide an excellent means of directly studying the role of NF alterations in axonopathy in diabetes. The advent of transgene-mediated expression of NFs specifically in the adult CNS and PNS will allow the impact of developmental changes in NF phenotype to be isolated and set apart from pathogenic mechanisms occurring in the adult.

References

Acheson, A., Conover, J. C., Fandl, J. P., DeChiara, T. M., Russell, M., Thadani, A., Squinto, S. P., Yancopoulos, G. D., and Lindsay, R. M. (1995). A BDNF autocrine loop in adult sensory neurons prevents cell death. *Nature* **374**, 450–453.

Ackerley, S., Grierson, A. J., Brownlees, J., Thornhill, P., Anderton, B. H., Leigh, P. N., Shaw, C. E., and Miller, C. C. (2000). Glutamate slows axonal transport of neurofilaments in transfected neurons. *J. Cell Biol.* **150**, 165–176.

Archer, D. R., Watson, D. F., and Griffin, J. W. (1994). Phosphorylation-dependent immunoreactivity of neurofilaments and the rate of slow axonal transport in the central and peripheral axons of the rat dorsal root ganglion. *J. Neurochem.* **62**, 1119–1125.

Awazu, M., Ishikura, K., Hida, M., and Hoshiya, M. (1999). Mechanisms of mitogen-activated protein kinase activation in experimental diabetes. *J. Am. Soc. Nephrol.* **10**, 738–745.

Banner, L. R., and Patterson, P. H. (1994). Major changes in the expression of the mRNAs for cholinergic differentiation factor/leukemic inhibitory factor and its receptor after injury to adult peripheral nerves and ganglia. *Proc. Natl. Acad. Sci. USA* **91**, 7109–7113.

Bajaj, N. P. S., and Miller, C. C. J. (1997). Phosphorylation of neurofilament heavy-chain side-arm fragments by cyclin-dependent kinase-5 and glycogen synthase kinase-3α in transfected cells. *J. Neurochem.* **69**, 737–743.

Bondar, F. L., and Roots, B. I. (1977). Neurofibrillar changes in goldfish (*Carassius auratus L.*) brain in relation to environmental temperature. *Exp. Brain Res.* **30**, 577–585.

Brady, S. T. (1993). Motor neurons and NFs in sickness and in health. *Cell* **73**, 1–3.

Brownlee, M., Cerami, A., and Vlassara, H. (1988). Advanced glycosylation endproducts in tissue and the biochemical basis of diabetic complications. *N. Engl. J. Med.* **318**, 1315–1321.

Brownlees, J., Yates, A., Bajaj, N. P., Davis, D., Anderton, B. H., Leigh, P. N., Shaw, C. E., and Miller, C. C. J. (2000). Phosphorylation of neurofilament heavy chain side-arms by stress activated protein kinase-1b/Jun N-terminal kinase-3. *J. Cell Sci.* **113**, 401–407.

Canete-Soler, R., Schwartz, M. L., Hua, Y., and Schlaepfer, W. W. (1998). Stability determinants are localized to the 3′-untranslated region and 3'-coding region of the neurofilament light subunit mRNA using a tetracycline-inducible promoter. *J. Biol. Chem.* **273**, 12650–12654.

Canete-Soler, R., Wu, J., Zhai, J., Shamim, M., and Schlaepfer, W. W. (2001). p190RhoGEF binds to a destabilizing element in the 3′ untranslated region of light neurofilament subunit mRNA and alters the stability of the transcript. *J. Biol. Chem.* **276**, 32046–32050.

Carletti, R., Tacconi, S., Bettini, E., and Ferraguti, F. (1995). Stress activated protein kinases, a novel family of mitogen- activated protein kinases, are heterogeneously expressed in the adult rat brain and differentially distributed from extracellular- signal-regulated protein kinases. *Neuroscience* **69**,1103–1110.

Cheng, H.-L., and Feldman, E. L. (1998). Bidirectional regulation of p38 kinase and c-Jun N-terminal protein kinase by insulin-like growth factor-I. *J. Biol. Chem.* **273**, 14560–14565.

Chou, S. M., Han, C. Y., Wang, H. S., Vlassara, H., and Bucala, R. (1999). A receptor for advanced glycosylation endproducts (AGEs) is colocalized with NF-bound AGEs and SOD1 in motoneurons of ALS: Immunohistochemical study. *J. Neurol. Sci.* **169**, 87–92.

Chou, S. M., Taniguchi, A., Wang, H. S., and Festoff, B. W. (1998). Serpin = serine protease-like complexes within NF conglomerates of motoneurons in amyotrophic lateral sclerosis. *J. Neurol. Sci.* **160**, S73–S79.

Chou, S. M., Wang, H. S., and Taniguchi, A. (1996). Role of SOD-1 and nitric oxide/cyclic GMP cascade on NF aggregation in ALS/MND. *J. Neurol. Sci.* **139**, 16–26.

Cohen, P. (1999). The Croonian Lecture 1998. Identification of a protein kinase cascade of major importance in insulin signal transduction. *Philos. Trans. R. Soc. Lond. B Biol. Sci.* **354**, 485–495.

Copray, J. C., Mantingh, I., Brouwer, N., Biber, K., Kust, B. M., Liem, R. S., Huitinga, I., Tilders, F. J., Van Dam, A. M., and Boddeke, H. W. (2001). Expression of interleukin-1 beta in rat dorsal root ganglia. *J. Neuroimmunol.* **118**, 203–211.

Cote, F., Collard, J.-F., and Julien, J.-P. (1993). Progressive neuronopathy in transgenic mice expressing the human NF heavy gene: A mouse model of amyotrophic lateral sclerosis. *Cell* **73**, 35–46.

Cross, D. A., Culbert, A. A., Chalmers, K. A., Facci, L., Skaper, S. D., and Reith, A. D. (2001). Selective small-molecule inhibitors of glycogen synthase kinase-3 activity protect primary neurones from death. *J. Neurochem.* **77**, 94–102.

Davis, B. M., Goodness, T. P., Soria, A., and Albers, K. M. (1998). Over-expression of NGF in skin causes formation of novel sympathetic projections to trkA-positive sensory neurons. *NeuroReport* **9**, 1103–1107.

Denlinger, R. H., Anthony, D. C., Amarnath, V., and Graham, D. G. (1992). Comparison of location, severity and dose response of proximal axonal lesions induced by 3,3′-iminodipropionitrile and deuterium substituted analogs. *J. Neuropathol. Exp. Neurol.* **51**, 569–576.

Deshmukh, M., and Johnson, E. M., Jr. (1997). Programmed cell death in neurons: Focus on the pathway of nerve growth factor deprivation-induced death of sympathetic neurons. *Mol. Pharmacol.* **51**, 897–906.

De Waegh, S. M., Lee, V. M. Y., and Brady, S. T. (1992). Local modulation of neurofilament phosphorylation, axonal caliber, and slow axonal transport by myelinating Schwann cells. *Cell* **68**, 451–463.

Diabetes Control and Complications Trial Research Group. (1993). The effect of intensive treatment of diabetes on the development and progression of long-term complications in insulin-dependent diabetes mellitus. *N. Engl. J. Med.* **329**, 977–986.

Fernyhough, P., Gallagher, A., Averill, S. A., Priestley, J. V., Hounsom, L., Patel, J., and Tomlinson, D. R. (1999). Aberrant NF phosphorylation in sensory neurons of rats with diabetic neuropathy. *Diabetes* **48**, 881–889.

Galer, B. S. (1995). Neuropathic pain of peripheral origin: Advances in pharmacologic treatment. *Neurology* **45**(Suppl. 9), S17–S25.

Gao, W.-Q., Dybdal, N., Shinsky, N., Murnane, A., Schmelzer, C., Siegel, M., Keller, G., Hefti, F., Phillips, H. S., and Winslow, J. W. (1995). Neurotrophin-3 reverses experimental cisplatin-induced peripheral sensory neuropathy. *Ann. Neurol.* **38**, 30–37.

Giasson, B. I., and Mushynski, W. E. (1997). Study of proline-directed protein kinases involved in phosphorylation of the heavy neurofilament subunit. *J. Neurosci.* **17**, 9466–9472.

Gibb, B. K., Brion, J. P., Brownlees, J., Anderton, B. H., and Miller, C. C. (1998). Neuropathological abnormalities in transgenic mice harbouring a phosphorylation mutant NF transgene. *J. Neurochem.* **70**, 492–500.

Gold, B. G., and Austin, D. R. (1991). Regulation of aberrant NF phosphorylation in neuronal perikarya. III. Alterations following single and continuous β,β'-iminodipropionitrile administrations. *Brain Res.* **563**, 151–162.

Gold, B. G., Mobley, W. C., and Matheson, S. F. (1991). Regulation of axonal caliber, neurofilament content, and nuclear localization in mature sensory neurons by nerve growth factor. *J. Neurosci.* **11**, 943–955.

Gold, B. G., Price, D. L., Griffin, J. W., Rosenfeld, J., Hoffman, P. N., Sternberger, N. H., and Sternberger, L. A. (1988). NF antigens in acrylamide neuropathy. *J. Neuropathol. Exp. Neurol.* **47**, 145–157.

Goldstein, M. E., Cooper, H. S., Bruce, J., Carden, M. J., Lee, V. M., and Schlaepfer, W. W. (1987). Phosphorylation of neurofilament proteins and chromatolysis following transection of rat sciatic nerve. *J. Neurosci.* **7**, 1586–1594.

Graham, D. G., Szakal-Quin, G., Priest, J. W., and Anthony, D. C. (1984). *In vitro* evidence that covalent crosslinking of NFs occurs in gamma-diketone neuropathy. *Proc. Natl. Acad. Sci. USA* **81**, 4979–4982.

Grant, P., and Pant, H. C. (2000). Neurofilament protein synthesis and phosphorylation. *J. Neurocytol.* **29**, 843–872.

Grant, P., Sharma, P., and Pant, H. C. (2001). Cyclin-dependent protein kinase 5 (Cdk5) and the regulation of neurofilament metabolism. *Eur. J. Biochem.* **268**, 1534–1546.

Griffin, J. W., Hoffman, P. N., Clark, A. W., Carroll, P. T., and Price, D. L. (1978). Slow axonal transport of NF proteins: Impairment by β,β'-iminodipropionitrile administration. *Science* **202**, 633–635.

Guidato, S., Tsai, L. H., Woodgett, J., and Miller, C. C. (1996). Differential cellular phosphorylation of neurofilament heavy side-arms by glycogen synthase kinase-3 and cyclin-dependent kinase-5. *J. Neurochem.* **66**, 1698–1706.

Guo, L., Degenstein, L., Dowling, J., Yu, Q.-C., Wollmann, R., Perman, B., and Fuchs, E. (1995). Gene targeting of BPAG1: Abnormalities in mechanical strength and cell migration in stratified epithelium and neurologic degeneration. *Cell* **81**, 233–243.

Guroff, G. (1993). Nerve growth factor as a neurotrophic agent. *Ann. N.Y. Acad. Sci.* **692**, 51–59.

Hammarberg, H., Piehl, F., Cullhellm, S., Fjell, J., Hokfelt, T., and Fried, K. (1996). GDNF mRNA in Schwann cells and DRG satellite cells after chronic sciatic nerve injury. *NeuroReport* **7**, 857–860.

Haneda, M., Araki, S., Togawa, M., Sugimoto, T., Isono, M., and Kikkawa, R. (1997). Mitogen-activated protein kinase cascade is activated in glomeruli of diabetic rats and glomerular mesangial cells cultured under high glucose conditions. *Diabetes* **46**, 847–853.

Heumann, R., Korshing, S., Bandtlow, C., and Thoenen, H. (1987). Changes of nerve growth factor synthesis in nonneuronal cells in response to sciatic nerve transection. *J. Cell Biol.* **104**, 1623–1632.

Hirokawa, N., and Takeda, S. (1998). Gene targeting studies begin to reveal the function of neurofilament proteins. *J. Cell Biol.* **143**, 1–4.

Hirokawa, N., Terada, S., Funakoshi, T., and Takeda, S. (1997). Slow axonal transport: the subunit transport model. *Trends Cell Biol.* **7**, 384–388.

Hoffman, P. N., Cleveland, D. W., Griffin, J. W., Landes, P. W., Cowan, N. J., and Price, D. L. (1987). Neurofilament gene expression: A major determinant of axonal caliber. *Proc. Natl. Acad. Sci. USA* **84**, 3472–3476.

Hoffman, P. N., and Lasek, R. J. (1975). The slow component of axonal transport. Identification of major structural polypeptides of the axon and their generality among mammalian neurons. *J. Cell Biol.* **66**, 351–366.

Hsieh, S. T., Kidd, G. J., Crawford, T. O., Xu, Z., Lin, W. M., Trapp, B. D., Cleveland, D. W., and Griffin, J. W. (1994). Regional modulation of neurofilament organization by myelination in normal axons. *J. Neurosci.* **14**, 6392–6401.

Igarashi, M., Wakasaki, H., Takahara, N., Ishii, H., Jiang, Z. Y., Yamauchi, T., Kuboki, K., Meier, M., Rhodes, C. J., and King, G. L. (1999). Glucose or diabetes activates p38 mitogen-activated protein kinase via different pathways. *J. Clin. Invest.* **103**, 185–195.

Ji, R.-R., Zhange, Q., Zhang, X., Piehl, F., Reilly, T., Pettersson, R. F., and Hokfelt, T. (1995). Prominent expression of bFGF in dorsal root ganglia after axotomy. *Eur. J. Neurosci.* **7**, 2458–2468.

Julien, J.-P. (1999). Neurofilament functions in health and disease. *Curr. Opin. Neurobiol.* **9**, 554–560.

Julien, J. P., and Mushynski, W. E. (1998). Neurofilaments in health and disease. *Prog. Nucleic Acid Res. Mol. Biol.* **61**, 1–23.

Kennedy, W. R., and Wendelschafer-Crabb, G. (1999). Utility of the skin biopsy method in studies of diabetic neuropathy. *Electroencephalogr. Clin. Neurophysiol.* **50**, 553–559.

Kenney, A. M., and Kocsis, J. D. (1998). Peripheral axotomy induces long-term c-Jun amino-terminal-kinase-1 activation and activator protein-1 binding activity by c-Jun and junD in adult rat dorsal root ganglia *in vivo. J. Neurosci.* **18**, 1318–1328.

King, R.H.M., Sarsilmaz, M., Thomas, P. K., Jacobs, J. M., Muddle, J. R., and Duncan, I. D. (1993). Axonal neurofilamentous accumulations: A comparison between human and canine giant axonal neuropathy and 2,5-HD neuropathy. *Neuropathol. Appl. Neurobiol.* **19**, 224–232.

Kuan, C. Y., Yang, D. D., Roy, D. R. S., Davis, R. J., Rakic, P., and Flavell, R. A. (1999). The Jnk1 and Jnk2 protein kinases are required for regional specific apoptosis during early brain development. *Neuron* **22**, 667–676.

Kultz, D., Garcia-Perez, A., Ferraris, J. D., and Burg, M. B. (1997). Distinct regulation of osmoprotective genes in yeast and mammals. Aldose reductase osmotic response element is induced independent of p38 and stress-activated protein kinase/Jun N-terminal kinase in rabbit kidney cells. *J. Biol. Chem.* **272**, 13165–13170.

Lauria, G., McArthur, J. C., Hauer, P. E., John W., Griffin, J. W., and Cornblath, D. R. (1998). Neuropathological alterations in diabetic truncal neuropathy: Evaluation by skin biopsy. *J. Neurol. Neurosurg. Psychiatry* **65**, 762–766

Lee, M. K., Xu, Z., Wong, P. C., and Cleveland, D. W. (1993). Neurofilaments are obligate heteropolymers *in vivo*. *J. Cell Biol.* **122**, 1337–1350.

Lee, V., Trojanowski, J. Q., and Schlaepfer, W. W. (1982). Induction of neurofilament triplet proteins in PC12 cells by nerve growth factor. *Brain Res.* **238**, 169–180.

Lewis, S. E., and Nixon, R. A. (1988). Multiple phosphorylated variants of the high molecular mass subunit of neurofilaments in axons of retinal cell neurons: Characterization and evidence for their differential association with stationary and moving neurofilaments. *J. Cell Biol.* **107**, 2689–2701.

Li, B. S., Daniels, M. P., and Pant, H. C. (2001). Integrins stimulate phosphorylation of neurofilament NF-M subunit KSP repeats through activation of extracellular regulated-kinases (Erk1/Erk2) in cultured motoneurons and transfected NIH 3T3 cells. *J. Neurochem.* **76**, 703–710.

Li, B.-S., Zhang, L., Gu, J., Amin, N. D., and Pant, H. C. (2000). Integrin $\alpha_1\beta_1$-mediated activation of cyclin-dependent kinase 5 activity is involved in neurite outgrowth and human neurofilament protein H Lys-Ser-Pro tail domain phosphorylation. *J. Neurosci.* **20**, 6055–6062.

Lindenbaum, M. H., Carbonetto, S., Grosveld, F., Flavell, D., and Mushynski, W. E. (1988). Transcriptional and post-transcriptional effects of nerve growth factor on expression of the three neurofilament subunits in PC-12 cells. *J. Biol. Chem.* **263**, 5662–5667.

Liuzzi, F. J., Bufton, S. M., and Vinik, A. I. (1998). Streptozotocin-induced diabetes mellitus causes changes in primary sensory neuronal cytoskeletal mRNA levels that mimic those caused by axotomy. *Exp. Neurol.* **154**, 381–388.

Liuzzi, F. J. (1990). Proteolysis is a critical step in the physiological stop pathway: Mechanisms involved in the blockade of axonal regeneration by mammalian astrocytes. *Brain Res.* **512**, 277–283.

Macioce, P., Filliatreau, G., Figliomeni, B., Hassig, R., Thiéry, J., and Di Giamberardino, L. (1989). Slow axonal transport impairment of cytoskeletal proteins in streptozocin-induced diabetic neuropathy. *J. Neurochem.* **53**, 1261–1267.

Makkinje, A., Quinn, D. A., Chen, A., Cadilla, C. L., Force, T., Bonventre, J. V., and Kyriakis, J. M. (2000). Gene 33/Mig-6, a transcriptionally inducible adapter protein that binds GTP-Cdc42 and activates SAPK/JNK: A potential marker transcript for chronic pathologic conditions, such as diabetic nephropathy. *J. Biol. Chem.* **275**, 17838–17847.

Marszalek, J. R., Williamson, T. L., Lee, M. K., Xu, Z. S., Hoffman, P. N., Becher, M. W., Crawford, T. O., and Cleveland, D. W. (1996). Neurofilament subunit NF-H modulates axonal diameter by selectively slowing neurofilament transport. *J. Cell Biol.* **135**, 711–724.

Mattson, M. P. (2001). Neuronal death and GSK-3beta: A tau fetish? *Trends Neurosci.* **24**, 255–256.

McLachlan, E. M. and Hu, P. (1998). Axonal sprouts containing calcitonin gene-related peptide and substance P form pericellular baskets around large diameter neurons after sciatic nerve transection in the rat. *Neuroscience* **84**, 961–965.

McLachlan, E. M., Janig, W., Devor, M., and Michaelis, M. (1993). Peripheral nerve injury triggers noradrenergic sprouting within dorsal root ganglia. *Nature* **363**, 543–546.

Medori, R., Autilio-Gambetti, L., Jenich, H., and Gambetti, P. (1988a). Changes in axon size and slow axonal transport are related in experimental diabetic neuropathy. *Neurology* **38**, 597–601.

Medori, R., Jenich, H., Autilio-Gambetti, L., and Gambetti, P. (1988b). Experimental diabetic neuropathy: Similar changes of slow axonal transport and axonal size in different animal models. *J. Neurosci.* **8**, 1814–1821.

Mizisin, A. P., Bache, M., DiStefano, P. S., Acheson, A., Lindsay, R. M., and Calcutt N. A. (1997). BDNF attenuates functional and structural disorders in nerves of galactose-fed rats. *J. Neuropathol. Exp. Neurol.* **56**, 1290–1301.

Mizisin, A. P., Calcutt, N. A., Tomlinson, D. R., Gallagher, A., and Fernyhough, P. (1999). Neurotrophin-3 reverses nerve conduction velocity deficits in streptozotocin-diabetic rats. *J. Periph. Nerv. Sys.* **4**, 211–221.

Mizisin, A. P., Kalichman, M. W., Bache, M., Dines, K. C., and DiStefano, P. S. (1998). NT-3 attenuates functional and structural disorders in sensory nerves of galactose-fed rats. *J. Neuropathol. Exp. Neurol.* **57**, 803–813.

Mohiuddin, L., Fernyhough, P., and Tomlinson, D. R. (1995). Reduced levels of mRNA encoding endoskeletal and growth-associated proteins in sensory ganglia in experimental diabetes mellitus. *Diabetes* **44**, 25–30.

Mohiuddin, L., and Tomlinson, D. R. (1997). Impaired molecular regenerative responses in sensory neurones of diabetic rats: Gene expression changes in dorsal root ganglia after sciatic nerve crush. *Diabetes* **46**, 2057–2062.

Muma, N. A., and Cork, L. C. (1993). Alterations in NF mRNA in hereditary canine spinal muscular atrophy. *Lab. Invest.* **69**, 436–442.

Murphy, P. G., Grondin, J., Altares, M., and Richardson, P. M. (1995). Induction of interleukin-6 in axotomized sensory neurons. *J. Neurosci.* **15**, 5130–5138.

Nakagawa, T., Chen, J., Zhang, Z., Kanai, Y., and Hirokawa, N. (1995). Two distinct functions of the carboxyl-terminal tail domain of NF-M upon neurofilament assembly: Cross-bridge formation and longitudinal elongation of filaments. *J. Cell Biol.* **129**, 411–429.

Nixon, R. A. (1993). The regulation of neurofilament protein dynamics by phosphorylation: Clues to neurofibrillary pathobiology. *Brain Pathol.* **3**, 29–38.

Nixon, R. A., Clarke, J. F., Logvinenko, K. B., Tan, M.K.H., Hoult, M., and Grynspan, F. (1990). Aluminium inhibits calpain-mediated proteolysis and induces human NF proteins to form protease-resistant high molecular weight complexes. *J. Neurochem.* **55**, 1950–1959.

Nixon, R. A., Paskevich, P. A., Sihag, R. K., and Thayer, C. Y. (1994). Phosphorylation on carboxyl terminus domains of neurofilament proteins in retinal ganglion cell neurons *in vivo*: Influences on regional neurofilament accumulation, interneurofilament spacing, and axon caliber. *J. Cell Biol.* **126**, 1031–1046.

Nukada, H., Dyck, P. J., Low, P. A., Lais, A. C., and Sparks, M. F. (1986). Axonal caliber and neurofilaments are proportionately decreased in galactose neuropathy. *J. Neuropathol. Exp. Neurol.* **45**, 140–150.

Okonkwo, D. O., Pettus, E. H., Moroi, J., and Povlishock, J. T. (1998). Alteration of the NF sidearm and its relation to NF compaction occurring with traumatic axonal injury. *Brain Res.* **784**, 1–6.

Pant, H. C. (1988). Dephosphorylation of NF proteins enhances their susceptibility to degradation by calpain. *Biochem. J.* **256**, 665–668.

Parysek, L. M., McReynolds, M. A., Goldman, R. D., and Ley, C. A. (1991). Some neural intermediate filaments contain both peripherin and the NF proteins. *J. Neurosci. Res.* **30**, 80–91.

Pekiner, C., and McLean, W. G. (1991). NF protein phosphorylation in spinal cord of experimentally diabetic rats. *J. Neurochem* **56**, 1362–1367.

Pettus, E. H., Christman, C. W., Giebel, M. I., and Povlishock, J. T. (1994). Traumatically induced altered membrane permeability: Its relationship to traumatically induced reactive axonal change. *J. Neurotrauma* **11**, 507–522.

Pettus, E. H., and Povlishock, J. T. (1996). Characterization of a distinct set of intraaxonal ultrastructural changes associated with traumatically induced alteration in axolemmal permeability. *Brain Res.* **722**, 1–11.

Properzi, G., Francavilla, S., and Liu, Y.-F. (1993). Early increase precedes a depletion of VIP and PGP 9.5 in the skin of insulin dependent diabetics: Correlation between quantitative immunohistochemistry and clinical assessment of peripheral neuropathy. *J. Pathol.* **169**, 269–277.

Purves, T., Middlemas, A., Agthong, S., Jude, E. B., Boulton, A., Fernyhough, P., and Tomlinson, D. R. (2001). A role for mitogen-activated protein kinases in the aetiology of diabetic neuropathy. *FASEB J.* **15**, 2508–2514.

Rao, M. V., Houseweart, M. K., Williamson, T. L., Crawford, T. O., Folmer, J., and Cleveland, D. W. (1998). Neurofilament-dependent radial growth of motor axons and axonal organization of neurofilaments does not require the neurofilament heavy subunit (NF-H) or its phosphorylation. *J. Cell Biol.* **143**, 171–181.

Ratner, N., Bloom, G. S., and Brady, S. T. (1998). A role for cyclin-dependent kinase(s) in the modulation of fast anterograde axonal transport: effects defined by olomoucine and the APC tumor suppressor protein. *J. Neurosci.* **18**, 7717–7726.

Richardson, P. M. (1991). Neurotrophic factors in regeneration. *Curr. Opin. Neurobiol.* **1**, 401–406.

Roots, B. I. (1983). NF accumulation induced in synapses by leupeptin. *Science* **221**, 971–972.

Rotshenker, S., Aamar, S., and Barak, V. (1992). Interleukin-1 activity in lesioned peripheral nerve. *J. Neuroimmunol.* **39**, 75–80.

Roy, S., Coffee, P., Smith, G., Liem, R. K. H., Brady, S. T., and Black, M. M. (2000). Neurofilaments are transported rapidly but intermittently in axons: Implications for slow axonal transport. *J. Neurosci.* **20**, 6849–6861.

Ryle, C., Leow, C. K., and Donaghy, M. (1997). Nonenzymatic glycation of peripheral and central nervous system proteins in experimental diabetes mellitus. *Muscle Nerve* **20**, 577–584.

Sasaki, H., Schmelzer, J. D., Zollman, P. J., and Low, P. A. (1997). Neuropathology and blood flow of nerve, spinal roots and dorsal root ganglia in longstanding diabetic rats. *Acta Neuropathol.* **93**, 118–128.

Schlaepfer, W. W. (1987). NFs: Structure, metabolism and implications in disease. *J. Neuropathol. Exp. Neurol.* **46**, 117–129.

Schmidt, R. E. (1996). Synaptic dysplasia in sympathetic autonomic ganglia. *J. Neurocytol.* **25**, 777–791.

Schmidt, R. E., Beaudet, L. N., Plurad, S. B., and Roth, K. A. (1997). Axonal cytoskeletal pathology in aged and diabetic human sympathetic autonomic ganglia. *Brain Res.* **769**, 375–383.

Schmidt, R. E., Dorsey, D. A., Parvin, C. A., Beaudet, L. N., Plurad, S. B., and Roth, K. A. (1997). Dystrophic axonal swellings develop as a function of age and diabetes in human dorsal root ganglia. *J. Neuropathol. Exp. Neurol.* **56**, 1028–1043.

Schmidt, R. E., Plurad, S. B., and Modert, C. W. (1983). Neuroaxonal dystrophy in the autonomic ganglia of aged rats. *J. Neuropathol. Exp. Neurol.* **42**, 376–390.

Schmidt, R. E., Plurad, S. B., Parvin, C. A., and Roth, K. A. (1993). Effect of diabetes and aging on human sympathetic autonomic ganglia. *Am. J. Pathol.* **136**, 1327–1338.

Schmidt, R. E., Plurad, S. B., Sherman, W. R., Williamson, J. R., and Tilton, R. G. (1989). Effects of aldose reductase inhibitor sorbinil on neuroaxonal dystrophy and levels of myo-inositol and sorbitol in sympathetic autonomic ganglia of streptozocin-induced diabetic rats. *Diabetes* **38**, 569–579.

Schmidt, R. E., and Scharp, D. W. (1982). Axonal dystrophy in experimental diabetic autonomic neuropathy. *Diabetes* **31**, 761–770.

Schmidt, R. E., Spencer, S. A., Coleman, B. D., and Roth, K. A. (1991). Immunohistochemical localization of GAP-43 in rat and human sympathetic nervous system: Effects of aging and diabetes. *Brain Res.* **552**, 190–197.

Schroer, J. A., Plurad, S. B., and Schmidt, R. E. (1992). Fine structure of presynaptic axonal terminals in sympathetic autonomic ganglia of aging and diabetic human subjects. *Synapse* **12**, 1–13.

Schwartz, M. L., Hua, Y., and Schlaepfer, W. W. (1997). *In vitro* activation of the mouse mid-sized neurofilament gene by an NF-1-like transcription factor. *Mol. Brain Res.* **48**, 305–314.

Schwartz, M. L., Shneidman, P. S., Bruce, J., and Schlaepfer W. W. (1992). Actinomycin prevents the destabilization of neurofilament mRNA in primary sensory neurons. *J. Biol. Chem.* **267**, 24596–24600.

Scott, J. N., Clark, A. W., and Zochodne, D. W. (1999). NF and tubulin gene expression in progressive experimental diabetes: failure of synthesis and export by sensory neurons. *Brain* **122**, 2109–2118.

Shaw, G., Wialski, D., and Reier, P. (1988). The effect of axotomy and deafferentation on phosphorylation dependent antigenicity of NFs in rat superior cervical ganglion neurons. *Brain Res.* **460**, 227–234.

Shubayev, V. I., and Myers, R. R. (2001). Axonal transport of TNF-alpha in painful neuropathy: distribution of ligand tracer and TNF receptors. *J. Neuroimmunol.* **114**, 48–56.

Sidenius, P., and Jakobsen, J. (1980). Reduced perikaryal volume of lower motor and primary sensory neurons in early experimental diabetes. *Diabetes* **29**, 182–186.

Sima, A. A. F., Lorusso, A. C., and Thibert, P. (1982). Distal symmetric polyneuropathy in the spontaneously diabetic BB Wistar rat: An ultrastructural and teased fiber study. *Acta Neuropathol.* **58**, 39–47.

Sima, A. A. F., and Robertson, D. M. (1979). Peripheral neuropathy in the mutant mouse [C57/BL/KS(db/db)]: An ultrastructural study. *Lab. Invest.* **40**, 627–632.

Sima, A. A. F., and Yagihashi, S. (1986). Central-peripheral distal axonopathy in the spontaneously diabetic BB rat: Ultrastructural and morphometric findings. *Diabet. Res. Clin. Pract.* **1**, 289–298.

Simm, A., Munch, G., Seif, F., Schenk, O., Heidland, A., Richter, H., Vamvakas, S., and Schinzel, R. (1997). Advanced glycation endproducts stimulate the MAP-kinase pathway in tubulus cell line LLC-PK1. *FEBS Lett.* **410**, 481–484.

Smith, M. D., Morris, P. J., Dawson, S. J., Schwartz, M. L., Schlaepfer, W. W., and Latchman, D. S. (1997). Coordinate induction of the three neurofilament genes by the Brn-3a transcription factor. *J Biol Chem.* **272**, 21325–21333.

Taniuchi, M., Clark, H. B., and Johnson, E. M., Jr. (1986). Induction of nerve growth factor receptor in Schwann cells after axotomy. *Proc. Natl. Acad. Sci. USA* **83**, 4094–4098.

Terada, M., Yasuda, H., and Kikkawa, R. (1998). Delayed Wallerian degeneration and increased NF phosphorylation in sciatic nerves of rats with streptozotocin-induced diabetes. *J. Neurol. Sci.* **155**, 23–30.

Thomas, P. K., and Tomlinson, D. R. (1992). Diabetic and hypoglycaemic neuropathy. *In* "Peripheral Neuropathy" (P. J. Dyck, P. K. Thomas, J. W. Griffin, P. A. Low, and J. F. Poduslo, eds.), pp. 1219–1250. Saunders, Philadelphia.

Treisman, R. (1985). Transient accumulation of c-fos RNA following serum stimulation requires a conserved 5′ element and c-fos 3′ sequences. *Cell* **42**, 889–902.

Troncoso, J. C., Hoffman, P. N., Griffin, J. W., Hess-Kozlow, K. M., and Price, D. L. (1985). Aluminum intoxication: A disorder of NF transport in motoneurons. *Brain Res.* **342**, 172–175.

Veeranna, Amin, N. D., Ahn, N. G., Jaffe, H., Winters, C. A., Grant, P., and Pant, H. C. (1998). Mitogen-activated protein kinases (Erk1,2) phosphorylate Lys-Ser-Pro (KSP) repeats in neurofilament proteins NF-H and NF-M. *J. Neurosci.* **18**, 4008–4021.

Verge, V. M. K., Tetzlaff, W., Bisby, M. A., and Richardson P. M. (1990). Influence of nerve growth factor on neurofilament gene expression in mature primary sensory neurons. *J. Neurosci.* **10**, 2018–2025.

Vickers, J. C., Costa, M., Vitadello, M., Dahl, D., and Marotta, C. A. (1990). NF protein triplet immunoreactivity in distinct subpopulations of peptide-containing neurons in the guinea pig coeliac ganglion. *Neuroscience* **39**, 743–759.

Vinik, A. I., Park, T. S., Stansberry, K. B., and Pittenger, G. L. (2000). Diabetic neuropathies. *Diabetologia* **43**, 957–973.

Vlassara, H., Bucala, R., and Striker, L. (1994). Pathogenetic effects of advanced glycosylation: Biochemical, biologic and clinical implications for diabetes and aging. *Lab. Invest.* **70**, 138–151.

Wang, C., Li Y., Wible, B., Angelides, K. J., and Ishii, D. N. (1992). Effects of insulin and insulin-like growth factors on neurofilament mRNA and tubulin mRNA content in human neuroblastoma SH-SY5Y cells. *Mol. Brain Res.* **13**, 289–300.

Wang, L., Ho, C-L., Sun, D., Liem, R. K. H., and Brown, A. (2000). Rapid movements of neurofilaments interrupted by prolonged pauses. *Nature Cell Biol.* **2**, 137–141.

Watson, D. F., Glass, J. D., and Griffin, J. W. (1993). Redistribution of cytoskeletal proteins in mammalian axons disconnected from their cell bodies. *J. Neurosci.* **13**, 4354–4360.

Whitmarsh, A. J., and Davis, R. J. (1996). Transcription factor AP-1 regulation by mitogen-activated protein kinase signal transduction pathways. *J. Mol. Med.* **74**, 589–607.

Wong, P. C., Marszalek, J., Crawford, T. O., Xu, Z. S., Hsieh, S. T., Griffin, J. W., and Cleveland, D. W. (1995). Increasing neurofilament subunit NF-M expression reduces axonal NF-H, inhibits radial growth, and results in neurofilamentous accumulation in motor neurons. *J. Cell Biol.* **130**, 1413–1422.

Xia, Z., Dickens, M., Raingeaud, J., Davis, R. J., and Greenberg, M. E. (1995). Opposing effects of ERK and JNK-p38 MAP kinases on apoptosis. *Science* **270**, 1326–1331.

Xu, X., Raber, J., Yang, D., Su, B., and Mucke, L. (1997). Dynamic regulation of c-Jun N-terminal kinase activity in mouse brain by environmental stimuli. *Proc. Natl. Acad. Sci. USA* **94**, 12655–12660.

Xu, Z. S., Marszalek, J. R., Lee, M. K., Wong, P. C., Folmer, J., Crawford, T. O., Hsieh, S. T., Griffin, J. W., and Cleveland, D. W. (1996). Subunit composition of neurofilaments specifies axonal diameter. *J. Cell Biol.* **133**, 1061–1069.

Yagihashi, S. (1997). Nerve structural defects in diabetic neuropathy: Do animals exhibit similar changes? *Neurosci. Res. Commun.* **21**, 25–32.

Yagihashi, S., Kamijo, M., and Watanabe, K. (1990). Reduced myelinated fiber size correlates with loss of axonal neurofilaments in peripheral nerve of chronically streptozotocin diabetic rats. *Am. J. Pathol.* **136**, 1365–1373.

Yagihashi, S., and Sima, A.A.F. (1986). Neuroaxonal and dendritic dystrophy in diabetic autonomic neuropathy: Classification and topographic distribution in the BB-rat. *J. Neuropathol. Exp. Neurol.* **45**, 545–565.

Yang, D. D., Kuan, C. Y., Whitmarsh, A. J., Rincon, M., Zheng, T. S., Davis, R. J., Rakic, P., and Flavell, R. A. (1997). Absence of excitotoxicity-induced apoptosis in the hippocampus of mice lacking the Jnk3 gene. *Nature* **389**, 865–870.

Zhou, X.-F., Rush, R. A., and McLachlan, E. M. (1996). Differential expression of the p75 nerve growth factor receptor in glia and neurons of the rat dorsal root ganglia after peripheral nerve transection. *J. Neurosci.* **16**, 2901–2911.

APOPTOSIS IN DIABETIC NEUROPATHY

Aviva Tolkovsky

Department of Biochemistry, University of Cambridge,
Cambridge CB2 1QW, United Kingdom

It has been proposed that apoptotic death of some sensory neurons and Schwann cells occurs in, and may be causal for, diabetic neuropathy (DN). Some tantalizing but incomplete evidence for this has emerged from studies of rat models of diabetes and *in vitro* studies of sensory and sympathetic neurons, and Schwann cells exposed to very high concentrations of glucose. This article reviews the evidence and suggests that most studies to date are far from being conclusive. Hence there is room for proper studies of apoptosis in DN, as such studies may reveal hitherto unexplored drug targets that may improve management of the disease. © 2002, Elsevier Science (USA).

I. Introduction

Diabetic neuropathy (DN) is characterized by a heterogeneous range of abnormalities affecting proximal and distal peripheral sensory and motor nerves, as well as some parts of the autonomic nervous system. A hallmark is peripheral nerve dysfunction, which can be clearly measured as a deficit in nerve conduction (see chapter by Arezzo). Underlying this dysfunction there is clear evidence for axonal atrophy/loss and demyelination/remyelination. Biochemically, these differences have been shown to be preceded by metabolic impairment, increases in flux through aldose reductase, oxidative stress, impaired neurovascular blood flow, and impaired traffic of components of axoplasm by both retrograde and anterograde axonal transport and synthesis of growth factors such as nerve growth factor (NGF) (for reviews, see Vinik *et al.*, 2000; Thomas and Tomlinson, 1992). As is described by Arezzo, there is an early conduction deficit in both patients with badly controlled diabetes and in diabetic rats. This early deficit is clearly metabolic in origin, is without potentially causative structural abnormalities, and is reversible by improvement of glycemic control (Gregersen, 1968; Greene *et al.*, 1975). As the condition develops, so the conduction velocity deficit worsens and acquires a clear structural component. Clearly, basic causal elements that might be responsible could include either or both axonal impairment leading to distal shrinkage and collapse and/or Schwann cell disease leading to demyelination. Indeed, there is evidence for both of these components, to an extent that varies between patients. In the case on nonmyelinated fibers, the contribution of Schwann cell disease is still possible, but the mechanism remains covert (Thomas and Tomlinson, 1992).

Among the observations that have been suggested to account for the loss of neuronal function has been the simplistic suggestion of neuronal and Schwann cell apoptotic cell death (Russell *et al.*, 1999). This subject has received scant attention in most of the major reviews on DN, although loss of neurons and/or Schwann cells, if it were shown to occur, would have profound implications for understanding the causes of DN and developing appropriate therapies. One of the main problems in studying this question is that apoptotic cells are thought to be cleared very rapidly by professional phagocytes and neighboring cells (Savill and Fadok, 2000). Hence finding/documenting cells undergoing apoptosis in a chronic disease is not straightforward, even in diseases where profound cell loss is observed. As an extreme example, even in cases where optic nerve myelination during development involves apoptotic death of 50% of all of the estimated 50,000 newborn cells during every 24 h, only a few apoptotic profiles can be observed at any time (Barres and Raff, 1994). If apoptosis does contribute to the pathology of DN, several questions must be addressed. What are its causes? Why does not the process destroy all the affected neurons/Schwann cells? Is it likely that inhibition of apoptosis will ameliorate some of the pathology associated with DN? This review aims to summarize the current studies relating to the question whether cell death/apoptosis is a contributing factor to the pathology observed in DN.

II. Apoptotic Pathways

Before reviewing the literature pertaining to apoptosis in DN, it is worth reviewing briefly the current view of the apoptotic process so as to highlight several potential markers that can be used to diagnose apoptosis. For further information, readers are referred to Hengartner (2000); and Yuan and Yankner (2000).

A. DEATH RECEPTOR PATHWAY

The are two major pathways through which apoptosis occurs in vertebrate cells. One pathway is initiated by death receptors of the Fas/TNFa family. Upon ligand binding, these trimeric receptors recruit a set of death domain-interacting proteins, which form the death-inducing signaling complex (DISC). The DISC recruits procaspases 8/10, which then undergo cleavage and activation (Kruidering and Evan, 2000). The activation process occurs possibly because of the close proximity between the procaspase and/or due to a conformational change induced by binding to the death-effector domains of DISC proteins. Caspases 8/10 can in turn cleave and thereby activate effector caspases such as caspases 3/6/7, which cleave the numerous substrates that promote the orderly disintegration of the cell. One of the hallmarks of apoptosis is the condensation and fragmentation of DNA. This condensation is mediated by the cleavage and activation of the protein Acinus by caspase 3 (Sahara *et al.*, 1999), while fragmentation is caused by the release of DNase known as CAD/DFF40 from its inhibitory subunit ICAD/DFF45 by the latter's cleavage by caspase 3 (Enari *et al.*, 1998; Liu *et al.*, 1997).

The DR family currently consists of at least seven members, not all of which are well characterized in terms of ligands and proteins recruited to the DISC. Allied to the DR family is the p75 low-affinity neurotrophin receptor (p75^{NTR}), which is highly expressed on Schwann cells and peripheral neurons. p75^{NTR} shares some homology in its intracellular domain with the death domain of the DR family (Barrett 2000).

Binding of ligands to death receptors does not necessarily induce apoptosis. In many circumstances, the so-called death receptors can also recruit molecules to the DISC which promote survival, one of the best examples being the activation of the NF-κB signaling pathway by tumor necrosis factor (TNF)α and Fas. Indeed, protective effects of TNF in the nervous system have been noted (Lipton, 1997). The same duality is observed for p75^{NTR}. In Schwann cells, activation of p75^{NTR} by NGF can also lead to apoptotic death or to survival; it appears that the pathways involved are dependent on which proteins are recruited to the death domains of the p75^{NTR}. Recruitment of NRAGE, for example, will promote apoptotic death (Salehi *et al.*, 2000), perhaps via neutralization of IAPs (Jordan *et al.*, 2001), but recruitment of the RIP2 kinase will facilitate the activation of NF-κB and promote survival (Khursigara *et al.*, 2001). Moreover, in DRG neurons where the p75^{NTR} is coexpressed with the high-affinity NGF receptor (TrkA), NGF binding to p75^{NTR} is not able to activate cell death (Casaccia-Bonnefil *et al.*, 1999). However, it has been proposed that withdrawal of NGF leads to activation of apoptosis via p75^{NTR} in primary neurons (e.g., Bamji *et al.*, 1998). In addition to signaling via alternative survival and death-inducing pathways being active simultaneously at the DISC, some proteins can act as dominant-negative or antagonistic molecules to caspase 8 at the DISC (Hu *et al.*, 1997; Irmler *et al.*, 1997; Srinivasula *et al.*, 1997). Interesting in this regard is the observation that interferon (IFN)-γ (in combination with TNFα) can elevate FasL on Schwann cells but reduce the expression of Fas. This has been suggested to underlie the ability of Schwann cells to reject invading lymphocytes during inflammation without being in danger of undergoing simultaneous apoptosis (Wohlleben *et al.*, 2000). Hence receptor repertoire and mechanisms of apoptotic induction need to be evaluated carefully in studies of apoptosis, as mechanisms of apoptosis during DN in adult animals may not be mimicked by studying embryonic neurons or early postnatal Schwann cells.

B. MITOCHONDRIAL PATHWAY

The second mechanism of apoptotic induction is initiated by numerous stimuli, is regulated by the Bcl2 protein family, and is mediated by the release of proapoptotic factors from the mitochondria. Stimuli include withdrawal of survival factors [such as IGF-1 for Schwann cells (Campana *et al.*, 1999; Delaney *et al.*, 1999) or NGF for sensory neurons], DNA-damaging agents, prooxidants, inflammatory cytokines, and perturbation of the cytoskeleton. All of these can activate death signaling pathways that recruit proapoptotic members of the Bcl2 family to the mitochondria. The BH3-only members of this family bind to the antiapoptotic members (e.g., Bcl2, BclxL, Mcl1) and enable the activation of two key players—Bax and Bak—which adopt an active conformation in the mitochondrial membrane and promote the opening of a pore. It is still unclear whether the Bax/Bak form a channel composed of self-multimers or whether they regulate other pore-forming channels such as the voltage-activated anion channel (VDAC) (Adams and Cory, 2001). Pore formation is a critical step in the mitochondrial route of apoptosis, as this causes the release of cytochrome *c*. Cytochrome *c*, together with dATP, activates the holoenzyme composed of Apaf-1 and procaspase 9, causing caspase 9 to become cleaved and activated. Caspase 9 serves a similar role to caspases 8/10 in that it cleaves effector caspases, thereby initiating the execution of apoptosis. The only caspase that does not fit neatly into this mechanism is caspase 2—the mechanisms of its activation is still not well understood. Other proapoptotic factors may also be released from mitochondria, some of which contribute to apoptotic death; SMAC/DIABLO (Du *et al.*, 2000; Verhagen *et al.*, 2000), an inhibitor of IAPs—proteins that inhibit caspases, endonuclease G, a mitochondrial DNase that translocates to the nucleus (Li *et al.*, 2001), and SM-20 (Lipscomb *et al.*, 2001), whose precise role is still not known). Some mitochondrial proteins that are released upon induction of apoptosis promote cell death without involving caspases [AIF, a flavoprotein that also acts as a nuclease (Lorenzo *et al.*, 1999)]. Death receptors may also enhance their killing effects through the mitochondrial route, by caspase 8-mediated cleavage of the BH3-only protein Bid, and recruitment of the truncated [and myristoylated (Zha *et al.*, 2000)] BID to the mitochondrial membrane (Gross *et al.*, 1999; Li *et al.*, 1998). Moreover, there may be feedback between death signaling induced by various insults and activation of the death receptor pathway, as kinases such as JNK, which is implicated in cell death induced by trophic factor deprivation (Ham *et al.*, 2000), can promote the upregulation of FasL (Le-Niculescu *et al.*, 1999). In embryonic motoneurons, for example, it has been suggested that there is a window of time during which the withdrawal of survival factors causes upregulation of FasL, which promotes apoptotic death (Raoul *et al.*, 1999). The window may be closed after 2–3 days of culture in the presence of BDNF, possibly by upregulation of FLIP (Irmler *et al.*, 1997).

Figure 1 (see also color insert) reviews the current consensus regarding apoptosis pathways in primary neurons. Apoptotic death is characterized by several morphological and biochemical features. In studies of tissues, it is still common to characterize apoptosis by electron microscopy and by demonstrating DNA condensation, margination, or clumping. Chromatin clumping is often observed along with vacuole formation in the cytoplasm attributed to mitochondrial swelling, and darkening/shrinkage of the cytoplasm possibly through loss of movements of K^+ and water (Bortner and Cidlowski, 1999). The most common microscopic techniques for detecting apoptosis are those that label fragmented DNA by incorporating labeled DNA bases using DNA polymerase [ISEL, (Wijsman *et al.*, 1992)] or terminal d-transferase [TUNEL (Gavrieli *et al.*, 1992)] (*Methods in Cell Biology*, Volume 66 covers most of these techniques in detail). However, it is now clear that there are occasions where necrosis also promotes DNA fragmentation and therefore the criterion of TUNEL cannot be used alone to diagnose apoptotic death. A similar argument can be made when a DNA dye (propidium iodide) is used to indicate loss of plasma membrane integrity indicative of necrosis. Because apoptosis is so well characterized biochemically, it is now possible to study the process using several markers: cells can be probed with antibodies to cytochrome *c* (to examine whether the protein has been released from the mitochondria); specific antibodies are available against the active form of caspase 3 or the conformationally active Bax/Bak; and, in some cases, staining with annexin V will reveal externalization of phosphatidylserine to the outer leaflet of the plasma membrane. In tissue sections, anti-active caspase 3 antibodies have proved

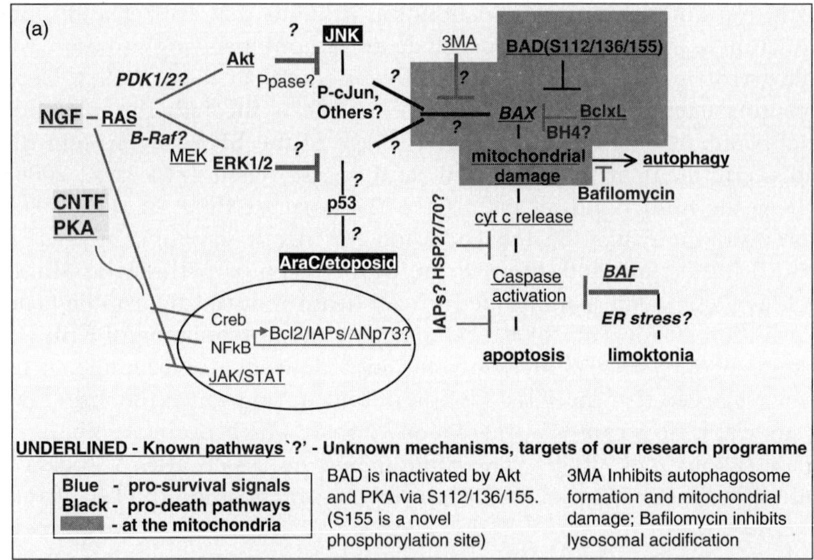

FIG. 1. (a) Pro- and anti-apoptotic pathways in primary peripheral NGF-dependent neurons. NGF activates at least two survival signalling pathways via stimulation of Ras. The Akt (also known as PKB) pathway suppresses signalling via the pro-apoptotic JNK (stress kinase) pathway. The ERK (MAPK) pathway suppresses pro-apoptotic signals induced by p53. The precise targets of the survival kinases, and the mechanisms by which the pro-apoptotic signals induce apoptosis are still not well understood. One target is the protein BAD which is multiply phosphorylated by Akt and its downstream target RSK, and by protein kinase A. This phosphorylation keeps BAD from binding to anti-apoptotic Bcl2 family members. In addition, the transcription factors NFkB, CREB, and STATs, regulate the expression of several pro- and anti-apoptotic factors, illustrated by the examples of Bcl2 family members, IAPs, and a proposed dominant interfering inhibitor of the p53 pathways, the N-truncated splice variant of p73α. The orange box depicts mitochondria, where active Bax and Bak form pores that promote the release of pro-apoptotic factors, whose activity culminate in caspase activation and orderly demolition of the cells. Possible inhibitors are noted, for example, caspase inhibitors such as BAF (Boc.Asp(O-methyl).fluoromethylketone) and Heat shock proteins (HSP70/27). In the mitochondrial compartment, overexpression of Bcl2 will antagonise the actions of Bax/Bak and other BH3-only members of the Bcl-2 family. BH4 denotes peptides based on the protective regions of Bcl-2/Bcl-xL. The activation of autophagy, a bi-product of apoptotic signalling, is suppressed by agents such as 3MA (which is an inhibitor of autophagosome formation, but is quite nonspecific) and Bafilomycine A1 which is an inhibitor of the lysosomal H^+-ATPase. (b) Death receptor-mediated signalling pathways. The main motifs are the recruitment of caspase 8/10 to the receptor through a series of intermediate proteins recruited to the receptors upon their activation. A link into the mitochondrial pathway is depicted via cleavage of Bid, a BH3-only protein members of the Bcl-2 family. Another motif to note is the link to survival signals via JNK (which can be pro- or anit-apoptotic depending on cell context) and NFkB. (b) Courtesy of Malcolm I. Roberts. (See also color insert.)

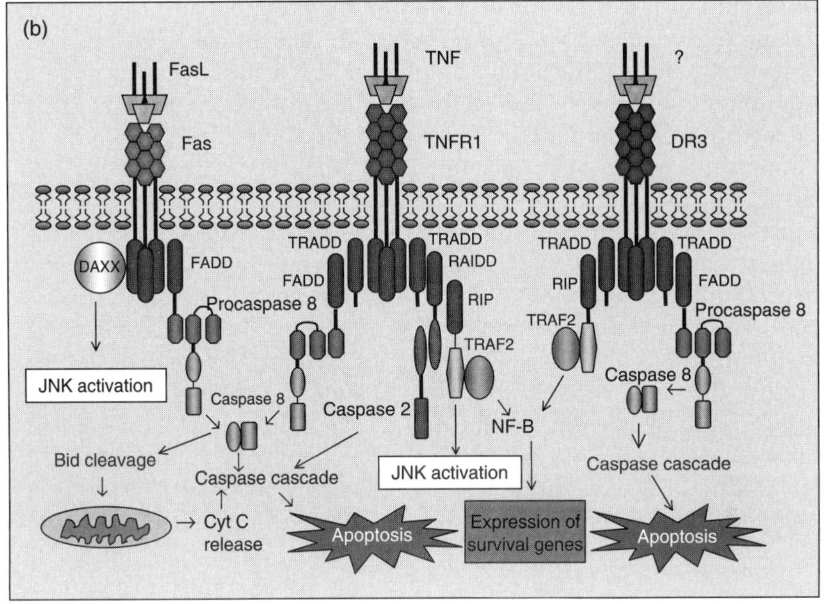

FIG. 1. (*continued*)

to be very useful. *In vitro* studies of apoptosis may be complemented by use of caspase inhibitors that eliminate most of the morphological features of apoptosis (although they do not eliminate cytochrome *c* release unless the death was initiated via the DR and initiated by caspase 8). There are now also knockout animals for various Bcl2 family members, as well as for various caspases. Hence it should not be difficult to assess the role of apoptotic death in rodent models of DN and in human biopsy material.

Both the DR pathway and the inductive pathway of apoptosis are under tight regulation by survival factors. These factors suppress apoptosis at many levels through transcriptional and posttranslational regulation. Moreover, many of the signals that have been shown to inhibit apoptosis (the PI3K pathway, NF-κB, MAPK) are also involved in other aspects of cell function, such as the control of protein synthesis, and regeneration, and therefore their use may provide a better prognosis for survival than simply inhibiting apoptosis. It is clear that inhibition of caspases may not be sufficient to inhibit apoptosis, as the mitochondrial dysfunction induced by cytochrome *c* release will still remain a limiting factor for survival and function (Fletcher *et al.*, 2000). Only if caspase 8 works upstream of mitochondrial dysfunction will its inhibition prove a good target. Hence all treatments that aspire to inhibit apoptotic death must be geared toward the preservation of mitochondrial function. It is instructive to examine the requirements of sensory and sympathetic neurons from adult animals in this regard. In rodent DRGs, the levels of Bcl-xL are upregulated by NGF following the period of programmed cell death (Vogelbaum *et al.*, 1998). This upregulation coincides with an increased independence from NGF. In adult SCGs, which also show an independence from neurotrophins, it appears that the PI3K/Akt pathway is constitutively active, which acts in concert with a requirement for anti-apoptotic Bcl2 proteins (Orike *et al.*, 2001). When neurons are studied *in vitro*, the axotomy that is needed to obtain single cells may change the order of events in the death pathway and may shift the death commitment point (Fletcher *et al.*, 2000).

It is useful to examine briefly some of the targets of the survival signals as they illuminate the risk factors involved in apoptosis. Some of the signals (such as the PI3K/Akt pathway) inhibit the JNK/p38 pathways implicated in apoptosis in numerous systems, including neurons. A key target may be the upstream activator of JNK, the kinase ASK-1 (Kim *et al.*, 2001). Other targets include BAD (one of the BH3-only proapoptotic proteins) and transcription factors belonging to the Forkhead family, which are known to induce Fas ligand in several neuronal systems (Brunet *et al.*, 2001). The role of the ERK/MAPK pathway is more controversial, causing an inhibition of p53-dependent apoptosis in SCG neurons (Anderson and Tolkovsky 1999), as well as possibly an inhibition of ROS production (Dugan *et al.*, 1997). NF-κB has been proposed to inhibit apoptosis by upregulating Bcl-xL and IAPs, inhibitors of caspases whose inhibitory action is neutralized by the binding of SMAC/DIBLO. Despite some broad themes that underlie these signaling modules, one should be cautious in attributing the survival effects each signal plays to a common mechanism because the signals play their roles in a cell- and age-specific manner. Moreover, in *in vitro* studies, while neurons are postmitotic a priori and will therefore only undergo changes induced by their isolation and culture conditions, Schwann cells are usually amplified in culture using forskolin before being studied. These Schwann cells may acquire very different properties to primary Schwann cells. Hence the best way of probing rodent models for the importance of apoptotic death in DN is to examine the fates of cells *in vivo* in a time-dependent manner using a range of techniques that will reveal whether the death is apoptotic and whether the cells being studied were there originally during the period when the induction of diabetes took place. If insulin is to be administered, time points just before, as well as after the induction, should be taken so as to provide information as to whether true reversal took place in the neurons that would have otherwise undergone apoptosis.

III. Apoptotic Indicators in Diabetic Neuropathy

Little work has been performed on mouse models of DN in relation to neuronal or Schwann cell apoptosis and so no use has been made of the many transgenic and knockout mice that could be used to examine the role of apoptosis in DN. However, a few studies have been performed in rats; the largest body of work has come from the laboratories of Feldman and colleagues. Russell *et al.* (1999) examined whether neurons and Schwann cells undergo apoptosis in animal and cell culture models of DN. Using young adult rats 4 weeks after treatment with streptozotocin (STZ), in which mean blood glucose concentration was elevated to about 15 mM, the authors showed evidence by EM of cell body and nuclear degeneration within the DRG. Of 4500 neurons counted after TUNEL staining, 34% of the DRG were TUNEL positive compared to 0.14% in controls. After 3 months, there was a loss of 22% of neurons in sections of lumbar L5 DRGs in diabetic animals (slightly less than the number of apoptotic neurons). Vacuoles in the cytoplasm, which appear to be derived from mitochondria, occur in cells in which chromatin condensation is apparently only beginning. This kind of swelling also occurs in necrosis but is rare in intact neurons. The authors further pointed to the cell autonomous nature of the event, as the satellite cell shown is normal. It would have been interesting to examine whether there was any loss in axon diameter in these animals and whether this change was observed in myelinated or unmyelinated fibers, but thus far this information is lacking. To show that the loss of neurons was not a result of STZ, the authors infused a glucose solution (100%) into adult rats for 6–10 h whereby average glucose levels were 31 mM. In this reference, an image of one neuron and one Schwann cell are shown with typical apoptotic morphology, but there is no indication as to the total number of cells that underwent apoptosis in this manner.

To analyze whether glucose induces apoptosis in a cell autonomous fashion, the authors cultured DRG neurons from E15 embryonic animals. At this stage the neurons are acutely dependent on NGF for their survival and have very low levels of Bcl2/BclxL compared to adult neurons. This is not the best model for studying the complexities of DN that manifest under chronic conditions of hyperglycemia, as the neurons are very sensitive to cell death inducers. Nevertheless, under the culture conditions used, it was found that neurite outgrowth was maximal when the explants were cultured in the presence of 30 mM glucose, which was defined as the optimal dose for growth and survival in the culture medium used. Normally this concentration is of course well into the diabetic hyperglycemic range. However, addition of another 20 mM glucose, but not mannitol, was sufficient to reduce neurite outgrowth. The effect was increased to 25%

inhibition when using 150 mM glucose. When the authors looked for signs of apoptosis, they found that treatment with 50 mM glucose promoted a significant increase in the percentage of apoptotic neurons, measured using TUNEL, and this observation is backed up using EM studies showing characteristic changes in nuclear condensation and fragmentation and an increase in immunoreactivity of the p17 subunit of active caspase 3. The death is very rapid compared to that induced by the withdrawal of survival factors, with measurements being taken 6 h after high glucose application.

Finally, the authors provided evidence that this apoptotic process can be controlled by the addition of IGF-1, which can prevent the apoptotic effects induced by glucose (added up to 180 mM). Based on the observation of swollen profiles of mitochondria in neurons *in vivo* in STZ-treated rats, the authors proposed the hypothesis that glucose causes oxidative stress, which in turn leads to apoptosis. That high glucose can cause apoptosis was previously claimed for human umbilical vein endothelial cells (HUVECs) (Baumgartner-Parzer *et al.*, 1995), but the differences in the percentage of DNA fragmentation between treated and control cells were very small (13.7 ± 6.5 vs 10.7 ± 5.6). The large number of stable TUNEL-positive cells found that might have been expected to have been phagocytosed is explained by the requirement for an additional signal. The authors suggested that the number of dying neurons is equivalent to the numbers of those small myelinated and unmyelinated fibers that are lost in DN, although this loss is not observed in rat models of the disease. Another interesting feature is that IGF-1 protects these neurons from apoptotic death whereas NGF does not. However, these neurons are NGF dependent for survival. Moreover, the authors proposed that IGF mediates its effects through activation of the PI3K pathway, but because this pathway is also activated by NGF, it is unclear why the former but not the latter can mediate protection against glucose. The signals that IGF1 induces in preference to NGF that inhibit this death remain to be explored.

Russell and Feldman (1999) have also shown that high concentrations of glucose induce apoptosis in SCG neurons (in the presence of NGF) and that this death is prevented by IGF-1. The authors proposed that glucose inhibits attachment of the neurons and thereby induces apoptosis. NGF, however, is capable of inducing survival in suspensions of SCG neurons. Clearly, there is much more work to be done to identify the causes of glucose-induced apoptosis and why IGF-1 but not NGF is protective against this effect. In neuroblastoma cells, the authors have shown that mannitol induces very rapid apoptosis that is blocked by IGF-1 (van Golen *et al.*, 2000), but the effects of high glucose were not tested. Hence it seems the hyperosmotic effects are just as potent as high glucose in inducing apoptosis. While it seems that the authors controlled for hyperosmolarity in their experiments on DRGs using mannitol and high NaCl, the fact that high glucose must be used to elicit maximal survival makes the data suggestive but difficult to interpret unequivocally.

In another series of papers, Schwann cell function was probed in relation to reversal of apoptosis by IGF1. Delaney *et al.* (1999) showed that the survival induced by IGF-1 is dependent on the PI3K/Akt (Cheng *et al.*, 2000) pathway, but in this instance the cells were withdrawn of serum and the ability of high concentrations of glucose to induce apoptosis in this paradigm is not described.

Srinivasan *et al.* (2000) used a slightly different model and set of parameters to demonstrate apoptosis in the DRG of STZ-treated rats. In addition, the rats were provided with insulin after confirmation that blood glucose was elevated. Although it is a little unclear when insulin was administered in relation to the time when the STZ treatment was begun, it appears that hyperglycemia was maintained 1–4 weeks before 2 weeks of insulin treatment. The level of glucose during hyperglycemia was 14–28 mM at the time of insulin treatment. The authors first showed that STZ causes the reduction in nerve conduction, which is reversed (or annulled, if there was no effect before insulin treatment) by insulin. They then derived DRG from adult rats treated with STZ, or STZ and insulin, and measured the mitochondrial membrane potential in these neurons. They show a twofold increase in the ratio of JC-1 green to red fluorescence to indicate a reduction in mitochondrial membrane potential. However, the effect is small relative to the sixfold increase in dye ratio induced by glutamate. Insulin treatment reverses both the STZ-induced basal increase and the stimulated increase in the dye ratiometric readings. The absolute magnitude of depolarization is not known. TUNEL was used to measure the number of neurons with DNA fragmentation in DRGs isolated from the three groups of animals. There was a twofold increase in the relative amount of TUNEL-positive neurons (out of 200 neurons counted), although the actual percentage of apoptosis was low, rising from about 4.5 to 9%. Immunostaining for Bcl2 in tissue sections shows a global reduction in intensity in STZ-treated DRGs. The authors write that Bcl2 levels were reduced by 85%, although the results of a blot are not shown. Whether insulin reversed this effect is also not clear. No change was observed in BclxL or Bax levels by Western blotting (a "positive control" appears to be "negative"). Finally, the authors showed a blot of cytochrome c staining derived from cytoplasmic and mitochondrial fractions, which, as discussed earlier, is released from mitochondria during apoptosis. However, data show the molecular mass of mitochondrial cytochrome c at about 40 Da, whereas that of cytochrome c is about 14 kDa. Moreover, in cytoplasmic fractions which should contain cytochrome c, there is a new band at 28 kDa, whereas there is no increase in the 40-kDa band intensity. These may be dimers and trimers of cytochrome c, but this issue is not discussed. The results are consistent with apoptotic death in DRG neurons caused by hyperglycemia induced by STZ, but the causal connections are not clear. Because apoptosis (and the decrease in nerve conduction) is prevented by a 2-week treatment with insulin in this model, yet all of the neurons appeared to be deficient in Bcl2 staining, it is not clear why apoptosis was only detected in 5% of the neurons. Moreover, it is unclear how it was possible to see a change in localization of cytochrome c if only 5% of the neurons were affected. If the apoptotic neurons are cleared from the ganglion, this would contradict the *in situ* studies performed by Russell and colleagues. Perhaps apoptotic neurons are preferentially lost during their isolation. Further careful time course studies will have to be conducted to reconcile these observations and to demonstrate a causal connection.

IV. Conclusions

While there has been an enormous emphasis on the reduction in axonal function in studies of DN, there has been very little examination of whether neuron and Schwann cell activity is compromised by apoptosis. The current evidence is too fragmented to be able to discount an involvement of apoptosis in the pathology of DN, although it is hard to understand why only certain neurons would be affected unless the apoptosis occurred as a result of a coincidence of two factors, such as hyperglycemic induction of reactive oxygen species (Greene *et al.*, 1999; Hounsom *et al.*, 2001) and loss of a protective mechanism [such as NGF availability or Bcl2/Bcl-xL or heat shock protein 27 (Lewis *et al.*, 1999) expression]. Any study of apoptosis as it relates to DN must first investigate the effects *in vivo* and relate these carefully to the loss of nerve conduction. Use could be made of the numerous mouse models in which components of the apoptotic pathway have been knocked out so that a dissection of the factors that contribute to apoptosis could be assessed. Such an approach, combined with carefully timed studies in which primary proapoptotic events are imposed *in vitro*, might define the necessary markers and sequencing for studies *in vivo*. Ultimately, this might not only confirm or refute the relevance of apoptosis in the pathogenesis of this condition, but it might also reveal hitherto ignored drug targets for its prevention or management.

References

Adams, J. M., and Cory, S. (2001). Life-or-death decisions by the Bcl-2 protein family. *Trends Biochem. Sci.* **26**, 61–66.

Anderson, C. N., and Tolkovsky, A. M. (1999). A role for MAPK/ERK in sympathetic neuron survival: Protection against a p53-dependent, JNK-independent induction of apoptosis by cytosine arabinoside. *J. Neurosci.* **19**, 664–673.

Bamji, S. X., Majdan, M., Pozniak, C. D., Belliveau, D. J., Aloyz, R., *et al.* (1998). The p75 neurotrophin receptor mediates neuronal apoptosis and is essential for naturally occurring sympathetic neuron death. *J. Cell. Biol.* **140**, 911–923.

Barres, B. A., and Raff, M. C. (1994). Control of oligodendrocyte number in the developing rat optic nerve. *Neuron* **12**, 935–942.

Barrett, G. L. (2000). The p75 neurotrophin receptor and neuronal apoptosis. *Prog. Neurobiol.* **61**, 205–229.

Baumgartner-Parzer, S. M., Wagner, L., Pettermann, M., Grillari, J., Gessl, A., and Waldhausl, W. (1995). High-glucose–triggered apoptosis in cultured endothelial cells. *Diabetes* **44**, 1323–1327.

Bortner, C. D., and Cidlowski, J. A. (1999). Caspase independent/dependent regulation of K(+), cell shrinkage, and mitochondrial membrane potential during lymphocyte apoptosis. *J. Biol. Chem.* **274**, 21953–21962.

Brunet, A., Datta, S. R., and Greenberg, M. E. (2001). Transcription-dependent and—independent control of neuronal survival by the PI3K-Akt signaling pathway. *Curr. Opin. Neurobiol.* **11**, 297–305.

Campana, W. M., Darin, S. J., and O'Brien, J. S. (1999). Phosphatidylinositol 3-kinase and Akt protein kinase mediate IGF-I- and prosaptide-induced survival in Schwann cells. *J. Neurosci. Res.* **57**, 332–341.

Casaccia-Bonnefil, P., Gu, C., Khursigara, G., and Chao, M. V. (1999). p75 neurotrophin receptor as a modulator of survival and death decisions. *Microsc. Res. Tech.* **45**, 217–224.

Cheng, H. L., Steinway, M., Delaney, C. L., Franke, T. F., and Feldman, E. L. (2000). IGF-I promotes Schwann cell motility and survival via activation of Akt. *Mol. Cell. Endocrinol.* **170**, 211–215.

Delaney, C. L., Cheng, H. L., and Feldman, E. L. (1999). Insulin-like growth factor-I prevents caspase-mediated apoptosis in Schwann cells. *J. Neurobiol.* **41**, 540–548.

Du, C., Fang, M., Li, Y., Li, L., and Wang, X. (2000). Smac, a mitochondrial protein that promotes cytochrome c-dependent caspase activation by eliminating IAP inhibition. *Cell* **102**, 33–42.

Dugan, L. L., Creedon, D. J., Johnson, E. M., Jr., and Holtzman, D. M. (1997). Rapid suppression of free radical formation by nerve growth factor involves the mitogen-activated protein kinase pathway. *Proc. Natl. Acad. Sci. USA* **94**, 4086–4091.

Enari, M., Sakahira, H., Yokoyama, H., Okawa, K., Iwamatsu, A., and Nagata, S. (1998). A caspase-activated DNase that degrades DNA during apoptosis, and its inhibitor ICAD. *Nature* **391**, 43–50.

Fletcher, G. C., Xue, L., Passingham, S. K., and Tolkovsky, A. M. (2000). Death commitment point is advanced by axotomy in sympathetic neurons. *J. Cell. Biol.* **150**, 741–754.

Gavrieli, Y., Sherman, Y., and Ben-Sasson, S. A. (1992). Identification of programmed cell death in situ via specific labeling of nuclear DNA fragmentation. *J. Cell. Biol.* **119**, 493–501.

Greene, D. A., De Jesus, P. V., Jr., and Winegrad, A. I. (1975). Effects of insulin and dietary myoinositol on impaired peripheral motor nerve conduction velocity in acute streptozotocin diabetes. *J. Clin. Invest.* **55**, 1326–1336.

Greene, D. A., Stevens, M. J., Obrosova, I., and Feldman, E. L. (1999). Glucose-induced oxidative stress and programmed cell death in diabetic neuropathy. *Eur. J. Pharmacol.* **375**, 217–223.

Gregersen, G. (1968). Variations in motor conduction velocity produced by acute changes of the metabolic state in diabetic patients. *Diabetologia* **4**, 273–277.

Gross, A., Yin, X. M., Wang, K., Wei, M. C., Jockel, J., *et al.* (1999). Caspase cleaved BID targets mitochondria and is required for cytochrome *c* release, while BCL-XL prevents this release but not tumor necrosis factor-R1/Fas death. *J. Biol. Chem.* **274**, 1156–1163.

Ham, J., Eilers, A., Whitfield, J., Neame, S. J., and Shah, B. (2000). c-Jun and the transcriptional control of neuronal apoptosis. *Biochem. Pharmacol.* **60**, 1015–1021.

Hengartner, M. O. (2000). The biochemistry of apoptosis. *Nature* **407**, 770–776.

Hounsom, L., Corder, R., Patel, J., and Tomlinson, D. R. (2001). Oxidative stress participates in the breakdown of neuronal phenotype in experimental diabetic neuropathy. *Diabetologia* **44**, 424–428.

Hu, S., Vincenz, C., Ni, J., Gentz, R., and Dixit, V. M. (1997). I-FLICE, a novel inhibitor of tumor necrosis factor receptor-1- and CD- 95-induced apoptosis. *J. Biol. Chem.* **272**, 17255–17257.

Irmler, M., Thome, M., Hahne, M., Schneider, P., Hofmann, K., *et al.* (1997). Inhibition of death receptor signals by cellular FLIP. *Nature* **388**, 190–195.

Jordan, B. W., Dinev, D., LeMellay, V., Troppmair, J., Gotz, R., *et al.* (2001). NRAGE is an inducible IAP-interacting protein that augments cell death. *J. Biol. Chem.* **6**, 6.

Khursigara, G., Bertin, J., Yano, H., Moffett, H., DiStefano, P. S., and Chao, M. V. (2001). A prosurvival function for the p75 receptor death domain mediated via the caspase recruitment domain receptor-interacting protein 2. *J. Neurosci.* **21**, 5854–5863.

Kim, A. H., Khursigara, G., Sun, X., Franke, T. F., and Chao, M. V. (2001). Akt phosphorylates and negatively regulates apoptosis signal-regulating kinase 1. *Mol. Cell. Biol.* **21**, 893–901.

Kruidering, M., and Evan, G. I. (2000). Caspase-8 in apoptosis: The beginning of "the end?" *IUBMB Life* **50**, 85–90.

Le-Niculescu, H., Bonfoco, E., Kasuya, Y., Claret, F. X., Green, D. R., and Karin, M. (1999). Withdrawal of survival factors results in activation of the JNK pathway in neuronal cells leading to Fas ligand induction and cell death. *Mol. Cell. Biol.* **19**, 751–763.

Lewis, S. E., Mannion, R. J., White, F. A., Coggeshall, R. E., Beggs, S., Costigan, M., Martin, J. L., Dillmann, W. H., and Woolf, C. J. (1999). A role for HSP27 in sensory neuron survival. *J. Neurosci.* **19**, 8945–8953.

Li, H., Zhu, H., Xu, C. J., and Yuan, J. (1998). Cleavage of BID by caspase 8 mediates the mitochondrial damage in the Fas pathway of apoptosis. *Cell* **94**, 491–501.

Li, L. Y., Luo, X., and Wang, X. (2001). Endonuclease G is an apoptotic DNase when released from mitochondria. *Nature* **412**, 95–99.

Lipscomb, E. A., Sarmiere, P. D., and Freeman, R. S. (2001). SM-20 is a novel mitochondrial protein that causes caspase-dependent cell death in nerve growth factor-dependent neurons. *J. Biol. Chem.* **276**, 5085–5092.

Lipton, S. A. (1997). Janus faces of NF-kappa B: Neurodestruction versus neuroprotection. *Nature Med.* **3**, 20–22.

Liu, X., Zou, H., Slaughter, C., and Wang, X. (1997). DFF, a heterodimeric protein that functions downstream of caspase-3 to trigger DNA fragmentation during apoptosis. *Cell* **89**, 175–184.

Lorenzo, H. K., Susin, S. A., Penninger, J., and Kroemer, G. (1999). Apoptosis inducing factor (AIF): A phylogenetically old, caspase- independent effector of cell death. *Cell Death Differ.* **6**, 516–524.

Orike, N., Middleton, G., Borthwick, E., Buchman, V., Cowen, T., and Davies, A. M. (2001). Role of PI 3-kinase, Akt and Bcl-2-related proteins in sustaining the survival of neurotrophic factor-independent adult sympathetic neurons. *J. Cell. Biol.* **154**, 995–1005.

Raoul, C., Henderson, C. E., and Pettmann, B. (1999). Programmed cell death of embryonic motoneurons triggered through the Fas death receptor. *J. Cell. Biol.* **147**, 1049–1062.

Russell, J. W., and Feldman, E. L. (1999). Insulin-like growth factor-I prevents apoptosis in sympathetic neurons exposed to high glucose. *Horm. Metab. Res.* **31**, 90–96.

Russell, J. W., Sullivan, K. A., Windebank, A. J., Herrmann, D. N., and Feldman, E. L. (1999). Neurons undergo apoptosis in animal and cell culture models of diabetes. *Neurobiol. Dis.* **6**, 347–363.

Sahara, S., Aoto, M., Eguchi, Y., Imamoto, N., Yoneda, Y., and Tsujimoto, Y. (1999). Acinus is a caspase-3-activated protein required for apoptotic chromatin condensation. *Nature* **401**, 168–173.

Salehi, A. H., Roux, P. P., Kubu, C. J., Zeindler, C., Bhakar, A., *et al.* (2000). NRAGE, a novel MAGE protein, interacts with the p75 neurotrophin receptor and facilitates nerve growth factor-dependent apoptosis. *Neuron* **27**, 279–288.

Savill, J., and Fadok, V. (2000). Corpse clearance defines the meaning of cell death. *Nature* **407**, 784–788.

Srinivasan, S., Stevens, M., and Wiley, J. W. (2000). Diabetic peripheral neuropathy: Evidence for apoptosis and associated mitochondrial dysfunction. *Diabetes* **49**, 1932–1938.

Srinivasula, S. M., Ahmad, M., Ottilie, S., Bullrich, F., Banks, S., *et al.* (1997). FLAME-1, a novel FADD-like anti-apoptotic molecule that regulates Fas/TNFR1-induced apoptosis. *J. Biol. Chem.* **272**, 18542–18545.

Thomas, P.K., and Tomlinson, D.R. (1992). Diabetic and hypoglycaemic neuropathy. *In* "Peripheral Neuropathy" (P. J. Dyck, P. K. Thomas, J. W. Griffin, P. A. Low, and J. F. Poduslo eds.), (pp. 1219–1250). Saunders, Philadelphia.

van Golen, C. M., Castle, V. P., and Feldman, E. L. (2000). IGF-I receptor activation and BCL-2 overexpression prevent early apoptotic events in human neuroblastoma. *Cell Death Differ.* **7**, 654–665.

Verhagen, A. M., Ekert, P. G., Pakusch, M., Silke, J., Connolly, L. M., *et al.* (2000). Identification of DIABLO, a mammalian protein that promotes apoptosis by binding to and antagonizing IAP proteins. *Cell* **102**, 43–53.

Vinik, A. I., Park, T. S., Stansberry, K. B., and Pittenger, G. L. (2000). Diabetic neuropathies. *Diabetologia* **43**, 957–973.

Vogelbaum, M. A., Tong, J. X., and Rich, K. M. (1998). Developmental regulation of apoptosis in dorsal root ganglion neurons. *J. Neurosci.* **18**, 8928–8935.

Wijsman, J., Atkison, P., Mazaheri, R., Garcia, B., Paul, T., *et al.* (1992). Histological and immunopathological analysis of recovered encapsulated allogeneic islets from transplanted diabetic BB/W rats. *Transplantation* **54**, 588–592.

Wohlleben, G., Ibrahim, S. M., Schmidt, J., Toyka, K. V., Hartung, H. P., and Gold, R. (2000). Regulation of Fas and FasL expression on rat Schwann cells. *Glia* **30**, 373–381.

Yuan, J., and Yankner, B. A. (2000). Apoptosis in the nervous system. *Nature* **407**, 802–809.

Zha, J., Weiler, S., Oh, K. J., Wei, M. C., and Korsmeyer, S. J. (2000). Posttranslational N-myristoylation of BID as a molecular switch for targeting mitochondria and apoptosis. *Science* **290**, 1761–1765.

NERVE AND GANGLION BLOOD FLOW IN DIABETES: AN APPRAISAL

Douglas W. Zochodne

Department of Clinical Neurosciences
University of Calgary, Calgary, Alberta, Canada T2N 4N1

Vasa nervorum, the vascular supply to peripheral nerve trunks, and their associated cell bodies in ganglia have unique anatomical and physiological characteristics. Several different experimental approaches toward understanding the changes in vasa nervorum following injury and disease have been used. Quantitative techniques most widely employed have been microelectrode hydrogen clearance polarography and [^{14}C]iodoantipyrine autoradiographic distribution, whereas estimates of red blood cell flux using a fiber-optic laser Doppler probe offer real time data at different sites along the nerve trunk. There are important caveats about the use of these techniques, their advantages, and their limitations. Reports of nerve blood flow require careful documentation of physiological variables, including mean arterial pressure and near nerve temperature during the recordings. Several ischemic models of the peripheral nerve trunk have addressed the ischemic threshold below which axonal degeneration ensues (<5 ml/100 g/min). Following injury, rises in local blood flow reflect actions of vasoactive peptides, nitric oxide, and the development of angiogenesis. In experimental diabetes, a large number of studies have documented reductions in nerve blood flow and tandem corrections of nerve blood flow and conduction slowing. A significant proportion, however, of the work can be criticized on the basis of methodology and interpretation. Similarly, not all work has confirmed that reductions of nerve blood flow are an invariable feature of experimental or human diabetic polyneuropathy. Therefore, while there is disagreement as to whether early declines in nerve blood flow "account" for diabetic polyneuropathy, there is unquestioned evidence of early microangiopathy. Abnormalities of vasa nervorum and microvessels supplying ganglia at the very least develop parallel to and together with changes in neurons, Schwann cells, and axons.   © 2002, Elsevier Science (USA).

I. Introduction

Prior to 1984, direct measurements of blood flow in peripheral nerves had been described in sporadic reports, with widely varying estimations provided, depending on the approach. In that year, Low, Tuck, and colleagues brought order to the field and new interest in its potential importance. These investigators published seminal work describing the use of microelectrode hydrogen clearance polarography to make direct measurements of nerve blood flow in rats (Low and Tuck, 1984). They demonstrated feasibility and reproducibility using the technique with blood flow values in the range of 15 ml/100 g/min, closely comparable with those achieved using autoradiographic [14C]iodioantipyrine endoneurial distribution measurements. These investigators went on to describe the lack of autoregulation in peripheral nerve blood flow and suggested that it was reduced in a rat model of diabetic neuropathy (Tuck *et al.*, 1984). While the latter finding has generated some disagreement, measurements of nerve blood flow have subsequently provided interesting data on the responses of peripheral nerve to ischemia, injury, inflammation, denervation, and other facets of peripheral nerve biology. This chapter reviews some of these ideas about nerve blood flow, in particular its relationship to experimental and human diabetes.

II. Anatomy and Physiology of Peripheral Nerve and Spinal Ganglia Vascular Supply

A. Peripheral Nerve

The vascular supply of the peripheral nerve trunk, or vasa nervorum, can be considered as two interdependent compartments: the epineurial/perineurial vascular plexus and the endoneurial blood supply. Further upstream, these vascular beds are supplied by regional feeding vessels and this supply accounts for watershed zones of susceptibility to ischemia, such as the proximal tibial nerve of the rat (McManis and Low, 1988; Zochodne *et al.*, 1992; Sladky *et al.*, 1985). Despite their common arterial source, the two nerve vascular compartments have distinct anatomical and physiological characteristics. The epineurial plexus consists of large arterioles and venules and prominent arteriovenous shunts. Its distribution over the surface of the nerve trunk is highly variable, but it ultimately supplies the endoneurial vascular compartment by direct penetrating branches or remote penetrating branches that supply longitudinal endoneurial feeding vessels (Adams, 1942; Bell and Weddell, 1984). The epineurial circulation does not have a tight barrier, in contrast to endoneurial capillaries, and Evans' blue or other macromolecular markers easily diffuse from it over the surface of the nerve (Rechthand and Rapoport, 1987). Epineurial vessels have been postulated to exert regulatory control over the downstream endoneurial circulation because, unlike most endoneurial vessels, they are innervated with both sympathetic noradrenergic terminals and sensory peptidergic terminals (Appenzeller *et al.*, 1984; Dhital *et al.*, 1986; Dhital and Appenzeller, 1988; Kihara and Low, 1990; Zochodne and Low, 1990; Zochodne and Ho, 1991b). Moreover, these terminals appear to arise from axons traveling within the parent peripheral nerve trunk, a form of "self-innervation" of vasa nervorum (Rechthand *et al.*, 1986). Rechthand and colleagues (1986) demonstrated that α-adrenoceptors on epineurial (not observed on endoneurial vessels) disappear with proximal section of the sciatic nerve trunk in rats, and we have observed similar findings with innervation from nerves storing (and presumably releasing) calcitonin gene-related peptide (CGRP) (Zochodne and Cheng, unpublished findings).

Both sympathetic and peptidergic innervation of vasa nervorum appear to act tonically. Thus, for example, acute doses of phentolamine, an α-adrenoceptor antagonist, transiently increased nerve blood flow, whereas chronic destruction of the sympathetic nervous system with guanethidine (that generates a rat species-specific autoimmune sympathectomy) chronically increased local blood flow (Zochodne *et al.*, 1990; Zochodne and Low, 1990; Zochodne and Ho, 1994b). Such local increases in flow occurred despite falls in mean arterial pressure (MAP) rendered by sympathetic blockade and represented a fall in local microvascular resistance, or true vasodilation. This is important because local blood flow of the peripheral nerve does not appear to autoregulate (Low and Tuck, 1984; Zochodne and Ho, 1991a). Instead, there is a mild, possibly curvilinear relationship between mean arterial pressure and nerve blood flow. Nerve blood flow therefore gradually declines with falls in mean arterial pressure, in contrast to ganglia (see later). As an expected feature of sympathetic adrenergic innervation, norepinephrine administered to the arterial supply of the sciatic nerve, or applied topically, reduced nerve blood flow through vasoconstriction (Zochodne and Low, 1990; Kihara and Low, 1990). Videoangiology, a technique allowing live video recordings of epineurial vessels following norepinephrine application, suggested that noradrenergic vasoconstriction was confined to specific vascular segments, portions of vessels with higher concentrations of α-adrenoceptors that may have roles in controlling downstream blood flow (Zochodne and Low, 1990).

That peptidergic fibers innervate vasa nervorum was initially demonstrated by Appenzeller and Dhital, who identified fibers containing CGRP, substance P (SP), VIP, NPY, and serotonin associated with epineurial vessels (Appenzeller *et al.*, 1984; Dhital and Appenzeller, 1988). However, it is important to point out here that not all fibers self-innervating the peripheral nerve, known as nervi nervorum, are perivascular (Hromada, 1963). Perivascular fibers comprise one subset of nervi nervorum, a form of innervation that may play a role in the generation of neuropathic pain (Asbury and Fields, 1984; Bove and Light, 1997). CGRP released from perivascular fibers acts as a potent vasodilator (Brain *et al.*, 1985), whereas SP induces plasma extravasation and is a somewhat less potent vasodilator (Lembeck and Holzer, 1979). Capsaicin, an agent acting through vanilloid receptors, acutely releases peptides from perivascular afferent terminals at their neuroeffector sites (Holzer, 1988, 1991). Experimental studies applying capsaicin to the epineurial plexus generated abrupt vasodilation of vasa nervorum, in turn blocked by antagonists of CGRP or SP (Zochodne and Ho, 1991b). In keeping with these findings, topical application of SP or CGRP alone to vasa nervorum lowered microvascular resistance, indicating a local vasodilating action mediated by NK-1 (SP receptors) and α CGRP receptors, respectively, on epineurial vessels. Alternatively, application of a SP or CGRP antagonist to the epineurial plexus lowered "resting" nerve blood flow—evidence that peptidergic input to vasa nervorum, as with sympathetic innervation, is tonic (Zochodne and Ho, 1993a). In summary, removal of a tonic peptide action using an antagonist resulted in vasoconstriction in vasa nervorum, likely by the sympathetic input, but also probably by circulating vasoconstrictors.

That other vasoactive molecules influence vasa nervorum has been established in additional work examining prostaglandin metabolites (Kihara and Low, 1995a), endothelin (ET) and other molecules (Zochodne *et al.*, 1992). Among the prostaglandins, vasodilation was observed with topical application in the order of $PGE_1 > PGI_2$ (prostacyclin)$\geq PGF_{2\alpha}$. Endothelin is a very potent constrictor of vasa nervorum, exerting an action approximately one log order of magnitude greater than that of norepinephrine (Zochodne *et al.*, 1992). ETa receptors that mediate vasoconstriction are present on epineurial arterioles and probably act as a substrate for circulating levels of ET (Dashwood and Thomas, 1997). Angiotensin II similarly reduces nerve blood flow (Kihara *et al.*, 1999). Acetylcholine acts as a vasodilator (Thomsen *et al.*, 2000). While measurements of endoneurial blood flow made using microelectrode hydrogen clearance polarography were not influenced by hypercarbia [that increases blood flow in the central nervous system (CNS)], measurements emphasizing epineurial circulation using a laser Doppler probe suggested that hypercarbia induces a mild rise in perfusion (Rechthand *et al.*, 1988; Low and Tuck, 1984).

Whether perineurial vessels should be considered separately from those of the epineurium is uncertain. During perineurial transit, vessels from the epineurium acquire a blood–nerve barrier. Myers and colleagues (1986) have postulated that perineurial passage can influence vessel caliber by mechanical compression of the vessel when the nerve is edematous.

Endoneurial circulation is considered to be largely passive and capillary; its flow is determined by upstream vasoresponsive arterioles in the epineurium. As discussed earlier, it may be that specific segments of such upstream vessels are critical determinants of what the endoneurial blood flow will be. The concept that endoneurial vessels are purely capillary, however, is likely oversimplified, with some evidence that there may be endoneurial arterioles and arteriovenous shunts, and that some perivascular innervating fibers may follow these vessels into this compartment (Lagerlund and Low, 1994). Endoneurial capillaries are larger than those of other tissue beds, have occasional investment of pericytes, and constitute one component of the blood–nerve barrier (Bell and Weddell, 1984; Olsson and Kristensson, 1971, 1973).

Endoneurial blood flow is much higher in newborn rats than adults (Kihara *et al.*, 1991b). In humans, quantitative measurements of nerve blood flow, other than laser Doppler flowmetry (LDF) (see later) or estimates of fluorescein plasma transit, have not been carried out.

B. Spinal Sensory and Autonomic Ganglia

Spinal sensory ganglia are supplied by segmental spinal artery branches that similarly supply cord, nerve roots, and proximal peripheral nerve trunks. There is a direct radicular branch and indirect anastomosis from the spinal artery (Adams, 1942). Comparatively few measurements of blood flow have been made in spinal sensory or autonomic ganglia. Measurements of sensory ganglia blood flow using microelectrode hydrogen clearance polarography and [^{14}C]idioantipyrine have given similar estimates of blood flow: approximately 30–40 ml/100 g/min, or two to three times that of the peripheral nerve (Zochodne and Ho, 1991a; McManis *et al.*, 1997; Sasaki *et al.*, 1997). In autonomic ganglia, somewhat higher values have been recorded, up to 70 ml/100 g/min (Cameron and Cotter, 2001; Sasaki *et al.*, 1997), but such values, higher than spinal cord or brain, seem difficult to explain. Because such recordings carried out in the superior cervical ganglia are complicated by the proximity of nearby high flow at the carotid bifurcation, it is possible that these values may be spuriously elevated.

While higher than that of peripheral nerve, blood flow in ganglia exhibited physiological features of partial autoregulation, not seen in peripheral nerve trunks. This was suggested by evidence of relatively preserved flow down to a mean arterial pressure (MAP) level of 60 mm Hg and a flattened relationship between flow and MAP between 60 and 120 mm Hg. The finding of partial autoregulation is of interest because it may indicate that the higher metabolic demands of perikarya and lower known oxygen tensions in this structure entrain higher flow and protection from ambient fluctuations in MAP. Although the innervation of ganglion blood vessels has not been explored histologically, ganglion blood flow, unlike that of the peripheral nerve trunk, was not influenced by sympathectomy (Zochodne and Ho, 1994b). Like CNS, and unlike peripheral nerve trunks, it may be that stronger autoregulatory control of flow in ganglia and lesser sympathetic influence protects neurons and preserves flow during episodic hypotension and generalized sympathetic vasoconstriction. Ganglion blood flow was not altered by hypercarbia, unlike CNS (Zochodne and Ho, 1991a).

There is no information to determine whether blood flow in ganglia, like that in nerve trunks, can be divided into subcompartments; e.g., subcapsular, supplying subcapsular ganglion cells (expected to be higher, analogous to gray matter), and centroganglionic, supplying mainly axons (perhaps lower, resembling that of white matter). It is also uncertain what the ischemic threshold of ganglia is and how ganglion neurons are targeted by either acute or low grade ischemia.

III. Measurements of Local Blood Flow

Methodology and pitfalls in the recording of nerve blood flow have been reviewed in detail (Low *et al.*, 1989). The principal techniques include microelectrode hydrogen clearance polarography, distribution of [^{14}C]iodoantipyrine, laser Doppler flowmetry, and microsphere embolization. Ancillary approaches include videoangiology microscopic visualization of the epineurial plexus, measurements of indicator transit times, and morphological studies of vessel numbers and area. In view of the advantages and disadvantages of each of the techniques, it may be advisable to use more than one technique (e.g., hydrogen clearance and LDF) in specific studies of blood flow.

Microelectrode hydrogen clearance polarography (HC) involves insertion of a fine tipped (3- to 5-μm tip ideally) platinum microelectrode, linearly sensitive to hydrogen, into the tissue of interest. In laboratories using this approach, such microelectrodes have been constructed on site, and the suitability of commercial versions (e.g., platinum and iridium blends) has not been established. In the case of peripheral nerve, a small epineurial window is made in an area relatively free of vessels, through which the microelectrode is inserted (puncturing the perineurium). Two reports of tip sizes smaller than 1 μm were apparently in error (Obrosova *et al.*, 2000; Stevens *et al.*, 2000; Stevens, World Precision Instruments, personal communications). The tissue is saturated by adding hydrogen to the inspiratory gas mixture (approximately 10%) of a paralyzed (curarized) and ventilated animal, and blood flow is calculated using a washout clearance curve after the hydrogen is turned off. The washout curve is fitted to a monoexponential or biexponential curve, and flow is calculated according to the approach of Aukland and colleagues (1964) as applied to nerve by Low, Lagerlund, and colleagues (Low and Tuck, 1984; Lagerlund and Low, 1994). Biexponential washout curves are thought to include a fast component contributed to by the epineurial plexus, arteriovenous shunts (some are found in the endoneurium), and a slow component contributed by the endoneurial nutritive vascular compartment. At a near nerve temperature of 37°C, normal nerve blood flow in the endoneurial compartment (slow washout component) is calculated as 15–20 ml/100 g/min. Low, Lagerlund, and colleagues (Day *et al.*, 1989a; Lagerlund and Low, 1994) have also described a composite blood flow measurement that takes into account weighting of both slow and fast components of the washout curve and is thought to reflect a composite of both endoneurial nutritive blood supply and arteriovenous shunting (particularly from the epineurial plexus). This value is usually calculated as 25–30 ml/100 g/min. This is an important issue because some indicator approaches that calculate uptake of an isotope in whole nerve without autoradiographic subcompartment localization probably reflect a similar composite flow value.

The advantages of HC include its quantitative reproducibility, comparable with autoradiographic [14C]iodoantipyrine distribution, and its provision of selective endoneurial flow data. Moreover, it allows serial measurements in a preparation that can be monitored physiologically. Disadvantages include the invasiveness of the procedure, its technical challenges (only a few laboratories have used the technique routinely), and the difficulties sometimes in assigning best fit to a monoexponential or biexponential washout curve. In normal intact nerve, the first washout curve after microelectrode insertion usually yields blood flow measurements higher than subsequent washouts that in turn remain stable for two to three more washouts (each lasting 30–60 min). Commercially developed setups (only a single one in development to this author's knowledge) to carry out HC in peripheral nerve or ganglia have not demonstrated their validity in experimental use to date. Hotta and colleagues (1995) have used a variation of the HC technique involving local electrolysis and generation of local concentrations of hydrogen.

Laser Doppler flowmetry involves placement of a laser Doppler probe containing efferent and afferent fiber-optic ends separated by a fixed distance so that they are just touching or above the tissue of interest. The probe bombards tissue and vessels with an efferent laser light that is reflected back to the afferent probe by moving erythrocytes. The velocity of movement determines the amount of Doppler shift of wavelength of the backscattered laser signal. LDF thus measures erythrocyte flux, a value that depends on the velocity of individual erythrocytes, and the overall number of erythrocytes that contribute to the amplitude of the signal. Perfusion measurements at varying levels of MAP in rats using LDF correlate well with those measured by HC (Takeuchi and Low, 1987). While LDF is relatively much easier to perform than HC, there are several important caveats that govern its use. While some devices have sought to generate a quantitative blood flow value from the signal in ml/100 g/min, this is problematic in nonuniform tissues such as peripheral nerve and the results may be inaccurate. The LDF signal can vary by a factor of two to three over a few millimeters in the epineurial plexus, depending on the relationship of the probe to plexus vessels. Because it may be very difficult to prevent slight movement of the probe during respiration of the animal, signals from single sites along the nerve may also be variable and inaccurate. While the probe can provide a "real time" signal, allowing the examiner to measure acute interventions (e.g., application of a vasoconstrictor), the most accurate quantitative approach is that reported by Kalichman and Lalonde (1991), requiring a minimum of 10 measurements (later calculated as a mean) along the nerve (e.g., at 1-mm intervals). LDF measurements must be made with all ambient lights (including heating lamps) turned off (because they also influence the signal). The probe should be mounted in a micromanipulator (microtremor from an examiner's hand influences the signal drastically) at a fixed angle to the nerve, preferably close to 90°. Near nerve temperature can be maintained (and should be controlled) by warming the local environment to 37°C just prior to measurement. Separation of afferent and efferent fibers within the probe determines the depth of sampling, and we have generally sampled at a fiber separation of 250 μm. Smaller interfiber distances deepen the sampling. A variety of separations is available from commercial suppliers. Finally, signals generated are in arbitrary units, reflecting erythrocyte flux, and can be considered quantitative provided the same conditions are used with each measurement (and controls are included). Complete circulatory arrest is associated with a small but not negligible residual signal from the LDF that may have to be considered (e.g., in assessing the severity of vasoconstriction).

Advantages of LDF include its ease of use, its "real time" measurement, and its ability to serially sample and geographically sample in the same animal. Disadvantages include its inability to provide true quantitative blood flow measurement results and its limited penetration into peripheral nerve. Because the outer surface of the peripheral nerve trunk directly beneath the LDF probe is covered by the epineurial plexus and because of limited laser penetration into the endoneurium, LDF signals are dominated by erythrocyte flux in the epineurial plexus (approaches to sample epineurium "free of vessels" may introduce sampling bias). We have considered the LDF technique as predominantly providing information about events in this plexus. The issue is important because changes assessed by LDF compared, for example, with HC may differ. This is the case for hypercarbic vasodilation detected by LDF but not by HC. Mild changes (in the range of 10–15%) could be identified in "real" time by the LDF probe in vasoactive epineurial vessels, whereas downstream, largely capillary endoneurial flow was unchanged, at least as detected by HC washout curves. HC may not be sensitive to small (e.g., less than 25%) or transient changes in blood flow. Another example of differing HC and LDF findings was shown in studies by Hoke *et al.* (2001) of vasa nervorum after complete nerve transections that were immediately resutured. Whereas endoneurial flow (HC) was well preserved in such lesions when examined 6 months later, there were substantial reductions in LDF erythrocyte flux at the transection and suture site. Findings suggested that the epineurial plexus may be susceptible to chronic scarring and devascularization, whereas endoneurial flow is compensated by maintained longitudinal feeding vessels.

[^{14}C]Iodoantipyrine, an isotope with relatively high tissue penetration, has been used as an autoradiographic indicator of local blood flow in the peripheral nerve (Sladky *et al.*, 1985, 1991; Rundquist *et al.*, 1985). This technique provides high-quality selective information about epineurial/perineurial and endoneurial blood flow. Rundquist *et al.* (1985) demonstrated that local blood flow with [^{14}C]iodoantipyrine measured 13 ml/100 g/min in nerve stripped of epineurium (reflecting endoneurial blood flow) and 27 ml/100 g/min of whole nerve, a value close to the composite values estimated using microelectrode hydrogen clearance polarography. There was a close correlation between estimates of blood flow and recordings using LDF. Disadvantages of the technique include the need for isotope handling, the delay in generating autoradiographic images, and the inability to perform serial measurements in a single animal.

Nerve blood flow has been calculated from the distribution of labeled microspheres into nerve. Because this technique generally involves removal of the complete nerve trunk, without separation of the epineurium and endoneurium, values calculated reflect whole trunk blood flow, as discussed earlier. Most estimates of blood flow using labeled microspheres, however, are lower than those provided by microelectrode hydrogen clearance polarography or [^{14}C]iodoantipyrine distribution (Pugliese *et al.*, 1989; Sutera *et al.*, 1992; Tilton *et al.*, 1995). One explanation for such findings has been that arteriovenous shunting vessels that are relatively large fail to trap microspheres (that are, in turn, trapped in the pulmonary vasculature, preventing redistribution), underestimating their distribution. One report (Chang *et al.*, 1997) sought to circumvent this difficult by injecting a "particulate microsphere" [^{3}H] desmethylimipramine, unlikely to be shunted or held up in pulmonary vessels. As with other isotope techniques, however, only a single measurement in a given preparation is possible.

Other approaches to measure nerve blood flow have been reported infrequently and include [^{14}C]-butanol distribution (Sugimoto *et al.*, 1986) and videoangiology (Zochodne and Low, 1990) in rats and fluroescein angiography in humans (Tesfaye *et al.*, 1993). The latter two techniques do not measure blood flow directly. Videoangiology is a live visualization of epineurial plexus vessels to address responses of specific vascular segments to vasoactive agents. The highly variable choice of where to visualize and the confinement of all of these choices to the epineurial plexus are drawbacks.

Morphological approaches to studying vasa nervorum have included counting and sizing vessels under light microscopy (LM) or EM using glutaraldehyde-fixed (by perfusion, *in situ* superfusion or immersion), epon-embedded, and toluidine blue-stained transverse nerve sections, counting immunohistochemistically labeled vascular profiles (e.g., factor VIII), or perfusing vessels with Indian ink or a similar indicator (Sugimoto and Yagihashi, 1997; Zochodne and Nguyen, 1997, 1999; Cameron *et al.*, 1991a; Sugimoto and Yagihashi, 1997). Studies using fixed sections can provide relatively accurate estimates of the numbers of vessels (there is a small risk of missing some small capillary profiles) and accurate information about vascular wall caliber. Measurements of luminal area, however, may not accurately reflect physiological status because of distortions during euthanasia and fixation. Despite this concern, this approach did identify rises in vessel caliber associated with early nerve injury, previously and subsequently verified using physiological blood flow measurements and an Indian ink perfusion approach (see later) (Podhajsky and Myers, 1993). Immunohistochemical approaches may be used to estimate vessel numbers (assuming that nonspecific labeling of the marker can be confidently excluded) but not their caliber (Cameron *et al.*, 1991a). Indian ink-labeled vascular profiles can be used to measure numbers and estimate the caliber of vessels. One disadvantage has been that Indian ink escapes from relatively leaky epineurial vessels, limiting analysis to the endoneurial compartment (Zochodne *et al.*, 1990).

In our laboratory, adopting an approach originally described by Bray *et al.* (1990) examining vessels in rat knee ligaments, we studied microvascular morphometry in unfixed rat sciatic nerves and ganglia after an infusion with a combination of Indian ink, mannitol, and gelatin (Zochodne and Nguyen, 1997, 1999). The perfusate (approximately 20 ml) was slowly (1 ml/min) injected into a catheter inserted into one femoral artery so that the tip lay close to the bifurcation of the aorta. By this approach, the contralateral limb and its sciatic nerve were directly perfused while the rat was alive. Later during the infusion, recirculation of Indian ink and perfusion of other organs were associated with eventual cardiopulmonary arrest. At this stage, the carcass was placed in a freezer at $-20°C$ for 1 h, so that the perfused microvessels solidify forming luminal casts and the nerve was removed, placed in an OCT block without fixation, and then later sectioned on a cryostat. The use of mannitol and gelatin followed by cooling prevents vascular leakage, allowing morphometric study of the epineurial plexus, as well as the endoneurial compartment. Vessels were counted and sized using an image analysis program.

Overall issues that emerge then in the measurement of blood flow are whether studies that apply specific techniques rigorously keep in mind the limitations of a particular approach. Physiological properties of the preparation are critical, such as near nerve temperature, core body temperature, mean arterial pressure, and arterial blood gas oxygenation, particularly for the more demanding HC technique. For the LDF technique, similar considerations apply: near nerve and core temperature, how many measurements along the nerve are sampled, whether a micromanipulator is used, whether ambient lights are shut off during the measurements, and whether mean arterial pressure is monitored. In reporting nerve or ganglia perfusion, all of the aforementioned data should be included, including a calculation of microvascular resistance (mean arterial pressure/blood flow) or conductance (blood flow/mean arterial pressure).

IV. Changes in Nerve Blood Flow with Ischemia and Injury

A. Ischemia Models

Changes in nerve blood flow have been reported in several experimental models of peripheral nerve ischemia. Sladky *et al.* (1985) measured regional sciatic-posterior tibial nerve blood flow in rats using [14C]iodoantipyrine autoradiographic distribution 90 min following acute occlusion of the femoral artery. A zone of ischemia (lowest NBF 4.0 ml/100 g/min) was maximum in a proximal tibial centrofascicular "watershed" zone. McManis and Low (1988) studied longitudinal and radial changes in NBF (HC) following femoral artery ligation, hypogastric arterial trunk ligation, or hypotension and compared these changes with alterations in endoneurial oxygen tension. A watershed zone of susceptibility to ischemia was encountered in the proximal tibial nerve, but there was no evidence of centrofascicular ischemia. In contrast, hypotension was associated with relative centrofascicular hypoxia. Day *et al.* (1989b) studied hindlimb and sciatic nerve ischemia in rats undergoing abdominal aortic occlusion and reperfusion. Despite interruption of flow in this major vessel, only mild reperfusion abnormalities were observed after 60 min of occlusion without an impact on nerve excitability. The findings illustrated the rich anastomotic vascular supply to the hindlimb in general and peripheral nerves in particular. Peripheral nerves are relatively difficult to make ischemic. Schmelzer *et al.* (1989) studied nerve ischemia following extensive ligations to arteries supplying the rat hindlimb, including the abdominal aorta, common iliac arteries, and anastomotic vessels. A temporary tie over aorta allowed investigation of reperfusion. Release of aortic occlusion and reperfusion was associated with a profound but also progressive decline in NBF (HC), suggesting reperfusion injury to vessels. After 1 h of ischemia during ligations, axonal excitability of sciatic-tibial motor fibers recovered with reperfusion, but after 3 h, there was persistent conduction block across the previously ischemic zone and loss of distal excitability, indicating early axonal degeneration. Zochodne *et al.* (1992) examined a model of segmental endoneurial ischemia (HC) of the rat sciatic nerve induced by exposure of the epineurial plexus to topical endothelin-1. In this model, there was a temporary but reversible motor conduction block across the ischemic zone in some treated nerves. This contrasted with work in diabetes, described later, where similar exposure to endothelin vasoconstriction resulted in nerve infarction. Endothelin studies have helped verify the conclusion that vasoactive events in the epineurial plexus have a significant bearing on downstream endoneurial flow and viability. Sladky *et al.* (1991) generated a model of chronic sciatic nerve endoneurial ischemia by creating a proximal limb arteriovenous shunt. Endoneurial blood flow, measured by [^{14}C]iodoantipyrine autoradiographic distribution 6 weeks after creation of the model, was reduced in an ischemic zone that began just proximal to the origin of the posterior tibial nerve and declined distally to a value of 3.6 ml/100 g/min. The model was associated with slowing of conduction velocity, abnormalities at nodes of Ranvier, and mild axonal atrophy.

Kihara *et al.* (1993) studied the impact of a microsphere embolization model on NBF (HC) and axonal degeneration. This model was
originally described by Nukada and Dyck (1984). Higher doses of microspheres ($>1 \times 10^6$) led to more severe ($>75\%$ degenerating fibers) axonal
degeneration with associated reductions in NBF of <3 ml/100 g/min at
7 days following embolization. Using the same model, hypothermia during
ischemia (Kihara *et al.*, 1996) protected against axonal degeneration from
microsphere embolization. NBF, however, at 7 days was reduced to a greater
extent in nerves that had been exposed to greater hypothermia, despite
the benefits of the intervention. A further paper by the same group (Kihara
et al., 1995) rescued similarly ischemic nerves using hyperbaric oxygenation
but no improvement of NBF at 7 days accompanied this. Overall, microsphere embolization studies have provided important information on what
the relationship between embolization dose and persistent ischemia might
be. Whether these levels of NBF, however, can be strictly correlated with
fiber degeneration is uncertain. What may be more important is the impact
of acute ischemia during the 3-h epoch of ischemia, when permanent
axonal damage can occur. Ischemia after this time may be important in
how regenerative and reparative events are marshaled.

Studies of experimental ischemia have therefore provided important
correlation with what levels of NBF are associated with fiber damage.
Overall, reductions of NBF below 5 ml/100 g/min appear to be required
to cause axonal degeneration. Interestingly, these levels, including those
of the chronic limb arterial ischemia model of Sladky and colleagues
discussed earlier, are considerably lower than values described in experimental diabetes.

B. Injury Models

Acute ischemia of the peripheral nerve might be an expected consequence of blunt or penetrating injury from physical disruption of vasa
nervorum or microthrombosis. Indeed, studies by Lundborg (1975) suggested that there was significant vascular disruption of vasa nervorum following
injury when examining perfused vascular profiles. These earlier studies,
however, did not include measurements of nerve blood flow. Somewhat
surprisingly then, quantitative measurements failed to identify ischemia,
even if measurements were made directly at sites of nerve crush or in
distal and proximal stumps after transection (Zochodne *et al.*, 1995, 2001;
Zochodne and Ho, 1990, 1992). Even longitudinally extensive crush zones,
associated with declines in epineurial plexus flow (LDF), had preserved
endoneurial (HC) blood flow in the center of the crush zone (Xu and
Zochodne, unpublished data). On the contrary, not only was blood flow

preserved, e.g., immediately at the site of an acute nerve crush, but there was a trend toward higher levels by 3 h after crush (Zochodne and Ho, 1990). By 24–48 h postcrush or proximal and distal to transection there were substantial rises in blood flow, up to twice normal values that were maintained for up to 14 days following the injury (Zochodne *et al.*, 1995; Zochodne and Ho, 1992). Parallel studies reported by Podhajski and Myers (1993) of morphological changes of endoneurial microvessels following crush identified an early wave of endoneurial microvessel dilatation, corresponding to the same time points that hyperemia was identified using physiological recordings (Podhajsky and Myers, 1993, 1994). Subsequent work linked such rises to CGRP and nitric oxide at the injury site (Zochodne *et al.*, 1995, 1999). CGRP and NO both appeared to arise from axonal end bulbs, swollen structures at the proximal end of severed axons that accumulate peptides, proteins, and other constituents from arrested axoplasmic transport. Although not explored, it is likely these molecules act synergistically on local vasa nervorum arterioles to account for vasodilation and rises in blood flow. Rises in local blood flow could be reversed by blocking α CGRP receptors or inhibiting the synthesis of NO by nitric oxide synthase (NOS). Later (7–14 days), persistent rises in nerve blood flow at the site of nerve injury were not linked to CGRP, but were linked to persistent NO action (possibly from a different isoform) and accompanied prominent mast cell accumulation, mast cell degranulation, and angiogenesis (Zochodne and Nguyen, 1997; Zochodne *et al.*, 1994). Podhajsky and Myers (1993, 1994) also linked a second wave of vascular change at similar time points with a rise in endoneurial vessel numbers. Hoke *et al.* (2001) confirmed that rises in epineurial vessel numbers detected in distal nerve stumps after transection accompanied rises in vascular endothelial growth factor (VEGF) mRNA. Chronic constriction injury (CCI) is a model of neuropathic pain induced by the placement of four loose chromic gut sutures around the sciatic nerve in rats or mice. Between the sutures, there is ischemic strangulation of the nerve with degeneration of large fibers (Sommer and Myers, 1996; Levy and Zochodne, 1998). Proximal to the sutures, local hyperemia related to CGRP and NO also developed (Levy and Zochodne, 1998). Distal to the sutures, there was a later rise in blood flow, likely related to angiogenesis, similar to other injuries described earlier. Shupeck *et al.* (1989) compared blood flow in 15-mm nerve autografts (transection at two sites then resuture) made with and without retaining a segmental vascular supply. At 24 h to 7 days following injury, endoneurial blood flow measured by HC was similar (mild trend toward reductions in nonvascularized grafts at 24, 48, and 72 h, but toward higher levels at 7 days) whether the vascular supply had been preserved or not, whereas nonnutritive (composite or fast component blood flow) was higher in the vascularized grafts.

In summary then, acute hyperemia of vasa nervorum followed by later angiogenesis has been identified following nerve injuries that include crush (within the crush zone), transection (both proximal and distal to the transection site), CCI, and placement of a ligature through one fascicle of the nerve. NBF is even preserved in nonvascularized grafts within 24-h.

Changes in local blood flow after later intervals following nerve injury have had less attention. Hoke *et al.* (2001) studied chronically denervated distal nerve stumps of transected rat sciatic nerves. Hyperemia gradually subsided by 1 month after the injury and by 6 months, blood flow was reduced significantly in both the endoneurial compartment (HC) and the epineurial plexus (LDF). Interestingly, the decline in flow developed despite retention of new vessels formed earlier after the injury. Xu and Zochodne (2002) studied local blood flow in chronic experimental neuromas of the proximal stump of the transected rat sciatic nerve. By 6 months, there was a substantial decline in stump tip epineurial plexus blood flow (LDF) but highly variable endoneurial blood flow readings (including zones with very low blood flow). Histologically, such lesions were similarly variable with zones containing populations of fibers associated with blood vessels and other areas, presumed ischemic, lacking both.

V. Nerve and Ganglia Blood Flow in Experimental and Human Diabetes

A. Nerve Blood Flow (NBF) and Experimental Diabetes

1. *Evidence That NBF Is Reduced in Experimental Diabetes*

In 1984, Tuck and colleagues reported that experimental streptozotocin (STZ)-induced diabetes in rats of 4 months duration had substantial reductions in endoneurial blood flow (measured by HC), rises in microvascular resistance, and lowered endoneurial oxygenation. The findings supported the hypothesis that early, e.g., electrophysiological, changes of diabetic polyneuropathy could be accounted for by functional microangiopathy and hypoxia. Subsequent studies from the same laboratory or investigators later established in their own laboratories using the same approach yielded similar findings (Cameron *et al.*, 1991; Kihara and Low, 1995b; Kihara *et al.*, 1991a, 1995). In addition, a number of like studies from the same investigators suggested that specific approaches to increase nerve blood flow, e.g., the use of sympathectomy or calcium channel antagonists, restored conduction abnormalities in diabetes in tandem with repairing deficits in blood flow (Cameron *et al.*, 1991a,b, 1993; Cameron and Cotter, 1996c;

Robertson *et al.*, 1992; Obrosova *et al.*, 2000). The list of such approaches using several techniques has grown and now includes beraprost, a prostacyclin analog (Ueno *et al.*, 1996; Hotta *et al.*, 1995a,b, 1996), antagonists of α_1-adrenergic receptors (doxazosin) (Cotter and Cameron, 1998), β_2-adrenergic agonists (salbutamol) (Cotter and Cameron, 1998), antagonists of both endothelin receptors (Stevens and Tomlinson, 1995) or of the endothelin-1 ETa receptor (Cameron and Cotter, 1996c), antagonists of the angiotensin II receptor (Cameron and Cotter, 1996c), cilostazol, a vasodilator (Kihara *et al.*, 1995; Hotta *et al.*, 1996b), carvedilol, a vasodilator (Cotter and Cameron, 1995), aldose reductase inhibitors (Hotta *et al.*, 1996; Cameron *et al.*, 1996; Cotter *et al.*, 1998; Yamamoto *et al.*, 2001; Yoshida *et al.*, 1998; Nakamura *et al.*, 1997, 1999), propionyl-L-carnitine, that may alter lipid balance or improve energy flux in cells (Hotta *et al.*, 1996a,b), evening primrose oil. (Stevens *et al.*, 1993; Cameron *et al.*, 1996), γ-linolenic acid or ascorbyl γ-linolenic acid (Cameron and Cotter, 1996a,b), aminoguanidine (Kihara *et al.*, 1991; Cameron and Cotter, 1996d), pentoxifylline (Flint *et al.*, 2000), cremophor, a diacylglycerol complexing agent (Jack *et al.*, 1999), a protein kinase C-β-selective inhibitor (Nakamura *et al.*, 1999), cilazapril, an angiotensin-converting enzyme inhibitor (Kihara *et al.*, 1999), or combinations of agents (Cameron *et al.*, 1996).

Related studies have improved NBF at the same time as conduction abnormalities, suggesting a reduction in oxidative stress, perhaps responsible for both. This argument has been applied to lipoic acid (Nagamatsu *et al.*, 1995; van Dam *et al.*, 2001; Stevens *et al.*, 2000; Cameron *et al.*, 1998). McManis and Low (1986) reported reductions in NBF in experimental galactose neuropathy. This is an important metabolic model of diabetic neuropathy that involves the excessive accumulation of polyols (galacticol) in nerve through accelerated aldose reductase activity. In half of the rats selected with the most prominent nerve edema, more striking reductions were observed in the subperineurial zone, postulated to be a result of increased subperineurial intercapillary distance.

In examining this large body of data, however, one is struck with the relatively invariable benefits of virtually every agent applied, both on blood flow and on nerve conduction. There are exceptions, with one study showing absolutely no effects of aldose reductase inhibition, under conditions when flux through the sorbitol pathway was probably completely blocked (Tomlinson *et al.*, 1998). This implies either that "negative" results are not being reported by some investigators or that all agents applied alter an easily modified parameter that has an impact on NBF recordings and conduction velocity. Another concern with vasodilator studies, while acceptable as basic work examining microangiopathy, is that they may be inappropriate approaches for later clinical trials. Many diabetic patients already take such vasoactive compounds for the control of hypertension, yet continue to suffer from polyneuropathy. Superimposing vasodilation on preexisting diabetic autonomic neuropathy also runs the serious risk of exacerbating postural hypotension, a significant cause of head injury, cerebral infarction, and myocardial ischemia in diabetic patients.

The inference in a number of pharmacological studies is that reversal of conduction abnormalities and changes in blood flow in tandem indicate that oligemia of nerve accounts for such abnormalities. Not usually documented in like studies have been other changes in experimental diabetic nerves, such as axon atrophy, axon loss, declines in neurofilament content, and other perhaps more relevant structural or molecular indices of nerve damage in diabetes. Finally, it is also uncertain whether reported benefits of any of the interventions described are attributable to an action on vasa nervorum or could operate at the level of dorsal root ganglia, spinal cord, or autonomic ganglia.

Surrogate measures of NBF support the argument that NBF is central to the pathogenesis of early experimental diabetic polyneuropathy. The most important such measurement has been that of endoneurial oxygen tension, reduced in both experimental and human diabetic polyneuropathy in studies thus far carried out (Newrick *et al.*, 1986; Tuck *et al.*, 1984; Zochodne and Ho, 1992a,b). An additional marker may be a decreased $NAD^+/NADH$ ratio and phosphocreatine levels in diabetes (Obrosova *et al.*, 2000). These findings are important evidence arguing that microangiopathy is indeed important and relevant. Such findings, however, do not necessarily imply that they arise from an overall decline in nerve blood flow but could, for instance, reflect more subtle alterations in oxygen delivery related to deficits in capillary (only) transit by nondeformable erythrocytes (Kowluru *et al.*, 1989), increased blood viscosity (Simpson, 1988), lowered 2,3-diphosphoglycerate levels (Nakamura *et al.*, 1995), or other factors.

2. *Evidence against Changes in NBF in Experimental Diabetes*

Not all investigators have identified reductions in nerve blood flow in experimental diabetes using physiological approaches. Williamson and colleagues, using measurements of microsphere distribution (limitations discussed earlier), consistently demonstrated early rises in nerve blood flow in experimental diabetes (Sutera *et al.*, 1992; Tilton *et al.*, 1995; Pugliese *et al.*, 1989). In a later paper, the same group used "particulate microsphere" [^{3}H]desmethylimipramine to address the criticisms of the microsphere approach and to demonstrate similar 1.5- to 2-fold rises in NBF in STZ-diabetic rats of 2–4 weeks duration (Chang *et al.*, 1997). Dines *et al.* (1999) reported that correction of a near nerve temperature from 32°

to 37°C changed nerve blood flow recording results, using laser doppler flowmetry (LDF), in diabetic [streptozotocin (STZ); 8 weeks] compared to nondiabetic rats. At 32°C, nerve blood flow was reduced in the diabetics compared to controls, but at 37°C, there was no difference between the two groups. Kalichman *et al.* (1998) also reported, in contrast to previous work, that NBF (LDF) in experimental galactose neuropathy was unchanged.

Keeping in mind the aforementioned studies, our laboratory has had findings differing from a substantial number of reports by other investigators, by failing to identify reductions in nerve blood flow in experimental diabetes. To summarize our studies, they have included measurements of NBF in STZ diabetes, compared to citrate-injected littermates after durations of hyperglycemia of 6 weeks, 16 weeks (two studies), 6 months, and 8 months, all using microelectrode (3- to 5-μm tip) hydrogen clearance polarography (Zochodne and Ho, 1992a,b, 1994a; Xu, Zochodne, and Sun, unpublished data; Kennedy and Zochodne, 2002) (Fig. 1). Specific projects using recordings of hydrogen clearance in diabetics and controls were carried out by four specifically trained experimentalists (Ho, Sun, Xu, and Kennedy; each project involving diabetics and their own controls, and each project was handled by one experimentalist) over different periods of time. While our initial studies were carried out expecting reductions to be identified (indicating a bias against the results identified), in the most recent study carried out at 8 months duration of diabetes, all curve analysis was carried out with the experimentalist (Kennedy) blinded to the status of the rats being studied. In additional

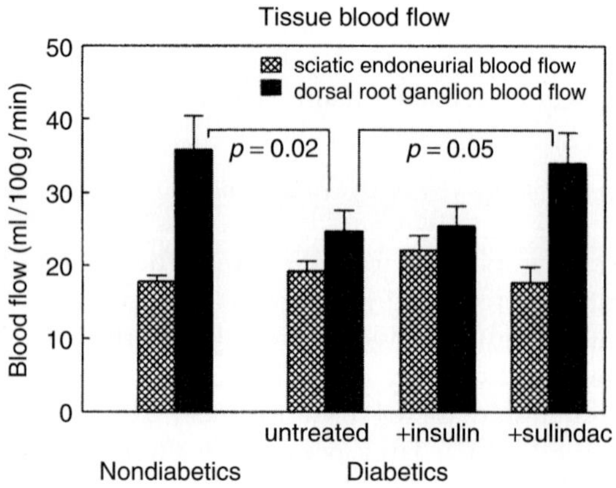

FIG. 1. Reductions in dorsal root ganglion (solid bars) but not sciatic endonerium blood flow (hatch bars) were identified in experimental STZ diabetes of 4 months duration. Reproduced with permission from Zochodne and Ho (1994a).

work, inherited diabetes in the BB Wistar model was studied 7–11 and 17–23 weeks after diabetes onset without evidence of oligemic NBF, also using HC (Zochodne *et al.*, 1994). In all models examined, concurrent electrophysiological recordings have identified slowing of conduction velocities identifying experimental neuropathy (Fig. 2). All recordings were carried out at a near nerve temperature of 37°C with recording of mean arterial pressure and arterial blood gases to ensure physiological stability. Examples of microelectrode HC curves are given in Fig. 3. Despite using an approach sensitive to peptidergic vasodilatation (Fig. 3b), peptide antagonism (Fig. 3b), or endothelin vasoconstriction (Fig. 3c), curves in diabetics have been routinely similar to nondiabetics (Fig. 3a). Calculations of microvascular resistance and composite blood flow (that includes a component from the epineurial plexus and arteriovenous shunts; see earlier discussion) were also similar between all diabetic models and nondiabetic controls. Measurements carried out using LDF, but applying the Kalichman and Lalonde approach of multiple sampling, also at 37°C, and after 6–8 weeks or 8 months (Zochodne *et al.*, 1996) (Kennedy and Zochodne, 2002) of STZ diabetes, did not identify differences in erythrocyte flux between diabetics and controls.

Similarly, our experience with approaches to treat experimental diabetic neuropathy have differed from other reported work. Our laboratory has explored the impact of guanethidine sympathectomy on experimental diabetic neuropathy in two separate investigations (Zochodne and Ho,

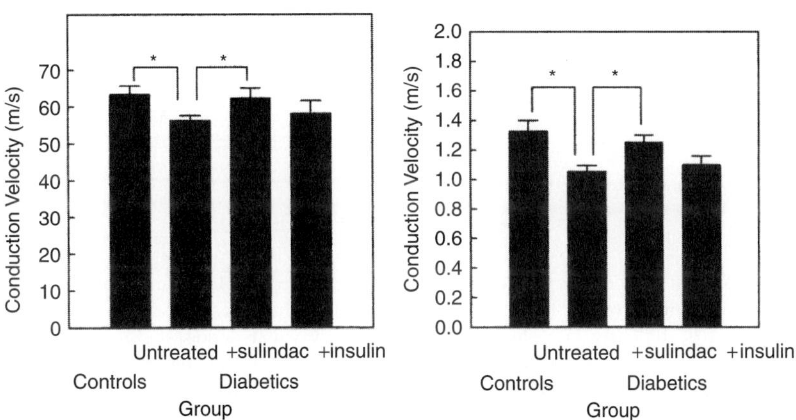

FIG. 2. Conduction velocity studies *in vitro* of myelinated (left) and unmyelinated (right) fibers in 4-month experimental STZ diabetes indicating reduction in conduction velocities in both fiber classes despite the preservation in endoneurial blood flow given in Fig. 1. Reproduced with permission from Zochodne and Ho (1994a).

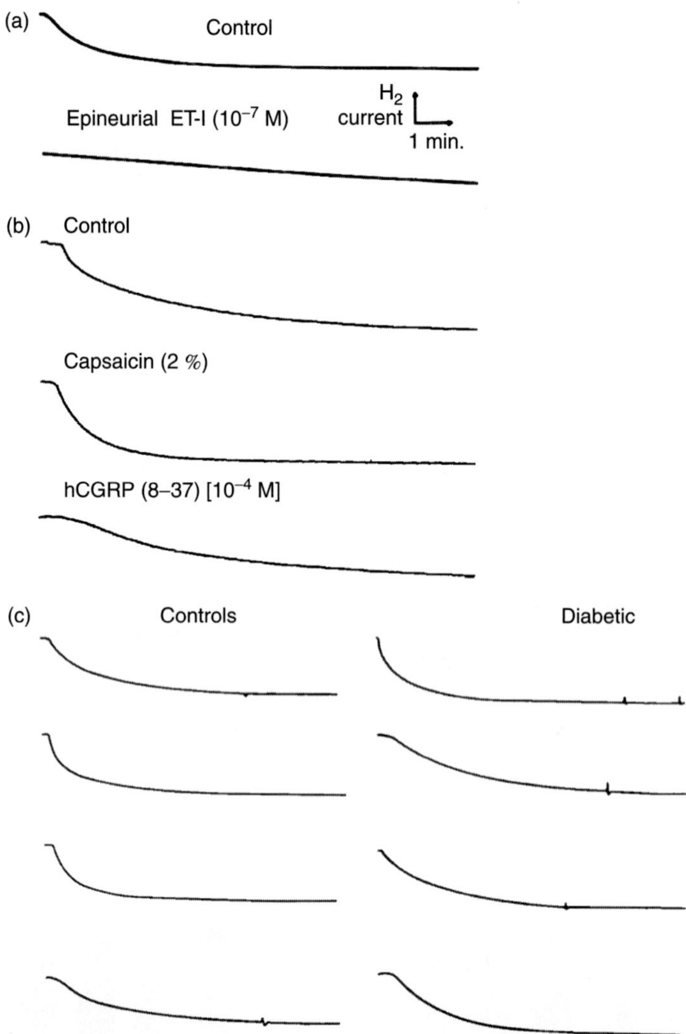

FIG. 3. Examples of microelectrode hydrogen clearance polarography washout curves. In normal sciatic nerves of the rat, topical endothelin induces vasoconstriction indicated by a flattening of the curve (a), whereas pepdigeric perivascular depolarization with capsaicin induces hyperemia, or a rise in nerve blood flow (b). In intact nerves, application of a CGRP antagonist causes vasoconstriction by removing tonic peptide vasodilation (b). No such changes in nerve blood flow were observed in experimental STZ diabetes of 6 months duration (c) illustrated in several individual rats. Reproduced with permission from Zochodne *et al.*, (1992) and Zochodne and Ho (1991b, 1992c).

1992a,b). Our rationale for proceeding was based on prior work carried out by the author when in the Low laboratory, which initially identified guanethidine sympathectomy in nondiabetic rats as generating chronic rises in NBF and dilation of nerve vessels (Zochodne *et al.*, 1990). This experiment, in turn, had been prompted by yet prior findings, also in the Low and Dyck laboratories, by the author that guanethidine sympathectomy apparently dilated microvessels in the cervical sympathetic trunk, being examined at that time for evidence of fiber loss (Zochodne *et al.*, 1989). After setting up an independent laboratory in 1988, we followed up the work on guanethidine hyperemia by applying it to experimental diabetes, an approach coincidentally adopted in parallel by Cameron *et al.* (1991b) at that time having arrived in the Low laboratory. Results of these parallel investigations, despite the use of apparently identical NBF-recording approaches, were quite different. Cameron *et al.* (1991) reported substantial deficits in NBF in experimental diabetes soon after the onset of hyperglycemia, and guanethidine sympathectomy not only restored this deficit, but corrected early slowing of nerve conduction velocity. In contrast, work in our hands failed to identify a reduction in NBF at a near nerve temperature of 37°C, and guanethidine sympathectomy (in two separate experiments) failed to correct conduction velocity, but instead worsened it (Zochodne and Ho, 1992a,b). Sympathectomized diabetic animals were chronically unwell and underweight, a feature of the work that did not encourage us to continue with it. In further work, we explored the impact of indomethacin and sulindac in experimental diabetes, both agents that improved conduction velocity abnormalities, but without meaningful impact on NBF (Zochodne and Ho, 1992a, 1994a). Finally, in a more recent experiment, Singhal *et al.* (1997) in our laboratory demonstrated that intermittent (three times weekly) near nerve injections of low-dose subhypoglycemic (not altering systemic hyperglycemia) insulin unilaterally corrected nerve conduction velocity, despite ongoing slowing in territories not directly exposed to it. This finding was important because it implied that conduction velocity could be corrected by brief intermittent therapy not altering the ongoing exposure of the treated nerve and its vessels to high concentrations of delivered glucose. Because reductions in NBF are thought to chronically contribute to conduction abnormalities in experimental diabetes, such intermittent treatment would not be expected to impact NBF. This finding was therefore difficult to explain on a microvascular basis. Similarly, acute insulin administration had been demonstrated to acutely worsen endoneurial hypoxia and might have been expected to worsen, not improve neuropathy (Kihara *et al.*, 1994). In addition to its impact on conduction velocity, near nerve insulin treatment was associated with an increase in the numbers of local small myelinated fibers, perhaps suggesting a local trophic action by insulin.

Acute axotomy of a peripheral nerve generates a change in perikaryal gene expression with upregulation of regeneration associated genes (RAGS) and declines in genes subserving maintenance or neurotransmitter roles. If axons in diabetes are targeted directly by diabetes, there should be associated RAG induction. No such induction, however, has been identified (Scott *et al.*, 1999; Zochodne *et al.*, 2000, 2001b; Zochodne and Verge, 1999). Distal axons are either not primarily targeted by diabetes or diabetes impairs RAG expression directly in sensory perikarya.

3. *Explaining Discrepant Findings*

For over a decade it has been difficult to reconcile diverse reports addressing the direction of change and importance of changes in NBF in diabetes. Unfortunately, some of this difficulty has not been addressed by a willingness of laboratories to explore these apparent differences and understand them. Dines, Kalichman, and colleagues in San Diego have, however, been among the few to directly test why such differences might occur. The role of near nerve temperature in changing the physiological conditions associated with measurements of NBF is an obvious choice for explaining different results. Pharmacological studies have provided good evidence that diabetes is associated with impaired vascular reactivity, with several studies identifying blunted vasodilation but excessive vasoconstriction. These findings have been attributed to mechanisms that include unresponsiveness of vessels to NO (Calver *et al.*, 1992; Elliott, *et al.*, 1993; Diederich *et al.*, 1994; Durante *et al.*, 1988; Hill *et al.*, 1990; Kihara and Low, 1995b; Lawrence and Brain, 1992; Tolins *et al.*, 1993), accelerated consumption of intraluminal NO by advanced glycosylation end products (Bucala *et al.*, 1991), unresponsiveness to peptide vasodilation (Zochodne and Ho, 1993b), loss of perivascular peptidergic fibers (Willars *et al.*, 1989), a shift in the thromboxane to prostacyclin ratio in blood favoring vasoconstriction (Ward *et al.*, 1989), increased sensitivity to angiotensin II (Kihara *et al.*, 1999), and increased levels of circulating endothelin, a potent vasoconstrictor (Takahashi *et al.*, 1990). In vasa nervorum specifically, impaired NO action, loss of peptidergic "neurogenic" vasodilation (Zochodne and Ho, 1993b), decreased prostacyclin (Ward *et al.*, 1989), and excessive responsiveness to endothelin vasoconstriction (resulting in nerve infarction) have all been demonstrated (Zochodne and Cheng, 1999; Zochodne *et al.*, 1996). In view of such changes in diabetes, excessive vasoconstriction of hindlimb arterioles supplying vasa nervorum could be a consequence of lower near nerve temperatures in physiological preparations, through heightened sympathetic activation, or other mechanisms. Dines and colleagues (1999) have suggested that hindlimb atrophy in diabetic models might account for

in experimental diabetes, might do so through direct actions on axon excitability.

Unfortunately, some of the published studies also pose problems in interpretation. For example, a report reversing abnormalities in experimental diabetes of rabbits and rats used a VEGF gene transfer protocol

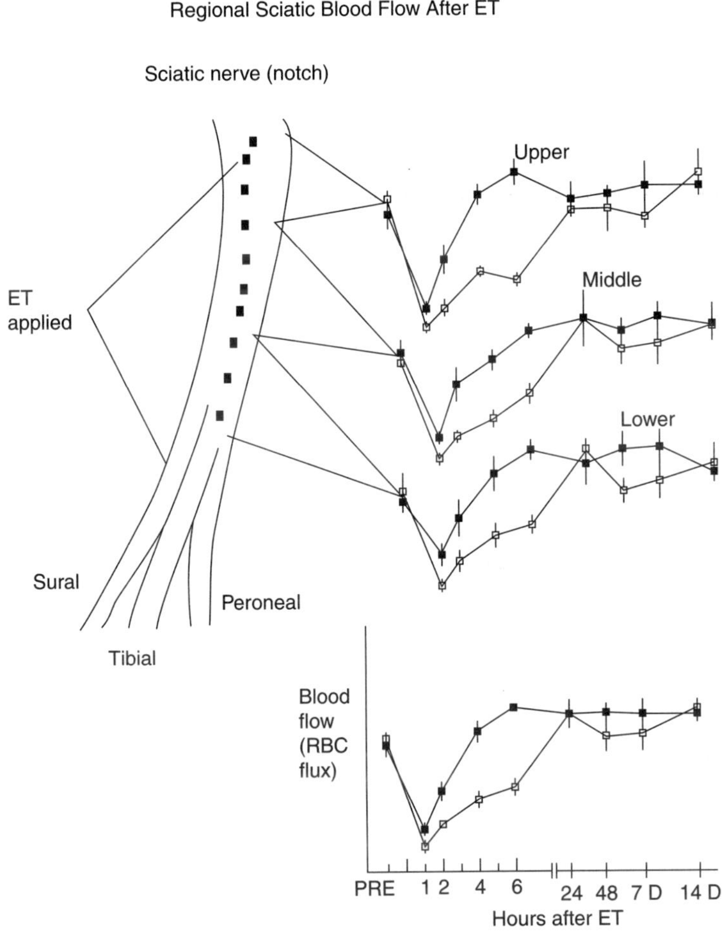

FIG. 4. Experimental work identifying prolonged reductions in diabetic NBF following topical administration of the vasoconstrictor endothelin. (a) Studies using LDF at several regions along the sciatic nerve identified more marked and prolonged vasoconstriction in the diabetics (lower curves, open symbols). (b) Acute ischemic conduction block is illustrated 2 h after ET application followed by distal loss of excitability that is persistent and indicates nerve infarction. Reproduced with permission from Zochodne *et al.* (1996).

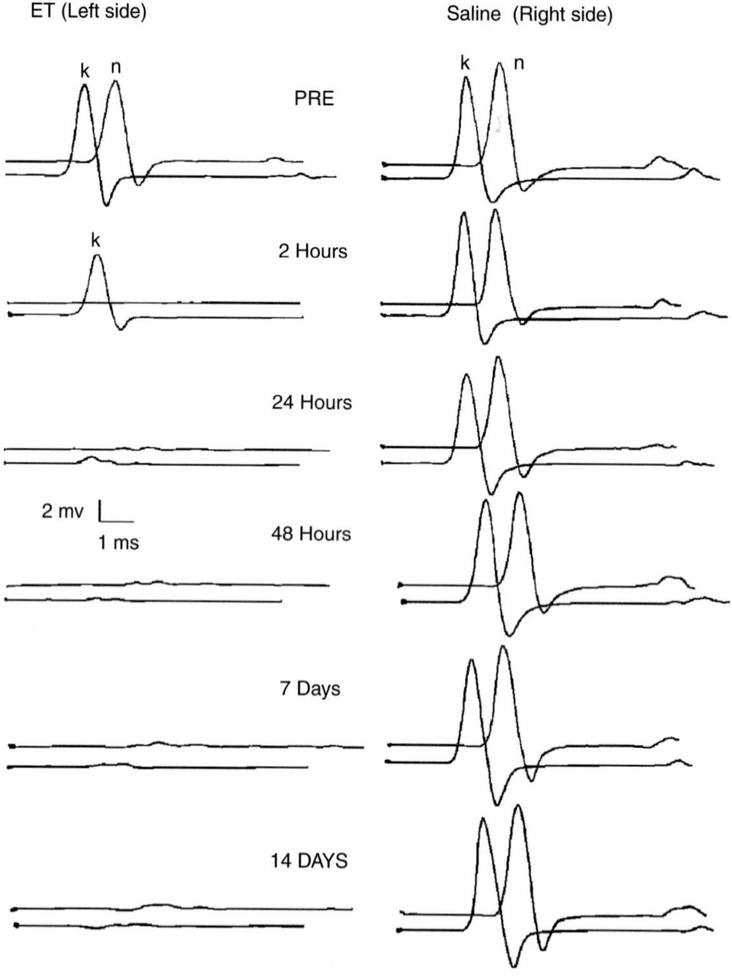

FIG. 4. (*continued*)

(Schratzberger *et al.*, 2001). While a novel approach, several questions have arisen from the approach. Given the failure of most investigators to identify loss of nerve vessels in diabetes, the need to supplement with VEGF is unclarified. VEGF may have direct actions on neurons (Sondell *et al.*, 1999). Excessive production of VEGF in the diabetic retina has been implicated

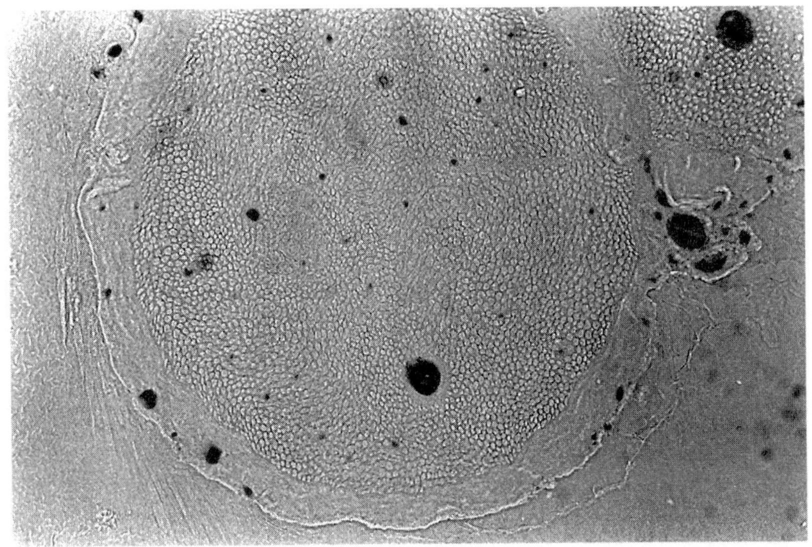

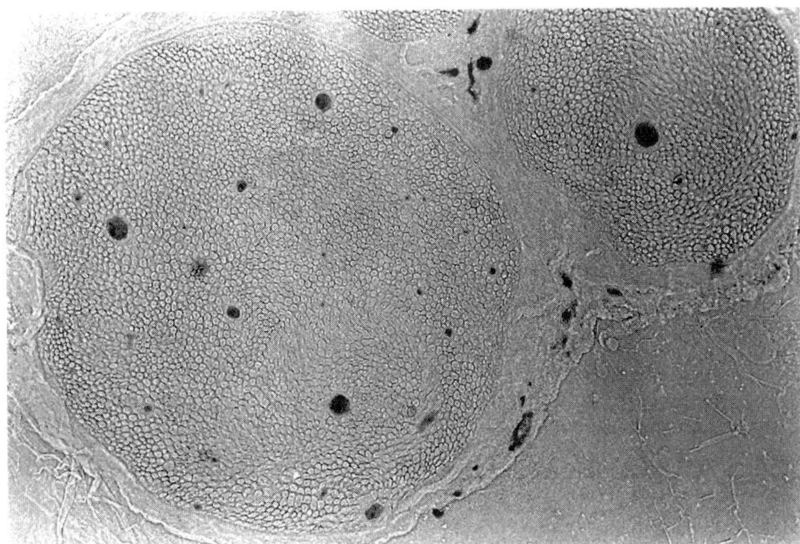

FIG. 5. Examples of epineurial and endoneurial vasa nervorum outlined with an India ink perfusion technique. (Top) Sciatic nerve of a rat with experimental STZ diabetes of 3 months duration. (Bottom) A similar section from a nondiabetic littermate. Quantitative studies identified a rise in the number of epineurial vessels in diabetics, suggesting angiogenesis. Note that there is no evidence of vascular loss or obliteration to suggest ongoing ischemia of the diabetic nerves. Reproduced from Zochodne and Nguyen (1997).

as a response to hypoxia that leads to retinal neovascularity, vessel rupture, and blindness (Hofman *et al.*, 2001). Nonvalidated approaches toward estimating vessel numbers (fluorescein perfusion without subcompartment analysis; approach to eliminate nonspecific fluorescence not documented) and blood flow (laser Doppler imager) were used in the work, and temperature maintenance during blood flow measurements or nerve conduction studies was not documented.

4. *Evidence of Functional Peripheral Nerve Trunk Microangiopathy without Oligemia*

Despite issues raised about the role of NBF in initiating experimental diabetic neuropathy, diabetic microangiopathy and endothelial dysfunction remain important features of diabetes (Hill *et al.*, 1990). Early functional microangiopathy may precede later structural abnormalities of vessels (Bohlen and Hankins, 1982). In some cases, these functional abnormalities comprise abnormally dilated microvessels or vessels susceptible to excessive vasoconstriction, as discussed earlier. That early functional microangiopathy occurs in experimental diabetes seems clear enough, particularly if one considers all prior reports of changes in NBF, irrespective of direction, as some evidence for this. The question arises as to whether ischemia of the peripheral nerve trunk, implying a mismatch between the relatively low metabolic demands of intact nerve and its blood supply, accounts for early changes of experimental diabetic neuropathy. For this, the evidence is debated, as discussed previously.

We encountered evidence of early functional microangiopathy in vasa nervorum of experimental diabetes despite preserved baseline NBF recordings. Diabetic nerves failed to develop inflammatory hyperemia from local capsaicin (Zochodne and Ho, 1993b). The failure of diabetic nerves to generate peptide-mediated rises in blood flow parallels deficits in neurogenic flare response in the skin of diabetic patients, a risk factor for foot ulcer development (Parkhouse and Le Quesne, 1988). Kennedy and Zochodne (2002) have provided evidence that microvascular changes that accompany peripheral nerve trunk injury are similarly impaired in diabetes. In this work, diabetes significantly attenuated early nerve trunk hyperemia in both proximal and distal stumps of transected sciatic nerves at early time points after injury and was associated with a delay in the development of injury-related angiogenesis. The findings have provided a potential mechanism for impaired diabetic nerve regeneration when the metabolic demands of proliferating and sprouting cells and axons may be high.

Finally, work addressing endothelin-induced epineurial vasoactivity (Zochodne *et al.*, 1997; Zochodne and Cheng, 1999) identified sensitivity of diabetic vasa nervorum to prolonged and intense vasoconstriction, resulting in multifocal axon damage (Figs. 4 and 6).

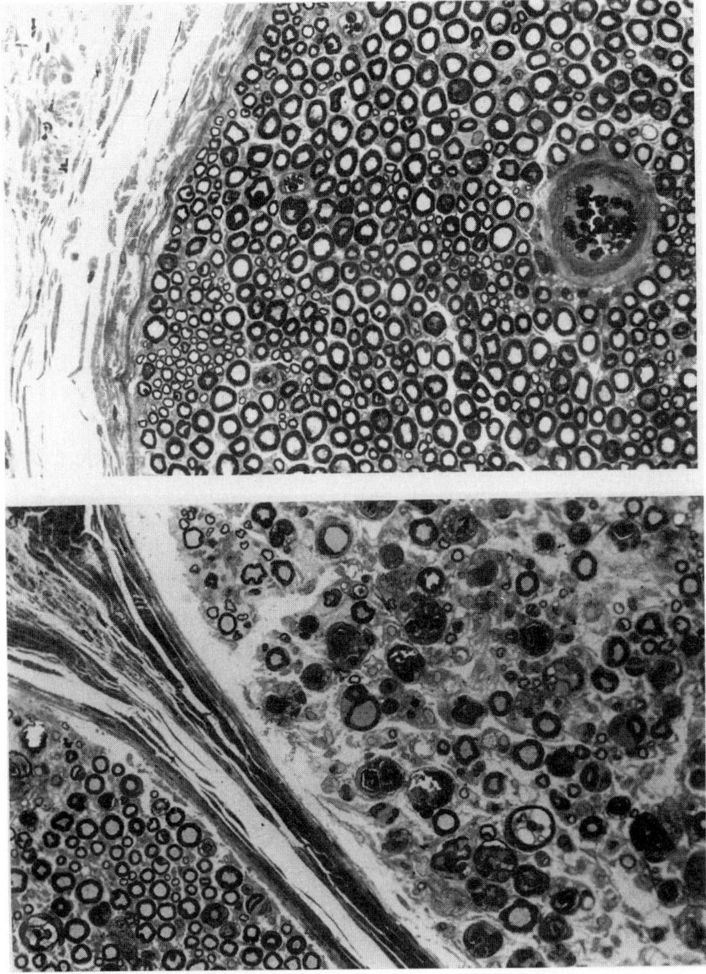

FIG. 6. An example of multifocal axon damage in a diabetic nerve exposed to topical endothelin. (Top) A section of an intact nondiabetic nerve exposed to endothelin. (Bottom) Severe axonal degeneration 14 days after ET application in one fascicle with preservation in an adjacent fascicle. Such a multifocal pattern of fiber loss is a feature of human diabetic polyneuropathy. Reproduced with permission from Zochodne and Cheng (1999).

B. NBF AND HUMAN DIABETES

Studies of nerve blood flow in human diabetes have been obviously difficult to come by. Histological work by Malik and colleagues (2001) suggests that microangiopathy can occur even in early human disease. This adds important information to the already considerable evidence that more established diabetic patients have a number of structural vessel abnormalities, including basement membrane thickening, endothelial reduplication, microthrombosis, capillary closure, smooth muscle proliferation, pericyte degeneration, and other changes (Yasuda and Dyck, 1987; Dyck and Giannini, 1996; Dyck *et al.*, 1985; Giannini and Dyck, 1995; Korthals *et al.*, 1988; Malik *et al.*, 1989, 1993; Malik, 1997; Timperley *et al.*, 1985; Williams *et al.*, 1980). The multifocal pattern of fiber loss in established patients with established diabetic neuropathy has also suggested that such nerves are ischemic (Dyck *et al.*, 1984, 1986a,b, Johnson *et al.*, 1986). The relationship, however, of this pattern of fiber loss to ischemia was questioned by Thomas and colleagues, who demonstrated similar multifocal fiber loss and microvascular abnormalities in chronic inherited neuropathies, where microangiopathy is not thought to contribute to axon loss (Bradley *et al.*, 1990; Llewelyn *et al.*, 1988). Tesfaye and colleagues (1993) demonstrated a delayed appearance of injected fluorescein visualized in exposed nerves of patients with chronic diabetic polyneuropathy, suggesting nerve ischemia. While not a direct measure of nerve blood flow, the findings suggested epineurial microangiopathy and there was also evidence of excessive arteriovenous shunting. Diabetic subjects without neuropathy had normal studies. Newrick *et al.* (1986) made direct measurements of oxygen tension in sural nerves of patients with chronic polyneuropathy and identified significant reductions in oxygen tension compared to nondiabetics. Similar reductions in oxygen saturation, accompanied by a delayed fluorescein appearance rise time, were observed by Ibrahim *et al.* (1999) using direct microlight-guide spectrophotometry in patients with mild to moderate sensory motor polyneuropathy. Young *et al.* (1995) noted that slowing of peroneal conduction velocity in diabetic patients undergoing limb revascularization procedures was improved. As in the experimental neuropathy studies discussed earlier, however, such changes in conduction velocity might also be accounted for by limb warming after revascularization.

Theriault *et al.* (1997) failed to identify declines in blood flow in diabetic patients in comparison to nondiabetic patients scheduled for sural nerve biopsy. Direct sural blood flow measurements were made using a laser Doppler flowmetry probe with multiple measurements taken along the exposed nerve using a micromanipulator, with the operating theater lights shut off, prior to nerve resection. The diabetic subjects were recruited as part of a clinical trial in which subjects agreed to a sural nerve biopsy in one leg prior to entry and then a contralateral sural biopsy 1 year later after treatment. The study specifically addressed patients that had early diabetic polyneuropathy (with a preserved sural potential). Nondiabetic sural nerve biopsies included those taken for routine diagnostic purposes for a variety of conditions. Diabetic erythrocyte flux was comparable to that of nondiabetics with a nonsignificant trend toward higher values. One year after the initial measurement and biopsy, erythrocyte flux in their contralateral sural nerve again trended toward higher rather than lower values within individual patients, despite evidence of progressive fiber loss over the year (Fig. 7). Patients with necrotizing vasculitis and one patient exposed to epineurial epinephrine during the nerve exposure had substantial reductions in nerve blood flow. A single patient with diabetic lumbosacral plexopathy and severe fiber loss (as well as capillary closure observed on the nerve biopsy) had a marked reduction in erythrocyte flux. One drawback of the study was the inability to control near nerve temperature (for ethical reasons) and the inability to sample the endoneurial vascular compartment of the nerve.

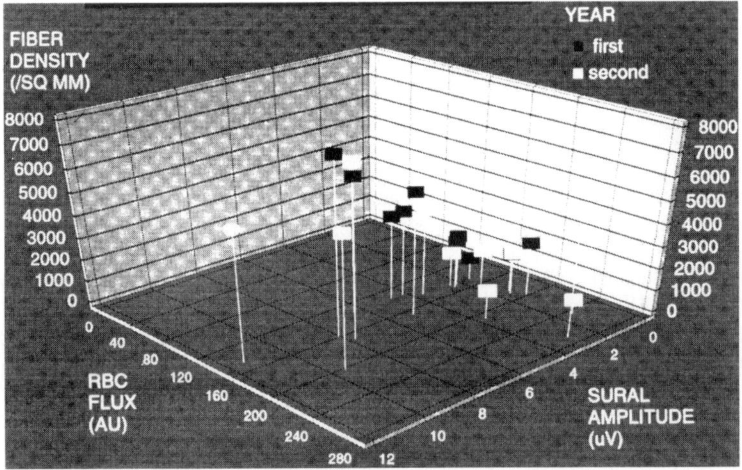

FIG. 7. Human nerve blood flow measurements using LDF in patients with mild diabetic polyneuropathy undergoing nerve biopsy as part of a clinical trial. RBC (erythrocyte) flux did not correlate with reductions in sural amplitude or fiber density, i.e., there were trends toward rises in flux with declines in amplitude and density, suggesting increased nerve blood flow despite loss of axons. Dark flags indicate an initial biopsy, and white flags indicate a second measurement in the contralateral sural nerve of the same patient 1 year later. Despite trends toward declines in sural potentials and fiber density over the year, RBC flux values did not decline. Reproduced with permission from Theriault *et al.* (1997).

C. GANGLION BLOOD FLOW AND EXPERIMENTAL DIABETES

Four published studies have examined local ganglion blood flow in experimental diabetes. Zochodne and Ho (1994a,b) observed selective reductions in ganglion, but not sciatic nerve blood flow in two separate models of experimental diabetes. Sixteen weeks after induction of diabetes using STZ, there was a reduction in ganglion blood flow by approximately 30% corrected by the use of sulindac. Reductions of DRG blood flow, but not sciatic nerve blood flow, were observed after longer (17–23 weeks) but not shorter (7–11) durations of diabetes in female BB Wistar rats with genetic diabetes. Sasaki *et al.* (1997) described reductions of sciatic, DRG, and superior sympathetic ganglia blood flow in STZ-diabetic rats of 12–15 months duration using [^{14}C]iodoantipyrine autoradiography to measure blood flow. Cameron and Cotter (2001) reported a reduction of superior cervical ganglion blood flow using HC in 24-week STZ-diabetic rats.

Acknowledgments

Brenda Boake provided expert secretarial assistance. DWZ is a Senior Medical Scholar of the Alberta Heritage Foundation for Medical Research.

References

Adams, W. E. (1942). The blood supply of nerves. I. Historical review. *J. Anat.* **76**, 323–341.

Appenzeller, O., Dhital, K. K., Cowen, T., and Burnstock, G. (1984). The nerves to blood vessels supplying blood to nerves: the innervation of vasa nervorum. *Brain Res.* **304**, 383–386.

Asbury, A. K., and Fields, H. L. (1984). Pain due to peripheral nerve damage: An hypothesis. *Neurology* **34**, 1587–1590.

Aukland, K., Bower, B. F., and Berliner, R. W. (1964). Measurement of local blood flow with hydrogen gas. *Circ. Res.* **14**, 164–187.

Bell, M. A., and Weddell, A. G. (1984). A descriptive study of the blood vessels of the sciatic nerve in the rat, man and other mammals. *Brain* **107**, 871–898.

Biessels, G. J., Stevens, E. J., Mahmood, S. J., Gispen, W. H., and Tomlinson, D. R. (1996). Insulin partially reverses deficits in peripheral nerve blood flow and conduction in experimental diabetes. *J. Neurol. Sci.* **140**, 12–20.

Bohlen, H. G., and Hankins, K. D. (1982). Early arteriolar and capillary changes in streptozotocin-induced diabetic rats and intraperitoneal hyperglycaemic rats. *Diabetologia* **22**, 344–348.

Bove, G. M., and Light, A. R. (1997). The nervi nervorum, missing link for neuropathic pain? *Pain Forum* **6**, 181–190.

Bradley, J., Thomas, P. K., King, R. H., Llewelyn, J. G., Muddle, J. R., and Watkins, P. J. (1990). Morphometry of endoneurial capillaries in diabetic sensory and autonomic neuropathy. *Diabetologia* **33**, 611–618.

Brain, S. D., Williams, T. J., Tippins, J. R., Morris, H. R., and MacIntyre, I. (1985). Calcitonin gene-related peptide is a potent vasodilator. *Nature* **313**, 54–56.

Bray, R. C., Fisher, A. W., and Frank, C. B. (1990). Fine vascular anatomy of adult rabbit knee ligaments. *J. Anat.* **172**, 69–79.

Bucala, R., Tracey, K. J., and Cerami, A. (1991). Advanced glycosylation products quench nitric oxide and mediate defective endothelium-dependent vasodilatation in experimental diabetes. *J. Clin. Invest.* **87**, 432–438.

Calcutt, N. A., Mizisin, A. P., and Kalichman, M. W. (1994). Aldose reductase inhibition, Doppler flux and conduction in diabetic rat nerve. *Eur. J. Pharmacol.* **251**, 27–33.

Calver, A., Collier, J., and Vallance, P. (1992). Inhibition and stimulation of nitric oxide synthesis in the human forearm arterial bed of patients with insulin-dependent diabetes. *J. Clin. Invest.* **90**(6), 2548–2554.

Cameron, N. E., and Cotter, M. A. (1994). Effects of evening primrose oil treatment on sciatic nerve blood flow and endoneurial oxygen tension in streptozotocin-diabetic rats. *Acta Diabetol.* **31**, 220–225.

Cameron, N. E., and Cotter, M. A. (1996a). Comparison of the effects of ascorbyl gamma-linolenic acid and gamma-linolenic acid in the correction of neurovascular deficits in diabetic rats. *Diabetologia* **39**, 1047–1054.

Cameron, N. E., and Cotter, M. A. (1996b). Interaction between oxidative stress and gamma-linolenic acid in impaired neurovascular function of diabetic rats. *Am. J. Physiol.* **271**, E471–E476.

Cameron, N. E., and Cotter, M. A. (1996c). Effects of a nonpeptide endothelin-1 ETA antagonist on neurovascular function in diabetic rats: Interaction with the renin-angiotensin system. *J. Pharmacol. Exp. Ther.* **278**, 1262–1268.

Cameron, N. E., and Cotter, M. A. (1996d). Rapid reversal by aminoguanidine of the neurovascular effects of diabetes in rats: Modulation by nitric oxide synthase inhibition. *Metabolism* **45**, 1147–1152.

Cameron, N. E., and Cotter, M. A. (2001). Diabetes causes an early reduction in autonomic ganglion blood flow in rats. *J. Diabet. Complicat.* **15**, 198–202.

Cameron, N. E., Cotter, M. A., Dines, K. C., and Maxfield, E. K. (1993). Pharmacological manipulation of vascular endothelium function in non-diabetic and streptozotocin-diabetic rats: Effects on nerve conduction, hypoxic resistance and endoneurial capillarization. *Diabetologia* **36**, 516–522.

Cameron, N. E., Cotter, M. A., Dines, K. C., Maxfield, E. K., Carey, F., and Mirrlees, D. J. (1994a). Aldose reductase inhibition, nerve perfusion, oxygenation and function in streptozotocin-diabetic rats: Dose-response considerations and independence from a myo-inositol mechanism. *Diabetologia* **37**, 651–663.

Cameron, N. E., Cotter, M. A., Ferguson, K., Robertson, S., and Radcliffe, M. A. (1991a). Effects of chronic alpha-adrenergic receptor blockade on peripheral nerve conduction, hypoxic resistance, polyols, Na$^+$-k$^+$-ATPase activity, and vascular supply in STZ-D rats. *Diabetes* **40**, 1652–1658.

Cameron, N. E., Cotter, M. A., and Hohman, T. C. (1996). Interactions between essential fatty acid, prostanoid, polyol pathway and nitric oxide mechanisms in the neurovascular deficit of diabetic rats. *Diabetologia* **39**, 172–182.

Cameron, N. E., Cotter, M. A., Horrobin, D. H., and Tritschler, H. J. (1998). Effects of alpha-lipoic acid on neurovascular function in diabetic rats: Interaction with essential fatty acids. *Diabetologia* **41**, 390–399.

Cameron, N. E., Cotter, M. A., and Low, P. A. (1991b). Nerve blood flow in early experimental diabetes in rats: relation to conduction deficits. *Am. J. Physiol.* **261**, E1–E8.

Cameron, N. E., Dines, K. C., and Cotter, M. A. (1994b). The potential contribution of endothelin-1 to neurovascular abnormalities in streptozotocin-diabetic rats. *Diabetologia* **37**, 1209–1215.

Chang, K., Ido, Y., LeJeune, W., Williamson, J. R., and Tilton, R. G. (1997). Increased sciatic nerve blood flow in diabetic rats: Assessment by "molecular" vs. particulate microspheres. *Am. J. Physiol.* **273**, E164–E173.

Cotter, M. A., and Cameron, N. E. (1995). Neuroprotective effects of carvedilol in diabetic rats: Prevention of defective peripheral nerve perfusion and conduction velocity. *Naunyn Schmiedebergs Arch. Pharmacol.* **351**, 630–635.

Cotter, M. A., and Cameron, N. E. (1998). Correction of neurovascular deficits in diabetic rats by beta2-adrenoceptor agonist and alpha1-adrenoceptor antagonist treatment: Interactions with the nitric oxide system. *Eur. J. Pharmacol.* **343**, 217–223.

Cotter, M. A., Cameron, N. E., and Hohman, T. C. (1998). Correction of nerve conduction and endoneurial blood flow deficits by the aldose reductase inhibitor, tolrestat, in diabetic rats. *J. Peripher. Nerv. Syst.* **3**, 217–223.

Cotter, M. A., Love, A., Watt, M. J., Cameron, N. E., and Dines, K. C. (1995). Effects of natural free radical scavengers on peripheral nerve and neurovascular function in diabetic rats. *Diabetologia* **38**, 1285–1294.

Dashwood, M. R., and Thomas, P. K. (1997). Neurovascular [125I]-ET-1 binding sites on human peripheral nerve. *Endothelium* **5**, 119–123.

Day, T. J., Lagerlund, T. D., and Low, P. A. (1989a). Analysis of H2 clearance curves used to measure blood flow in rat sciatic nerve. *J. Physiol.* **414**, 35–54.

Day, T. J., Schmelzer, J. D., and Low, P. A. (1989b). Aortic occlusion and reperfusion and conduction, blood flow and the blood-nerve barrier of rat sciatic nerve. *Exp. Neurol.* **103**, 173–178.

Dhital, K., Lincoln, J., Appenzeller, O., and Burnstock, G. (1986b). Adrenergic innervation of vasa and nervi nervorum of optic, sciatic, vagus and sympathetic nerve trunks in normal and streptozotocin-diabetic rats. *Brain Res.* **367**, 39–44.

Dhital, K. K., and Appenzeller, O. (1988). Innervation of vasa nervorum. In "Nonadrenergic Innervation of Blood Vessels" (G. Burnstock and S. G. Griffith, eds.), pp. 191–211. CRC Press, Boca Raton, FL.

Diederich, D., Skopec, J., Diederich, A., and Dai, F. X. (1994). Endothelial dysfunction in mesenteric resistance arteries of diabetic rats: Role of free radicals. *Am. J. Physiol.* **266**, H1153–H1161.

Dines, K. C., Calcutt, N. A., Nunag, K. D., Mizisin, A. P., and Kalichman, M. W. (1999). Effects of hindlimb temperature on sciatic nerve laser Doppler vascular conductance in control and streptozotocin-diabetic rats. *J. Neurol. Sci.* **163**, 17–24.

Dines, K. C., Cameron, N. E., and Cotter, M. A. (1995). Comparison of the effects of evening primrose oil and triglycerides containing y-linolenic acid on nerve conduction and blood flow in diabetic rats. *J. Pharmacol. Exp. Ther.* **273**, 49–55.

Durante, W., Sen, A. K., and Sunahara, F. A. (1988). Impairment of endothelium-dependent relaxation in aortae from spontaneously diabetic rats. *Br. J. Pharmacol.* **94**, 463–468.

Dyck, P. J., and Giannini, C. (1996). Pathologic alterations in the diabetic neuropathies of humans: a review. *J. Neuropathol. Exp. Neurol.* **55**, 1181–1193.

Dyck, P. J., Hansen, S., Karnes, J., O'Brien, P., Yasuda, H., Windebank, A., and Zimmerman, B. (1985). Capillary number and percentage closed in human diabetic sural nerve. *Proc. Natl. Acad. Sci. USA* **82**, 2513–2517.

Dyck, P. J., Karnes, J., O'Brien, P., Nukada, H., Lais, A., and Low, P. (1984). Spatial pattern of nerve fiber abnormality indicative of pathologic mechanism. *Am. J. Pathol.* **117**, 225–238.

Dyck, P. J., Karnes, J. L., O'Brien, P., Okazaki, H., Lais, A., and Engelstad, J. (1986a). The spatial distribution of fiber loss in diabetic polyneuropathy suggests ischemia. *Ann. Neurol.* **19**, 440–449.

Dyck, P. J., Lais, A., Karnes, J. L., O'Brien, P., and Rizza, R. (1986b). Fiber loss is primary and multifocal in sural nerves in diabetic polyneuropathy. *Ann. Neurol.* **19**, 425–439.

Elliott, T. G., Cockcroft, J. R., Groop, P. H., Viberti, G. C., and Ritter, J. M. (1993). Inhibition of nitric oxide synthesis in forearm vasculature of insulin-dependent diabetic patients: Blunted vasoconstriction in patients with microalbuminuria. *Clin. Sci.* **85**, 687–693.

Flint, H., Cotter, M. A., and Cameron, N. E. (2000). Pentoxifylline effects on nerve conduction velocity and blood flow in diabetic rats. *Int. J. Exp. Diabet. Res.* **1**, 49–58.

Giannini, C., and Dyck, P. J. (1995). Basement membrane reduplication and pericyte degeneration precede development of diabetic polyneuropathy and are associated with its severity. *Ann. Neurol.* **37**(4), 498–504.

Hill, M. A., Meininger, G. A., and Larkins, R. G. (1990). Alterations in microvascular reactivity in experimental diabetes mellitus: Contribution of the endothelium? *In* "Endothelial Cell Function in Diabetic Microangiopathy: Problems in Methodology and Clinical Aspects" (G. M. Molinatti, R. S. Bar, F. Belfiore, and M. Porta, eds.), pp. 118–126. Karger, Basel.

Hofman, P., Blaauwgeers, H. G., Vrensen, G. F., and Schlingemann, R. O. (2001). Role of VEGF-A in endothelial phenotypic shift in human diabetic retinopathy and VEGF-A-induced retinopathy in monkeys. *Ophthalmic Res.* **33**, 156–162.

Hoke, A., Sun, H., Gordon, T., and Zochodne, D. W. (2001). Do denervated peripheral nerve trunks become ischemic? *Exp. Neurol.* **172**, 398–406.

Holzer, P. (1988). Local effector functions of capsaicin-sensitive sensory nerve endings: Involvement of tachykinins, calcitonin gene-related peptide and other neuropeptides. *Neuroscience* **24**, 739–768.

Holzer, P. (1991). Capsaicin: Cellular targets, mechanisms of action, and selectivity for thin sensory neurons. *Pharmacol. Rev.* **43**, 143–201.

Hotta, N., Kakuta, H., Fukasawa, H., Koh, N., Sakakibara, F., Komori, H., and Sakamoto, N. (1992). Effect of niceritrol on streptozocin-induced diabetic neuropathy in rats. *Diabetes* **41**, 587–591.

Hotta, N., Koh, N., Sakakibara, F., Nakamura, J., Hamada, Y., Hara, T., Mori, K., Nakashima, E., Naruse, K., Fukasawa, H., Kakuta, H., and Sakamoto, N. (1996a). Effects of beraprost sodium and insulin on the electroretinogram, nerve conduction, and nerve blood flow in rats with streptozotocin-induced diabetes. *Diabetes* **45**, 361–366.

Hotta, N., Koh, N., Sakakibara, F., Nakamura, J., Hamada, Y., Hara, T., Mori, K., Naruse, K., Fukasawa, H., Kakuta, H., and Sakamoto, N. (1996b). Nerve function and blood flow in Otsuka Long-Evans Tokushima Fatty rats with sucrose feeding: Effect of an anticoagulant. *Eur. J. Pharmacol.* **313**, 201–209.

Hotta, N., Koh, N., Sakakibara, F., Nakamura, J., Hamada, Y., Wakao, T., Hara, T., Mori, K., Naruse, K., Nakashima, E., and Sakamoto, N. (1996c). Effect of propionyl-L-carnitine on motor nerve conduction, autonomic cardiac function, and nerve blood flow in rats with streptozotocin-induced diabetes: Comparison with an aldose reductase inhibitor. *J. Pharmacol. Exp. Ther.* **276**, 49–55.

Hotta, N., Koh, N., Sakakibara, F., Nakamura, J., Hamada, Y., Wakao, T., Hara, T., Mori, K., Naruse, K., Nakashima, W., *et al.* (1995a). Prevention of abnormalities in motor nerve conduction and nerve blood-flow by a prostacyclin analog, beraprost sodium, in streptozotocin-induced diabetic rats. *Prostaglandins* **49**, 339–349.

Hotta, N., Koh, N., Sakakibara, F., Nakamura, J., Hamara, Y., Hara, T., Nakashima, E., Sasaki, H., Fukasawa, H., Kakuta, H., and Sakamoto, N. (1996d). Effects of propionyl-L-carnitine and insulin on the electroretinogram, nerve conduction and nerve blood flow in rats with streptozotocin-induced diabetes. *Pflug. Arch.* **431**, 564–570.

Hotta, N., Koh, N., Sakakibara, F., Nakamura, J., Hara, T., Yamada, H., Hamada, Y., and Takeuchi, N. (1995b). Neurotropin prevents neurophysiological abnormalities and ADP-induced hyperaggregability in rats with streptozotocin-induced diabetes. *Life Sci.* **57**, 2101–2111.

Hromada, J. (1963). On the nerve supply of the connective tissue of some peripheral nervous system components. *Acta Anat.* **55**, 343–351.

Ibrahim, S., Harris, N. D., Radatz, M., Selmi, F., Rajbhandari, S., Brady, L., Jakubowski, J., and Ward, J. D. (1999). A new minimally invasive technique to show nerve ischaemia in diabetic neuropathy. *Diabetologia* **42**, 737–742.

Ido, Y., Chang, K., LeJeune, W., Tilton, R. G., Monafo, W. W., and Williamson, J. R. (1997). Diabetes impairs sciatic nerve hyperemia induced by surgical trauma: implications for diabetic neuropathy. *Am. J. Physiol.* **273**, E174–E184.

Jack, A. M., Cameron, N. E., and Cotter, M. A. (1999). Effects of the diacylglycerol complexing agent, cremophor, on nerve-conduction velocity and perfusion in diabetic rats. *J. Diabet. Complicat.* **13**, 2–9.

Johnson, P. C., Doll, S. C., and Cromey, D. W. (1986). Pathogenesis of diabetic neuropathy. *Ann. Neurol.* **19**, 450–457.

Kalichman, M. W., Dines, K. C., Bobik, M., and Mizisin, A. P. (1998). Nerve conduction velocity, laser Doppler flow, and axonal caliber in galactose and streptozotocin diabetes. *Brain Res.* **810**, 130–137.

Kalichman, M. W., and Lalonde, A. W. (1991). Experimental nerve ischemia and injury produced by cocaine and procaine. *Brain Res.* **565**, 34–41.

Kennedy, J. M., and Zochodne, D. W. (2002). The influence of experimental diabetes mellitus on the microcirculation of injured peripheral nerve: Functional and morphological aspects. *Diabetes*, in press.

Kihara, M., and Low, P. A. (1990). Regulation of rat nerve blood flow: Role of epineurial α-receptors. *J. Physiol.* **422**, 145–152.

Kihara, M., and Low, P. A. (1995a). Vasoreactivity to prostaglandins of rat peripheral nerve. *J. Physiol.* **484**, 463–467.

Kihara, M., and Low, P. A. (1995b). Impaired vasoreactivity to nitric oxide in experimental diabetic neuropathy. *Exp. Neurol.* **132**, 180–185.

Kihara, M., McManis, P. G., Schmelzer, J. D., Kihara, Y., and Low, P. A. (1995a). Experimental ischemic neuropathy: Salvage with hyperbaric oxygenation. *Ann. Neurol.* **37**, 89–94.

Kihara, M., Mitsui, M. K., Mitsui, Y., Okuda, K., Nakasaka, Y., Takahashi, M., and Schmelzer, J. D. (1999). Altered vasoreactivity to angiotension II in experimental diabetic neuropathy: Role of nitric oxide. *Muscle Nerve* **22**, 920–925.

Kihara, M., Schmelzer, J. D., Kihara, Y., Smithson, I. L., and Low, P. A. (1996). Efficacy of limb cooling on the salvage of peripheral nerve from ischemic fiber degeneration. *Muscle Nerve* **19**(2), 203–209.

Kihara, M., Schmelzer, J. D., and Low, P. A. (1995b). Effect of cilostazol on experimental diabetic neuropathy in the rat. *Diabetologia* **38**, 914–918.

Kihara, M., Schmelzer, J. D., Poduslo, J. F., Curran, G. L., Nickander, K. K., and Low, P. A. (1991a). Aminoguanidine effects on nerve blood flow, vascular permeability, electrophysiology, and oxygen free radicals. *Proc. Natl. Acad. Sci. USA* **88**, 6107–6111.

Kihara, M., Weerasuriya, A., and Low, P. A. (1991b). Endoneurial blood flow in rat sciatic nerve during development. *J. Physiol.* **439**, 351–360.

Kihara, M., Zollman, P. J., Schmelzer, J. D., and Low, P. A. (1993). The influence of dose of microspheres on nerve blood flow, electrophysiology and fiber degeneration of rat peripheral nerve. *Muscle Nerve* **16**, 1383–1389.

Kihara, M., Zollman, P. J., Smithson, I. L., Lagerlund, T. D., and Low, P. A. (1994). Hypoxic effect of exogenous insulin on normal and diabetic peripheral nerve. *Am. J. Physiol.* **266**, E980–E985.

Korthals, J. K., Gieron, M. A., and Dyck, P. J. (1988). Intima of epineurial arterioles is increased in diabetic polyneuropathy. *Neurology* **38**, 1582–1586.

Kowluru, R., Bitensky, M. W., Kowluru, A., Dembo, M., Keaton, P. A., and Buican, T. (1989). Reversible sodium pump defect and swelling in the diabetic rat erythrocyte: Effects on filterability and implications for microangiopathy. *Proc. Natl. Acad. Sci. USA* **86**, 3327–3331.

Lagerlund, T. D., and Low, P. A. (1994). Mathematical modeling of hydrogen clearance blood flow measurements in peripheral nerve. *Comput. Biol. Med.* **24**, 77–89.

Lawrence, E., and Brain, S. D. (1992). Altered microvascular reactivity to endothelin-1, endothelin-3 and NG-nitro-L-arginine methyl ester in streptozotocin-induced diabetes mellitus. *Br. J. Pharmacol.* **106**, 1035–1040.

Lembeck, F., and Holzer, P. (1979). Substance P as neurogenic mediator of antidromic vasodilation and neurogenic plasma extravasation. *Naunyn Schmiedebergs Arch. Pharmacol.* **310**, 175–183.

Levy, D., and Zochodne, D. W. (1998). Local nitric oxide synthase activity in a model of neuropathic pain. *Eur. J. Neurosci.* **10**, 1846–1855.

Llewelyn, J. G., Thomas, P. K., Gilbey, S. G., Watkins, P. J., and Muddle, J. R. (1988). Pattern of myelinated fibre loss in the sural nerve in neuropathy related to type 1 (insulin-dependent) diabetes. *Diabetologia* **31**, 162–167.

Low, P. A., Lagerlund, T. D., and McManis, P. G. (1989). Nerve blood flow and oxygen delivery in normal, diabetic, and ischemic neuropathy. *Int. Rev. Neurobiol.* **31**, 355–438.

Low, P. A., and Tuck, R. R. (1984). Effects of changes of blood pressure, respiratory acidosis and hypoxia on blood flow in the sciatic nerve of the rat. *J. Physiol.* **347**, 513–524.

Lundborg, G. (1975). Structure and function of the intraneural microvessels as related to trauma, edema formation, and nerve function. *J. Bone Joint Surg. AM.* **57**, 938–948.

Malik, R. A. (1997). The pathology of human diabetic neuropathy. *Diabetes* **46**, S50–S53.

Malik, R. A., Newrick, P. G., Sharma, A. K., Jennings, A., Ah-See, A. K., Mayhew, T. M., Jakubowski, J., Boulton, A. J., and Ward, J. D. (1989). Microangiopathy in human diabetic neuropathy: Relationship between capillary abnormalities and the severity of neuropathy. *Diabetologia* **32**, 92–102.

Malik, R. A., Tesfaye, S., Thompson, S. D., Veves, A., Sharma, A. K., Boulton, A. J. M., and Ward, J. D. (1993). Endoneurial localisation of microvascular damage in human diabetic neuropathy. *Diabetologia* **36**, 454–459.

Malik, R. A., Veves, A., Walker, D., Siddique, I., Lye, R. H., Schady, W., and Boulton, A. J. (2001). Sural nerve fibre pathology in diabetic patients with mild neuropathy: Relationship to pain, quantitative sensory testing and peripheral nerve electrophysiology. *Acta Neuropathol.* **101**, 367–374.

McManis, P. G., and Low, P. A. (1988). Factors affecting the relative viability of centrifascicular and subperineurial axons in acute peripheral nerve ischemia. *Exp. Neurol.* **99**, 84–95.

McManis, P. G., Low, P. A., and Yao, J. K. (1986). Relationship between nerve blood flow and intercapillary distance in peripheral nerve edema. *Am. J. Physiol.* **251**, E92–E97.

McManis, P. G., Schmelzer, J. D., Zollman, P. J., and Low, P. A. (1997). Blood flow and autoregulation in somatic and autonomic ganglia: Comparison with sciatic nerve. *Brain* **120**(Pt. 3), 445–449.

Monafo, W. W., Eliasson, S. G., Shimazaki, S., and Sugimoto, H. (1988). Regional blood flow in resting and stimulated sciatic nerve of diabetic rats. *Exp. Neurol.* **99**, 607–614.

Myers, R. R., Murakami, H., and Powell, H. C. (1986). Reduced nerve blood flow in edematous neuropathies: a biomechanical mechanism. *Microvasc. Res.* **32**, 145–151.

Nagamatsu, M., Nickander, K. K., Schmelzer, J. D., Raya, A., Wittrock, D. A., Trischler, H., and Low, P. A. (1995). Lipoic acid improves nerve blood flow, reduces oxidative stress, and improves distal nerve conduction in experimental diabetic neuropathy. *Diabet. Care* **18**, 1160–1167.

Nakamura, J., Kato, K., Hamada, Y., Nakayama, M., Chaya, S., Nakashima, E., Naruse, K., Kasuya, Y., Mizubayashi, R., Miwa, K., Yasuda, Y., Kamiya, H., Ienaga, K., Sakakibara, F., Koh, N., and Hotta, N. (1999). A protein kinase C-beta-selective inhibitor ameliorates neural dysfunction in streptozotocin-induced diabetic rats. *Diabetes* **48**, 2090–2095.

Nakamura, J., Koh, N., Sakakibara, F., Hamada, Y., Wakao, T., Hara, T., Mori, K., Nakashima, E., Naruse, K., and Hotta, N. (1995). Polyol pathway, 2,3-diphosphoglycerate in erythrocytes and diabetic neuropathy in rats. *Eur. J. Pharmacol.* **294**, 207–214.

Nakamura, J., Koh, N., Sakakibara, F., Hamada, Y., Wakao, T., Sasaki, H., Mori, K., Nakashima, E., Naruse, K., and Hotta, N. (1997). Diabetic neuropathy in sucrose-fed Otsuka Long-Evans Tokushima fatty rats: effect of an aldose reductase inhibitor, TAT. *Life Sci.* **60**, 1847–1857.

Newrick, P. G., Wilson, A. J., Jakubowski, J., Boulton, A. J., and Ward, J. D. (1986). Sural nerve oxygen tension in diabetes. *Br. Med. J.* **293**, 1053–1054.

Nukada, H., and Dyck, P. J. (1984). Microsphere embolization of nerve capillaries and fiber degeneration. *Am. J. Pathol.* **115**, 275–287.

Obrosova, I. G., Van Huysen, C., Fathallah, L., Cao, X., Stevens, M. J., and Greene, D. A. (2000). Evaluation of alpha(1)-adrenoceptor antagonist on diabetes-induced changes in peripheral nerve function, metabolism, and antioxidative defense. *FASEB J.* **14**, 1548–1558.

Olsson, Y., and Kristensson, K. (1971). Permeability of blood vessels and connective tissue sheaths in the peripheral nervous system to exogenous proteins. *Acta Neuropathol.* (Suppl. 5), 61–69.

Olsson, Y., and Kristensson, K. (1973). The perineurium as a diffusion barrier to protein tracers following trauma to nerves. *Acta Neuropathol.* **23**, 105–111.

Parkhouse, N., and Le Quesne, P. M. (1988). Impaired neurogenic vascular response in patients with diabetes and neuropathic foot lesions. *N. Engl. J. Med.* **318**, 1306–1309.

Podhajsky, R. J., and Myers, R. R. (1993). The vascular response to nerve crush: Relationship to Wallerian degeneration and regeneration. *Brain Res.* **623**, 117–123.

Podhajsky, R. J., and Myers, R. R. (1994). The vascular response to nerve transection: Neovascularization in the silicone nerve regeneration chamber. *Brain Res.* **662**, 88–94.

Pugliese, G., Tilton, R. G., Speedy, A., Chang, K., Santarelli, E., Province, M. A., Eades, D., Sherman, W. R., and Williamson, J. R. (1989). Effects of very mild versus overt diabetes on vascular haemodynamics and barrier function in rats. *Diabetologia* **32**, 845–857.

Rechthand, E., Hervonen, A., Sato, S., and Rapoport, S. I. (1986). Distribution of adrenergic innervation of blood vessels in peripheral nerve. *Brain Res.* **374**, 185–189.

Rechthand, E., and Rapoport, S. I. (1987). Regulation of the microenvironment of peripheral nerve: Role of the blood nerve barrier. *Prog. Neurobiol.* **28**, 303–343.

Rechthand, E., Sato, S., Oberg, P. A., and Rapoport, S. I. (1988). Sciatic nerve blood flow response to carbon dioxide. *Brain Res.* **446**, 61–66.

Robertson, S., Cameron, N. E., and Cotter, M. A. (1992). The effect of the calcium antagonist nifedipine on peripheral nerve function in streptozotocin-diabetic rats. *Diabetologia* **35**, 1113–1117.

Rundquist, I., Smith, Q. R., Michel, M. E., Ask, P., Oberg, P. A., and Rapoport, S. I. (1985). Sciatic nerve blood flow measured by laser Doppler flowmetry and [^{14}C]iodoantipyrine. *Am. J. Physiol.* **248**, H311–H317.

Sasaki, H., Schmelzer, J. D., Zollman, P. J., and Low, P. A. (1997). Neuropathology and blood flow of nerve, spinal roots and dorsal root ganglia in longstanding diabetic rats. *Acta Neuropathol. (Berl.)* **93**, 118–128.

Schmelzer, J. D., Zochodne, D. W., and Low, P. A. (1989). Ischemic and reperfusion injury of rat peripheral nerve. *Proc. Natl. Acad. Sci. USA* **86**, 1639–1642.

Schratzberger, P., Walter, D. H., Rittig, K., Bahlmann, F. H., Pola, R., Curry, C., Silver, M., Krainin, J. G., Weinberg, D. H., Ropper, A. H., and Isner, J. M. (2001). Reversal of experimental diabetic neuropathy by VEGF gene transfer. *J. Clin. Invest.* **107**, 1083–1092.

Scott, J. N., Clark, A. W., and Zochodne, D. W. (1999). Neurofilament and tubulin gene expression in progressive experimental diabetes: Failure of synthesis and export by sensory neurons. *Brain* **122**, 2109–2117.

Shupeck, M., Ward, K. K., Schmelzer, J. D., and Low, P. A. (1989). Comparison of nerve regeneration in vascularized and conventional grafts: Nerve electrophysiology, norepinephrine, prostacyclin, malondialdehyde, and the blood-nerve barrier. *Brain Res.* **493**, 225–230.

Simpson, L. O. (1988). Altered blood rheology in the pathogenesis of diabetic and other neuropathies. *Muscle Nerve* **11**, 725–744.

Singhal, A., Cheng, C., Sun, H., and Zochodne, D. W. (1997). Near nerve local insulin prevents conduction slowing in experimental diabetes. *Brain Res.* **763**, 209–214.

Sladky, J. T., Greenberg, J. H., and Brown, M. J. (1985). Regional perfusion in normal and ischemic rat sciatic nerves. *Ann. Neurol.* **17**, 191–195.

Sladky, J. T., Tschoepe, R. L., Greenberg, J. H., and Brown, M. J. (1991). Peripheral neuropathy after chronic endoneurial ischemia. *Ann. Neurol.* **29**, 272–278.

Sommer, C., and Myers, R. R. (1996). Vascular pathology in CCI neuropathy: A quantitative temporal study. *Exp. Neurol.* **141**, 113–119.

Sondell, M., Lundborg, G., and Kanje, M. (1999). Vascular endothelial growth factor has neurotrophic activity and stimulates axonal outgrowth, enhancing cell survival and Schwann cell proliferation in the peripheral nervous system. *J. Neurosci.* **19**, 5731–5740.

Stevens, E. J., Lockett, M. J., Carrington, A. L., and Tomlinson, D.R. (1993). Essential fatty acid treatment prevents nerve ischaemia and associated conduction anomalies in rats with experimental diabetes mellitus. *Diabetologia* **36**, 397–401.

Stevens, E. J., and Tomlinson, D. R. (1995). Effects of endothelin receptor antagonism with bosentan on peripheral nerve function in experimental diabetes. *Br. J. Pharmacol.* **115**, 373–379.

Stevens, M. J., Obrosova, I., Cao, X., Van Huysen, C., and Greene, D.A. (2000). Effects of DL-alpha-lipoic acid on peripheral nerve conduction, blood flow, energy metabolism, and oxidative stress in experimental diabetic neuropathy. *Diabetes* **49**, 1006–1015.

Sugimoto, H., Monafo, W. W., and Eliasson, S. G. (1986). Regional sciatic nerve and muscle blood flow in conscious and anesthetized rats. *Am. J. Physiol.* **251**, H1211–H1216.

Sugimoto, K., and Yagihashi, S. (1997). Effects of aminoguanidine on structural alterations of microvessels in peripheral nerve of streptozotocin diabetic rats. *Microvasc. Res.* **53**, 105–112.

Sutera, S. P., Chang, K., Marvel, J., and Williamson, J. R. (1992). Concurrent increases in regional hematocrit and blood flow in diabetic rats: Prevention by sorbinil. *Am. J. Physiol.* **263**, H945–H950.

Takahashi, K., Ghatei, M. A., Lam, H.-C., O'Halloran, D. J., and Bloom, S. R. (1990). Elevated plasma endothelin in patients with diabetes mellitus. *Diabetologia* **33**, 306–310.

Takeuchi, M., and Low, P. A. (1987). Dynamic peripheral nerve metabolic and vascular responses to exsanguination. *Am. J. Physiol.* **253**, E349–E353.

Tesfaye, S., Harris, N., Jakubowski, J. J., Mody, C., Wilson, R. M., Rennie, I. G., and Ward, J. D. (1993). Impaired blood flow and arterior-venous shunting in human diabetic neuropathy:

A novel technique of nerve photography and fluorescein angiography. *Diabetologia* **36**, 1266–1274.

Theriault, M., Dort, J., Sutherland, G., and Zochodne, D. W. (1997). Local human sural nerve blood flow in diabetic and other polyneuropathies. *Brain* **120**, 1131–1138.

Thomsen, K., Rubin, I., and Lauritzen, M. (2000). *In vivo* mechanisms of acetylcholine-induced vasodilation in rat sciatic nerve. *Am. J. Physiol. Heart Circ. Physiol.* **279**, H1044–H1054.

Tilton, R. G., Chang, K., Nyengaard, J. R., Van Den Enden, M., Ido, Y., and Williamson, J. R. (1995). Inhibition of sorbitol dehydrogenase: Effects on vascular and neural dysfunction in streptozocin-induced diabetic rats. *Diabetes* **44**, 234–242.

Timperley, W. R., Boulton, A. J., Davies-Jones, G. A., Jarratt, J. A., and Ward, J. D. (1985). Small vessel disease in progressive diabetic neuropathy associated with good metabolic control. *J. Clin. Pathol.* **38**, 1030–1038.

Tolins, J. P., Schultz, P. J., Raij, L., Brown, D. M., and Mauer, S. M. (1993). Abnormal renal hemodynamic response to reduced renal perfusion pressure in diabetic rats: Role of NO. *Am. J. Physiol.* **265**, F886–F895.

Tomlinson, D. R., Dewhurst, M., Stevens, E. J., Omawari, N., Carrington, A. L., and Vo, P. A. (1998). Reduced nerve blood flow in diabetic rats: Relationship to nitric oxide production and inhibition of aldose reductase. *Diabet. Med.* **15**, 579–585.

Tuck, R. R., Schmelzer, J. D., and Low, P. A. (1984). Endoneurial blood flow and oxygen tension in the sciatic nerves of rats with experimental diabetic neuropathy. *Brain* **107**, 935–950.

Ueno, Y., Koike, H., Nakamura, Y., Ochi, Y., Annoh, S., and Nishio, S. (1996). Effects of beraprost sodium, a prostacyclin analogue, on diabetic neuropathy in streptozotocin-induced diabetic rats. *Jap. J. Pharmacol.* **70**, 177–182.

Van Buren, T., Kasbergen, C. M., Gispen, W. H., and De Wildt, D. J. (1996). Presynaptic deficit of sympathetic nerves: A cause for disturbed sciatic nerve blood flow responsiveness in diabetic rats. *Eur. J. Pharmacol.* **296**, 277–283.

van Dam, P. S., van Asbeck, B. S., Van Oirschot, J. F., Biessels, G. J., Hamers, F. P., and Marx, J. J. (2001). Glutathione and alpha-lipoate in diabetic rats: Nerve function, blood flow and oxidative state. *Eur. J. Clin. Invest.* **31**, 417–424.

Ward, K. K., Low, P. A., Schmelzer, J. D., and Zochodne, D. W. (1989). Prostacyclin and noradrenaline in peripheral nerve of chronic experimental diabetes in rats. *Brain* **112**, 197–208.

Willars, G. B., Calcutt, N. A., Compton, A. M., Tomlinson, D. R., and Keen, P. (1989). Substance P levels in peripheral nerve, skin, atrial myocardium and gastrointestinal tract of rats with long-term diabetes mellitus: Effects of aldose reductase inhibition. *J. Neurol. Sci.* **91**, 153–164.

Williams, E., Timperley, W. R., Ward, J. D., and Duckworth, T. (1980). Electron microscopical studies of vessels in diabetic peripheral neuropathy. *J. Clin. Pathol.* **33**, 462–470.

Wright, A., and Nukada, H. (1994). Sciatic nerve morphology and morphometry in mature rats with streptozocin-induced diabetes. *Acta Neuropathol.* **88**, 571–578.

Xu, Q.-G., and Zochodne, D. W. (2002). Ischemia and failed regeneration in chronic experimental neuromas. *Brain Res.*, in press.

Yamamoto, T., Takakura, S., Kawamura, I., Seki, J., and Goto, T. (2001). The effects of zenarestat, an aldose reductase inhibitor, on minimal F-wave latency and nerve blood flow in streptozotocoin-induced diabetic rats. *Life Sci.* **68**, 1439–1448.

Yasuda, H., and Dyck, P. J. (1987). Abnormalities of endoneurial microvessels and sural nerve pathology in diabetic neuropathy. *Neurology* **37**, 20–28.

Yasuda, H., Sonobe, M., Yamashita, M., Terada, M., Hatanaka, I., Huitian, Z., and Shiegeta, Y. (1989). Effect of prostaglandin E1 analogue TFC 612 on diabetic neuropathy in streptozocin-induced didabetic rats Comparison with aldose reductase inhibitor ONO 2235. *Diabetes* **38**, 832–838.

Yoshida, M., Sugiyama, Y., Akaike, N., Ashizawa, N., Aotsuka, T., Ohbayashi, S., and Matsuura, A. (1998). Amelioration of neurovascular deficits in diabetic rats by a novel aldose reductase inhibitor, GP-1447: Minor contribution of nitric oxide. *Diabet. Res. Clin. Pract.* **40**, 101–112.

Young, M. J., Veves, A., Smith, J. V., Walker, M. G., and Boulton, A. J. (1995). Restoring lower limb blood flow improves conduction velocity in diabetic patients. *Diabetologia* **38**, 1051–1054.

Zochodne, D. W. (1996). Is early diabetic neuropathy a disorder of the dorsal root ganglion? A hypothesis and critique of some current ideas on the etiology of diabetic neuropathy. *J. Peripher. Nerv. Syst.* **1**, 119–130.

Zochodne, D. W., Allison, J. A., Ho, W., Ho, L. T., Hargreaves, K., and Sharkey, K. A. (1995). Evidence for CGRP accumulation and activity in experimental neuromas. *Am. J. Physiol.* **268**, H584–H590.

Zochodne, D. W., and Cheng, C. (1999). Diabetic peripheral nerves are susceptible to multifocal ischemic damage from endothelin. *Brain Res.* **838**, 11–17.

Zochodne, D. W., Cheng, C., Miampamba, M., Hargreaves, K., and Sharkey, K. A. (2001a). Peptide accumulations in proximal endbulbs of transected axons. *Brain Res.* **902**, 40–50.

Zochodne, D. W., Cheng, C., and Sun, H. (1996). Diabetes increases sciatic nerve susceptibility to endothelin-induced ischemia. *Diabetes* **45**, 627–632.

Zochodne, D. W., and Ho, L. T. (1990). Endoneurial microenvironment and acute nerve crush injury in the rat sciatic nerve. *Brain Res.* **535**, 43–48.

Zochodne, D. W., and Ho, L. T. (1991a). Unique microvascular characteristics of the dorsal root ganglion in the rat. *Brain Res.* **559**, 89–93.

Zochodne, D. W., and Ho, L. T. (1991b). Influence of perivascular peptides on endoneurial blood flow and microvascular resistance in the sciatic nerve of the rat. *J. Physiol.* **444**, 615–630.

Zochodne, D. W., and Ho, L. T. (1992a). Hyperemia of injured peripheral nerve: Sensitivity to CGRP antagonism. *Brain Res.* **598**, 59–66.

Zochodne, D. W., and Ho, L. T. (1992b). Normal blood flow but lower oxygen tension in diabetes of young rats: Microenvironment and the influence of sympathectomy. *Can. J. Physiol. Pharmacol.* **70**, 651–659.

Zochodne, D. W., and Ho, L. T. (1992c). The influence of indomethacin and guanethidine on experimental streptozotocin diabetic neuropathy. *Can. J. Neurol. Sci.* **19**, 433–441.

Zochodne, D. W., and Ho, L. T. (1993a). Vasa nervorum constriction from substance P and calcitonin gene- related peptide antagonists: Sensitivity to phentolamine and nimodipine. *Regul. Pept.* **47**, 285–290.

Zochodne, D. W., and Ho, L. T. (1993b). Diabetes mellitus prevents capsaicin from inducing hyperaemia in the rat sciatic nerve. *Diabetologia* **36**, 493–496.

Zochodne, D. W., and Ho, L. T. (1994a). The influence of sulindac on experimental streptozotocin-induced diabetic neuropathy. *Can. J. Neurol. Sci.* **21**, 194–202.

Zochodne, D. W., and Ho, L. T. (1994b). Neonatal guanethidine treatment alters endoneurial but not dorsal root ganglion perfusion in the rat. *Brain Res.* **649**, 147–150.

Zochodne, D. W., Ho, L. T., and Allison, J. A. (1994a). Dorsal root ganglia microenvironment of female BB Wistar diabetic rats with mild neuropathy. *J. Neurol. Sci.* **127**, 36–42.

Zochodne, D. W., Ho, L. T., and Gross, P. M. (1992). Acute endoneurial ischemia induced by epineurial endothelin in the rat sciatic nerve. *Am. J. Physiol.* **263**, H1806–H1810.

Zochodne, D. W., Huang, Z. X., Ward, K. K., and Low, P. A. (1990). Guanethidine-induced adrenergic sympathectomy augments endoneurial perfusion and lowers endoneurial microvascular resistance. *Brain Res.* **519**, 112–117.

Zochodne, D. W., Levy, D., Zwiers, H., Sun, H., Rubin, I., Cheng, C., and Lauritzen, M. (1999). Evidence for nitric oxide and nitric oxide synthase activity in proximal stumps of transected peripheral nerves. *Neuroscience* **91**, 1515–1527.

Zochodne, D. W., and Low, P. A. (1990). Adrenergic control of nerve blood flow. *Exp. Neurol.* **109**, 300–307.

Zochodne, D. W., Low, P. A., and Dyck, P. J. (1989). Adrenergic sympathectomy ablates unmyelinated fibers in the rat 'preganglionic' cervical sympathetic trunk. *Brain Res.* **498**, 221–228.

Zochodne, D. W., Misra, M., Cheng, C., and Sun, H. (1997). Inhibition of nitric oxide synthase enhances peripheral nerve regeneration in mice. *Neurosci. Lett.* **228**, 71–74.

Zochodne, D. W., and Nguyen, C. (1997). Angiogenesis at the site of neuroma formation in transected peripheral nerve. *J. Anat.* **191**, 23–30.

Zochodne, D. W., and Nguyen, C. (1999). Increased peripheral nerve microvessels in early experimental diabetic neuropathy: Quantitative studies of nerve and dorsal root ganglia. *J. Neurol. Sci.* **166**, 40–46.

Zochodne, D. W., Nguyen, C., and Sharkey, K. A. (1994b). Accumulation and degranulation of mast cells in experimental neuromas. *Neurosci. Lett.* **182**, 3–6.

Zochodne, D. W., Verge, V. M. K., Cheng, C., Sun, H., and Johnston, J. (2001b). Does diabetes target ganglion neurones? Progressive sensory neurone involvement in long term experimental diabetes. *Brain* **124**, 2319–2334.

Zochodne, D. W., Verge, V. M., Cheng, C., Hoke, A., Jolley, C., Thomsen, K., Rubin, I., and Lauritzen, M. (2000). Nitric oxide synthase activity and expression in experimental diabetic neuropathy. *J. Neuropathol. Exp. Neurol.* **59**, 798–807.

Zochodne, D. W., and Verge, V. M. K. (1999). Diabetes is associated with degenerative changes in sensory neuron gene expression: Implications for neurotrophin therapy. *Am. Acad. Neurol.* **52**, A552–A552. [Abstract]

PART III
MANIFESTATIONS

POTENTIAL MECHANISMS OF NEUROPATHIC PAIN IN DIABETES

Nigel A. Calcutt

Department of Pathology University of California, San Diego
La Jolla, California 92093

Abnormal sensations and pain are features of approximately 10% of all cases of diabetic neuropathy and can cause marked diminution in the quality of life for these patients. The quality and distribution of pain are variable, although descriptions of burning pain in the hands and feet are commonly reported. Like other neuropathic pain states, painful diabetic neuropathy has an unknown pathogenesis and, in many cases, is not alleviated by nonsteroidal anti-inflammatory drugs or opiates. In the last decade, a number of behavioral and physiologic studies have revealed indices of sensory dysfunction in animal models of diabetes. These include hyperalgesia to mechanical and noxious chemical stimuli and allodynia to light touch. Animal models of painful diabetic neuropathy have been used to investigate the therapeutic potential of a range of experimental agents and also to explore potential etiologic mechanisms. There is relatively little evidence to suggest that the peripheral sensory nerves of diabetic rodents exhibit spontaneous activity or increased responsiveness to peripheral stimuli. Indeed, the weight of evidence suggests that sensory input to the spinal cord is decreased rather than increased in diabetic rodents. Aberrant spinal or supraspinal modulation of sensory processing may therefore be involved in generating allodynia and hyperalgesia in these models. Studies have supported a role for spinally mediated hyperalgesia in diabetic rats that may reflect either a response to diminished peripheral input or a consequence of hyperglycemia on local or descending modulatory systems. Elucidating the effects of diabetes on spinal sensory processing may assist development of novel therapeutic strategies for preventing and alleviating painful diabetic neuropathy. © 2002, Elsevier Science (USA).

I. Diabetic Neuropathy and Painful Diabetic Neuropathy

Degeneration of the peripheral nervous system is one of the most common of the secondary complications associated with diabetes. Diabetic neuropathy encompasses a diverse range of nerve disorders that may be broadly divided into focal and symmetrical neuropathies. Focal neuropathies are usually a consequence of compression or ischemic lesions to a specific nerve, whereas symmetrical neuropathy suggests a widespread metabolic or physiologic etiology. The most frequently described manifestation of diabetic neuropathy is of a distal symmetrical polyneuropathy, which, as the name suggests, can involve sensory, motor, and autonomic nerves and tends to present first in the most distal extremities of the hands and feet. Neuropathy can be confirmed by biopsy of the sural or other distal nerves, and pathologic changes have been noted in most cells present. Schwann cells show an accumulation of lipid granules, intermediate filaments, and π granules of Reich, cell death, and subsequent breakdown of myelin that can lead to segmental demyelination (Thomas and Lascelles, 1965; Kalichman *et al.*, 1998). Neurons of all classes also show pathology ranging from axonal dwindling to the dystrophic and degenerative changes that precede axonal degeneration (Behse *et al.*, 1977). Clusters of regenerating fibers are noted in diabetic nerve (Behse *et al.*, 1977; Archer *et al.*, 1983), but any recovery clearly cannot keep pace with degeneration so that the inevitable consequence is progressive fiber loss. Pathologic changes are not confined to nerve fibers, and endoneurial blood vessels show evidence of microangiopathy with hypertrophy and hyperplasia of endothelial cells and pericytes along with thickening of basal lamina (Johnson *et al.*, 1981; Powell *et al.*, 1985) that has lead to the speculation that diabetic neuropathy has an ischemic origin. The perineurial cells that form the blood–nerve barrier also exhibit pathologic changes (Johnson *et al.*, 1981; Powell *et al.*, 1985). Unfortunately, it is not possible to deduce the primary location of the lesion that leads to such a diverse cellular pathology from these biopsy studies, and while some have argued that diabetic neuropathy is a primary axonopathy, a primary Schwannopathy, or a consequence of systemic microvascular disease, it is also plausible that diabetes affects all cell types via a common and fundamental pathogenic mechanism.

While nerve biopsies provide an appreciation of the widespread cellular pathology associated with diabetic neuropathy, they are not usually required for diagnostic purposes, as other manifestations of nerve dysfunction can be measured in a less invasive manner. These include progressive conduction velocity slowing of large myelinated sensory and motor fibers (Downie and Newell, 1961; Lawrence and Locke, 1961) and sensory loss (Gregersen,

1968; Apfel *et al.*, 2000). Around 10–20% of neuropathic diabetics also describe abnormal sensations (Boulton *et al.*, 1983; Partanen *et al.*, 1995) than can include spontaneous pain, paresthesias, dysesthesias, exaggerated pain perception to mildly noxious stimuli (hyperalgesia), or pain in response to innocuous stimuli (allodynia). A number of these signs may be present at the same time and they can coexist with elevated sensory thresholds to give the condition of hyperpathia. Painful diabetic neuropathy can be extremely distressing and disruptive to the patient and diminishes the quality of life greatly (Ahroni *et al.*, 1994).

II. Clinical Clues to Potential Mechanisms of Painful Diabetic Neuropathy

Painful neuropathy occurs in both insulin-deficient and insulin-replete diabetics (Ahroni *et al.*, 1994), suggesting that it is unlikely to be a result of insulin deficiency *per se*. In some patients, implementation of glycemic control can alleviate pain relatively quickly (Archer *et al.*, 1983) and normal subjects can show significant decreases in pain threshold following glucose infusions (Morley *et al.*, 1984). These findings suggest that hyperglycemia or metabolic consequences of increasing blood sugar levels may provide an etiologic mechanism for painful diabetic neuropathy. However, pain may persist for years after improving glycemic control, and painful episodes are not associated with fluctuations in blood sugar levels (Chan *et al.*, 1990). Modulation of pain thresholds by ambient glucose levels cannot therefore be the sole explanation for all of the manifestations of painful diabetic neuropathy.

Diabetics with chronic painful diabetic neuropathy are often considered to be refractory, within tolerable dose limits, to treatment with nonsteroidal anti-inflammatory drugs or opioids, the two classes of analgesic currently in most widespread use. While newer opiates may be tolerated better (Harati *et al.*, 2000), tricyclic antidepressants are usually the first line of treatment for painful diabetic neuropathy. Unfortunately, these agents are also of limited efficacy and can have problematic side effects (Max *et al.*, 1992; Watson, 2000). Pain in diabetic neuropathy therefore somewhat resembles neuropathic pain secondary to physical or viral nerve injury. This observation has led to the speculation that some aspect of ongoing nerve degeneration is involved in the pathogenesis of a wide range of neuropathic pain states, including painful diabetic neuropathy (Dyck *et al.*, 1976). The presence of frustrated regenerating fibers has also been evoked as a potential mechanism for producing pain (Asbury and Fields, 1984; Brown *et al.*, 1976) in a manner similar to the formation of painful neuromas that can occur after nerve transection (Matzner and Devor, 1987). Unfortunately, a number of biopsy studies conducted over the last decade have been unable to draw any clear morphologic associations between the degree of fiber degeneration and/or regeneration and the presence or absence of pain in diabetic patients (Lleweleyn *et al.*, 1991; Britland *et al.*, 1992; Malik *et al.*, 2001). While it is plausible that these sural nerve biopsy studies have examined the wrong location for any pertinent structural correlate to be found, clinical studies to date provide no supportable hypothesis to explain painful diabetic neuropathy. In the absence of a mechanism that can be clearly established from clinical observations, animal models of diabetes have been studied to allow exploration of potential mechanisms that could underlie painful diabetic neuropathy.

III. Sensory Dysfunction in Diabetic Animals

The majority of studies of sensory function in diabetic animals have used rodents that either develop spontaneous diabetes, such as the BB Wistar rat and db/db mouse, or become hyperglycemic following chemical ablation of pancreatic β cells by streptozotocin or alloxan. These animal models can survive for periods of weeks to months but rarely follow the normal life span of the species without insulin supplementation. Thus, most published reports of sensory dysfunction in diabetic animals have examined short periods (4–12 weeks) of relatively extreme (25 mmol/liter or greater) hyperglycemia. This is significant because such animals rarely show evidence of overt neuropathy, such as demyelination, axonal degeneration, fiber loss, or axonal regeneration in their peripheral nerves (Sharma and Thomas, 1974). While this makes diabetic rodents poor models for investigating the contribution of peripheral nerve degeneration or regeneration to pain, it does allow potential associations between the behavioral indices of sensory dysfunction seen in such animals, and the accompanying early neurochemical and functional disorders to be explored.

It is not possible to quantify spontaneous pain in animals, and the majority of experimental studies have therefore measured defined behavioral responses to external stimuli. Some tests, such as those that measure response latencies to noxious thermal stimuli, may be equated to the quantitative sensory tests used in diabetic patients and which generally demonstrate a progressive loss of thermal perception and thermal pain perception. There is less agreement when equivalent sensory tests have been performed in diabetic rodents. The tail flick test, where the time to movement of the tail from a noxious heat source is measured, has been reported to show unchanged (Kamei *et al.*, 1991, 1992; 1993a, 1994) or increased (Levine *et al.*, 1982) response times in diabetic mice, whereas diabetic rats have shown increased (Apfel *et al.*, 1994; Pertovaara *et al.*, 2001) or decreased (Courteix *et al.*, 1993a, 1998; Lee and McCarty 1990, 1992) response times. The causes of these disparities are unclear.

The tail flick test reflects activity of a simple spinal reflex arc and provides information on peripheral nerve and spinal function in isolation from higher nociceptive processing and cognitive systems. In contrast, measuring the paw withdrawal time from noxious thermal stimuli includes supraspinal sensory processing and may more closely reflect the quantitative sensory tests used in clinical studies. As with diabetic patients, the majority of studies using diabetic rats have found thermal hypoalgesia (Akunne and Soliman, 1987; Calcutt *et al.*, 1994, 1998b, 1999; 2000a,b; Chu *et al.*, 1986; Fox *et al.*, 1999), indicating loss of sensation rather than any exaggerated pain state. There has been a report of thermal hyperalgesia in the paw thermal response test (Forman *et al.*, 1986), but this appears to be a transient phenomenon that progresses to hypoalgesia (Kolta *et al.*, 1996). Thermal hypoalgesia can be prevented in diabetic rats by insulin therapy to achieve protracted normoglycemia (Chu *et al.*, 1986) and by prolonged treatment with an aldose reductase inhibitor (Fig. 1). The latter finding argues against the speculation that apparent hypoalgesia in diabetic animals reflects behavioral depression due to general ill health (Fox *et al.*, 1999),

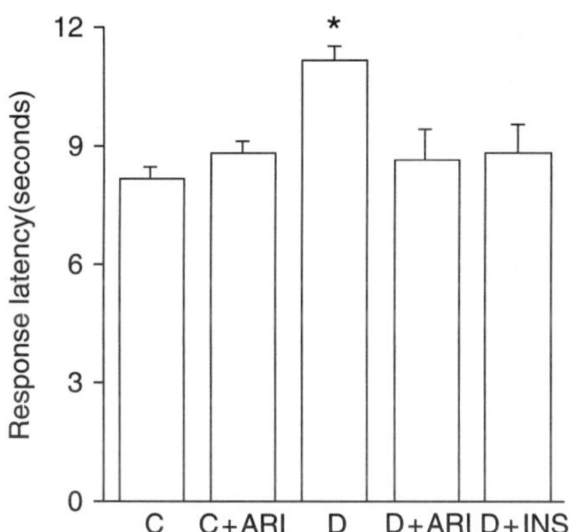

FIG. 1. Hind paw thermal response latency in control (C) and diabetic (D) rats treated daily with the aldose reductase inhibitor (ARI) Ponalrestat (50 mg/kg/day po) or slow-release (2 U/day) insulin implants (INS) for 8 weeks. Data are means ± SEM of $N = 14–17$/group. Statistical analysis by one-way ANOVA with Dunnets posthoc test is used to compare untreated diabetic rats against all other groups. $*p < 0.05$ versus all other groups. From N.A. Calcutt and J. D. Freshwater, unpublished observations.

as aldose reductase inhibitors do not alter hyperglycemia, polyurea, muscle
wasting, or other general metabolic indices of severe diabetes in rats. It also
suggests that metabolism of excess glucose through the polyol pathway is
involved in the pathogenesis of thermal hypoalgesia, although at present
the precise etiologic mechanism remains to be established.

Responses to high threshold mechanical stimuli have also been
measured in diabetic animals. These tests involve application of increasing
pressure to the paw or pinching the tail with forceps, although there is no
equivalent clinical test used in diabetic patients. The majority of such studies
have found hyperalgesia following stimulation of either the paw (Ahlgren
and Levine, 1993, Coudoré-Civiale *et al.*, 1998, 2000a,b, 2001; Courteix
et al., 1993a,b, 1996, 1998; Fox *et al.*, 1999; Klampt, 1998; Malcangio and
Tomlinson, 1998; Pertovaara *et al.*, 2001; Wuarin-Bierman *et al.*, 1987) or
the tail (Kamei *et al.*, 1991, 1992). Taken together with the studies of
thermal sensitivity described earlier, it appears that diabetes induces a state
of thermal hypoalgesia and mechanical hyperalgesia. This could arise from
distinct effects on nociceptors of different modalities in the periphery or
altered nociceptive processing at more central sites.

In addition to spontaneous pain, another of the manifestations of
painful diabetic neuropathy is the development of tactile allodynia, where
light touch is perceived as painful. A similar phenomenon occurs in diabetic
rats, where the light touch (less than 15 g) of von Frey filaments (Calcutt
et al., 1996) or light stroking of the paw (Field *et al.*, 1999) induces a
withdrawal response from the stimulus. Tactile allodynia is also a feature
of nerve injury models of neuropathic pain (Seltzer *et al.*, 1990; Kim and
Chung, 1991; Chaplan *et al.*, 1994), and both diabetes and spinal nerve
ligation produce allodynia of a similar magnitude (Calcutt *et al.*, 1996).
Tactile allodynia is not alleviated by rapid normalization of blood sugar,
but can be prevented and reversed in diabetic rats by protracted insulin
therapy (Calcutt *et al.*, 1996) or nitecarpone, an inhibitor of catechol-*O*-
methyltransfersae and antioxidant (Pertovaara *et al.*, 2001). The etiologic
mechanism is presently unclear. Nevertheless, a number of studies have
used tactile allodynia in diabetic rats to screen potential therapeutic agents
and a variety of compounds have produced temporary alleviation of the
condition (see later).

Injection of the chemical irritant formalin into the paw produces a
biphasic injury response characterized by flinching and guarding of the
afflicted limb (Dubuisson and Dennis, 1977). This test has been widely
studied as a model of pain-associated behavior that is triggered (phase 1)
and maintained (phase 2) by primary afferent activity from the lesion
site, but which is also modulated during phase 2 by a developing state
of spinal sensitization that sustains a behavioral hyperalgesia relative to

the primary afferent input. The model may have a clinical correlate in the conditions of postoperative and postinjury pain. Responses of diabetic rodents to paw formalin injection have been examined, not because there is an equivalent clinical test, but because it allows examination of the effects of hyperglycemia on protracted activation of nociceptive pathways and processing in both peripheral and central portions of the nervous system.

In diabetic mice, there is suppression of formalin-evoked behavior (Kamei *et al.*, 1993b; Takeshita *et al.*, 1995, 1997, 1998). This is particularly evident during phase 2, the period where behavior is associated with spinal amplification of nociceptive input from primary afferents. Potential explanations for this finding in diabetic mice include the absence of spinal mechanisms that generate locally produced hyperalgesia or changes in the balance of endogenous inhibitory systems (Kamei *et al.*, 1997). In contrast, diabetic rats exhibit increased behavioral responses following paw formalin injection, such as an increase in the frequency of paw flinching (Courteix *et al.*, 1993a; Malmberg *et al.*, 1993). When high concentrations of formalin are used, this hyperalgesia is noted only during the quiescent period between phases 1 and 2 of the behavioral response (Malmberg *et al.*, 1993), probably because flinching during phases 1 and 2 is at a maximal frequency. Lower concentrations of formalin, which reduce the frequency of flinching in phases 1 and 2 of the response in control rats, reveal hyperalgesia in diabetic rats that is most notable in both the quiescent phase and phase 2 (Calcutt *et al.*, 1995). Hyperalgesia during the formalin test is prevented in diabetic rats by insulin treatment to maintain euglycemia (Calcutt *et al.*, 1996) and also by aldose reductase inhibition (Calcutt *et al.*, 1994, 1995), indicating an etiologic mechanism involving glucose metabolism by the polyol pathway. Further stages of the pathogenic sequence have yet to be resolved.

IV. Using Animal Models to Predict Therapeutic Efficacy of Experimental Agents

The description of behavioral indices of sensory dysfunction in diabetic rodents has prompted their use to screen potential agents for treating painful diabetic neuropathy. However, there are a number of caveats that must be considered when adopting this approach. Concerns that hypoalgesic responses of diabetic animals may be related to the general behavioral depression seen in diabetic rats (Fox *et al.*, 1999) or altered thermal properties of the skin (Pertovaara *et al.*, 2001), rather than being directly related to nerve dysfunction, are valid but can be overcome by demonstrating the efficacy of neuroactive agents that do not affect the general metabolic status of diabetic animals (Calcutt *et al.*, 1999; Pertovaara *et al.*, 2001; Fig. 1). The apparent demonstration of hyperalgesia must also be treated with caution, particularly in tests that require restraint of diabetic animals with markedly reduced body weight or those that involve repeated testing and thus the potential incorporation of learning behaviors (Calcutt *et al.*, 2000a). The majority of studies to date have used systemic delivery of the drug in question, and this approach may be of some benefit for investigating clinical efficacy. However, systemic delivery of drugs does not provide information regarding the site of action of the agents. More importantly, the different distribution kinetics and clearance rates that occur in polydypsic diabetic animals (Courteix *et al.*, 1998) can mislead by suggesting apparently different relative potencies of drug between control and diabetic animals. Care should therefore be taken in interpreting *in vivo* pharmacological studies in diabetic rats and before any extrapolation to a clinical setting.

Hyperalgesia to high threshold mechanical stimuli and tactile allodynia to low threshold mechanical stimuli has been most widely used to demonstrate the transient efficacy of a range of agents delivered either systemically or spinally (Tables I and II). That some of these agents also have shown a

TABLE I
AGENTS SHOWING EFFICACY AGAINST TACTILE ALLODYNIA IN DIABETIC RATS

Agent (route)	Drug class	Reference
Dexmedetomidine (it)	α2 agonist	Calcutt and Chaplan (1997)
Morphine (it)	μ-opioid agonist	Calcutt and Chaplan (1997)
Morphine (ip)	μ-opioid agonist	Lynch *et al.* (1999)
Morphine (sc)	μ-opioid agonist	Field *et al.* (1999)
Lidocaine (ip)	Local anesthetic	Calcutt *et al.* (1996)
SNX 239 (it)	N-type Ca channel antagonist	Calcutt and Chaplan (1997)
CI-1021 (sc)	NK-1 antagonist	Field *et al.* (1998)
RP-67580 (it)	NK-1 antagonist	Coudoré-Civiale *et al.* (2000a)
SR-48968 (it)	NK-2 antagonist	Coudoré-Civiale *et al.* (2000a)
AP5 (it)	NMDA antagonist	Calcutt and Chaplan (1997)
Prazosin (sc)	α1 antagonist	Lee *et al.* (2000)
NBQX (it)	Non-NMDA antagonist	Calcutt and Chaplan (1997)
5'd-5IT (ip)	Adenosine kinase inhibitor	Lynch *et al.* (1999)
ABT-702 (ip)	Adenosine kinase inhibitor	Kowaluk *et al.* (2000)
ABT 627 (ip/po)	Endothelin A antagonist	Jarvis *et al.* (2000)
Gabapentin/pregabalin (po)	Antiepileptic	Field *et al.* (1999)
Amitryptyline (po)	Tricyclic antidepressant	Field *et al.* (1999)
TX14(A) (ip)	Prosaposin fragment	Calcutt *et al.* (2000a)
RB 101 (iv)	Inhibitor of enkephalin catabolism	Coudoré-Civiale *et al.* (2001)

TABLE II

AGENTS SHOWING EFFICACY AGAINST MECHANICAL HYPERALGESIA (PAW PRESSURE TEST) IN
DIABETIC RATS

Agent (route)	Drug class	Reference
RP-67580 (sc)	NK-1 antagonists	Courteix *et al.* (1993b)
CP-96345 (it)		Coudoré-Civiale *et al.* (1998, 2000a)
SR-48968 (it)	NK-2 antagonist	Coudoré-Civiale *et al.* (1998)
Clonidine (sc)	$\alpha2$ antagonist	Courteix *et al.* (1994)
MK 801 (ip)	NMDA receptor antagonist	Malcangio and Tomlinson (1998)
		Begon *et al.* (2000)
CI-988 (it)	CCK-B receptor antagonist	Coudoré-Civiale *et al.* (2000b)
MgSO$_4$ (ip)	Magnesium ion donor	Begon *et al.* (2000)
Lamotrigine (it)	Glutamate release inhibitor	Klamt (1996)
CPA (intraplantar)	$\alpha1$ agonist	Ahlgren and Levine (1993)
Morphine (iv)	μ-opioid receptor agonist	Courteix *et al.* (1994, 1998)
Morphine (ip)	μ-opioid receptor agonist	Malcangio and Tomlinson (1998)
Morphine (sc)	μ-opioid receptor agonist	Fox *et al.* (1999)
RB 101 (iv)	Inhibitor of enkephalin catabolism	Coudoré-Civiale *et al.* (2001)
Baclofen (ip)	GABA$_B$ agonist	Malcangio and Tomlinson (1998)
Baclofen (it)	GABA$_B$ agonist	Fox *et al.* (1999)
5-HT (it)	Neurotransmitter	Bardin *et al.* (2000)
1DMe (it)	Neuropeptide FF analog	Courteix *et al.* (1999)
DDA (intraplantar)	Adneylate cyclase inhibitor	Ahlgren and Levine (1993)
Staurosporine (intradermal)	Protein kinase C inhibitor	Ahlgren and Levine (1994)
Desipramine (iv)	Tricyclic antidepressant	Courteix *et al.* (1994)
Lidocaine (iv)	Local anesthetic	Courteix *et al.* (1994)

degree of efficacy during clinical trials in diabetic patients (Bach *et al.*, 1990; Backonja *et al.*, 1998; Byas-Smith *et al.*, 1995; Dejgaard *et al.*, 1988; Harati *et al.*, 2000; Kastrup *et al.*, 1987; Max *et al.*, 1992; Oskarsson *et al.*, 1997) suggests that these animal models may have some validity as preclinical screening tools. However, the transient nature of the pain relief achieved in these experimental studies suggests that these agents are unlikely to protect against the progression of diabetic neuropathy, whereas any effect on mechanical and tactile thresholds of control animals raises the possibility of analgesia and sedation as clinical side effects.

The formalin test in diabetic rats has also been used to examine the efficacy of potential therapeutics in alleviating established hyperalgesia, although the nature of the test does not allow for repeated measurements in the same animal and it is more suited to preventative studies, such as those examining the pathogenic role of the polyol pathway (Calcutt *et al.*, 1994, 1995). Systemic delivery of morphine (Courteix *et al.*, 1998), a synthetic ganglioside (Malmberg *et al.*, 1993), or gabapentin (Ceseña and Calcutt, 1999) has been shown to rapidly and transiently reduce hyperalgesia in diabetic rats. Because these agents were also able to reduce responses in control rats, it is likely that they act to generally suppress mechanisms of hyperalgesia initiated by paw formalin injection. In contrast, the prosaposin-derived peptide TX14(A) transiently and selectively suppresses the exaggerated hyperalgesia of diabetic rats during the formalin test without influencing the responses of normal rats (Calcutt *et al.*, 2000a). If these properties translate to the clinical setting, the drug might be less likely to be associated with side effects or analgesia arising from actions on normal nociceptive processing. These findings also imply that the exaggerated hyperalgesia of diabetic rats is not a simple amplification of mechanisms operational in normal rats following formalin stimulation. TX14(A) may therefore provide a tool to identify novel hyperalgesic mechanisms that are induced by diabetes.

V. Contribution of Peripheral Nerves to Hyperalgesia in Diabetic Animals

In humans, the most dramatic structural and functional effects of diabetes are found in the distal regions of the peripheral nervous system, and pain is often initially localized to the hands and feet before spreading more proximally. This presentation has prompted the assumption that the site of pain generation in diabetes is likely to be in the peripheral nerves. However, there is scant evidence of increased sensitivity of primary afferents in diabetic patients, and most studies indicate diminished nerve function. Thus, there is slowing of large sensory fiber conduction (Downie and Newell, 1961) and reduced levels of sensory neuropeptides in peripheral nerve and skin (Anand *et al.*, 1996; Tsigos *et al.*, 1993). Neurogenic inflammation, which reflects a combination of small fiber activity and vascular responsiveness, is also impaired (Aronin *et al.*, 1987; Boolell and Tooke, 1990). The apparent depression of peripheral sensory nerve function and the progressive fiber loss seen in diabetic patients seem more likely to parallel the hypoalgesia revealed by quantitative sensory testing than peripherally generated pain.

The peripheral sensory nerves of diabetic rodents have been studied extensively and also present a picture largely consistent with decreased, rather than increased activity. Large myelinated sensory fibers have slowed conduction velocities (Moore *et al.*, 1980), accompanied by decreased volume of their cell bodies in the dorsal root ganglia (Sidenius and Jakobsen, 1980) and decreased axonal caliber of both peripheral and central projections of primary afferents (Mizisin *et al.*, 1999). These large myelinated fibers mediate light touch, but there is no electrophysiologic evidence of altered peripheral excitability of these fibers that could contribute to behavioral tactile allodynia discussed earlier unless it is generated at the level of the dorsal roots, where there is evidence of myelin splitting (Tamura and Parry, 1994).

The structure of unmyelinated fibers has been studied less frequently in diabetic rodents, although there is a report of dystrophic changes at the peripheral terminals (Jirminova, 1993), which could plausibly initiate spontaneous activity of nociceptors. A report of spontaneous activity in C fibers of diabetic rats (Burchiel *et al.*, 1985) has not been confirmed by subsequent studies (Ahlgren *et al.*, 1992; Russell and Burchiel, 1993). Thus, there does not appear to be any electrophysiological indication of any peripherally derived spontaneous pain in these animals. Alterations in the terminals of nociceptive fibers could also modify response thresholds or firing patterns and thus participate in the allodynic and hyperalgesic behavioral hyperalgesia seen after paw stimulation of diabetic rats. Electrophysiologic studies have been unable to demonstrate changes in thermal or mechanical thresholds of C fibers (Ahlgren *et al.*, 1992; Ahlgren and Levine, 1994). An increase in firing after sustained suprathreshold mechanical stimulation of C fibers has been noted (Ahlgren *et al.*, 1992, 1997; Ahlgren and Levine, 1994) and could contribute to the behavioral hyperalgesia that occurs in the paw pressure tests. However, this phenomenon does not extend to postsynaptic wide dynamic range neurons in the spinal cord (Pertovaara *et al.*, 2001) and how it relates to the spontaneous pain or allodynia seen in diabetic patients is unclear.

Other functional indices of the peripheral sensory nerves of diabetic rodents also suggest depressed rather than increased function. Sensory nerve regeneration is impaired after injury (Bisby, 1980; Ekström and Tomlinson, 1989; Whitworth *et al.*, 1995). This occurs in a setting of reduced uptake of amino acids into sensory cell bodies (Thomas *et al.*, 1984), attenuated synthesis of proteins by sensory neurons following nerve injury (Calcutt *et al.*, 1993; McLean *et al.*, 1987), and a reduced rate of slow axonal transport of structural proteins in sensory neurons (Jakobsen and Sidenius, 1980; Medori *et al.*, 1985).

Neurochemical disorders in sensory nerves of diabetic rats are not restricted to the postinjury state and can resemble those that follow physical injury of normal nerves. For example, levels of the sensory neuropeptides substance P and CGRP are reduced in the dorsal root ganglia and sciatic nerve of diabetic rats and this is associated with a decrease in the amount undergoing axonal transport (Diemel *et al.*, 1992; Robinson *et al.*, 1987; Tomlinson *et al.*, 1988). Consistent with reduced delivery to the terminals, there is also a reduction in the evoked release of neuropeptides from both peripheral and central terminals of primary afferents (Calcutt *et al.*, 1998a, 2000b; Garret *et al.*, 1997). This may have functional consequences, as neuropeptides modulate central nociceptive processing (Hylden and Wilcox, 1981; Malmberg and Yaksh, 1992; Yashpal and Henry, 1984). However, hypoalgesia rather than allodynia or hyperalgesia is the most obvious predicted consequence of diminished neuropeptide release from central terminals, and an association between neuropeptide depletion and sensory loss is supported by reports showing that agents that prevent neuropeptide depletion in diabetic rats, such as aldose reductase inhibitors, nerve growth factor (NGF), and TX14(A), also prevent thermal hypoalgesia (Fig. 1; Robinson *et al.*, 1987; Apfel *et al.*, 1994; Diemel *et al.*, 1994; Calcutt *et al.*, 1999).

The decrease in the synthesis and axonal transport of substance P following nerve injury (Bisby and Keen, 1985) is caused by loss of target organ-derived NGF, which directly regulates gene expression of these neuropeptides (Lindsay and Harmar, 1989). Loss of neurotrophic support also underlies this phenomenon in diabetic animals. mRNA for NGF is reduced in peripheral target organs, and sensory axons contain less of the p75 subunit of the NGF receptor (Delcroix *et al.*, 1997; Fernyhough *et al.*, 1995; Roriguez-Pena *et al.*, 1995). Consequently, less NGF is captured by C fiber terminals and undergoes retrograde axonal transport to the cell body (Hellweg and Hartung, 1990; Jakobsen *et al.*, 1981). This pathogenic mechanism has been supported by NGF therapy studies in diabetic rats in which systemic NGF restored peripheral nerve neuropeptide levels (Diemel *et al.*, 1994).

Loss of neurotrophic support in diabetic animals is not selective for NGF and extends to other members of the neurotrophin family (Fernyhough *et al.*, 1995, 1998; Ihara *et al.*, 1996; Mizisin *et al.*, 1999b; Roriguez-Pena *et al.*, 1995) and also other neuroactive factors (Calcutt *et al.*, 1992; Wuarin *et al.*, 1994), suggesting the potential for profound effects on expression of the target genes. While the associations between depletion of neurotrophic factors and expression of specific genes have yet to be explored in detail, sensory neuron cell bodies in the dorsal root ganglia of diabetic rodents show altered gene expression for a number of proteins associated with peripheral nociceptive processing, including a 25% decrease in mRNA for the sensory nerve specific tetrodotoxin-resistant sodium channel (Okuse *et al.*, 1997).

Despite the widespread reports of behavioral allodynia and hyperalgesia in diabetic animals, there is little evidence to suggest that the peripheral sensory nerves of diabetic animals show exaggerated spontaneous or stimulus-evoked activity. Rather, the weight of evidence points toward depression of sensory nerve function. This paradox can be further illustrated by spinal microdialysis studies that measure neurotransmitter release following paw stimulation with formalin and allow concurrent behavioral assessment of formalin-evoked flinching. In such studies, the formalin-evoked spinal release of the excitatory transmitters substance P and glutamate is reduced in diabetic animals, yet the hyperalgesic behavior induced following paw formalin injection is exaggerated (Calcutt and Malmberg, 1995; Calcutt *et al.*, 2000b). Because the formalin test incorporates both peripheral and central responses, one plausible explanation is that changes in spinal or supraspinal processing are involved in the hyperalgesia of diabetic animals.

VI. A Potential Role for Altered Spinal Processing in Diabetic Hyperalgesia

The effect of diabetes on the spinal cord and higher central nervous system has been relatively unexplored, probably because nerve dysfunction is most apparent in the periphery of diabetic patients. Pathologic changes have been reported in spinal cord taken at autopsy from diabetic patients (Reske-Neilsen *et al.*, 1968; Slager, 1978), and a more recent study using magnetic resonance imaging has shown that spinal cord volume is reduced in neuropathic diabetics compared to nonneuropathic diabetics or control subjects (Eaton *et al.*, 2001). There is some electrophysiologic evidence of spinal nerve conduction slowing in both patients (Cracco *et al.*, 1984) and rodents (Biessels *et al.*, 1999). At present, little is known about how diabetes might alter sensory processing in the spinal cord. However, emerging evidence from animal studies, including electrophysiologic evidence of spontaneous spinal activity that is not driven by peripheral input (Perto-vaara *et al.*, 2001) and protracted behavioral hyperalgesia following spinal delivery of substance P (Calcutt *et al.*, 2000a), supports the suggestion that neuropathic pain may involve aberrant spinal modulation of sensory input.

The behavioral response of normal rats to paw formalin injection has been widely studied as a model of hyperalgesia that is mediated at the spinal level via the local release of molecules, such as prostaglandins and nitric oxide (Malmberg and Yaksh 1993, 1995). That diabetic rats show exaggerated hyperalgesia during the formalin test, despite reduced spinal release of primary afferent neurotransmitters (Calcutt and Malmberg 1995; Calcutt *et al.*, 2000b), is consistent with spinal or supraspinal amplification

of sensory input in these animals. One possible explanation for this amplification could be an increase in the mediators of hyperalgesic behavior.
Nitric oxide synthase (NOS) has been examined in diabetic rats, and while
there was an approximate doubling of mRNA for nNOS in the dorsal root
ganglia of diabetic rats (Luo *et al.*, 1999), no marked increases in NOS
protein or activity were detected in a time frame consistent with the onset
of hyperalgesia (Zochodne *et al.*, 2000). In contrast, another study found
that protein levels of the prostaglandin-synthesizing enzyme COX-2 were
increased threefold in the spinal cord of diabetic rats and that formalin-
evoked release of prostaglandin E_2 (PGE_2) in the spinal cord was enhanced
in diabetic rats (Calcutt *et al.*, 2001). These findings raise the possibility of
a hyperglycemia-induced tonic increase in local mechanisms that promote
spinally mediated hyperalgesia following peripheral nociceptive input.

Another potential mechanism that could contribute to spinally mediated
hyperalgesia in diabetes is development of a central hypersensitivity in
response to the neurochemical denervation that results from the reduced
release of neurotransmitters from primary afferents (Calcutt *et al.*, 1998a,
2000b; Garret *et al.*, 1997). The concept of a denervation hypersensitivity
following peripheral nerve injury that is mediated by increased receptor
production is well established at the neuromuscular junction (Fambrough,
1979). Postsynaptic changes at the central terminals of primary afferents
also occur following nerve trauma, which causes a depletion of peripheral
nerve substance P and a corresponding increase in spinal NK-1 receptors
(Abbadie *et al.*, 1996; Malmberg and Basbaum, 1998). A similar response
also appears to occur at the spinal level in diabetic animals where the
peripheral nerve damage is of a metabolic rather than physical origin
and is initially neurochemical in nature. Thus, measurements of receptor
levels using ligand binding have shown an increase in the peptidergic NK-1
receptor (Kamei *et al.*, 1990) and also the glutamatergic NMDA and AMPA
receptors (Li *et al.*, 1999) in the spinal cord of diabetic rodents, while there
is pharmacological evidence suggesting novel expression of the bradykinin
B1 receptor in the spinal cord of diabetic rats (Cloutier and Couture,
2000). The extended period of thermal hyperalgesia seen in diabetic rats
after intrathecal delivery of substance P (Calcutt *et al.*, 2000a) may also
be a behavioral correlate to changes in spinal receptor populations and
indicates an alteration of sensory processing that is more central to primary
afferent input to the spinal cord.

Other than amplification of sensory input, spinally- mediated hyperalgesia may also occur secondary to changes in the local or descending
pathways that modulate spinal nociceptive processing. To date, studies of
these systems have been largely restricted to direct assays of neurotransmitters or receptors and a clear picture has yet to emerge. Spinal 5-HT

release evoked by the stimulation of descending pathways is reduced despite normal tissue levels (Suh *et al.*, 1996) and could contribute to spinally mediated hyperalgesia. The situation for other descending inhibitory systems is not clear. Thus, spinal levels of met-enkephalin has been reported as increased (Kolta *et al.*, 1996) or decreased (Di Guilio *et al.*, 1989) by diabetes. Descending noradrenergic inhibitory activity may also be depressed in diabetic animals (Bitar *et al.*, 1999; Bitar and Pilcher, 2000), although others have suggested that it is increased (Stewart *et al.*, 1994). Consensus has also yet to be reached regarding the effects of diabetes on spinal adrenoceptors. While mRNA and binding sires for α_2-adrenoceptors have been reported as reduced (Bitar *et al.*, 1999; Bitar and Pilcher, 2000), a report of an increase in α_1-adrenoceptor-binding sites (Bitar *et al.*, 1999) is balanced by another showing no change in either mRNA or binding sites for α_1-adrenoceptors (Lee *et al.*, 2000). Further studies are clearly required to establish the nature of diabetes-induced changes in the spinal cord and to identify their pathogenesis.

VII. Summary

Painful diabetic neuropathy presents an array of manifestations that may reflect either a number of unrelated pathogenic mechanisms or a common mechanism beyond hyperglycemia that can trigger diverse responses. In the absence of clear clinical evidence to explain the etiology of this pain, animal models may be useful to allow both screening of potential therapeutic agents and investigation of pathogenic mechanisms. There is growing evidence that hyperglycemia induces a range of neurochemical changes in the spinal cord of rodents that may play a role in the behavioral hyperalgesia and allodynia seen in these animals. Understanding how spinal and supraspinal sensory processing is altered in diabetic animals will allow rational design and delivery of therapeutic agents that target selective mechanisms with the goal of either preventing the development of pain or providing pain relief without side effects.

References

Abbadie, C., Brown, J. L., Mantyh, P. W., and Basbaum, A. I. (1996). Spinal cord substance P receptor immunoreactivity increases in both inflammatory and nerve injury models of persistent pain. *Neuroscience* **70**, 201–209.

Ahlgren, S. C., and Levine, J. D. (1993). Mechanical hyperalgesia in streptozotocin-diabetic rats. *Neuroscience* **52**, 1049–1055.

Ahlgren, S. C., and Levine, J. D. (1994). Protein kinase C inhibitors decrease hyperalgesia and C-fiber hyperexcitability in the streptozotocin-diabetic rat. *J. Neurophysiol.* **72**, 684–692.

Ahlgren, S. C., White, D. M., and Levine, J. D. (1992). Increased responsiveness of sensory neurons in the saphenous nerve of the streptozotocin-diabetic rat. *J. Neurophysiol.* **68**, 2077–2085.

Ahlgren, S. C., Wang, J. F., and Levine, J. D. (1997). C-fiber mechanical stimulus-response functions are different in inflammatory versus neuropathic hyperalgesia in the rat. *Neuroscience* **76**, 285–290.

Ahroni, J. H., Boyko, E. J., Davignon, D. R., and Pecoraro, R. E. (1994). The health and functional status of veterans with diabetes. *Diabet. Care* **17**, 318–321.

Akunne, H.C., and Soliman, K.F. (1987). The role of opioid receptors in diabetes and hyperglycemia-induced changes in pain threshold in the rat. *Psychopharmacology* **93**, 167–712.

Anand, P., Terenghi, G., Warner, G., Kopelman, P., Williams-Chestnut, R. E., and Sinicropi, D. V. (1996). The role of endogenous nerve growth factor in human diabetic neuropathy. *Nature Med.* **2**, 703–707.

Apfel, S. C., Arezzo, J. C., Brownlee, M., Federoff, H., and Kessler, J. A. (1994). Nerve growth factor administration protects against experimental diabetic sensory neuropathy. *Brain Res.* **634**, 7–12.

Apfel, S. C., Schwartz, S., Adornato, B. T., Freeman, R., Biton, V., Rendell, M., Vinik, A., Giuliani, M., Stevens, J. C., Barbano, R., and Dyck, P. J. (2000). Efficacy and safety of recombinant human nerve growth factor in patients with diabetic polyneuropathy: A randomized controlled trial. *JAMA* **284**, 2215–2221.

Archer, A. G., Watkins, P. J., Thomas, P. K., Sharma, A. K., and Payan, J. (1983). The natural history of acute painful neuropathy in diabetes mellitus. *J. Neurol. Neurosurg. Psych.* **46**, 491–499.

Aronin, N., Leeman, S. E., and Clements, R. S., Jr. (1987). Diminished flare response in neuropathic diabetic patients: Comparison of effects of substance P, histamine, and capsaicin. *Diabetes* **36**, 1139–1143.

Asbury, A. K., and Fields, H. L. (1984). Pain due to peripheral nerve damage: An hypothesis. *Neurology* **34**, 1587–1590.

Bach, F. W., Jensen, T. S., Kastrup, J., Stigsby, B., and Dejgard, A. (1990). The effect of intravenous lidocaine on nociceptive processing in diabetic neuropathy. *Pain* **40**, 29–34.

Backonja, M., Beydoun, A., Edwards, K. R., Schwartz, S. L., Fonseca, V., Hes, M., LaMoreaux, L., and Garofalo, E. (1998). Gabapentin for the symptomatic treatment of painful neuropathy in patients with diabetes mellitus: A randomized controlled trial. *JAMA* **280**, 1831–1836.

Bardin, L., Schmidt, J., Alloui, A., and Eschalier, A. (2000). Effect of intrathecal administration of serotonin in chronic pain models in rats. *Eur. J. Pharmacol* **409**, 37–43.

Begon, S., Pickering, G., Eschalier, A., and Dubray, C. (2000). Magnesium and MK801 have a similar effect in two experimental models of neuropathic pain. *Brain Res.* **887**, 436–439.

Behse, F., Buchthal, F., and Carlsen, F. (1977). Nerve biopsy and conduction studies in diabetic neuropathy. *J. Neurol. Neurosurg. Psych.* **40**, 1072–1082.

Biessels, G. J., Cristino, N. A., Rutten, G. J., Hamers, F. P., Erkelens, D. W., and Gispen, W. H. (1999). Neurophysiological changes in the central and peripheral nervous system of streptozotocin-diabetic rats. Course of development and effects of insulin treatment. *Brain* **122**, 757–768.

Bisby, M. A. (1980). Axonal transport of labeled protein and regeneration rate in nerves of streptozocin-diabetic rats. *Exp. Neurol.* **69**, 74–84.

Bisby, M. A., and Keen, P. (1985). Axonal transport of substance P-like immunoreactivity in regenerating rat sciatic nerve. *Brain Res.* **361**, 396–399.

Bitar, M. S., Bajic, K. T., Farook, T., Thomas, M. I., and Pilcher, C. W. (1999). Spinal cord noradrenergic dynamics in diabetic and hypercortisolaemic states. *Brain Res.* **830**, 1–9.

Bitar, M. S., and Pilcher, C. W. (2000). Diabetes attenuates the response of the lumbospinal noradrenergic system to idazoxan. *Pharmacol. Biochem. Behav.* **67**, 247–255.

Boolell, M., and Tooke, J. E. (1990) The skin hyperaemic response to local injection of substance P and capsaicin in diabetes mellitus. *Diabet. Med.* **7**, 898–901.

Boulton, A. J., Armstrong, W. D., Scarpello, J. H., and Ward, J. D. (1983). The natural history of painful diabetic neuropathy: A 4-year study. *Postgrad. Med. J.* **59**, 556–559.

Britland, S. T., Young, R. J., Sharma, A. K., and Clarke, B. F. (1992). Acute and remitting painful diabetic polyneuropathy: A comparison of peripheral nerve fibre pathology. *Pain* **48**, 361–370.

Brown, M. J., Martin, J. R., and Asbury, A. K. (1976). Painful diabetic neuropathy: A morphometric study. *Arch. Neurol.* **33**, 164–171.

Burchiel, K. J., Russell, L. C., Lee, R. P., and Sima, A. A. (1985). Spontaneous activity of primary afferent neurons in diabetic BB/Wistar rats: A possible mechanism of chronic diabetic neuropathic pain. *Diabetes* **34**, 1210–1213.

Byas-Smith, M. G., Max, M. B., Muir, J., and Kingman, A. (1995). Transdermal clonidine compared to placebo in painful diabetic neuropathy using a two-stage "enriched enrollment" design. *Pain* **60**, 267–274.

Calcutt, N. A., Campana, W. M., Eskeland, N. L., Mohiuddin, L., Dines, K. C., Mizisin, A. P., and O'Brien, J. S. (1999). Prosaposin gene expression and the efficacy of a prosaposin-derived peptide in preventing structural and functional disorders of peripheral nerve in diabetic rats. *J. Neuropathol. Exp. Neurol.* **58**, 628–636.

Calcutt, N. A., and Chaplan, S. R. (1997). Spinal pharmacology of tactile allodynia in diabetic rats. *Br. J. Pharmacol.* **122**, 1478–1482.

Calcutt, N. A., Chen, P., and Hua, X. Y. (1998a). Effects of diabetes on tissue content and evoked release of calcitonin gene-related peptide-like immunoreactivity from rat sensory nerves. *Neurosci. Lett.* **254**, 129–132.

Calcutt, N. A., Dines, K. C., and Cesena R. M. (1998b). Effects of the peptide HP228 on nerve disorders in diabetic rats. *Metabolism* **47**, 650–656.

Calcutt, N. A., Freshwater, J. D., and O'Brien, J. S. (2000a). Protection of sensory function and antihyperalgesic properties of a prosaposin-derived peptide in diabetic rats. *Anesthesiology* **93**, 1271–1278.

Calcutt, N. A., Jorge, M. C., Yaksh, T. L., and Chaplan, S. R. (1996). Tactile allodynia and formalin hyperalgesia in streptozotocin-diabetic rats: Effects of insulin, aldose reductase inhibition and lidocaine. *Pain* **68**, 293–299.

Calcutt, N. A., Li, L., Yaksh, T. L., and Malmberg, A. B. (1995). Different effects of two aldose reductase inhibitors on nociception and prostaglandin E. *Eur. J. Pharmacol.* **285**, 189–197.

Calcutt, N. A., and Malmberg, A. B. (1995). Basal and formalin-evoked spinal levels of amino acids in conscious diabetic rats. *Soc. Neurosci. Abs.* **21**, 650.

Calcutt, N. A., Malmberg, A. B., Yamamoto, T., and Yaksh, T. L. (1994). Tolrestat treatment prevents modification of the formalin test model of prolonged pain in hyperglycemic rats. *Pain* **58**, 413–420.

Calcutt, N. A., Mizisin, A. P., and Yaksh, T. L. (1993). Impaired induction of vasoactive intestinal polypeptide after sciatic nerve injury in the streptozotocin-diabetic rat. *J. Neurol. Sci.* **119**, 154–161.

Calcutt, N. A., Muir, D., Powell, H. C., and Mizisin, A. P. (1992). Reduced ciliary neuronotrophic factor-like activity in nerves from diabetic or galactose-fed rats. *Brain Res.* **575**, 320–324.

Calcutt, N. A., Stiller, C., Gustafsson, H., and Malmberg, A. B. (2000b). Elevated substance-P-like immunoreactivity levels in spinal dialysates during the formalin test in normal and diabetic rats. *Brain Res.* **856**, 20–27.

Calcutt, N. A., Svensson, C. I., and Freshwater, J. D. (2001). NSAID-sensitive hyperalgesia in diabetic rats is associated with increased spinal COX-2. *J. Pain* **2**, 39.

Ceseña, R. M., and Calcutt, N. A. (1999). Gabapentin prevents hyperalgesia during the formalin test in diabetic rats. *Neurosci. Lett.* **262**, 101–104.

Chaplan, S. R., Bach, F. W., Pogrel, J. W., Chung, J. M., and Yaksh, T. L. (1994). Quantitative assessment of tactile allodynia in the rat paw. *J. Neurosci. Methods* **53**, 55–63.

Chan, A. W., MacFarland, I. A., Bowsher, D. R., and Wells, J. C. D. (1990). Does acute hyperglycaemia influence heat pain threshold? *J. Neurosurg.* **57**, 688–690.

Chu, P. C., Lin, M. T., Shian, L. R., and Leu, S. Y. (1986). Alterations in physiologic functions and in brain monoamine content in streptozocin-diabetic rats. *Diabetes* **35**, 481–485.

Cloutier, F., and Couture, R. (2000). Pharmacological characterization of the cardiovascular responses elicited by kinin B(1) and B(2) receptor agonists in the spinal cord of streptozotocin-diabetic rats. *Br. J. Pharmacol.* **130**, 375–385.

Coudoré-Civiale, M. A., Courteix, C., Eschalier, A., and Fialip, J. (1998). Effect of tachykinin receptor antagonists in experimental neuropathic pain. *Eur. J. Pharmacol.* **361**, 175–184.

Coudoré-Civiale, M., Courteix, C., Boucher, M., Fialip, J., and Eschalier, A. (2000a). Evidence for an involvement of tachykinins in allodynia in streptozocin-induced diabetic rats. *Eur. J. Pharmacol.* **401**, 47–53.

Coudoré-Civiale, M. A., Courteix, C., Fialip, J., Boucher, M., and Eschalier, A. (2000b). Spinal effect of the cholecystokinin-B receptor antagonist CI-988 on hyperalgesia, allodynia and morphine-induced analgesia in diabetic and mononeuropathic rats. *Pain* **88**, 15–22.

Coudoré-Civiale, M. A., Meen, M., Fournie-Zaluski, M. C., Boucher, M., Roques, B. P., and Eschalier, A. (2001). Enhancement of the effects of a complete inhibitor of enkephalin-catabolizing enzymes, RB 101, by a cholecystokinin-B receptor antagonist in diabetic rats. *Br. J. Pharmacol.* **133**, 179–185.

Courteix, C., Bardin, M., Chantelauze, C., Lavarenne, J., and Eschalier, A. (1994). Study of the sensitivity of the diabetes-induced pain model in rats to a range of analgesics. *Pain* **57**, 153–160.

Courteix, C., Bardin, M., Massol, J., Fialip, J., Lavarenne, J., and Eschalier, A. (1996). Daily insulin treatment relieves long-term hyperalgesia in streptozocin-diabetic rats. *Neuroreport* **7**, 1922–1924.

Courteix, C., Bourget, P., Caussade, F., Bardin, M., Coudore, F., Fialip, J., and Eschalier, A. (1998). Is the reduced efficacy of morphine in diabetic rats caused by alterations of opiate receptors or of morphine pharmacokinetics? *J. Pharmacol. Exp. Ther.* **285**, 63–67.

Courteix, C., Coudoré-Civiale, M. A., Privat, A. M., Zajac, J. M., Eschalier, A., and Fialip, J. (1999). Spinal effect of a neuropeptide FF analogue on hyperalgesia and morphine-induced analgesia in mononeuropathic and diabetic rats. *Br. J. Pharmacol.* **127**, 1454–1462.

Courteix, C., Eschalier, A., and Lavarenne, J. (1993a). Streptozocin-induced diabetic rats: Behavioural evidence for a model of chronic pain. *Pain* **53**, 81–88.

Courteix, C., Lavarenne, J., and Eschalier, A. (1993b). RP-67580, a specific tachykinin NK1 receptor antagonist, relieves chronic hyperalgesia in diabetic rats. *Eur. J. Pharmacol.* **241**, 267–270.

Cracco, J., Castells, S., and Mark, E. (1984). Spinal somatosensory evoked potentials in juvenile diabetes. *Ann. Neurol.* **15**, 55–58.

Dejgard, A., Petersen, P., and Kastrup, J. (1988). Mexiletine for treatment of chronic painful diabetic neuropathy. *Lancet* **1**, 9–11.

Delcroix, J. D., Tomlinson, D. R., and Fernyhough, P. (1997). Diabetes and axotomy-induced deficits in retrograde axonal transport of nerve growth factor correlate with decreased levels of p75LNTR protein in lumbar dorsal root ganglia. *Mol. Brain. Res.* **51**, 82–90.

Diemel, L. T., Stevens, E. J., Willars, G. B., and Tomlinson, D. R. (1992). Depletion of substance *P* and calcitonin gene-related peptide in sciatic nerve of rats with experimental diabetes; effects of insulin and aldose reductase inhibition. *Neurosci. Lett.* **30**, 253–256.

Diemel, L. T., Brewster, W. J., Fernyhough, P., and Tomlinson, D. R. (1994). Expression of neuropeptides in experimental diabetes; effects of treatment with nerve growth factor or brain-derived neurotrophic factor. *Mol. Brain. Res.* **21**, 171–175.

Di Giulio, A. M., Tenconi, B., La Croix, R., Mantegazza, P., Abbracchio, M. P., Cattabeni, F., and Gorio, A. (1989). Denervation and hyperinnervation in the nervous system of diabetic animals. II. Monoaminergic and peptidergic alterations in the diabetic encephalopathy. *J. Neurosci. Res.* **24**, 362–368.

Downie, A. W., and Newell, D. J. (1961). Sensory nerve conduction in patients with diabetes mellitus and controls. *Neurology* **11**, 876–882.

Dubuisson, D., and Dennis, S. G. (1977). The formalin test: A quantitative study of the analgesic effects of morphine, meperidine, and brain stem stimulation in rats and cats. *Pain* **4**, 161–174.

Dyck, P. J., Lambert, E. H., and O'Brien, P. C. (1976). Pain in peripheral neuropathy related to rate and kind of fiber degeneration. *Neurology* **26**, 466–471.

Eaton, S. E., Harris, N. D., Rajbhandari, S. M., Greenwood, P., Wilkinson, I. D., Ward, J. D., Griffiths, P. D., and Tesfaye, S. (2001). Spinal-cord involvement in diabetic peripheral neuropathy. *Lancet* **358**, 35–36.

Ekstrom, A. R., and Tomlinson, D. R. (1989). Impaired nerve regeneration in streptozotocin-diabetic rats: Effects of treatment with an aldose reductase inhibitor. *J. Neurol. Sci.* **93**, 231–237.

Fambrough, D. M. (1979). Control of acetylcholine receptors in skeletal muscle. *Physiol. Rev.* **59**, 165–227.

Fernyhough, P., Diemel, L. T., Brewster, W. J., and Tomlinson, D. R. (1995). Altered neurotrophin mRNA levels in peripheral nerve and skeletal muscle of experimentally diabetic rats. *J. Neurochem.* **64**, 1231–1237.

Fernyhough, P., Diemel, L. T., and Tomlinson, D. R. (1998). Target tissue production and axonal transport of neurotrophin-3 are reduced in streptozotocin-diabetic rats. *Diabetologia* **41**, 300–306.

Field, M. J., McCleary, S., Boden, P., Suman-Chauhan, N., Hughes, J., and Singh, L. (1998). Involvement of the central tachykinin NK1 receptor during maintenance of mechanical hypersensitivity induced by diabetes in the rat. *J. Pharmacol. Exp. Ther.* **285**, 1226–1232.

Field, M. J., McCleary, S., Hughes, J., and Singh, L. (1999). Gabapentin and pregabalin, but not morphine and amitriptyline, block both static and dynamic components of mechanical allodynia induced by streptozocin in the rat. *Pain* **80**, 391–398.

Forman, L. J., Estilow, S., Lewis, M., and Vasilenko, P. (1986). Streptozocin diabetes alters immunoreactive beta-endorphin levels and pain perception after 8 wk in female rats. *Diabetes* **35**, 1309–1313.

Fox, A., Eastwood, C., Gentry, C., Manning, D., and Urban, L. (1999). Critical evaluation of the streptozotocin model of painful diabetic neuropathy in the rat. *Pain* **81**, 307–316.

Garrett, N. E., Malcangio, M., Dewhurst, M., and Tomlinson, D. R. (1997). alpha-Lipoic acid corrects neuropeptide deficits in diabetic rats via induction of trophic support. *Neurosci. Lett.* **222**, 191–194.

Gregersen, G. (1968). Vibratory perception threshold and motor conduction velocity in diabetics and non-diabetics. *Acta Med. Scand.* **183**, 61–65.

Harati, Y., Gooch, C., Swenson, M., Edelman, S. V., Greene, D., Raskin, P., Donofrio, P., Cornblath, D., Olson, W. H., and Kamin, M. (2000). Maintenance of the long-term effectiveness of tramadol in treatment of the pain of diabetic neuropathy. *J. Diabet. Comp.* **14**, 65–70.

Hellweg, R., and Hartung, H. D. (1990). Endogenous levels of nerve growth factor (NGF) are altered in experimental diabetes mellitus: A possible role for NGF in the pathogenesis of diabetic neuropathy. *J. Neurosci. Res.* **26**, 258–267.

Hylden, J. L., and Wilcox, G. L. (1981). Intrathecal substance P elicits a caudally-directed biting and scratching behavior in mice. *Brain Res.* **217**, 212–215.

Ihara, C., Shimatsu, A., Mizuta, H., Murabe, H., Nakamura, Y., and Nakao, K. (1996). Decreased neurotrophin-3 expression in skeletal muscles of streptozotocin-induced diabetic rats. *Neuropeptides* **30**, 309–312.

Jakobsen, J., Brimijoin, S., Skau, K., Sidenius, P., and Wells, D. (1981). Retrograde axonal transport of transmitter enzymes, fucose-labeled protein, and nerve growth factor in streptozotocin-diabetic rats. *Diabetes* **30**, 797–803.

Jakobsen, J., and Sidenius, P. (1980). Decreased axonal transport of structural proteins in streptozotocin diabetic rats. *J. Clin. Invest.* **66**, 292–297.

Jarvis, M. F., Wessale, J. L., Zhu, C. Z., Lynch, J. J., Dayton, B. D., Calzadilla, S. V., Padley, R. J., Opgenorth, T. J., and Kowaluk, E. A. (2000). ABT-627, an endothelin ET(A) receptor-selective antagonist, attenuates tactile allodynia in a diabetic rat model of neuropathic pain. *Eur. J. Pharmacol.* **388**, 29–35.

Jirmanova, I. (1993). Giant axonopathy in streptozotocin diabetes of rats. *Acta Neuropathol.* **86**, 42–48.

Johnson, P. C., Brendel, K., and Meezan, E. (1981). Human diabetic perineurial cell basement membrane thickening. *Lab. Invest.* **44**, 265–270.

Kalichman, M. W., Powell, H. C., and Mizisin, A. P. (1998). Reactive, degenerative, and proliferative Schwann cell responses in experimental galactose and human diabetic neuropathy. *Acta Neuropathol.* **95**, 47–56.

Kamei, J., Hitosugi, H., and Kasuya, Y. (1993b). Formalin-induced nociceptive responses in diabetic mice. *Neurosci. Lett.* **149**, 161–164.

Kamei, J., Hitosugi, H., Kawashima, N., Aoki, T., Ohhashi, Y., and Kasuya, Y. (1992). Antinociceptive effect of mexiletine in diabetic mice. *Res. Commun. Chem. Pathol. Pharmacol.* **77**, 245–248.

Kamei, J., Kashiwazaki, T., Hitosugi, H., and Nagase, H. (1997). The role of spinal delta1-opioid receptors in inhibiting the formalin-induced nociceptive response in diabetic mice. *Eur. J. Pharmacol.* **326**, 31–36.

Kamei, J., Kawashima, N., and Kasuya, Y. (1993a). Serum glucose level-dependent and independent modulation of mu-opioid agonist-mediated analgesia in diabetic mice. *Life Sci.* **52**, 53–60.

Kamei, J., Kawashima, N., Narita, M., Suzuki, T., Misawa, M., and Kasuya, Y. (1994). Reduction in ATP-sensitive potassium channel-mediated antinociception in diabetic mice. *Psychopharmacology* **113**, 318–321.

Kamei, J., Ogawa, M., and Kasuya, Y. (1990). Development of supersensitivity to substance P in the spinal cord of the streptozotocin-induced diabetic rats. *Pharmacol. Biochem. Behav.* **35**, 473–475.

Kamei, J., Ohhashi, Y., Aoki, T., and Kasuya, Y. (1991). Streptozotocin-induced diabetes in mice reduces the nociceptive threshold, as recognized after application of noxious mechanical stimuli but not of thermal stimuli. *Pharmacol. Biochem. Behav.* **39**, 541–544.

Kastrup, J., Petersen, P., Dejgard, A., Angelo, H. R., and Hilsted, J. (1987). Intravenous lidocaine infusion: A new treatment of chronic painful diabetic neuropathy? *Pain* **28**, 69–75.

Kim, S. H., and Chung, J. M. (1992). An experimental model for peripheral neuropathy produced by segmental spinal nerve ligation in the rat. *Pain* **50**, 355–363.

Klamt, J. G. (1998). Effects of intrathecally administered lamotrigine, a glutamate release inhibitor, on short- and long-term models of hyperalgesia in rats. *Anesthesiology* **88**, 487–494.

Kolta, M. G., Ngong, J. M., Rutledge, L. P., Pierzchala, K., and Van Loon, G. R. (1996). Endogenous opioid peptide mediation of hypoalgesic response in long-term diabetic rats. *Neuropeptides* **30**, 335–344.

Kowaluk, E. A., Mikusa, J., Wismer, C. T., Zhu, C. Z., Schweitzer, E., Lynch, J. J., Lee, C. H., Jiang, M., Bhagwat, S. S., and Gomtsyan, A. (2000). ABT-702 (4-amino-5-(3-bromo-phenyl)-7-(6-morpholino-pyridin-3-yl)pyrido[2,3-d]pyrimidine), a novel orally effective adenosine kinase inhibitor with analgesic and anti-inflammatory properties. II. *In vivo* characterization in the rat. *J. Pharmacol. Exp. Ther.* **295**, 1165–1174.

Lawrence, D. G., and Locke S. (1961). Motor nerve conduction velocity in diabetes. *Arch. Neurol.* **5**, 483–489.

Lee, J. H., and McCarty, R. (1990). Glycemic control of pain threshold in diabetic and control rats. *Physiol. Behav.* **47**, 225–230.

Lee, J. H., and McCarty, R. (1992). Pain threshold in diabetic rats: Effects of good versus poor diabetic control. *Pain* **50**, 231–236.

Lee, Y. H., Ryu, T. G., Park, S. J., Yang, E. J., Jeon, B. H., Hur, G. M., and Kim, K. J. (2000). Alpha1-adrenoceptors involvement in painful diabetic neuropathy: A role in allodynia. *Neuroreport* **11**, 1417–1420.

Li, N., Young, M. M., Bailey, C. J., and Smith, M. E. (1999). NMDA and AMPA glutamate receptor subtypes in the thoracic spinal cord in lean and obese-diabetic ob/ob mice. *Brain Res.* **849**, 34–44.

Luo, Z. D., Chaplan, S. R., Scott, B. P., Cizkova, D., Calcutt, N. A., and Yaksh, T. L. (1999). Neuronal nitric oxide synthase mRNA upregulation in rat sensory neurons after spinal nerve ligation: Lack of a role in allodynia development. *J. Neurosci.* **19**, 9201–9208.

Levine, A. S., Morley, J. E., Wilcox, G., Brown, D. M., and Handwerger, B. S. (1982). Tail pinch behavior and analgesia in diabetic mice. *Physiol. Behav.* **28**, 39–43.

Lindsay, R. M., and Harmar, A. J. (1989). Nerve growth factor regulates expression of neuropeptide genes in adult sensory neurons. *Nature* **337**, 362–364.

Llewelyn, J. G., Gilbey, S. G., Thomas, P. K., King, R. H., Muddle, J. R., and Watkins, P. J. (1991). Sural nerve morphometry in diabetic autonomic and painful sensory neuropathy: A clinicopathological study. *Brain* **114**, 867–892.

Lynch, J. J., Jarvis, M. F., and Kowaluk, E. A. (1999). An adenosine kinase inhibitor attenuates tactile allodynia in a rat model of diabetic neuropathic pain. *Eur. J. Pharmacol.* **364**, 141–146.

Malcangio, M., and Tomlinson, D. R. (1998). A pharmacologic analysis of mechanical hyperalgesia in streptozotocin-diabetic rats. *Pain* **76**, 151–157.

Malik, R. A., Veves, A., Walker, D., Siddique, I., Lye, R. H., Schady, W., and Boulton, A. J. (2001). Sural nerve fibre pathology in diabetic patients with mild neuropathy: Relationship to pain, quantitative sensory testing and peripheral nerve electrophysiology. *Acta Neuropathol* **101**, 367–374.

Malmberg, A. B., and Basbaum, A. I. (1998). Partial sciatic nerve injury in the mouse as a model of neuropathic pain: Behavioral and neuroanatomical correlates. *Pain* **76**, 215–222.

Malmberg, A. B., and Yaksh, T. L. (1992). Hyperalgesia mediated by spinal glutamate or substance P receptor blocked by spinal cyclooxygenase inhibition. *Science* **257**, 1276–1279.

Malmberg, A. B., and Yaksh, T. L. (1993). Spinal nitric oxide synthesis inhibition blocks NMDA-induced thermal hyperalgesia and produces antinociception in the formalin test in rats. *Pain* **54**, 291–300.

Malmberg, A. B., and Yaksh, T. L. (1995). Cyclooxygenase inhibition and the spinal release of prostaglandin E2 and amino acids evoked by paw formalin injection: A microdialysis study in unanesthetized rats. *J. Neurosci.* **15**, 2768–2776.

Malmberg, A. B., Yaksh, T. L., and Calcutt, N. A. (1993). Anti-nociceptive effects of the GM1 ganglioside derivative AGF 44 on the formalin test in normal and streptozotocin-diabetic rats. *Neurosci. Lett.* **161**, 45–48.

Matzner, O., and Devor, M. (1987). Contrasting thermal sensitivity of spontaneously active A- and C-fibers in experimental nerve-end neuromas. *Pain* **30**, 373–384.

Max, M. B., Lynch, S. A., Muir, J., Shoaf, S. E., Smoller, B., and Dubner, R. (1992). Effects of desipramine, amitriptyline, and fluoxetine on pain in diabetic neuropathy. *N. Engl. J. Med.* **326**, 1250–1256.

McLean, W. G., Chapman, J. E., and Cullum, N. A. (1987). Impaired induction of ornithine decarboxylase activity following nerve crush in the streptozotocin-diabetic rat. *Diabetologia* **30**, 963–965.

Medori, R., Autilio-Gambetti, L., Monaco, S., and Gambetti, P. (1985). Experimental diabetic neuropathy: Impairment of slow transport with changes in axon cross-sectional area. *Proc. Natl. Acad. Sci. USA* **82**, 7716–7720.

Mizisin, A. P., Calcutt, N. A., Tomlinson, D. R., Gallagher, A., and Fernyhough, P. (1999a). Neurotrophin-3 reverses nerve conduction velocity deficits in streptozotocin-diabetic rats. *J. Periph. Nerv. Syst.* **4**, 211–221.

Mizisin, A. P., DiStefano, P. S., Liu, X., Garrett, D. N., and Tonra, J. R. (1999b). Decreased accumulation of endogenous brain-derived neurotrophic factor against constricting sciatic nerve ligatures in streptozotocin-diabetic and galactose-fed rats. *Neurosci. Lett.* **263**, 149–152.

Moore, S. A., Peterson, R. G., Felten, D. L., and O'Connor, B. L. (1980). A quantitative comparison of motor and sensory conduction velocities in short- and long-term streptozotocin- and alloxan-diabetic rats. *J. Neurol. Sci.* **48**, 133–152.

Morley, G. K., Mooradian, A. D., Levine, A. L., and Morley, J. E. (1984). Mechanisms of pain in diabetic peripheral neuropathy: Effects of glucose on pain in humans. *Am. J. Med.* **77**, 79–86.

Okuse, K., Chaplan, S. R., McMahon, S. B., Luo, Z. D., Calcutt, N. A., Scott, B. P., Akopian, A. N., and Wood, J. N. (1997). Regulation of expression of the sensory neuron-specific sodium channel SNS in inflammatory and neuropathic pain. *Mol. Cell. Neurosci.* **10**, 196–207.

Oskarsson, P., Ljunggren, J. G., and Lins, P. E. (1997). Efficacy and safety of mexiletine in the treatment of painful diabetic neuropathy. The Mexiletine Study Group. *Diabet. Care* **20**, 1594–1597.

Partanen, J., Niskanen, L., Lehtinen, J., Mervaala, E., Siitonen, O., and Uusitupa, M. (1995). Natural history of peripheral neuropathy in patients with non-insulin-dependent diabetes mellitus. *N. En. J. Med.* **333**, 89–94.

Pertovaara, A., Wei, H., Kalmari, J., and Ruotsalainen, M. (2001). Pain behavior and response properties of spinal dorsal horn neurons following experimental diabetic neuropathy in the rat: modulation by nitecapone, a COMT inhibitor with antioxidant properties. *Exp. Neurol.* **167**, 425–434.

Powell, H. C., Rosoff, J., and Myers, R. R. (1985). Microangiopathy in human diabetic neuropathy. *Acta Neuropathol.* **68**, 295–305.

Reske-Nielsen, E., and Lundbaek, K. (1968). Pathological changes in the central and peripheral nervous system of young long-term diabetics. II. The spinal cord and peripheral nerves. *Diabetologia* **4**, 34–43.

Robinson, J. P., Willars, G. B., Tomlinson, D. R., and Keen, P. (1987). Axonal transport and tissue contents of substance P in rats with long-term streptozotocin-diabetes. Effects of the aldose reductase inhibitor 'statil'. *Brain Res.* **426**, 339–348.

Rodriguez-Pena, A., Botana, M., Gonzalez, M., and Requejo, F. (1995). Expression of neurotrophins and their receptors in sciatic nerve of experimentally diabetic rats. *Neurosci. Lett.* **200**, 37–40.

Russell, L. C., and Burchiel, K. J. (1993). Abnormal activity in diabetic rat saphenous nerve. *Diabetes* **42**, 814–819.

Seltzer, Z., Dubner, R., and Shir, Y. (1990). A novel behavioral model of neuropathic pain disorders produced in rats by partial sciatic nerve injury. *Pain* **43**, 205–218.

Sharma, A. K., and Thomas, P. K. (1974). Peripheral nerve structure and function in experimental diabetes. *J. Neurol. Sci.* **23**, 1–15.

Sidenius, P., and Jakobsen, J. (1980). Reduced perikaryal volume of lower motor and primary sensory neurons in early experimental diabetes. *Diabetes* **29**, 182–186.

Slager, U. T. (1978). Diabetic myelopathy. *Arch. Pathol. Lab. Med.* **102**, 467–469.

Stewart, J. K., Campbell, T. G., Gbadebo, T. D., Narasimhachari, N., and Manning, J. W. (1994). Cardiovascular responses and central catecholamines in streptozocin-diabetic rats. *Neurochem. Int.* **24**, 183–189.

Suh, H. W., Song, D. K., Wie, M. B., Jung, J. S., Hong, H. E., Choi, S. R., and Kim, Y. H. (1996). The reduction of antinociceptive effect of morphine administered intraventricularly is correlated with the decrease of serotonin release from the spinal cord in streptozotocin-induced diabetic rats. *Gen. Pharmacol.* **27**, 445–450.

Takeshita, N., Ohkubo, Y., and Yamaguchi, I. (1995). Tiapride attenuates pain transmission through an indirect activation of central serotonergic mechanism. *J. Pharmacol. Exp. Ther.* **275**, 23–30.

Takeshita, N., and Yamaguchi, I. (1997). Insulin attenuates formalin-induced nociceptive response in mice through a mechanism that is deranged by diabetes mellitus. *J. Pharmacol. Exp. Ther.* **281**, 315–321.

Takeshita, N., and Yamaguchi, I. (1998). Antinociceptive effects of morphine were different between experimental and genetic diabetes. *Pharmacol. Biochem. Behav.* **60**, 889–897.

Tamura, E., and Parry, G. J. (1994). Severe radicular pathology in rats with longstanding diabetes. *J. Neurol. Sci.* **127**, 29–35.

Thomas P. K., and Lascelles R. G. (1965). Schwann cell abnormalities in diabetic neuropathy. *Lancet* **1**, 1355–1357.

Thomas, P. K., and Wright, D. W., and Tzebelikos, E. (1984). Amino acid uptake by dorsal root ganglia from streptozotocin-diabetic rats. *J. Neurol. Neurosurg. Psych.* **47**, 912–916.

Tomlinson, D. R., Robinson, J. P., Willars, G. B., and Keen, P. (1988). Deficient axonal transport of substance P in streptozocin-induced diabetic rats. Effects of sorbinil and insulin. *Diabetes* **37**, 488–493.

Tsigos, C., Diemel, L. T., White, A., Tomlinson, D. R., and Young, R. J. (1993). Cerebrospinal fluid levels of substance P and calcitonin-gene-related peptide: Correlation with sural nerve levels and neuropathic signs in sensory diabetic polyneuropathy. *Clin. Sci.* **84**, 305–311.

Watson, C. P. (2000). The treatment of neuropathic pain: Antidepressants and opioids. *Clin. J. Pain.* **16**, S49–55.

Whitworth, I. H., Terenghi, G., Green, C. J., Brown, R. A., Stevens, E., and Tomlinson, D. R. (1995). Targeted delivery of nerve growth factor via fibronectin conduits assists nerve regeneration in control and diabetic rats. *Eur. J. Neurosci.* **7**, 2220–2225.

Wuarin, L., Guertin, D. M., and Ishii, D. N. (1994). Early reduction in insulin-like growth factor gene expression in diabetic nerve. *Exp. Neurol.* **130**, 106–114.

Wuarin-Bierman, L., Zahnd, G. R., Kaufmann, F., Burcklen, L., and Adler, J. (1987). Hyperalgesia in spontaneous and experimental models of diabetic neuropathy. *Diabetologia* **30**, 653–658.

Yashpal, K., and Henry, J. L. (1984). Substance p analogue blocks sp-induced facilitation of a spinal nociceptive reflex. *Brain. Res. Bull.* **13**, 597–600.

Zochodne, D. W., Verge, V. M., Cheng, C., Hoke, A., Jolley, C., Thomsen, K., Rubin, I., and Lauritzen, M. (2000). Nitric oxide synthase activity and expression in experimental diabetic neuropathy. *J. Neuropathol. Exp. Neurol.* **59**, 798–807.

ELECTROPHYSIOLOGIC MEASURES OF DIABETIC NEUROPATHY: MECHANISM AND MEANING

Joseph C. Arezzo and Elena Zotova

Albert Einstein College of Medicine
Bronx, New York 10461

Whole nerve electrophysiologic procedures afford a battery of measures that can provide a noninvasive and objective index of the onset and progression of diabetic polyneuropathy (DPN). Advances in physiologic procedures, digital hardware, and mathematical models have allowed assessment of activity in slower conducting fibers, as well as measures that reflect changes in refractory periods and threshold excitability. These expanded options can augment standard measures of maximal conduction velocity and compound amplitude and greatly enhance the sensitivity of whole nerve measures to both structural (e.g., demyelination) and "nonstructural" (e.g., redistribution of ion channels) deficits associated with DPN. The mechanisms underlying the physiologic events in DPN are multifactorial and their sequence is complex, with different mechanisms contributing to change at overlapping, but distinct points in the progression. Factors influencing early change in velocity may differ from those contributing to chronic deficits and these mechanisms may also differ in their response to various putative therapies. This review attempts to summarize the pattern of whole nerve electrophysiologic change associated with DPN, outlines the strengths and limitations of the various measures that are feasible, and discusses the specific impact of known pathophysiologic mechanisms on these end points.   © 2002, Elsevier Science (USA).

I. Introduction

Diabetic polyneuropathy (DPN) is a complex disease process that reflects alterations in both the function and the structure of neurons. The onset of dysfunction can be insidious and is likely driven by multiple pathophysiologic mechanisms. Even a brief and incomplete review of putative sources of nerve damage includes deficits in endoneurial blood flow (e.g., Cameron and Cotter, 1997), modification of proteins by glycation and glycosylation (e.g., Brownlee, 1990), altered metabolism of fatty acids (e.g., Jamal, 1990), increased flux through the polyol pathway (e.g., Greene *et al.*, 1987), aberrant phosphorylation of neurofilaments (e.g., Fernyhough *et al.*, 1999), altered axonal transport (Abbate *et al.*, 1991), basement membrane replication and pericyte degeneration (e.g., Giannini and Dyck, 1995), reduction in neurotrophic factors (e.g., Tomlinson *et al.*, 1996), oxidative stress (e.g., Low *et al.*, 1997), and mitochondrial superoxide overproduction (e.g., Nishikawa *et al.*, 2000). Most of these hypotheses are reviewed in detail in this volume. In individual patients, the time course and relative weighting of these factors are likely due to a combination of genetics and diet. Because of the multiple manifestations of the deficits, no single measure has emerged as a sensitive, reliable, and valid index of the onset and progression of DPN. However, in recent years, whole nerve electrophysiologic procedures (e.g., conduction velocities and/or amplitude of the compound muscle action potential) have gained increased importance as objective measures of this condition. Multiple consensus panels have recommended the inclusion of electrophysiologic measures in the clinical evaluation of DPN, as well as the use of these procedures as surrogate measures in multicenter clinical trials (e.g., American Diabetic Association, 1988; American Diabetic Association and American Neurologic Association, 1992; Peripheral Nerve Society, 1995). In parallel, electrophysiologic procedures have been used extensively to explore the mechanisms of dysfunction and the value of various therapeutic interventions in chemical and genetic animal models of hyperglycemia [e.g., streptozotocin (STZ)]. A literature search cross-referencing diabetic neuropathy and nerve conduction yields more than 150 papers since 1998.

The strengths of electrophysiology are obvious (for a review, see Arezzo, 1997). An appropriate battery of tests supports the measurement of the speed of both sensory and motor conduction, the amplitude of the propagating neural signal, the density and synchrony of muscle fibers activated by maximal nerve stimulation, and the integrity of neuromuscular transmission. These are objective, parametric, and highly standardized measures. The limitations of electrophysiologic measures are also clear. "Standard" procedures, such as maximal nerve conduction velocity (maxNCV), reflect only a limited aspect of neural activity, and then only in a small subset of large-diameter and heavily myelinated axons. In many nerve segments, axons contributing to maxNCV represent less than 10% of the total population. Largely because of the ease of assessment and the consistency of the response, maxNCV has become the principal, if not the exclusive, index of DPN in many studies. However, this parameter is often assessed using a procedure developed in the early 1960s and essentially unmodified in the ensuing decades (see Downie and Newell, 1961). Even in large-diameter fibers, velocity, driven by a single pulse and measured at one or two points, is insensitive to many pathologic changes known to be associated with DPN. For example, strong evidence links DPN with a reduction in Na^+/K^+ adenosine triphosphatase activity (Calcutt *et al.*, 1998; Raccah *et al.*, 1998; Cameron *et al.*, 1999; Hohman *et al.*, 2000), which would primarily affect the ability of the responsive neural populations to rapidly reestablish select transmembrane ion gradients after activation. This deficit, and its possible alteration by effective therapy, would likely be entirely missed by traditional maxNCV procedures. In a similar vein, maxNCV would also be insensitive to the effects of therapeutic agents [e.g., nerve growth factor (NGF)] that target small fiber function (Apfel *et al.*, 1998). Thus, in many studies, decisions are being made about treatment options and about correlations of structure, biochemistry, and function using a principal measure (i.e., maxNCV) that is insensitive to some key aspects of neural activity. The obvious statement that "nerve conduction velocity is not synonymous with nerve function" has been ignored, with potentially negative consequences, in several recent publications and presentations.

II. Functional and Structural Deficits Associated with Diabetic Neuropathy

Although the etiology is controversial, there is general agreement as to the pattern of structural and functional neural defects associated with the most common form of DPN. DPN is generally considered a type of "length-dependent distal axonopathy" (Schaumburg *et al.*, 1992) in which the cell body remains relatively unaffected at a time point when the distal peripheral axon begins to undergo pathologic alterations. Initial changes may be limited to functional alterations in transmembrane ion gradients or to redistributions of specific ion channels in the nodal regions (for a review, see Waxman, 1995). Disturbances in transduction, with consequent inappropriate or chaotic firing, may be associated with hyperexcitability and with "positive symptoms" (e.g., pain, paresthesia). Deficits in multiple forms of axonal transport, with a consequent reduction in neurotrophic support, may be part of the sequelae leading to early, length-dependent changes in the cross-sectional diameter of existing axons (Tomlinson *et al.*, 1984; Abbate *et al.*, 1991). Axon caliber may also be reduced in DPN due to alterations in tissue osmolarity (Sugimura *et al.*, 1980) or hydration (Eaton *et al.*, 1996). Swelling or bubbling of the myelin accompanies early axonal pathology, particularly in the vicinity of the nodes of Ranvier, which may lead to paranodal regions of axoglial disjunction (Sima *et al.*, 1988). Frank demyelination, which can be distributed along several points on the peripheral neuraxis, is characteristic of later stages of DPN and may result in complete conduction block for select axons. As some fibers are lost due to degeneration, new regenerating fibers are often detected clustered within the basal laminae, or basal laminae remnants, of previously degenerated myelinated axons. The abundant regeneration seen in early DPN is replaced by a relative failure of regenerating fibers as the neuropathy progresses (Bradley *et al.*, 1995). Substantial "dying back" Wallerian degeneration, resulting in a loss of a portion of the population of both large (Dyck, 1986) and small (Said *et al.*, 1983) caliber axons and a consequent reduction in total fiber density, is the structural hallmark of severe DPN. These structural alterations are associated with "negative symptoms" (e.g., numbness, weakness) and with severe clinical consequences (e.g., Boulton, 1997).

III. Whole Nerve Neural Response

To explore the value of noninvasive electrophysiology as an index of DPN, we must first understand the nature of the signal that can be measured from the whole nerve. An electrophysiologic response recorded from a whole nerve represents the composite of activity from thousands of fibers and is therefore different from the neural signal generated and propagated in individual axons. Under normal circumstances, even the fastest neural impulses are conducted along the whole nerve at the remarkably slow speed of 60–80 m/s. The slowest fibers conduct at less than 1 m/s. When stimulated at muscle or skin, the whole nerve afferent volley is characterized by a series of overlapping waves that propagate at different speeds (Okajima *et al.*, 1994; Kimura, 1995). With increasing distance from the site of sensory

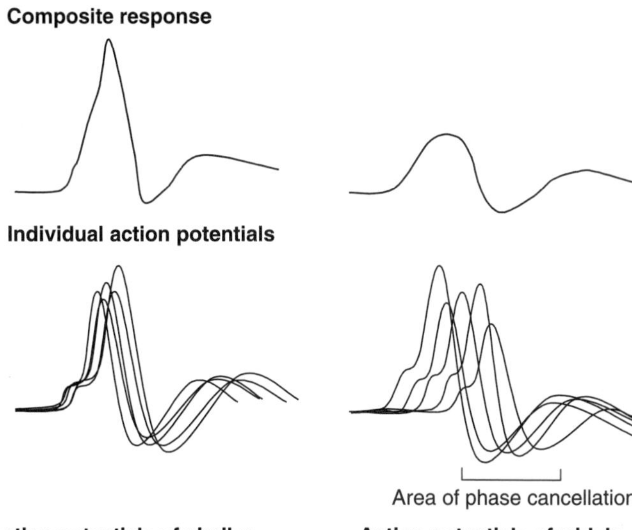

FIG. 1. Model of how action potentials reflecting the activity in a series of neurons sum to form the composite "whole nerve" response. The composite response for axons conducting at a uniform velocity has a peak amplitude approximately double the composite of the same number of axons, with approximately the same individual amplitudes, but conducting at disparate velocities. When measured at onset, maxNCV represents activity in only the fastest conducting fibers and can be similar for these two distributions.

transduction or from the ventral horn of the spinal cord for efferent nerves, the composite wave is progressively delayed in onset, reduced in peak amplitude, broadened in duration, and, with sufficient distance, begins to demonstrate phase cancellation (Olney *et al.*, 1987; vanDijk *et al.*, 1994; Schulte-Mattler and Zierz, 1999). This occurs because of spatial–temporal interactions even though the action potentials at each node basically behave according to the "all-or-none" rule and are approximately of equivalent amplitude at each point along the nerve (Fig. 1).

The goal of this review is to outline several whole nerve electrophysiologic parameters that are available and feasible and to explore how each of these measures is influenced by the structural and functional deficits associated with the onset and progression of DPN (Tables IA and IB).

TABLE 1A

Histopathologic Correlates of Early/Subtle Changes in Electrophysiologic Measures Associated with DPN

Electrophysiologic measures	Histopathologic measures															
	Frank demyelination		Axo-glia changes at nodes		Axonal atrophy (distal)		Axonal atrophy (proximal)		Reduction of fiber density		Denervation of muscle		Redistribution of ions at node		Reduction of Na^+/K^+ ATPase activity	
Axon diameter	Large	Small	Large	Small	Large	Small	Large	Small	Large	Small	Large	Small	Large	Small	Large	Small
MaxNCV			X		X								X			
Distribution of velocities				X	X									X		
Distal motor latency			X		X						X		X			
F-wave latency			X				X						X			
SNAP amplitude			X		X				X				X			
CMAP amplitude					X				X		X		X			
Area of depolarization							X			X						
Repetitive stimulation													X		X	
Axon excitability													X			

TABLE 1B

HISTOPATHOLOGIC CORRELATES OF CHRONIC/SEVERE CHANGES IN ELECTROPHYSIOLOGIC MEASURES ASSOCIATED WITH DPN

Electrophysiologic measures	Histopathologic measures															
	Frank demyelination		Axo-glia changes at nodes		Axonal atrophy (distal)		Axonal atrophy (proximal)		Reduction of fiber density		Denervation of muscle		Redistribution of ions at node		Reduction of Na$^+$/K$^+$ ATPase activity	
Axon diameter	Large	Small	Large	Small	Large	Small	Large	Small	Large	Small	Large	Small	Large	Small	Large	Small
MaxNCV	X		X		X		X		X				X			
Distribution of velocities		X		X		X				X				X		
Distal motor latency	X		X		X				X		X		X			
F-wave latency	X		X				X						X			
SNAP amplitude	X		X		X				X				X			
CMAP amplitude	X		X		X				X		X		X			
Area of depolarization	X	X	X	X	X	X			X	X			X	X		
Repetitive stimulation	X		X										X		X	
Axon excitability	X		X		X								X			

IV. Maximal Nerve Conduction Velocity

In whole nerve recordings, the principal factors that influence the speed of maxNCV are (1) the integrity and degree of myelination of the largest diameter fibers, (2) the mean cross-sectional diameter of the responding axons, (3) the representative internodal distance in the segment under study, and (4) the microenvironment at the nodes, including the distribution of ion channels. Figure 2 illustrates the progression of change in electrophysiologic measures, including maxNCV, associated with DPN, and lists putative underlying mechanisms.

A. Myelin

More than 120 years ago, Ranvier noted the lipid properties of myelin and suggested its role as an electrical insulator (Ranvier, 1878). In fact, in the peripheral nervous system, Schwann cell myelin is a rather poor insulator because of current leakage through transverse cytoplasmic belts and Schmidt–Lanterman incisures, as well as gaps along the periaxonal space (for review, see Ritchie, 1995). Myelin, however, has a profound influence on the speed of conduction by decreasing membrane capacitance (c_m) and therefore allowing threshold depolarization at a distant site along the axon cable without the need for threshold events (e.g., active ionic current flow) in the intervening axolemma. The signal thus "jumps" along the axon (i.e., saltatory conduction), saving an enormous amount of time and energy. As outlined later, a second critical factor that determines conduction velocity is the cross-sectional diameter of the axon, which is proportional to axial resistance (r_a). The ratio of axonal diameter divided by total diameter (i.e., axon and myelin) is termed the g ratio, which under normal conditions ranges from approximately 0.5 to 0.7. A shift of this ratio, due to a change in axon cytoplasm, myelin, or both, can strongly influence maxNCV.

Frank demyelination, consequent conduction block, and alterations in the g ratio of large-diameter axons have all been reported in late-stage DPN (Sima *et al.*, 1988). These changes unquestionably contribute to the observed slowing of maxNCV and to the high incidence of unobtainable signals from surface recording procedures (e.g., sural orthodromic conduction velocity) in chronic DPN. However, alterations in myelin appear to have little direct effect on the early changes in conduction evident in DPN. This is consistent with a reduction of maxNCV that is remarkably gradual in the early stages of DPN. While the precise rate of change in maxNVC for individual diabetic subjects is dependent on glycemic control

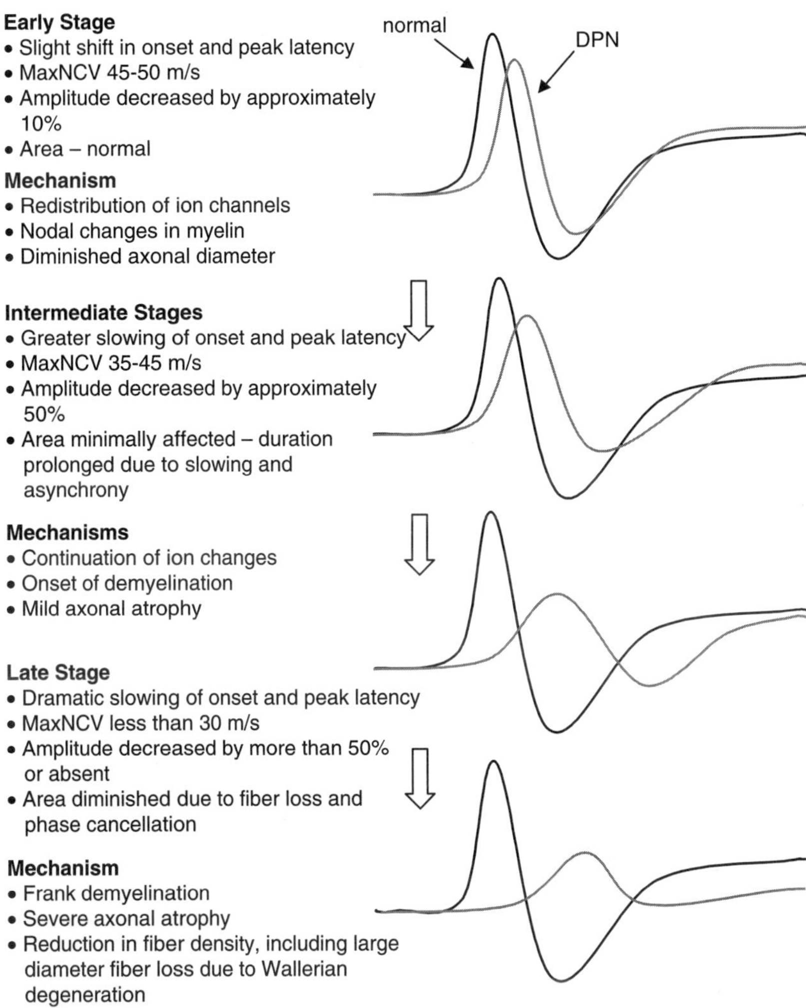

Early Stage
- Slight shift in onset and peak latency
- MaxNCV 45-50 m/s
- Amplitude decreased by approximately 10%
- Area – normal

Mechanism
- Redistribution of ion channels
- Nodal changes in myelin
- Diminished axonal diameter

Intermediate Stages
- Greater slowing of onset and peak latency
- MaxNCV 35-45 m/s
- Amplitude decreased by approximately 50%
- Area minimally affected – duration prolonged due to slowing and asynchrony

Mechanisms
- Continuation of ion changes
- Onset of demyelination
- Mild axonal atrophy

Late Stage
- Dramatic slowing of onset and peak latency
- MaxNCV less than 30 m/s
- Amplitude decreased by more than 50% or absent
- Area diminished due to fiber loss and phase cancellation

Mechanism
- Frank demyelination
- Severe axonal atrophy
- Reduction in fiber density, including large diameter fiber loss due to Wallerian degeneration
- Conduction block

FIG. 2. Model of the change in the whole nerve electrophysiologic response as a function of the progression of DPN, including a likely mechanism of dysfunction.

and a variety of anthropometric factors (e.g., age, height), population data suggest a mean loss of only 0.5 m/s/year (Partanen *et al.*, 1995). In contrast, in conditions characterized by primary demyelination (e.g., Guillain–Barré syndrome), maxNCV can be reduced by more than 10 m/s within weeks. The early change in maxNCV consistently seen in animal models of hyperglycemia (i.e., rats treated with STZ) is not related to frank demyelination. Although changes in myelin have been reported as early as 8 weeks after exposure to STZ (Wattig *et al.*, 1986), in most studies, there has been no alteration in myelin reported until the late stages of prolonged chemically induced hyperglycemia (Yagihashi *et al.*, 1990; Yagihashi, 1995). Despite this, maxNCV can be significantly reduced within 2–4 weeks of exposure to STZ (e.g., Hounsom and Tomlinson, 1997; Kato *et al.*, 1998; Mizuno *et al.*, 1999; Walker *et al.*, 1999). The dissociation of "acute" changes in maxNCV and alterations in myelin suggests metabolic factors altering membrane leakage, ion channel distributions, and axonal cytoskeleton proteins as the source of early slowing of conduction in DPN.

Axoglial disjunction in the paranodal region, which is a more restricted and subtle form of altered myelin, has been suggested as an early component of DPN (Jakobsen, 1979; Sima *et al.*, 1988), but the impact of this change, in isolation, on altered maxNCV remains undetermined.

B. Axon Diameter

In addition to myelin, maxNCV is also critically related to the "length constant" of the axon. The length constant (also termed the space constant) is the distance along the axon at which the transmembrane voltage decays to a specified percentage of its original value. The loss of potential is principally due to two factors: (1) "leakage" through passive channels along the membrane and (2) resistance to longitudinal current flow down the axon cable. The "fatter" the axon, the better the conductance of the inner core (i.e., less resistance within the cytoplasm). If the axon withers, the length constant is reduced and the magnitude of the current sink established at the next node is slightly diminished. Thus, the time to reach threshold for the opening of "active channels" at each node is increased and the velocity across multiple nodes is slowed.

Ultrastructural studies in alloxan or streptozotocin models of diabetic neuropathy have generally reported that axons are reduced in size compared to age-matched controls (Jakobsen, 1976, 1979; Sharma *et al.*, 1977; Yagihashi, 1995; Calcutt *et al.*, 1998). This reduction in cross-sectional diameter could reasonably be due to a diminished production of endoskeletal and growth-associated proteins (Fernyhough *et al.*, 1999), which may be secondary to a DPN-induced reduction of neurotrophic factors (Hellweg *et al.*, 1994; Mohiuddin *et al.*, 1995; Riaz and Tomlinson, 1996). The change in cross-sectional diameter of existing axons could occur rapidly and may be a structural component of the acute changes in maxNCV in animal models of DPN. Conversely, because a change in axonal diameter may antecede demyelination or Wallerian degeneration, this mechanism may contribute to relatively early changes in maxNCV.

In chronic DPN, changes in axonal diameter could influence maxNCV through a second mechanism. Data from *ex vivo* nerve biopsies (Sima *et al.*, 1988; Veves *et al.*, 1991; Greene and Brown, 1995; Greene *et al.*, 1999), postmortem studies (Dyck, 1987), and histopathology in chronic animal models (Yagihashi *et al.*, 1990) have all documented the loss of axons and a consequent reduction of fiber density as a component of long-term DPN. These changes reflect accelerated Wallerian degeneration. Regenerating fibers, which can partially compensate for the loss of axons, tend to be smaller in diameter with shorter internodal lengths and do not contribute at all to maxNCV. A gradual, but relentless shift in the fiber diameter histogram toward smaller diameter fibers undoubtedly is a significant component of the continued loss of maxNCV, even in the later stages of DPN.

C. INTERNODAL DISTANCE

A third mechanism that can strongly influence maxNCV is the mean internodal distance in large-diameter, myelinated fibers. In normal whole nerve recordings, maxNCV has been reported to slow between 6 and 10 m/s (depending on age) over the distal median versus the proximal median nerve (Davis-King *et al.*, 1992). This is in part due to a slight reduction in cross-sectional axon diameter (see earlier discussion), but it also likely reflects shorter mean internodal distances in the distal nerve segments. Several authors have evaluated teased nerve fibers and determined that hyperglycemia induced by STZ was associated with a reduction in mean internodal length along the entire nerve (e.g., Mattingly and Fischer, 1983). This factor could theoretically contribute to the slowing of maxNCV in DPN; however, the effect is usually seen principally in small-diameter axons and therefore may have little or no impact on maxNCV. In addition to length-dependent vulnerabilities, the shorter internodal distances in distal nerve segments may exacerbate the early nodal effects of DPN. Consistent with this suggestion, Patel and Tomlinson (1999) reported a proximal–distal gradient in the severity of slowing of maxNCV, with STZ-induced deficits of 14% in proximal sensory maxNCV coincident with deficits of 40% in distal sensory maxNCV.

D. DISTRIBUTION OF ION CHANNELS

The responsiveness of the neuron and its ability to propagate neuroelectric signals are ultimately dependent on the distribution of transmembrane ion channels. In myelinated axons, the nodal region is characterized by a high density of voltage-sensitive channels permeable to Na^+, K^+, or both, as well as a nonspecific channel likely responsible for nodal "leakage currents." The density of Na^+ channels in the nodal region is generally greater than 1000 per μm^2, whereas the density in the axolemma is generally less than 25 per μm^2 (Ritchie, 1995). Although multiple types of Na^+ channels have been identified in the peripheral nervous system (for a review, see Goldin *et al.*, 2000), only one type (i.e., $Na_v1.6$) is found in the nodes of Ranvier (Caldwell *et al.*, 2000). A number of theories have been proposed for the anchoring of specific Na^+ channels to the nodal region, including a barrier at the junction of the axon and myelin, suppression of Na^+ channels in the internodal regions, or select binding to the extracellular matrix (for a review, see Waxman, 1995). DPN has been reported to alter the nodal/paranodal distribution of ion channels (Cherian *et al.*, 1996) and to widen the nodal gap (Jakobsen, 1976; Mattingly and Fischer, 1983); each of these modifications in large-diameter fibers would likely contribute to the slowing of whole nerve maxNCV. Voltage clamp studies at the node have demonstrated reduced sodium current in the BB Wistar rat model of diabetes (Brismar and Sima, 1981) and associated slowing of maxNCV. Furthermore, it is likely that alteration in the density and distribution of specific ion channels could occur within days of a triggering event and antecede the structural deficits outlined earlier. These changes may reflect "metabolic" deficits and could underlie acute and rapidly reversible changes in maxNCV. Later, chronic changes in maxNCV may represent the addition of structural deficits such as demyelination, axonal atrophy, and loss of fiber density. This duel mechanism underlying the slowing of maxNCV may partially explain why many putative therapeutic agents [e.g., acetyl-L-carnitine, low doses of aldose reductase inhibitors (ARIs)] have demonstrated positive results on acute changes in maxNCV in the STZ rat model, but failed to alter the effects of chronic diabetic neuropathy in human trials.

V. F-Wave Responses

As generally measured, F-waves reflect the antidromic conduction of the compound neural volley to the ventral spinal cord, the activation of a subpopulation of spinal motor neurons, the orthodromic conduction of the newly established volley, and the postsynaptic activation of a portion of the muscle fibers in the innervated muscle (for a review, see Kimura, 1989). The latency of the induced muscle contraction is variable and generally more than 10 times that evoked by standard distal stimulation;

the amplitude is a small fraction of the direct muscle response. Because of its "long–loop" nature, this whole nerve measure is very sensitive to factors that alter the speed of conduction, especially those distributed widely along the nerve or those that have a significant proximal component. A subtle change affecting each node may not be detected in measures focused on an isolated distal segment, but may accumulate and become evident in the long latency F-wave response. Consistent with this suggestion, F-wave procedures have been reported as a more sensitive tool in patients with axonal polyneuropathy than standard motor conduction studies (Fraser and Olney, 1993; Shin *et al.*, 2000). The length of the conduction pathway also reduces the contribution of technical factors (i.e., errors in distance measurement) and makes the F-wave a remarkably reproducible measure of nerve conduction in DPN (Kohara *et al.*, 2000). The flip side of this coin is that changes limited to the distal segment of the axon, including possible therapeutic benefits, may be lost in a measure that is substantially driven by conduction along proximal nerve segments. The most frequent measures of F-wave activity (i.e., minimal latency, which can be converted to an estimate of velocity) reflect changes exclusively in fast conducting fibers. However, the addition of other measures, such as mean and maximal latency, chronodispersion and mean duration, persistence, and amplitude, can add sensitivity to slower conducting axons (Toyokura, 1998; Kohara *et al.*, 2000; Nobrega *et al.*, 2001).

VI. Distribution of Velocities

MaxNCV utilizes only the onset of the evoked depolarization at specific points along the peripheral nerve and is therefore entirely insensitive to functional or structural deficits in unmyelinated or thinly myelinated neurons. The axons from these cells are ubiquitous in cutaneous and mixed nerves and are uniquely associated with the modalities of pain, temperature, and diffuse touch. Using a horse race analogy, maxNCV provides the finishing time of the winning horse, but tells you nothing about the other horses in the race. In a whole nerve procedure, the speed of conduction of the slower fibers must be determined by analysis of the morphology of the composite response (e.g., distribution of velocities) (for a review, see Dorfman, 1984) or by techniques designed to cancel the activity from faster fibers (e.g., the collision technique) (for a review, see Kimura, 1976).

Several procedures have been developed to analyze the distribution of conduction velocities from a single fixed site as a means of measuring activity in small-diameter axons (Pollak *et al.*, 1992; Ruijten *et al.*, 1993; Papadopoulou and Panas, 1999; Tu *et al.*, 1999; Wells and Gozani, 1999). The original studies of an altered distribution of conduction velocities in DPN are approximately two decades old (Cummins and Dorfman, 1981), but the technique is rarely used in current clinical studies or in experimental animal models of DPN. Analysis of the distribution of conduction velocities can be enhanced by simultaneously recording from multiple sites along the nerve. Further, fusion of a collision technique and analysis of the distribution of velocities (Ingram *et al.*, 1987; Saito, 1991; Caccia *et al.*, 1993) are promising as practical and valuable electrophysiologic procedures to explore the effects of DPN on slower fibers.

Most neuropathies, including DPN, are characterized by a relatively uniform involvement of axons across the fiber diameter spectrum. However, the value of a separate analysis of fiber groups has been illustrated by several studies. Caccia *et al.* (1993) demonstrated that a computer-assisted collision procedure was both capable of examining velocities in slower conducting fibers and sensitive to the presence of subclinical neuropathy in insulin-dependent diabetic subjects. Bertora *et al.* (1998) compared the distribution of conduction velocities with standard maxNCV in 138 patients with subclinical DPN. For sensory nerves, deficits were detected in 58% of the subjects using distribution data compared to only 11% of subjects using standard procedures.

VII. Amplitude and Area

In whole nerve recordings, the amplitude of the compound sensory nerve action potential (SNAP) is principally a measure of the number of activated large- and moderate-diameter myelinated fibers. When measured at the peak of depolarization, SNAP amplitude is also influenced by the synchrony of activity and therefore affected by velocity (Fig. 1). A slowing of conduction, without fiber dropout, could alter peak amplitude (flatten response due to asynchrony). The amplitude of the compound muscle action potential (CMAP) following supramaximal stimulation is more complex, reflecting the integrity of the distal portion of motor axons inner- vating the target muscle, the status of the neuromuscular junction, and the electrical properties (including conduction velocity) of the muscle fibers. The dissociation of CMAP amplitude and neuropathy is well established in diseases isolating postsynaptic neuromuscular events (e.g., myasthenia gravis).

Theoretically, the amplitude of the whole nerve compound action potential or the amplitude of the CMAP following supramaximal nerve stimulation should be outstanding measures of the onset and progression of DPN. Because DPN, in its early stages, is generally considered an axonopathy, rather than a myelinopathy or a neuronopathy, changes in amplitude should precede, and perhaps be more consistent, than changes in velocity. In practice, the value of whole nerve amplitude measures is severely limited by the variability of the measurement and by the fact that nonneural factors (e.g., edema) can significantly influence recorded values. To compensate for the variability, changes in amplitude can be summed or averaged across multiple nerves (Russell *et al.*, 1996), or the ratio of amplitudes across different nerves can be calculated (Pastore *et al.*, 1999; Shin *et al.*, 2000).

Estimates of total area or duration of the depolarizing response should also be excellent markers of DPN, but again the variability of these measures in whole nerve recordings limits their value (Bril *et al.*, 2001). Area is especially sensitive to temporal dispersion, leading to phase cancellation (Figs. 1 and 2) or to conduction block (Olney *et al.*, 1987). Area alone, or in association with peak amplitude, can also be used to examine changes in the refractory period by the collision technique (Ruijten *et al.*, 1994) or by repetitive stimulation (Low and McLeon, 1977).

Although there is controversy regarding the weighting of underlying causes, there is no doubt that fiber loss, manifest as a reduction in axon density per square millimeter, is a significant component of chronic DPN and a factor in the reduced amplitude of the SNAP measured in the whole nerve and the CMAP measured in the innervated muscle when motor or mixed nerves are stimulated (for a review, see Kimura, 1995). There is a strong correlation ($r = 0.74$; $p < 0.001$) between myelinated fiber density and whole nerve sural amplitude (Veves *et al.*, 1991). Russell and coauthors (1996) calculated that a change of 1.0 μV in sural nerve SNAP amplitude is associated with a decrease of approximately 150 fibers/mm^2, whereas a loss of 200 fibers/mm^2 is associated with an approximate 1.0-mV reduction in the mean amplitude of the CMAP from the ulnar, peroneal, and tibial nerves. Longitudinal studies suggest an average loss of SNAP amplitude at a rate of approximately 5% per year in DPN over a 10-year period (Partanen *et al.*, 1995); however, there are no compelling data as to whether this fall-off is linear. In the early stages of DPN, amplitude, as is the case for velocity, is generally affected principally in the lower limbs (for a review, see Daube, 1987).

There have been few studies of the distal–proximal gradient of amplitude deficits, but studies in normal subjects have outlined new methods of recording from the distal extreme of the sensory nerve and reported that distal amplitude is approximately 50% of that recorded over the proximal sensory nerve (Oh *et al.*, 2001). Amplitude changes (both deterioration and

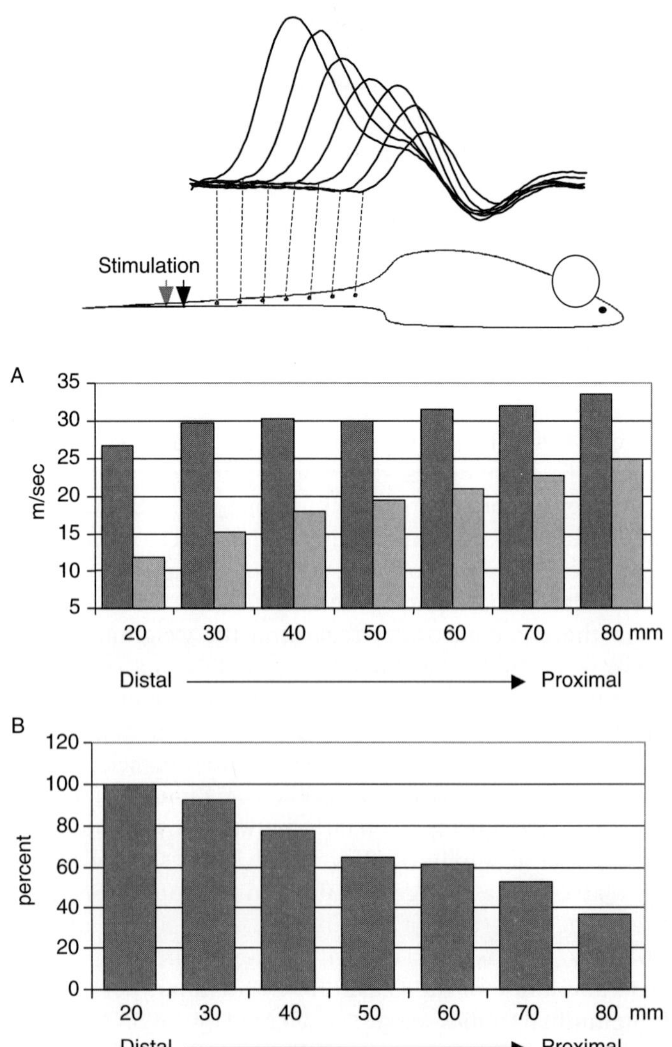

FIG. 3. Whole nerve response recorded at seven points (10 mm spaced) along the caudal nerve in a rat model. (A) MaxNCV (dark gray columns) reflects activity of the fastest conducting fibers and shows slight elevation at the proximal recording sites. NCV measured at the "peak" of the depolarization (light gray columns) includes activity in slower conducting axons and demonstrates a greater distal-to-proximal gradient. (B) Maximum amplitude consistently falls off with increased distance from the stimulating cathode due to a decrease in synchrony and a reduced contribution from slower fibers to the peak response.

possible improvement with therapy) over the distal sensory nerve of the lower limb may be an especially sensitive electrophysiologic procedure in charting the progression of DPN. The change in amplitude as a function of distance from the stimulating electrode is a second promising procedure to add to the sensitivity of whole nerve electrophysiologic measures. Although the signal is regenerated at each node, the amplitude of the composite wave systematically and predictably decreases with distance from the stimulating cathode due to progressive asynchronies (Fig. 3). In DPN, a conduction block of specific fibers and greater variability in propagation speed increase the temporal dispersion and thus the "fall-off" of amplitude per unit distance (Weber, 1997). Changes in the availability of amplifiers, inexpensive multiplexers, and high-capacity digital storage make it practical to simultaneously measure and average peripheral nerve activity from multiple sites. Simultaneous data from evenly spaced electrodes can be used to enhance resolution of the spatial properties and temporal dispersion of propagating neural signals, and the sensitivity of whole nerve amplitude measures to DPN (Fig. 3).

VIII. Refractory Periods and Neural Fatigue

A remarkable feature of the nervous system is that energy is principally utilized to establish an imbalance of transmembrane ions and voltage "at rest." These gradients are then preserved against both concentration and voltage forces by a series of conductance barriers and are maintained by a series of ion pumps to correct for leakage. The establishment and propagation of the action potential use little energy. Instead, energy is needed to reestablish the conditions necessary for the next pulse. In a standard maximal nerve conduction procedure, the stimulation rate is relatively low (e.g., 1 Hz) and there may be only one supramaximal stimulus presented to drive the response measured. Two key aspects of altered neural function associated with DPN may be a diminished capacity to follow high-frequency stimulation related to alterations in relative and absolute refractory periods and an increase in the rate of "neural fatigue" following prolonged periods of rapid stimulation, likely related to deficits in ATPase activity (Morita *et al.*, 1993; Chapman and Yeomans, 1994, Calcutt *et al.*, 1998). The previous reliance on slow stimulation rates appears to have seriously underestimated the incidence of conduction block in a variety of neuropathies (Cappelen-Smith *et al.*, 2000; Kaji *et al.*, 2000). In animal models, whole nerve responses can follow frequencies of up to 100 Hz with little change in velocity, amplitude, or duration. In contrast, animals with

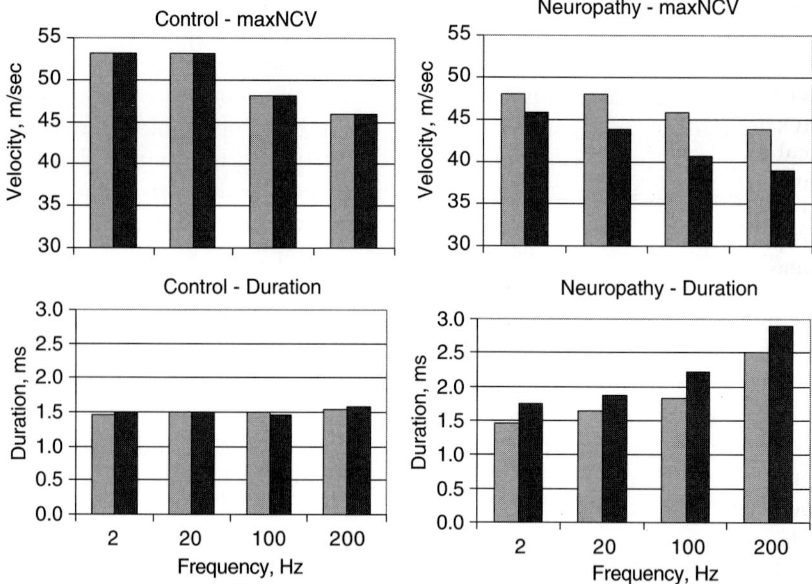

FIG. 4. MaxNCV and duration of depolarization (negative wave) as a function of stimulation frequency in a rat model. Data were recorded following either a 10-s period of prestimulation (light gray columns) or a 50-s period of prestimulation (dark gray columns) for both a control subject and a representative subject with a mild induced neuropathy. In the control subject, maxNCV is reduced at higher stimulation rates, but is not affected by the length of the prestimulation period; duration is not affected by either rate or prestimulation. As expected, neuropathy results in a slowing of maxNCV at all frequencies and in a further reduction of maxNCV following prolonged prestimulation. Duration is increased at higher frequencies (presumably due to increased asynchrony) and is altered by the characteristics of prestimulation (presumably due to energy-related deficits in the ability to rapidly reestablish optimal ion gradients).

induced neuropathy show marked alterations in each of these parameters at rapid stimulation rates (Fig. 4). These changes are exacerbated by prolonged periods of previous stimulation (i.e., neural fatigue) and clearly manifest using multiple simultaneous recording points (Fig. 5). Again, using a horse race analogy, a deficit may not be evident in the speed of the first race, but may become clear if the horses were forced to repeat the race with little or no rest, or to race for the entire day.

The deficits associated with rapid repetitive stimulation reflect the energy requirements of the neuron needed to reestablish a resting imbalance of ion distributions across the axonal membrane, which is in turn related to metabolic efficiency, the integrity of Na^+/K^+ ion pumps and

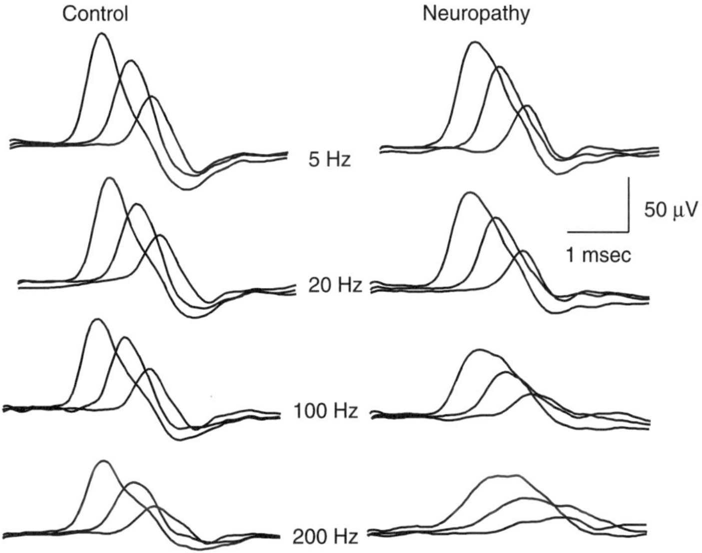

FIG. 5. Whole nerve responses recorded at three points (20 mm spaced) along the caudal nerve at different rates of repetitive stimulation in a normal subject and in a subject with induced peripheral neuropathy. Note the similarity of responses at slow rates of repetitive stimulation (5 and 20 Hz) and the clear changes of potentials in the subject with neuropathy at high rates (100 and 200 Hz). These deficits are exacerbated with increasing distance from the cathode along the nerve.

exchangers, energy reserves, the integrity and distribution of glial current sinks, and the distribution of ion channels within and surrounding the nodal gap (for a review, see Koh *et al.*, 1994). At very rapid rates, each of these mechanisms is likely impaired in DPN and the effects may antecede frank structural deficits evident in the associated axons or myelin. For instance, Greene *et al.* (1997) hypothesized that metabolic deficits in the efficiency of the transmembrane ion pumps related to DPN lead to an accumulation of Na^+ ions within the axon, an alteration in the Na^+ concentration gradient, and a slowing of conduction.

Most studies of repetitive stimulation have used the classical paradigm of a "conditioning" stimulus followed by a "test" stimulus using various interpulse delays (Low and McLeod, 1977; Schütt *et al.*, 1980, 1983; Braune, 1999). Generally, the magnitude of the response decrement or the minimal interstimulus interval needed to achieve a specific response decrement (e.g., 67% of response to conditioning stimulus) has been used to judge the integrity of nerve or neuromuscular transmission. Repetitive stimulation procedures have been used to document deficits in refractory periods (Low and McLeod, 1977; Schütt *et al.*, 1983) and in the tracking of rapid stimulation rates (Tackmann and Lehmann, 1980; Ruijten *et al.*, 1994). Repetitive stimulation procedures have been used to increase the sensitivity of whole nerve electrophysiologic measures to early and subtle deficits associated with DPN (Schütt *et al.*, 1980; Braune, 1997, 1999). At earlier stages of diabetes, when a decrease in maxNCV of motor and sensory nerves had an incidence of 10 and 25%, respectively, measurements sensitive to impaired refractory period were abnormal in 50% of subjects (Schütt *et al.*, 1980). In addition to an amplitude change, these studies documented alterations in the absolute latency of the response to the "test" stimuli, suggesting that conduction velocity may also be frequency sensitive. Consistent with this suggestion, Caccia *et al.* (1993) used paired stimulation combined with collision techniques to demonstrate a DPN-induced alteration of conduction in fibers of intermediate velocity.

IX. Measures of Excitability

With few exceptions (e.g., tactile stimulation), whole nerve electrophysiologic procedures rely on electrical stimulation of the bundle of axons placed either distal or proximal to the recording sites. In most studies, stimulation intensity is increased systematically until the outcome measures (e.g., SNAP or CMAP) are not further affected by increased current, and thus a supramaximal point is established. Supposedly, this procedure ensures the activation of all of the responsive axons, but as we have seen, peak amplitude principally reflects activity in large-diameter, heavily myelinated fibers. The excitability of specific axons, defined as the amount and type of current needed to just generate an action potential, is related to several elements, including transmembrane ion kinetics, nodal–paranodal ion distributions, and capacitance of paranodal space. In whole nerve procedures, excitability is also influenced by the ability to sustain a propagated signal, perineural ensheathments, the geometry of stimulation, and the impedance of the tissue intervening between the stimulating cathode and the active recording electrode. Because the largest fibers have the lowest thresholds, excitability generally reflects the relative density of large-diameter fibers and the mean cross-sectional diameter of the responding neurons.

There has been increased interest in using computer-driven programs to systematically vary stimulation parameters and explore the magnitude and nature of current necessary to drive whole nerve responses (for a review, see Burke *et al.*, 2001). Threshold tracking is a procedure that examines the pattern of stimulation necessary to achieve a predetermined target percentage of maximal whole nerve response (e.g., 50%). Because the absolute current necessary varies greatly across subjects, the analysis focuses on the relationship across stimulus parameters, such as the trade-off of strength and duration. The computer programs are capable of pairing multiple stimuli and adjusting parameters on the basis of the response obtained. One variant, termed threshold electrotonus, explores changes in excitability following stimulation with prolonged hyperpolarizing and depolarizing subthreshold currents (Kiernan *et al.*, 2000, 2001). This procedure, which is similar to the "threshold clamp" technique (Bostock and Baker, 1988), has been used to demonstrate differences in excitability between sensory and motor axons, fluctuation of excitability during the after activation period, and refractoriness, using a conditioning-testing stimuli paradigm (Kuwabara, *et al.*, 2000; Grosskreutz *et al.*, 2000; Burke *et al.*, 2001; Kiernan *et al.*, 2001). In the STZ model of DPN, a change in threshold electrotonus paralleled alterations in maxNCV (Yang *et al.*, 2001), but the mechanism underlying these changes may have differed. Excitation studies have also indicated that the diabetic nerve has less accommodation to hyperpolarization (i.e., inward rectification), which may limit its ability to follow rapid stimulus trains (Horn *et al.*, 1996; Yang *et al.*, 2001), and less of a modification of threshold as a function of ischemia (Weigl *et al.*, 1989; Grosskreutz *et al.*, 2000). At the cellular level, changes in excitability determined using ion-sensitive microelectrodes and the "threshold clamp" technique demonstrated that an elevation in threshold current preceded conduction block in demyelinated axons (Bostock and Grafe, 1985). Excitatory changes may be related to alterations in intracellular levels of cyclic adenosine monophosphate that have been associated with DPN (Yang *et al.*, 2001).

Studies of whole nerve excitability patterns are in their infancy, but they demonstrate considerable promise as an adjunct to conventional studies of velocity and amplitude in whole nerve procedures. This approach may add sensitivity to subtle aspects of membrane dynamics, which may be precursors to more obvious structural deficits.

X. Synopsis

To be valuable in the assessment of DPN, whole nerve electrophysiologic procedures must be noninvasive, sensitive to both early and chronic changes in nerve structure and function, reliable, and a valid measure of the underlying pathology. New procedures that meet these criteria are emerging;

others are being refined continuously. Computer assistance in programs exploring the distribution of velocities or parameters of excitability now makes these procedures feasible in a wide range of laboratories and clinics. Slowly the excessive reliance on maxNCV in both human and animal studies is being reduced as the limitations of this procedure are clarified and the options expanded. Perhaps the most important change is our growing appreciation that an accurate assessment of DPN requires information from multiple electrophysiologic measures, each sensitive to complementary but different aspects of nerve structure and function. To date, the accurate and early diagnosis of DPN has basically been an academic pursuit, but as treatment options develop, early and accurate assessment of DPN may become a paramount concern to physicians and patients.

References

Abbate, S. L., Atkinson, M. B., and Breuer, A. C. (1991). Amount and speed of fast axonal transport. *Diabetes* **40**, 111–117.

Apfel S. C., Kessler J. A., Adormato B. T., *et al.* (1998). Recombinant human nerve growth factor in the treatment of diabetic polyneuropathy. *Neurology* **51**, 695–702.

Arezzo, J. C. (1997). The use of electrophysiology for the assessment of diabetic neuropathy. *Neurosci. Res. Commun.* **21**, 13–23.

Bertora, P., Valla, P., Dezuanni, E., *et al.* (1998). Prevalence of subclinical neuropathy in diabetic patients: Assessment by study of conduction velocity distribution within motor and sensory nerve fibres. *J. Neurol.* **245**, 81–86.

Bostock, H., and Baker, M. (1988). Evidence for two types of potassium channel in human motor axons in vivo. *Brain Res.* **462**, 354–358.

Bostock, H., and Grafe, P. (1985). Activity-dependent excitability changes in normal and demyelinated rat spinal root axons. *J. Physiol.* **365**, 239–257.

Boulton, A. J. M. (1997). "Diabetic Neuropathy." Marius Press, UK.

Bradley, J. L., Thomas, P. K., King, R. H. M., *et al.* (1995). Myelinated nerve fibre regeneration in diabetic sensory polyneuropathy: Correlation with type of diabetes. *Acta Neuropathol.* **90**, 403–410.

Braune, H. J. (1997). Early detection of diabetic neuropathy: A neurophysiologic study on 100 patients. *Electromyogr. Clin. Neurophysiol.* **37**(7), 399–407.

Braune, H. J. (1999). Testing of the refractory period in sensory fibres is the most sensitive method to assess beginning polyneuropathy in diabetics. *Electromyogr. Clin. Neurophysiol.* **39**(6), 355–359.

Bril, V., Janzen, D., Gin, H., *et al.* (2001). Sensory nerve area measurements in patients with diabetic neuropathy. *Electromyogr. Clin. Neurophysiol.* **41**(1), 59–63.

Brismar, T., and Sima, A. A. F. (1981). Changes in nodal function in nerve fibres of the spontaneously diabetic BB-Wistar rat: Potential clamp analysis. *Acta. Physiol. Scand.* **113**, 499.

Brownlee, M. (1990). Advanced products of nonenzymatic glycosylation and the pathogenesis of diabetic complications. *In* "Diabetes Mellitus: Theory and Practice" (H. Rifkin and D. Porte, eds.), pp. 279–291. Elsevier Science, New York.

Burke, D., Kiernan, M. C., and Bostock, H. (2001). Excitability of human axons. *Clin. Neurophysiol.* **112**, 1575–1585.

Caccia, M. R., Salvaggio, A., Dezuanni, E., *et al.* (1993). An electrophysiological method to assess the distribution of the sensory propagation velocity of the digital nerve in normal and diabetic subjects. *Electroencephal. Clin. Neurophysiol.* **89**, 88–94.

Caldwell, J. H., Shaller, K. L., Lasher, R. S., *et al.* (2000). Sodium channel Na (v 1.6) is localized at nodes of Ranvier, dendrites, and synapses. *Proc. Natl. Acad. Sci. USA* **97**, 5616–5620.

Calcutt, N. A., Dines, K. C., and Ceseña, R. M. (1998). Effects of the peptide HP228 on nerve disorders in diabetic rats. *Metabolism* **47**(6), 650–656.

Cameron, N. E., and Cotter, M. A. (1997). Metabolic and vascular factors in the pathogenesis of diabetic neuropathy. *Diabetes* **46**, S31–S37.

Cameron, N. E., Cotter, M. A., Jack, A. M., *et al.* (1999). Protein kinase C effects on nerve function, perfusion, Na(+), K(+)-ATPase activity and glutathione content in diabetic rats. *Diabetologia* **42**(9) 1120–1130.

Cappelen-Smith, C., Kuwabara, S., Lin, C. S.-Y., *et al.* (2000). Activity-dependent hyperpolarization and conduction block in chronic inflammatory demyelinating polyneuropathy. *Ann. Neurol.* **48**, 826–832.

Chapman, C. A., and Yeomans, J. S. (1994). Motor cortex and pyramidal tract axons responsible for electrically evoked forelimb flexion: Refractory periods conduction velocities. *Neuroscience* **59**, 699–711.

Cherian, P. V., Kamijo, M., Angelides K. J., *et al.* (1996). Nodal Na$^+$-channel displacement is associated with nerve-conduction slowing in the chronically diabetic BB/W rat: Prevention by aldose reductase inhibition. *J. Diabet. Complicat.* **10**, 192–200.

Cummins, K. L., and Dorfman, L. J. (1981). Nerve fiber conduction velocity distributions: Studies of normal and diabetic human nerves. *Ann. Neurol.* **9**, 67–74.

Daube, J. R. (1987). Electrophysiologic testing in diabetic neuropathy. *In* "Diabetic Neuropathy" (P. J. Dyck, P. K. Thomas, A. K. Asbury, A. I. Winegrad, and D. Porte, Jr., eds.), pp. 162–176. Saunders, Philadelphia.

Davis-King, K. E., Sweeney, M. H., Willie, K. K., *et al.* (1992). Reference values for amplitudes and conduction velocities obtained from a cohort of middle-aged and retired workers. *Scand. J. Work Environ. Health* **18**(Suppl. 2), 24–26.

Dorfman, L. J. (1984). The distribution of conduction velocities (DVC) in peripheral nerves: A review. *Muscle Nerve* **7**, 2–11.

Downie, A. W., and Newell, D. J. (1961). Sensory nerve conduction in patients with diabetes mellitus and controls. *Neurology* **11**, 876.

Dyck, P. J. (1987). Pathology and pathophysiology: Human and experimental. *In* "Diabetic Neuropathy" (P. J. Dyck, P. K. Thomas, A. K. Asbury, A. I. Winegrad, and D. Porte, Jr., eds.), pp. 223–236. Saunders, Philadelphia.

Dyck, P. J., Lais, A., Karnes, J. L., *et al.* (1987). Fiber loss is primary and multifocal in sural nerves in diabetic polyneuropathy. *Ann. Neurol.* **19**, 425–439.

Eaton, R. P., Qualls, C., Bicknell, J., *et al.* (1996). Structure-function relationships within peripheral nerves in diabetic neuropathy: The hydration hypothesis. *Diabetologia* **39**, 439–446.

Fernyhough, P., Gallagher, A., and Averill, S. A. (1999). Aberrant neurofilament phosphorylation in sensory neurons of rats with diabetic neuropathy. *Diabetes* **48**, 881–889.

Fraser, J. L., and Olney, R. K. (1993). The relative diagnostic sensitivity of different F-wave parameters in various polyneuropathies. *Muscle Nerve* **16**(8), 877–878.

Giannini, C., and Dyck, P. J. (1995). Basement membrane reduplication and pericyte degeneration precede development of diabetic polyneuropathy and are associated with its severity. *Ann. Neurol.* **37**, 498–504.

Goldin, A. L., Barchi, R. L., Caldwell, J. H., *et al.* (2000). Nomenclature of voltage-gated sodium channels. *Neuron* **28**, 365–368.

Greene, D. A., Arezzo, J. C., Brown, M. B., *et al.* (1999). Effect of aldose reductase inhibition on nerve conduction and morphometry in diabetic neuropathy. *Neurology* **53**, 580–591.

Greene, D. A., and Brown, M. B. (1995). Validation of sural nerve fiber density and percent normal teased fibers as morphological endpoints in clinical trials of diabetic neuropathy. *In* "Diabetic Neuropathy: New Concepts and Insights" (N. Hotta, D. A. Greene, J. D. Ward, A. A. F. Sima, and A. J. M. Boulton, eds.) pp. 379–385. Elsevier Science, New York.

Greene, D. A., Feldman, E. L., Stevens, M. J., *et al.* (1997). *In* "Diabetic Neuropathy" (D. Porte and R. Sherwin, eds.), pp. 1009–1076. Appleton & Lange, Stamford, CO.

Greene, D. A., Lattimer, S. A., and Sima, A. A. (1987). Sorbitol, phosphoinositides, and sodium potassium-ATPase in the pathogenesis of diabetic complications. *N. Engl. J. Med.* **316**, 599–606.

Grosskreutz, J., Lin, C. S.-Y., Mogyoros, I., *et al.* (2000). Ischaemic changes in refractoriness of human cutaneous afferents under threshold-clamp conditions. *J. Physiol.* **523**, 807–815.

Hellweg, R., Raivich G., Hartung, H. D., *et al.* (1994). Axonal transport of endogenous nerve growth factor (NGF) and NGF receptor in experimental diabetic neuropathy *Exp. Neurol.* **130**, 24–30.

Hohman, T. C., Cotter, M. A., and Cameron, N. E. (2000). ATP-sensitive K(+) channel effects on nerve function, Na(+), K(+) ATPase, and glutathione in diabetic rats. *Eur. J. Pharmacol.* **3**;397(2–3), 335–341.

Horn, S., Quasthoff, S., Grafe, P., *et al.* (1996). Abnormal axonal inward rectification in diabetic neuropathy. *Muscle Nerve* **19**, 1268–1275.

Hounsom, L., and Tomlinson, D. R. (1997). Does neuropathy develop in animal models? *Clin. Neurosci.* **4**, 380–389.

Ingram, D. A., Davis, G. R., and Swash, M. (1987). Motor nerve conduction velocity distribution in man: Results of a new computer-based collision technique. *Electroencephal. Clin. Neurophysiol.* **66**, 235–243.

Jakobsen, J. (1976). Axonal dwindling in early experimental diabetes. II. A study of isolated nerve fibres. *Diabetologia* **12**, 547–553.

Jakobsen, J. (1979). Early and preventable changes of peripheral nerve structure and function in insulin-deficient diabetic rats. *J. Neurol. Neurosurg. Psychiat.* **42**, 509–518.

Jamal, G. A. (1990). Pathogenesis of diabetic neuropathy: The role of the n-6 essential fatty acids and their eicosanoid derivations. *Diabet. Med.* **7**, 574–579.

Kaji, R., Bostock, H., Kohara, N., *et al.* (2000). Activity-dependent conduction block in multifocal motor neuropathy. *Brain* **123**, 1602–1611.

Kato N., Makino M., Mizuno K., *et al.* (1998). Serial changes of sensory nerve conduction velocity and minimal F-wave latency in streptozocin-induced diabetic rats. *Neurosci. Lett.* **244**, 169–172.

Kiernan, M. C., Burke, D., Andersen, K. V., *et al.* (2000). Multiple measures of axonal excitability: A new approach in clinical testing. *Muscle Nerve* **23**, 399–409.

Kiernan, M. C., Lin, C. S.-Y., Andersen, K. V., *et al.* (2001). Clinical evaluation of excitability measures in sensory nerve. *Muscle Nerve* **24**, 883–892.

Kimura, J. (1976). Collision technique. *Neurology* **26**, 680–682.

Kimura, J. (1989). "Electrodiagnosis in Disease of Nerve and Muscle: Principles and Practice" (J. Kimura, ed.). F. A. Davis Company, Philadelphia.

Kimura, J. (1995). Clinical electrophysiology of peripheral nervous system axons. *In* "The Axon: Structure, Function and Pathophysiology" (S. G. Waxman, J. D. Kocsis, and P. K. Stys, eds.), pp. 590–628. Oxford Press, New York.

Koh, D.-S., Jonas, P., and Vogel, W. (1994). Na$^+$-activated K$^+$ channels localized in the nodal region of myelinated axons of *Xenopus. J. Physiol. (Lond.)* **479**, 183–197.

Kohara, N., Kimura, J., Kaji, R., *et al.* (2000). F-wave latency serves as the most reproducible measure in nerve conduction studies of diabetic polyneuropathy: Multicentre analysis in healthy subjects and patients with diabetic polyneuropathy. *Diabetologia* **43**, 915–921.

Kuwabara, S., Cappelen-Smith, C., Lin, C. S.-Y., *et al.* (2000). Excitability properties of median and peroneal motor axons. *Muscle Nerve* **23**, 1365–1373.

Low, P. A., and McLeod, J. G. (1977). Refractory period, conduction of trains of impulses and effect of temperature on conduction in chronic hypertrophic neuropathy. *J. Neurol. Neurosurg. Psychiat.* **40**, 434–447.

Low, P. A., Nicklander, K. K., and Tritschler, H. J. (1997). The roles of oxidative stress and antioxidant treatment in experimental diabetic neuropathy. *Diabetes* **46**(Suppl. 2), S38–S42.

Mattingly, G. E., and Fischer, V. W. (1983). Peripheral neuropathy following prolonged exposure to streptozocin-induced diabetes in rats: A teased nerve fiber study. *Acta Neuropathol.* **59**, 133–138.

Mizuno, K., Kato, N., Makino, M., *et al.* (1999). Continuous inhibition of extensive polyol pathway flux in peripheral nerves by aldose reductase inhibitor fidarestat leads to improvement of diabetic neuropathy. *J. Diabet. Comp.* **13**, 141–150.

Mohiuddin, L., Fernyhough, P., and Tomlinson, D. R. (1995). Reduced levels of mRNA encoding endoskeletal and growth-associated proteins in sensory ganglia in experimental diabetes. *Diabetes* **44**, 25.

Morita, H., Shindo, M., Yanagawa, S., *et al.* (1993). Neuromuscular response in man to repetitive nerve stimulation. *Muscle Nerve* **16**, 648–654.

Nishikawa, T., Edelstein, D., Du, X. L., *et al.* (2000). Normalizing mitochondrial superoxide production blocks three pathways of hyperglycaemic damage. *Nature* **404**, 787–790.

Nobrega, J. A. M., Manzano, G. M., and Montegudo, P. T. (2001). A comparison between different parameters in F-wave studies. *Clin. Neurophysiol.* **112**, 866–868.

Oh, S. J., Demirci, M., Dajani, B., *et al.* (2001). Distal sensory nerve conduction of the superficial peroneal nerve: New method and its clinical application. *Muscle Nerve* **24**, 689–694.

Okajima, Y., Chino, N., Tsubahara, A., *et al.* (1994). Waveform analysis of compound nerve action potentials: A computer simulation. *Arch. Phys. Med. Rehabil.* **75**, 960–964.

Olney, R. K., Budingen, H. J., and Miller, R. G. (1987). The effect of temporal dispersion on compound action potential area in human peripheral nerve. *Muscle Nerve* **10**, 728–733.

Papadopoulou, F. A., and Panas, S. M. (1999). Bispectral de-nosing of the compound action potential for estimation of the nerve conduction velocity distribution. *Med. Eng. Phys.* **21**, 499–505.

Partanen, J., Niskanen, L., Lehtinen, J., *et al.* (1995). Natural history of peripheral neuropathy in patients with non-insulin-dependent diabetes mellitus. *N. Engl. J. Med.* **333**, 89–94.

Pastore, C., Izura, V., Geijo-Barrientos, E., *et al.* (1999). A comparison of electrophysiological tests for the early diagnosis of diabetic neuropathy. *Muscle Nerve* **22**(12), 1667–1673.

Patel, J., and Tomlinson, D. R. (1999). Nerve conduction impairment in experimental diabetes-proximodistal gradient of severity. *Muscle Nerve* **22**, 1403–1411.

Peripheral Nerve Society. (1995). Diabetic polyneuropathy in controlled clinical trails: Consensus report of the peripheral nerve society. *Ann. Neurol.* **38**(3), 478–482.

Pollak, V. A., Ferbert, A., Cui, J., *et al.* (1992). Non-invasive determination of the distribution of the conductive velocity of the large-diameter fibers in peripheral nerves: Estimate based upon a single recording of the stimulus responses of the nerve. *Med. Prog. Technol.* **18**, 217–225.

Proceedings of the Consensus Development Conference on Standardized Measures in Diabetic Neuropathy. (1992). *Neurology* **42**(9), 1823–1839.

Raccah D., Coste T., Cameron N. E., *et al.* (1998). Effect of the aldose reductase inhibitor tolrestat on nerve conduction velocity, Na/K ATPase activity, and polyols in red blood cells, sciatic nerve, kidney cortex, and kidney medulla of diabetic rats. *J. Diabet. Comp.* **12**, 154–162.

Ranvier, M. L. (1878). "Legons sur l' Histologie du Systeme Neveux." Librairie F. Savy, Paris. Report and Recommendations of the San Antonio Conference on Diabetic Neuropathy. (1988). *Diabetes* **37**, 1000–1004.

Riaz, S. S., and Tomlinson, D. R. (1996). Neurotrophic factors in peripheral neuropathies: Pharmacological strategies. *Prog. Neurobiol.* **49**, 125–143.

Ritchie, J. M. (1995). Physiology of axons. *In* "The Axon" (S. G. Waxman, P. K. Stys, and J. D. Kocsis, eds.), pp. 68–96. Oxford Univ. Press, Oxford.

Russell, J. W., Karnes, J. L., and Dyck, P. J. (1996). Sural nerve myelinated fiber density differences associated with meaningful changes in clinical and electrophysiologic measurements. *J. Neurol. Sci.* **135**, 114–117.

Ruijten, M. W. M. M., Sallé, H. J. A., and Kingma, R. (1993). Comparison of two techniques to measure the motor nerve conduction velocity distribution. *Electroencephal. Clin. Neurophysiol.* **89**, 375–381.

Ruijten, M. W. M. M., De Haan, G.-J., Michels, R. P. J., *et al.* (1994). Motor nerve refractory period distribution assessed by two techniques in diabetic polyneuropathy. *Electroencephal. Clin. Neurophysiol.* **93**, 306–311.

Said, G., Slama, G., and Selva, J. (1983). Progressive centripetal degeneration of axons in small fibre diabetic neuropathy. *Brain* **106**, 791–807.

Saito, O. (1991). Electrophysiological study on pathology of Bell's palsy-distribution of nerve conduction velocities (DNCV) in facial nerve with collision method. *Nippon Jibiinkoka Gakkai Kiaiho* **94**, 906–914.

Schaumburg, H. H., Berger, A. R., and Thomas, P. K. (1992). Anatomical classification of peripheral nervous system disorders. *In* "Disorders of Peripheral Nerves" (H. H. Schaumburg, A. R. Berger, and P. K. Thomas, eds.), pp. 10–24. F. A. Davis Company, Philadelphia.

Schutt, P., Muche, H., Lehmann, H. J., *et al.* (1980). Sural nerve conduction velocity and refractory period in diabetics without clinical signs of neuropathy. *Horm. Metab. Res. Suppl.* 9, 39–42.

Schütt, P., Muche, H., and Lehmann, H. J. (1983). Refractory period impairment in sural nerves of diabetics. *J. Neurol.* **229**, 113–119.

Schülte-Mattler, W. J., and Zierz, M. J. S. (1999). Assessment of temporal dispersion in motor nerves with normal conduction velocity. *Clin. Neurophysiol.* **110**, 740–747.

Sharma, A. K., Thomas, P. K., and De Molina, A. F. (1977). Peripheral nerve fiber size in experimental diabetes. *Diabetes* **26**, 689–692.

Shin, J. B., Seong, Y. J., Lee, H. J., *et al.* (2000). The usefulness of minimal F-wave latency and sural/radial amplitude ratio in diabetic polyneuropathy. *Yonsei. Med. J.* **41**(3), 393–397.

Sirna, A. A. F., Nathaniel, V., Bril, V., *et al.* (1988). Histopathological heterogeneity of neuropathy in insulin-dependent and non-insulin dependent diabetes, and demonstration of axoglial dysfunction in human diabetic neuropathy. *J. Clin. Invest.* **81**, 349–364.

Sugimura, K., Windebank, A. J., Natarajan, V., *et al.* (1980). Interstitial hyperosmolarity may cause axis cylinder shrinkage in streptozotocin diabetic nerve. *J. Neuropathol. Exp. Neurol.* **30**, 710–721.

Tackmann, W., and Lehman, H. J. (1980). Conduction of electrically elicited impulses in peripheral nerves of diabetic patients. *Eur. Neurol.* **19**, 20–29.

Tomlinson, D. R., Moriarity, R. J., and Mayer, H. (1984). Prevention and reversal of defective axonal transport and motor nerve conduction velocity in rats with experimental diabetes by treatment with aldose reductase inhibitor sorbinil. *Diabetes* **33**, 470–476.

Tomlinson, D. R., Fernyhough, P., and Diemel, L. T. (1996). Neurotrophins and peripheral neuropathy. *Biol. Sci.* **351**, 455–462.

Toyokura, M. (1998). F-wave duration in diabetic polyneuropathy. *Muscle Nerve* **21**, 246–249.

Tu, Y., Honda, S., and Tomita, Y. (1999). Estimation of the conduction velocity distribution of peripheral nerve trunks. *Front. Med. Biol. Eng.* **9**, 198–197.

van Dijk, J. G., van der Kamp, W., Hilten, B. J., *et al.* (1994). Influence of recording site on CMAP amplitude on its variation over a length of nerve. *Muscle Nerve* **17**, 1286–1292.

Veves, A., Malik, R. A., Lye, R. H., *et al.* (1991). The relationship between sural nerve morphometric findings and measures of peripheral nerve function in mild diabetic neuropathy. *Diabet. Med.* **8**, 917–921.

Walker, D., Carrington, A., Cannan, S. A., *et al.* (1999). Structural abnormalities do not explain the early functional abnormalities in the peripheral nerves of the streptozocin diabetic rat. *J. Anat.* **195**, 419–427.

Wattig, B., Warzok, R., and Thomas, P. K. (1986). Experimental diabetic neuropathy. Morphometric studies on the rat N suralis in short-term streptozocin-induced diabetes. *Zentralbl. Allg. Pathol.* **131**, 451–458.

Waxman, S. G. (1995). Voltage-gated ion channels in axons: Localization, function, and development. *In* "The Axon: Structure, Function and Pathophysiology" (S. G. Waxman, J. D. Kocsis, and P. K. Stys, eds.), pp. 218–243. Oxford Press, New York.

Weber, F. (1997). Conduction block and abnormal temporal dispersion-diagnostic criteria. *Electromyogr. Clin. Neurophysiol.* **37**, 305–309.

Weigl, P., Bostock, H., Franz, P., *et al.* (1989). Threshold tracking provides a rapid indication of ischaemic resistance in motor axons of diabetic subjects. *Electroencephal. Clin. Neurophysiol.* **73**, 369–371.

Wells, M. D., and Gozani, S. N. (1999). A method to improve the estimation of conduction velocity distribution over a short segment of nerve. *IEEE Trans. Biomed. Eng.* **46**, 1107–1120.

Yagihashi, S., Kamijo, M., Ido, Y., *et al.* (1990). Effects of long-term aldose reductase inhibitor on development of experimental diabetic neuropathy: Ultrastructural and morphometric studies of sural nerve in streptozocin-induced diabetic rats. *Diabetes* **39**, 690–696.

Yagihashi, S. (1995). Pathology and pathogenetic mechanisms of diabetic neuropathy. *Diabet. Metab. Rev.* **11**, 193–225.

Yang, Q., Kaij, R., Takagi, T., *et al.* (2001). Abnormal axonal inward rectifier in streptozocin-induced experimental diabetic neuropathy. *Brain* **124**, 1149–1155.

NEUROPATHOLOGY AND PATHOGENESIS OF DIABETIC AUTONOMIC NEUROPATHY

Robert E. Schmidt

Department of Pathology and Immunology, Washington University School of Medicine
St. Louis, Missouri 63110

Autonomic neuropathy is a significant complication of diabetes resulting in increased patient morbidity and mortality. A number of studies, which have shown correspondence between neuropathologic findings in experimental animals and human subjects, have demonstrated that axonal and dendritic pathology in sympathetic ganglia in the absence of significant neuron loss represents a neuropathologic hallmark of diabetic autonomic neuropathy. A recurring theme in sympathetic ganglia, as well as in the postganglionic autonomic innervation of various end organs, is the involvement of distal portions of axons and nerve terminals by degenerative or dystrophic changes. In both animals and humans, there is a surprising selectivity of the diabetic process for subpopulations of autonomic ganglia, nerve terminals within sympathetic ganglia and end organs, from end organ to end organ, and between vascular and other targets within individual end organs. Although the involvement of autonomic axons in somatic nerves may reflect an ischemic pathogenesis, the selectivity of the diabetic process confounds simple global explanations of diabetic autonomic neuropathy as the result of diminished blood flow with resultant tissue hypoxia. A single unifying pathogenetic hypothesis has not yet emerged from clinical and experimental animal studies, and it is likely that diabetic autonomic neuropathy will be shown to have multiple causative mechanisms, which will interact to result in the variety of presentations of autonomic injury in diabetes. Some of these mechanisms will be shared with aging changes in the autonomic nervous system. The role of various neurotrophic substances and the polyol pathway in the pathogenesis and treatment of diabetic neuropathy likely represents a two-edged sword with both salutary and exacerbating effects. The basic neurobiologic processes underlying the diabetes-induced development of neuroaxonal dystrophy, synaptic dysplasia, defective axonal regeneration, and alterations in neurotrophic substances may be mechanistically related. © 2002, Elsevier Science (USA).

I. Introduction

Diabetic autonomic neuropathy, whose symptoms range from minor pupillary and sweating problems to significant disturbances in cardiovascular, alimentary, and genitourinary function, results in significantly increased patient morbidity and mortality (Hosking *et al.*, 1978; Vinik *et al.*, 1992; Ewing *et al.*, 1980). Gastrointestinal dysfunction in clinical diabetes, for example, may present acutely in response to a rapid increase in serum glucose or as a chronic syndrome characterized by hypomotility (e.g., gastroparesis, chronic constipation), hypermotility (diabetic diarrhea), or as alternating cycles of diarrhea and constipation (DePonti *et al.*, 1987; Sims *et al.*, 1995; Quigley, 1997). These symptoms result from the loss of coordinated alimentary reflexes (Battle *et al.*, 1980; Camilleri and Malagelada, 1984) and reflect the complex dysfunction of sympathetic and parasympathetic components with the possible contribution of visceral sensory and enteric neurons (Sampson *et al.*, 1990; Bellavere *et al.*, 1992; Kreiner *et al.*, 1995). Rather than resulting from a simple deficit in any single effector pathway, diabetic alimentary tract dysfunction may proceed from the inability to integrate portions of several complex pathways in gut function, which is to a significant degree accomplished in the prevertebral sympathetic ganglia (celiac, superior, and inferior mesenteric). Similar processes may compromise other autonomic functions.

II. Neuropathology of Clinical Diabetic Autonomic Neuropathy

Although autonomic function in human subjects reflects the simultaneous and integrated participation of neurons at various levels of the central and peripheral nervous systems, for simplicity, the sites of involvement are presented separately in the following sections.

A. CENTRAL NERVOUS SYSTEM

Loss of neurons in the intermediolateral nuclei of the spinal cord, in preganglionic sympathetic axons comprising communicating ("white") rami to paravertebral chain ganglia, and in the greater splanchnic nerve serving prevertebral ganglia have been reported (Olsson and Sourander, 1968; Appenzeller and Ogin, 1974; Low *et al.*, 1975; Low, 1984). Other central pathways are likely involved in the control of the autonomic nervous system, but have been incompletely studied in human diabetics.

B. SYMPATHETIC GANGLIA

Degeneration and significant loss of sympathetic neurons of prevertebral and paravertebral autonomic ganglia have been claimed (Appenzeller and Richardson, 1966; Duchen *et al.*, 1980) and disputed (Schmidt *et al.*, 1993) in previous studies. Duchen's often-quoted study of sympathetic ganglia of 5 patients with symptomatic diabetic autonomic neuropathy reported a variety of neuropathologic findings, including neuronal gigantism and necrosis, vacuolated neuronal endoplasmic reticulum, an intraganglionic inflammatory infiltrate, and argyrophilic swollen axons. The neuronal population density of the superior cervical (SCG) and superior mesenteric (SMG) ganglia of 345 human subjects (including 70 diabetics) in our extensive autopsy series (Schmidt *et al.*, 1993; Schmidt, 1996) was well maintained with diabetes and aging without the development of actively degenerating, apoptotic or chromatolytic neurons, or compact nodules of satellite cells (an indicator of remote neuronal loss). Rather, the neuropathologic hallmark of diabetes in our series was neuroaxonal dystrophy, i.e., markedly swollen (5–30 μm diameter) structures originating as a distal axonopathy involving presynaptic terminal axons. Neuroaxonal dystrophy preferentially involved the diabetic prevertebral SMG and celiac ganglia, largely sparing the paravertebral SCG. Dystrophic swellings (arrow, Fig. 1) contained disorganized neurofilamentous aggregates or an admixture of synaptic vesicles, tubulovesicular forms, and multivesicular bodies. Quantitative analysis determined that the frequency of neuroaxonal dystrophy

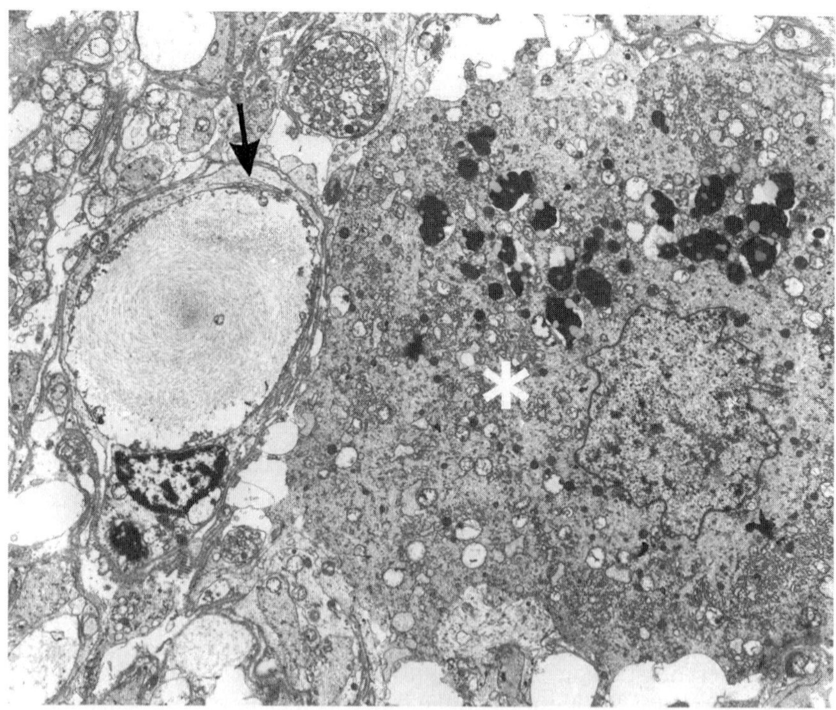

FIG. 1. Human diabetic autonomic neuropathy in the superior mesenteric ganglion. A neurofilament-laden dystrophic axon (arrow) is tightly apposed to an adjacent neuronal perikaryon (*). (Magnification: 2500×)

varied as a function of diabetes, location (SMG ≫ SCG), age, and gender (males 3-fold > females). Neuroaxonal dystrophy was immunohistochemically and ultrastructurally identical in aging and diabetes, developing in greater numbers and at a younger age in diabetics, which suggests shared pathogenetic mechanisms. Quantitative comparison of one of Duchen's original young symptomatic diabetic patients to nondiabetics of his approximate age and gender in our control population demonstrated a dramatically increased frequency of neuroaxonal dystrophy. In addition, although our studies of Duchen's patient showed large numbers of dystrophic axons in the celiac ganglion and almost none in the SCG, lumbar sympathetic chain ganglia showed an intermediate number of dystrophic axons (i.e., the severity of neuroaxonal dystrophy is not simply a function of pre- versus paravertebral ganglia). Some apparently vacuolated neuronal cell bodies actually represented dilated presynaptic axon terminals containing lucent proteinaceous material. Although the presence of lymphocytic infiltrates in postmortem sympathetic diabetic ganglia has been interpreted as evidence of an autoimmune process (Rabinowe *et al.*, 1990), nearly half of all examined SCG and SMG in our adult autopsy series exhibited comparable changes that were neither more severe nor frequent in diabetics and may largely reflect a common aspect of the terminal perimortem course or lymphocytic trafficking (Schmidt, 1996).

Prevertebral sympathetic ganglia integrate the contribution of neurons originating in the intermediolateral nuclei of the spinal cord, dorsal root ganglia, parasympathetic ganglia, sympathetic neurons (intrinsic or extrinsic to the SMG), and retrograde projections from enteric neurons, many of which have a distinctive neuropeptide and/or transmitter "signature." Dystrophic axons in the diabetic SMG were immunoreactive for tyrosine hydroxylase (TOH), dopamine-β-hydroxylase (DβH, Fig. 2, see also color insert), trkA, and p75^{NTR} (the high and low affinity neurotrophin receptors, respectively); however, immediately adjacent terminals containing substance P, VIP, GRP/bombesin, and met-enkephalin were typically uninvolved (Schmidt *et al.*, 1993, unpublished data). This immunophenotype is most consistent with the origin of the majority of dystrophic axons from noradrenergic neurons, most likely arising as intraganglionic sprouts from neighboring neurons.

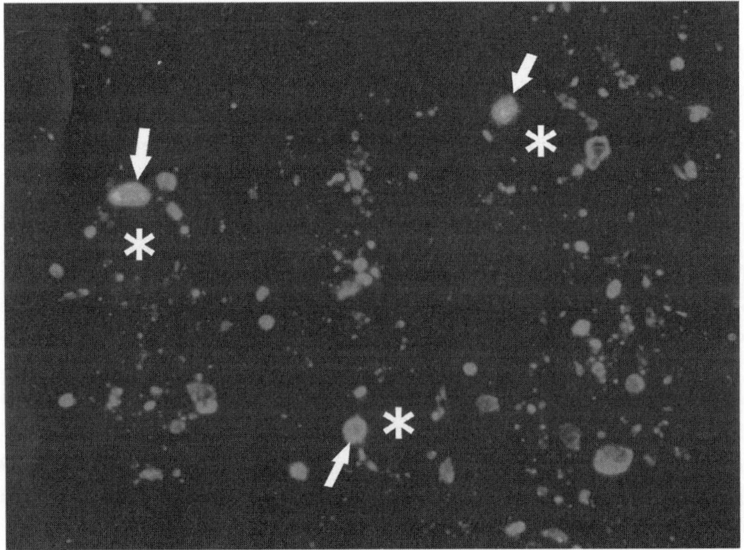

FIG. 2. DBH immunohistochemistry, diabetic human SMG. Swollen DBH immunoreactive dystrophic axons (arrows) cluster around relatively unlabeled perikarya (*). (Magnification: 300×) (See also color insert.)

C. Parasympathetic Ganglia and Projections

Degeneration and dropout of vagal axons have been described in diabetic autonomic neuropathy (Kristensson *et al.*, 1971; Guo *et al.*, 1987). In one case of diabetic gastroparesis, dramatic axon loss in the abdominal vagus was described (Guy *et al.*, 1984), a result that has been challenged by more recent studies (Yoshida *et al.*, 1988). Other studies of gastroparesis have shown degenerative changes in smooth muscle (Moscoso *et al.*, 1996). Preferential loss of presumably parasympathetic VIP-containing axons in the penile corpora cavernosa has been described (Gu *et al.*, 1984; Crowe *et al.*, 1983).

D. Peripheral Nerves and End Organ Innervation

Autonomic axons may be lost in somatic nerves as part of diabetic symmetrical sensorimotor neuropathy (Said *et al.*, 1992), which is thought to have an ischemic basis, resulting in local, distally accentuated autonomic symptoms. Degenerative changes in or a frank loss of autonomic innervation of the vasa nervorum of somatic nerves (Grover-Johnson *et al.*, 1981; Beggs *et al.*, 1993) may, in turn, significantly contribute to further nerve ischemia. Degenerative changes in the autonomic innervation of sweat glands of diabetic patients described in early studies (Faerman *et al.*, 1982) are being reinvestigated by elegant new studies of skin biopsies of patients with diabetic neuropathy (Kennedy and Wendelschafer-Crabb, 1997; Hirai *et al.*, 2000).

Distal visceral autonomic nerves have been infrequently characterized in diabetic patients with autonomic dysfunction. Although some studies have described a simple loss of autonomic axons in the diabetic human heart (Turpeinen *et al.*, 1996), positron emission tomography (PET) scanning with the sympathetic neurotransmitter analog [^{11}C]hydroxyephedrine demonstrated distal denervation of the heart ventricle in clinical diabetic autonomic neuropathy in the presence of more proximal hyperinnervation, a process that may enhance myocardial electrical instability (Stevens *et al.*, 1999). Meissner's and Auerbach's plexuses in patients with diabetic diarrhea have typically failed to demonstrate reproducible neuronal histopathology (Hensley and Soergel, 1968), although one ultrastructural study has demonstrated marked axonal swellings in intramural ganglia (Schmidt *et al.*, 1984). Immunolocalization of substance P is decreased in the rectal mucosa in diabetic patients, especially those with chronic constipation (Lysy *et al.*, 1993). Early studies reported the loss of autonomic innervation of the bladder and penis (Faerman *et al.*, 1973, 1974; Melman and Henry, 1979). Early selective interruption of autonomic innervation of pancreatic islets is proposed to interfere with glucagon secretion during hypoglycemia and to develop independently of generalized somatic or autonomic neuropathy (Taborsky *et al.*, 1998).

E. SUMMARY

Neuropathologic studies of human diabetic autonomic neuropathy suggest a predilection for nerve terminal damage or degeneration, both within sympathetic ganglia and involving autonomic nerve terminals in various end organs, in the absence of significant neuron loss.

III. Experimental Diabetic Autonomic Neuropathy

Animal models of diabetic neuropathy have been used to study the development of diabetic autonomic neuropathy, particularly its early phases, which presumably would be most amenable to therapy and provide insight into its pathogenetic mechanisms. Unlike many animal models of diabetic somatic sensory polyneuropathy, autonomic neuropathy in both streptozotocin (STZ)-diabetic and BB/W rats shows many of the neuropathologic features described in human subjects.

A. CENTRAL NERVOUS SYSTEM

Compared to the wealth of information on the peripheral nervous system of BB/W- and STZ-diabetic rats, relatively little is known concerning the diabetic rat central nervous system. Neuron loss, peptide alterations, and ultrastructural abnormalities have been described in the diabetic rat hypothalamus, cerebral cortex, brain stem, and cerebellum (Bestetti and Rossi, 1980; Dheen *et al.*, 1994; Ohtani *et al.*, 1997; Reagan *et al.*, 1999); however, the role of these alterations, if any, in the pathogenesis of diabetic autonomic neuropathy is unestablished.

B. SYMPATHETIC GANGLIA

A significant loss of sympathetic neurons is not a typical feature of diabetic autonomic neuropathy in the STZ rat SMG or SCG, even after 10 months of severe, untreated diabetes (Schmidt, 2001). As in human subjects, STZ- and BB/W-diabetic rats, as well as genetically diabetic Chinese hamsters, developed neuroaxonal dystrophy in prevertebral sympathetic ganglia (arrows, Figs. 3A and 3B) innervating the small bowel (Yagihashi

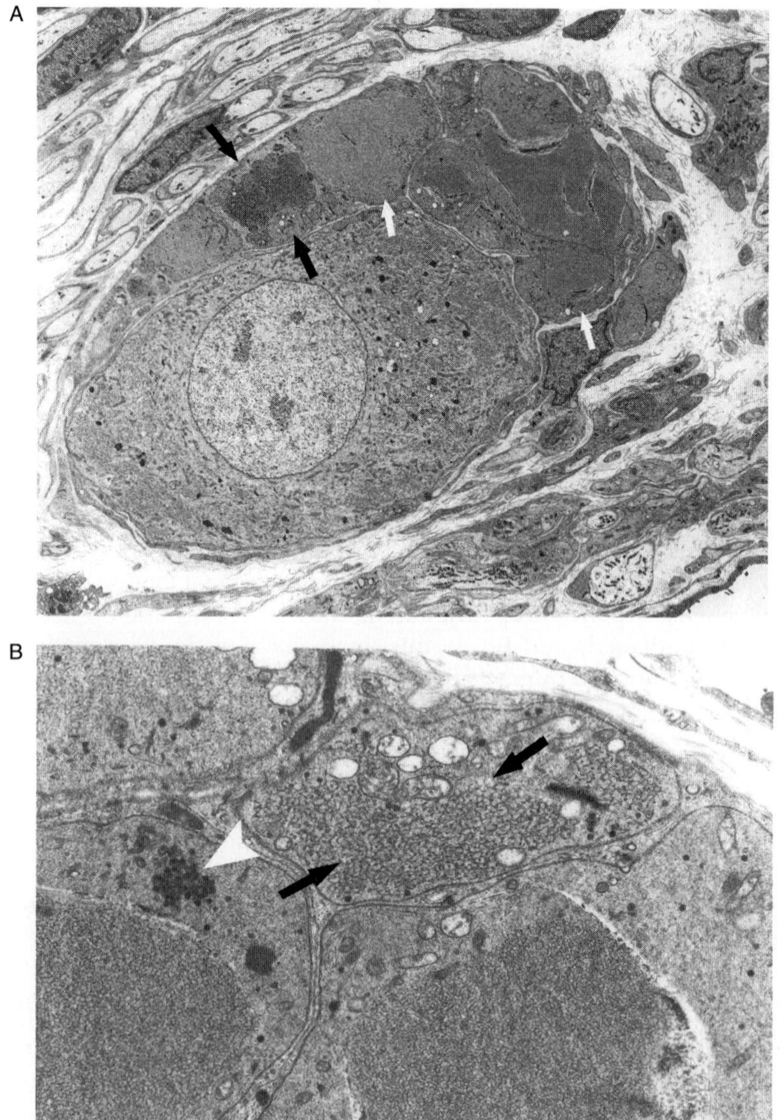

FIG. 3. Neuroaxonal dystrophy in STZ-diabetic rat SMG. (A–C) A single cell body is surrounded by numerous markedly enlarged dystrophic axons (white arrows, A) and a cluster of regenerative sprouts (black arrows, A), which are seen at higher magnification in B and C, respectively. Dystrophic axons contain delicate tubulovesicular elements (black arrows, B), which are compacted in adjacent axons. A collection of neurotransmitter granules (white arrowhead, B) and individual vesicles are noted. (Magnification A: 2000×; B: 12,000×) (C) Higher magnification of the area in A (black arrows, A) demonstrates large numbers of minute axonal sprouts (arrows, C). (Magnification: 12,000×)

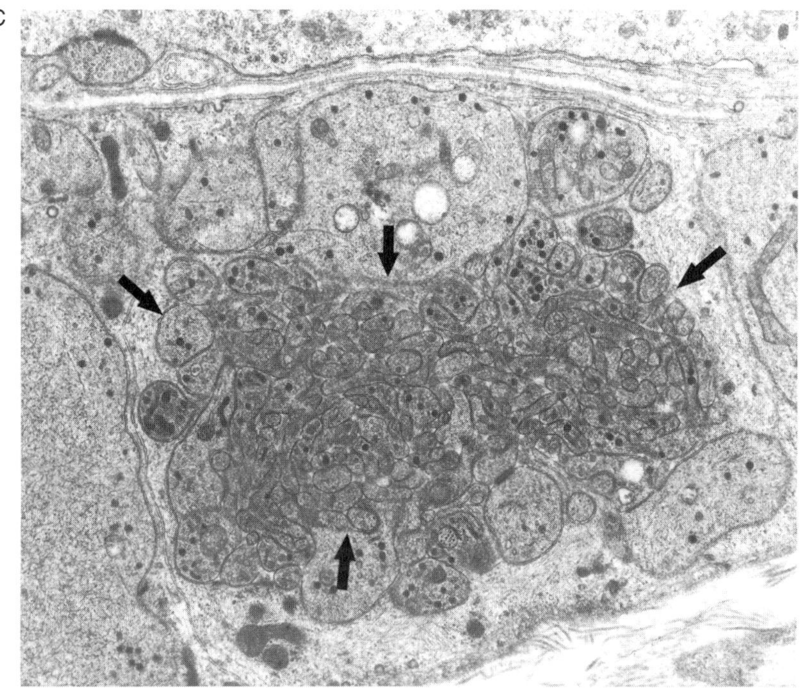

Fig. 3. (*continued*)

and Sima, 1985a,b, 1986a; Schmidt and Plurad, 1986; Schmidt *et al.*, 1989b). Neurofilament-laden neurites were found in the ganglionic neuropil (STZ rat) and in proximal axons and occasional neuronal perikarya (BB rat), whereas tubulovesicular elements were associated more frequently with axon termini in all three rodent models. Postsynaptic dendrites or somal spines in diabetic rat SMG were occasionally dilated by distinctive branched tubular aggregates (Yagihashi and Sima, 1986a; Schmidt and Plurad, 1986). Scattered neurons containing a few large apparently intracytoplasmic vacuoles appeared more frequently in the perikarya of diabetic rat SCG and SMG compared to nondiabetic controls, although their significance, if any, is undetermined (Schmidt *et al.*, unpublished data). Quantitative studies of rodent models demonstrated (1) that neuroaxonal dystrophy required 3–6 months of diabetes to develop significant numbers of lesions; (2) that neuroaxonal dystrophy preferentially involved the diabetic SMG and celiac ganglia, largely sparing the SCG (Schmidt and Plurad, 1986); (3) that the ultrastructural and immunohistologic appearance and anatomical distribution of neuroaxonal dystrophy in STZ-diabetic rat and Chinese hamster ganglia were identical to that which developed in aged rodents; (4) that insulin and pancreatic islet transplantation prevented the development of neuroaxonal dystrophy and substantially reversed established neuroaxonal dystrophy (Schmidt *et al.*, 1989a); (5) that aldose reductase inhibitors (ARI) exerted a salutary effect but did not normalize the frequency of neuroaxonal dystrophy in diabetic rat SMG and that sorbitol dehydrogenase inhibitors (SDI) actually dramatically worsened neuroaxonal dystrophy (see Section IVB); and (6) that IGF-I acting as a neurotrophic substance normalized neuroaxonal dystrophy (Schmidt *et al.*, 1999), whereas nerve growth factor (NGF) and neurotrophin-3 administered comparably failed to have a salutary effect (Schmidt *et al.*, 2001a) and NGF appeared to worsen neuroaxonal dystrophy in control ganglia (see Section IVD). Markedly swollen dystrophic axons involved a minority of intraganglionic axons; however, similar pathogenetic processes may be at work in nerve terminals before the emergence of dramatic neuropathologic changes. Significantly, Yagihashi and Sima (1985a,b) reported that neuroaxonal dystrophy correlated with the progressive loss of normal nondystrophic presynaptic terminals, resulting in deafferentation.

C. Parasympathetic Ganglia

Very few studies have concentrated on the neuropathology of parasympathetic ganglia in experimental diabetes. A study of the major pelvic ganglion in chronically diabetic rats showed an increased size of neuronal perikarya (Steers *et al.*, 1994), but failed to show compelling evidence of neuroaxonal dystrophy (Cai and Schmidt, unpublished data). Intrinsic cardiac ganglia in STZ rat have been reported to develop subtle accumulations of glycogen, small vacuoles, and swollen mitochondria in perisomal axon terminals (Kamal *et al.*, 1991).

D. Peripheral Nerves and End Organ Innervation

Many studies have examined the effect of experimental diabetes on the autonomic innervation of a variety of end organs and tissues.

1. *Alimentary Tract*

The regular occurrence of degenerating, regenerating, and pathologically distinctive dystrophic axons (Figs. 4 and 5) has been demonstrated in noradrenergic axons contained in mesenteric nerves innervating the distal alimentary tract of rats with chronic long-term STZ-induced

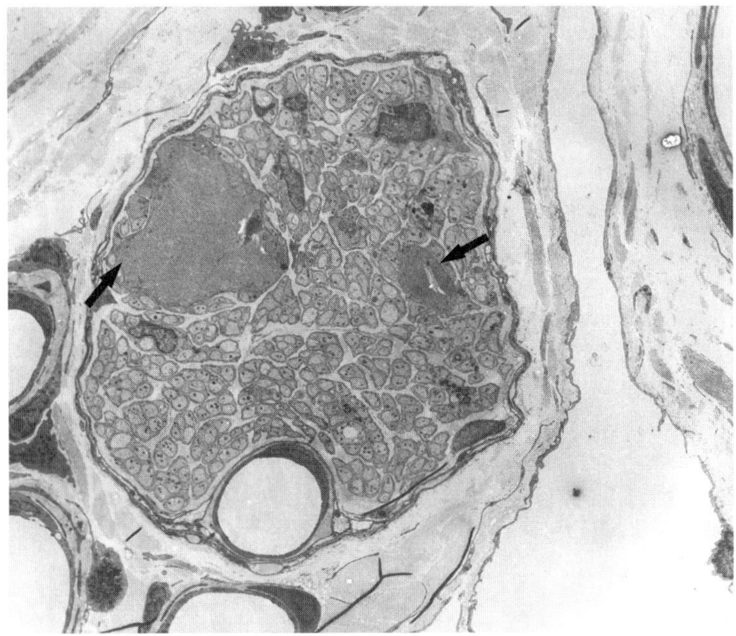

FIG. 4. Neuroaxonal dystrophy in ileal mesenteric nerve of STZ-diabetic rat. A single fascicle contains several dystrophic axons (arrows). (Magnification: 10,000×)

diabetes and in the BB rat (Schmidt and Scharp, 1982; Schmidt *et al.*, 1981,1986; Yagihashi and Sima, 1985a,b). The development of ileal mesenteric neuroaxonal dystrophy in STZ-diabetic rats began within 2–3 months of initiation of diabetes but required more than 6 months for the development of significant numbers of lesions (Schmidt and Plurad, 1986). Even in long-term diabetic rats, numbers of paravascular mesenteric nerve axons did not decrease; rather, axon numbers actually increased slightly, perhaps as the result of regenerative axonal sprouting (Schmidt and Plurad, 1986). Morphometric studies also failed to identify axonal atrophy in nondystrophic ileal mesenteric nerves (Schmidt and Plurad, 1986). Dystrophic axons involved lengthy (12–15 cm) ileal mesenteric axons but spared shorter (2–4 cm) jejunal mesenteric axons, as expected in a distal axonopathy (Schmidt and Plurad, 1986). In addition, examination of the proximal and distal portions of individual mesenteric pedicles, sites that may be separated by 3–4 cm, also showed distal worsening of neuroaxonal dystrophy. Ultrastructural and immunohistochemical studies demonstrated that neuroaxonal dystrophy selectively involved noradrenergic postganglionic axons within paravascular nerves,

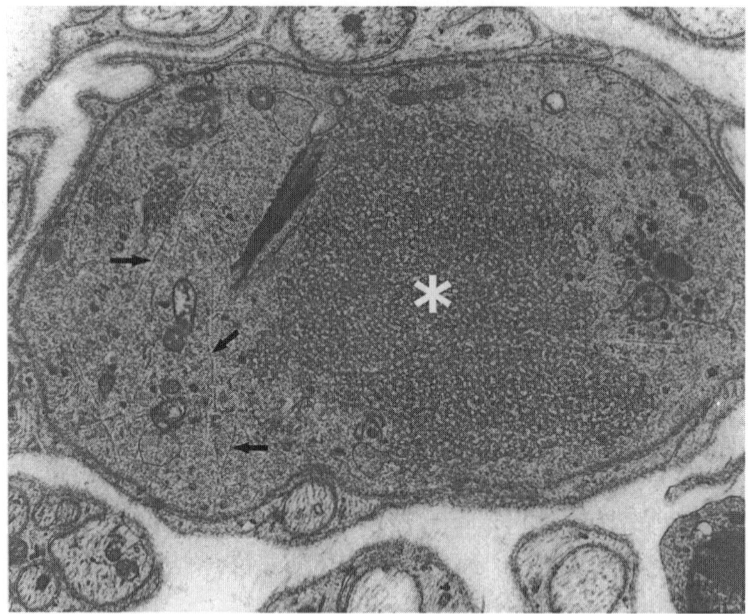

FIG. 5. Neuroaxonal dystrophy in ileal mesenteric nerve of a STZ-diabetic rat. A single dystrophic axon contains tubulovesicular elements (*), scattered dense core vesicles, and disoriented microtubules (arrows). Notice that the dystrophic axon shares a Schwann cell with a normal unmyelinated axon. (Magnification: 10,000×)

which provide much of the extrinsic sympathetic innervation of the intramural myenteric and submucosal ganglia, while sparing the equally lengthy immediately adjacent noradrenergic perivascular axons within the vascular adventitia (Schmidt *et al.*, 1983b; Clark and Schmidt, 1984). NPY and substance P containing axons within paravascular nerves (destined for vascular targets at various levels and serving as visceral sensory axons projecting into the gut wall, respectively) did not develop neuroaxonal dystrophy comparable to that involving nonvascular noradrenergic axons, even though all of these axons were located within the same fascicle and immediately adjacent to (or even within the same) Schwann cell unit. In further support of selectivity of damage to axon subpopulations, an increase in substance P and CGRP intensity of immunoreactivity and density of innervation of mesenteric arteries and veins (but no change in that of DβH or NPY) has been described in 8-week diabetic rats (Belai *et al.*, 1996). Pancreatic islet transplantation or chronic insulin therapy given in a preventative paradigm or to treat established dystrophy resulted in the regression of dystrophic axons (Schmidt and Scharp, 1982).

Diabetes-induced changes to the enteric nervous system have been further defined using electrophysiologic, immunohistologic, biochemical, and ultrastructural techniques. Both denervation and hyperinnervation have been described in the diabetic gut (DiGiulio *et al.*, 1989). Degenerative changes, but not neuroaxonal dystrophy, developed within the alimentary tract of short-term STZ-diabetic rats involving subpopulations of axons containing VIP (Loesch *et al.*, 1986) and CGRP (Belai and Burnstock, 1987) but not substance P. Measurement of neuropeptide content in diabetic rat ileum has demonstrated increased VIP and decreased substance P (variable) and met-enkephalin content (Ballman and Conlon, 1985; Gorio *et al.*, 1992), although changes varied with level of the alimentary tract sampled and duration of diabetes (Belai *et al.*, 1991). Simple measurement of neuropeptide content, however, may not adequately characterize the functional health of individual nerve terminals, as it has been reported that VIP and CGRP in the diabetic small bowel were not released appropriately in response to electrical stimuli (Belai *et al.*, 1987). Enteric ganglia responded to experimental diabetes in the rat by an apparent increase in NPY and VIP immunoreactivity in enteric neurons (Eaker *et al.*, 1996) and the superimposition of ultrastructural degenerative changes in the diabetic Chinese hamster (Diani *et al.*, 1979). Autonomic alterations in parasympathetic and sympathetic innervation of the pancreatic islets of diabetic rats and Chinese hamsters (Diani *et al.*, 1983; Luiten *et al.*, 1986) have been described.

Changes in the innervation of the diabetic rodent bowel may underlie gut dysfunction. Delayed small intestinal transit time has been reported in STZ-diabetic rats (Scott and Ellis, 1980; Chesta *et al.*, 1990) and in chronically diabetic Chinese hamsters (Diani *et al.*, 1979). Electrophysiologic studies of the alimentary tract in experimental diabetes have also established deficiencies of cholinergic transmission (Nowak *et al.*, 1986), muscarinic signal transduction (Lucas and Sardar, 1991), and prejunctional impairment of ileal sympathetic nerve function. Abnormal transmucosal ionic flux in response to tyramine, thought to reflect a deficiency of noradrenergic innervation (Chang *et al.*, 1985, 1986), has been identified in the ileum (but not jejunum) after 6 months of untreated diabetes (but not at 7 days), findings that parallel the time course and distribution of neuroaxonal dystrophy in this system. Therefore, determination of the functional significance of neuroaxonal dystrophy in the diabetic rat small intestine is complicated by changes in noradrenergic, sensory, and enteric nervous systems, a pathophysiologic response to hyperglycemia itself (Chang *et al.*, 1996) and, in the hyperphagic STZ rat, hypertrophic and hyperplastic changes in the gut. In other experimental and clinical paradigms (e.g., aged gracile nuclei, infantile neuroaxonal dystrophy, dystrophic neurites in Alzheimer's disease), neuroaxonal dystrophy is closely associated with pathophysiology. The enormous size of dystrophic axons may exert mechanical effects on adjacent nerve terminals and dendritic spines, release neurotransmitters continuously, resulting in post-synaptic tachyphylaxis, or fail to release neurotransmitter in response to an action potential.

Studies of gastric nonadrenergic noncholinergic (NANC) relaxation in the BB and STZ rat have identified impaired nitric oxide synthase (NOS) synthesis, a result supported by reduced NOS mRNA and decreased numbers of NOS neurons in the BB rat myenteric plexus and nodose ganglion (Takahashi *et al.*, 1997; Wrzos *et al.*, 1997). Genetically diabetic nonobese diabetic (NOD) mice, as well as STZ-treated mice and NOS knockout mice, also develop a delayed gastroparetic syndrome with a defect in nitrergic innervation of the pylorus (Watkins *et al.*, 2000). Decreased numbers of interstitial cells of Cajal in the distal stomach of NOD mice and their abnormal association with enteric intramuscular nerve terminals have also been reported (Ordog *et al.*, 2000). A complex pattern of alterations in myenteric neuron and intramural axons arising from neuropeptide- and neurotransmitter-containing subpopulations has been described for VIP, acetylcholine, and NOS, findings that are dependent on the level of gut examined (Spangeus *et al.*, 2000). Similarly, the anococcygeus muscle in diabetic rats showed impaired nitrergic transmission (Way *et al.*, 1999).

Although morphometric studies failed to show significant axon loss or dystrophic axons in the gut-directed abdominal vagus nerve of genet-ically diabetic Chinese hamsters (the sampled nerve was distant from its termination), evidence of significantly diminished numbers of axons comprising each Schwann cell unit and regenerative collections of Schwann cell processes devoid of axons have been interpreted as the residua of cycles of degeneration and regeneration (Schmidt *et al.*, 1989b).

2. *Genitourinary Tract*

Abnormal innervation of the urinary bladder (Paro *et al.*, 1991), vas deferens (Moss *et al.*, 1987), and penis, accompanied by sexual dysfunction (McVary *et al.*, 1997), have been described in several diabetic animal models. Axonal atrophy and increased numbers of axonal glycogenosomes (but not neuroaxonal dystrophy), as well as selective degeneration of nitrergic but not noradrenergic nerves, characterized the effect of diabetes on the rat penis (Yagihashi and Sima, 1986b; Cellek *et al.*, 1999). Prostate alveoli showed decreased density of noradrenergic innervation in the presence of increased noradrenergic, NPY, and VIP innervation of adjacent vasculature in 8-week diabetic rats (Crowe *et al.*, 1987).

3. *Cardiovascular System*

The diabetic rat heart shows blunting of the function of sympathetic, parasympathetic, and potentially afferent elements, resulting in altered cardiac function (Ustinova *et al.*, 2000; Sunagawa *et al.*, 1989; Zhang *et al.*, 1990; Fazan *et al.*, 1999). Neuropathologic studies have provided evidence for decreased cardiac sympathetic innervation in the db/db mouse (Tessari *et al.*, 1988) and atrophy, sequestration, and axon loss (after extended intervals) in the vagus nerve of chronic STZ- and BB-diabetic rats and db/db mouse, alterations that accompanied a decreased variability of heart rate (Robertson and Sima, 1980; McEwen *et al.*, 1987; Yagihashi and Sima, 1986b). The vasculature of the diabetic rat mystacial pad showed deficient CGRP in the presence of maintained sympathetic vascular innervation (Patel and Rice, 1999).

4. *Miscellaneous*

The VIPergic innervation of the diabetic rat iris is increased in the absence of a change in NPY, substance P, and noradrenergic innervation (Crowe and Burnstock, 1988). A decreased vagal sensory CGRP content is associated with a decreased stimulus-induced tracheal CGRP release in the absence of axon loss (Calcutt *et al.*, 1998).

E. COMPARISON OF THE EFFECT OF DIABETES ON HUMAN AND RODENT SYMPATHETIC GANGLIA

Chronically diabetic rodents and human diabetics (1) develop neuroaxonal dystrophy in sympathetic ganglia in the absence of the loss of significant numbers of sympathetic neurons; (2) prematurely develop individual dystrophic axons, which are immunohistochemically and ultrastructurally identical to those that eventually appear in aging nondiabetic sympathetic ganglia; (3) develop neuroaxonal dystrophy in the SMG and celiac ganglia, relatively sparing the SCG; and (4) exhibit a predilection of neuroaxonal dystrophy for selected subpopulations of nerve terminals.

IV. Proposed Pathogenetic Mechanisms of Diabetic Autonomic Neuropathy

The detailed pathogenesis of peripheral nervous system dysfunction in experimental and clinical diabetes is unknown. Proposed pathogenetic mechanisms (reviewed in Zochodne, 1999; Sugimoto *et al.*, 2000), based to a large degree on clinical and animal studies of somatic nerves and myelinated axons, include primary biochemical and metabolic alterations, ischemia due to microangiopathy, and deficiencies of a variety of neurotrophic substances. Although there is currently no consensus regarding which mechanism(s) is operative, multiple mechanisms may participate interactively and, indeed, may vary between different forms of diabetic neuropathy. The following sections focus on the potential role of some of these mechanisms in the diabetic autonomic nervous system.

A. Polyol Pathway and Deranged Phosphoinositide Metabolism

The reader is referred to the chapter by Oates for a full discussion of the mechanistics of the polyol pathway. Compared to the study of somatic nerves, much less is known about the polyol pathway in the autonomic nervous system. An early study of the diabetic BB rat SCG showed a salutary effect of an aldose reductase inhibitor (ARI) on decreased *myo*-inositol content, Na^+-K^+-ATPase activity, and muscarinic signal transduction (Greene and Mackway, 1986; Mackway *et al.*, 1986). ARI and sorbitol dehydrogenase inhibitors (SDI) produced comparable changes in sorbitol and fructose content in sympathetic ganglia as described in the sciatic nerve (Schmidt *et al.*, 1989c), and studies have demonstrated no compelling differences in glucose, fructose, sorbitol, and *myo*-inositol content in diabetic rat SCG and celiac ganglia (Schmidt *et al.*, 2001b). Previous studies (Schmidt *et al.*, 1989c, 1991,1998) of the effect of ARI on neuroaxonal dystrophy in the SMG and ileal mesenteric nerves of chronic STZ-diabetic rats showed significant amelioration, but not normalization, of neuropathologic findings. Altered VIP and galanin immunoreactivity in the STZ-diabetic rat myenteric plexus was normalized by ARI (Belai *et al.*, 1996). ARI treatment also prevented a diabetes-induced decrease in heart rate variability and atrophy of the vagus nerve (Nowak *et al.*, 1995). Surprisingly, SDI-treated STZ-diabetic rats (but not SDI-treated controls) demonstrated marked acceleration in the time of appearance and exaggeration of the frequency of otherwise typical neuroaxonal dystrophy in ileal mesenteric nerves and SMG compared to untreated diabetic rats (Schmidt *et al.*, 1998, 2001b). This SDI-accelerated STZ-diabetic rat model was faithful to the original "standard" STZ rat model and failed to develop neuroaxonal dystrophy in jejunal mesenteric nerves and in the SCG, while selectively targeting TOH-containing axons in the paravascular nerve bundles and myenteric ganglia and sparing the vascular innervation. The effect of SDI on neuroaxonal dystrophy in the diabetic SMG was completely prevented by concomitant administration of the ARI, sorbinil, evidence that the SDI effect was exerted through the sorbitol pathway rather than in a nonspecific toxic fashion (Schmidt *et al.*, 2001b).

The results of these experiments provide insight into the role of the sorbitol pathway in the pathogenesis of diabetic autonomic neuropathy. Although nerve *myo*-inositol content has been thought to be significant in the pathogenesis of diabetic somatic neuropathy in the rat (Greene *et al.*, 1985), sympathetic ganglionic *myo*-inositol content showed no consistent relationship to the frequency of neuroaxonal dystrophy in ARI- and SDI-treated diabetic rats (Schmidt *et al.*, 2001b). SDI treatment, presumably resulting in the improvement of nerve NAD levels (important in the prevention of oxidative toxicity), would be predicted to have a salutary effect (Lee *et al.*, 1985) rather than the observed worsening of sympathetic neuroaxonal dystrophy (see Oates, this volume). Treatment of diabetic rats with ARI or SDI has been shown to decrease or increase, respectively, neural markers of oxidative stress in the diabetic rat sciatic nerve (Obrosova *et al.*, 2000). A possible role for sorbitol-induced hyperosmolarity (Kinoshita and Nishimura, 1988) and associated depletion of antioxidant organic osmolytes (e.g., taurine; Oja and Kontro, 1990) in the pathogenesis of somatic neuropathy has been proposed. This hypothesis is not supported by the failure to produce neuroaxonal dystrophy in the SMG of nondiabetic control rats with the administration of galactose (Schmidt *et al.*, 1989c), which produces marked osmotic stress. In addition, SDI-treated control rats, whose ganglionic sorbitol level is comparable to that in untreated diabetics, fail to develop neuroaxonal dystrophy (Schmidt *et al.*, 1998).

B. Abnormal Axonal Transport

Transportation of intraxonal constituents from their sites of synthesis in the neuronal perikaryon to the most distal portions of the axon and, at least for some materials, their return to the cell body is accomplished by orthograde and retrograde axonal transport, respectively. Altered rapid transport of radiolabeled proteins and glycoproteins has been reported (Tomlinson, 1983) in the diabetic cat hypogastric nerve *in vitro*. Axonal transport in distal minute ($\leq$50 μm in diameter) rat mesenteric nerves has been measured by monitoring the retrograde axonal transport of intravenously administered ^{125}I-labeled NGF to the SMG (Schmidt *et al.*, 1983) and directly in the ileal mesenteric nerves themselves (Schmidt *et al.*, 1985). An apparent defect was identified in the retrograde ^{125}I-labeled NGF transport to prevertebral sympathetic ganglia serving the bowel, a site in which neuroaxonal dystrophy develops, but there was no change in ^{125}I-labeled NGF transport to the SCG, in whose peripheral territory comparable pathology does not develop (Schmidt *et al.*, 1983a). Measurement of DβH transport in diabetic rat mesenteric nerves (paravascular and perivascular together) provided no evidence of a generalized defect in orthograde or retrograde axonal transport in noradrenergic alimentary axons (Schmidt *et al.*, 1987). An abnormality in the reversal of transport polarity at the nerve ending ("turnaround transport") may result in the accumulation of membranous tubules, which are prominent constituents of neuroaxonal dystrophy (Sahenk and Brown, 1991).

C. Neurotrophic Substances

There have been prominent advances in the understanding of the role of neurotrophic substances in a wide variety of pathologic processes and they may play a role in the therapy or pathogenesis of diabetic neuropathy (Ishii, 1995; Vinik *et al.*, 1995; Apfel, 2001). Diabetes may alter the response of subpopulations of autonomic neurons to the neurotrophic milieu, possibly by upregulating or downregulating individual classes of receptors or interfering in subtle ways with neurotrophin synthesis, secretion, receptor binding, and uptake. These processes may be critical to the pathogenesis and/or therapy of experimental diabetic neuropathy and may provide important insights into its pathogenesis and therapy.

1. *Neurotrophins (NGF and NT-3)*

The sciatic nerves of STZ-diabetic rats are deficient in NGF content, its retrograde axonal transport, and levels of NGF-dependent substance P (Fernyhough *et al.*, 1994). Only a limited number of studies have addressed the role of neurotrophic substances in the pathogenesis or treatment of diabetic autonomic neuropathy. Early studies demonstrated a decrease in retrograde axonal transport of exogenously administered ^{125}I-labeled NGF in ileal mesenteric nerves (Schmidt *et al.*, 1983a, 1985); however, at that time, it was thought that the transport abnormality, which appeared to represent a lag in onset of transport of labeled exogenous NGF, might not result in a decrease in steady-state ganglionic NGF and, as a result, not reflect actual NGF deficiency. Subsequently, endogenous NGF in the SCG, sciatic nerve, and submandibular gland were reportedly decreased in short-term STZ-diabetic rats (Hellweg and Hartung, 1990). Other tissues (iris, heart, spleen, prostate) in chronically diabetic STZ rats contained increased or unchanged amounts of NGF, which could reflect decreased retrograde transport by dysfunctional autonomic or sensory axons.

Studies suggest that the effect of diabetes on endogenous NGF content in sympathetic ganglia and its functional significance may be complex. The diabetic rat SMG and celiac ganglia serving the bowel are preferentially targeted by neuroaxonal dystrophy; however, these ganglia fail to lose neurons (Schmidt, 2001), show perikaryal atrophy (Schmidt and Plurad, 1986), or decrease TOH and DβH activities (Schmidt and Cogswell, 1989), as would be expected in a true substantial NGF deficiency. To test the proposed mechanism that diabetic autonomic neuropathy represented an NGF deficiency disease, adult nondiabetic rats were chronically deprived of NGF for 1 year by an autoimmune paradigm (Gorin and Johnson, 1980), which failed to produce neuroaxonal dystrophy (Schroer *et al.*, 1995). Reanalysis of these ganglia using an updated quantitative scheme has determined that there is an actual significant decrease in age-related neuroaxonal dystrophy in NGF-deprived SMG (Schmidt *et al.*, unpublished data). Studies have found that ganglionic NGF is unchanged in the SCG and is increased markedly in the SMG and celiac ganglia of diabetic rats compared to controls, which probably represents increased retrograde transport of NGF from the hyperplastic diabetic bowel (Schmidt *et al.*, 2000). An increased NGF pool in the diabetic STZ rat bowel may have diluted tracer amounts of ^{125}I-labeled NGF in prior experiments, although tracer dilution would not explain the late arrival of ^{125}I-labeled NGF in greater amounts in diabetic SMG than control (Schmidt *et al.*, 1983a). Neither sympathetic ganglionic trkA nor p75^{NTR} was altered in long-term STZ diabetes (Schmidt *et al.*, 2000); however, heterogeneous upregulation of p75^{NTR} in a subpopulation of neurons or satellite cells may function to present increased NGF locally as a sprouting phenomenon, as proposed previously in sciatic nerve regeneration (Taniuchi *et al.*, 1986). Treatment of STZ-diabetic rats for 2–3 months with pharmacologic doses of NGF did not normalize established neuroaxonal dystrophy in the SMG and ileal mesenteric nerves (Schmidt *et al.*, 2001a). NGF treatment of control animals actually increased the frequency of neuroaxonal dystrophy in the SMG. PET scanning with [^{11}C]hydroxyephedrine showed distal loss and a proximal increase in sympathetic innervation of diabetic rat ventricle paralleling gradients of NGF distribution (Schmid *et al.*, 1999). Therefore, NGF may contribute to the pathogenesis of neuroaxonal dystrophy rather than its amelioration, perhaps as the result of inducing intraganglionic axonal sprouts in which dystrophic changes are superimposed, as proposed for fibroblast growth factor (FGF) in senile plaques of Alzheimer's disease (Cotman and Gomez-Pinilla, 1991). Neonatal sympathetic ganglia treated simultaneously with 6-OHDA and high doses of NGF *in vivo* developed large intraganglionic swellings reminiscent of neuroaxonal dystrophy, suggesting a possible pathogenetic role for increased ganglionic NGF and peripheral axotomy (Levi-Montalcini *et al.*, 1975). High concentrations of NGF produced reversible arrest of neurite outgrowth from DRG explants (Conti *et al.*, 1997), which may parallel the diabetic rat sympathetic ganglionic milieu.

Less attention has been directed to the role of neurotrophic substances in the diabetic parasympathetic nervous system. NGF content in the diabetic rat cervical vagus nerve (although most likely only the sensory axon subpopulation) was increased at 8 and 16 weeks of diabetes (but not at 24 weeks), and retrograde transport of NGF, chiefly to the nodose ganglion, was decreased at both later times in diabetic rats (Lee *et al.*, 2001). NGF treatment of STZ-diabetic rats normalized selected enteric neuropeptides and corrected degenerative changes in interstitial cells of Cajal (Park *et al.*, 1997).

The sciatic nerve content of NT-3 and its retrograde axonal transport is reportedly decreased in experimental somatic diabetic neuropathy (Fernyhough *et al.*, 1998) and in the vagus nerve (Lee *et al.*, 2001). Studies have suggested a possible role for NT-3 in the adult sympathetic nervous system (Zhou *et al.*, 1997); however, NT-3 administration did not alter the frequency of neuroaxonal dystrophy in the SMG or ileal mesenteric nerves of chronically diabetic animals (Schmidt *et al.*, 2001a).

2. *Insulin-like Growth Factors (IGFs)*

IGF-I has been reported to (1) support the development and growth of cultured sympathetic and sensory neurons (Ishii *et al.*, 1994; Recio-Pinto *et al.*, 1986); (2) influence synaptic development, nerve terminal, and axonal sprouting and regeneration (Ishii *et al.*, 1993, 1994; Cheng *et al.*, 1996; Engber *et al.*, 1995); (3) decrease in serum with aging and diabetes (Zhuang *et al.*, 1996; Bitar *et al.*, 1997), worsening with the development of somatic neuropathy in human subjects (Migdalis *et al.*, 1995) and rats (Ishii *et al.*, 1994); (4) promote growth cone reorganization and interfere with the glucose-induced degeneration of neuroblastoma cells *in vitro* (Feldman *et al.*, 1997); and (5) be deficient in diabetic somatic nerves and fail to be properly upregulated after axotomy (Sima *et al.*, 1997). In longitudinal studies, circulating IGF-I declined progressively as Rhesus monkeys aged, became diabetic, and developed neuropathy (Zhuang *et al.*, 1997). In addition, the IGF-I receptor protein was decreased in the SCG of diabetic rats (Bitar *et al.*, 1997).

IGF-I is thus poised to influence the pathogenesis or therapy of diabetic autonomic neuropathy. IGF-I, administered exogenously to chronically diabetic rats for several months, resulted in the nearly complete reversal of established neuroaxonal dystrophy in the SMG and ileal mesenteric nerves without affecting the metabolic severity of diabetes (Schmidt *et al.*, 1999). Local synthesis of IGF-I and perikaryal, axonal, and nerve terminal IGF-I receptors in the rat SCG also suggested an autocrine or paracrine role for IGF-I in sympathetic ganglia (Singhal *et al.*, 1997; Bitar *et al.*, 1996). A role for IGF-I in routine synaptic turnover and its deficiency in diabetes could underlie abnormal axonal regeneration (Ekstrom *et al.*, 1989) and synaptic turnover and contribute to the development of neuroaxonal dystrophy. The diabetic male rat major pelvic ganglion (a mixed parasympathetic and sympathetic ganglion) showed no significant change in IGF-I gene expression in the presence of increased expression of IGFBPs-3,4,5 and unchanged IGF-I receptor levels (Abdelbaky *et al.*, 1998), a constellation of findings that complicates evaluation of the role of individual elements in the IGF-I pathway.

3. *Glial Cell Line-Derived Neurotrophic Factor (GDNF) Family*

Studies have established a significant role for glial cell line-derived neurotrophic factor (GDNF) and neurturin, one of the GDNF family of neurotrophic substances, in the developing and adult parasympathetic and enteric nervous systems (Rossi *et al.*, 1999; Heuckeroth *et al.*, 1999; Enomoto *et al.*, 2000). Of particular significance, GDNF is expressed and retrogradely transported from the adult rat penis to the major pelvic parasympathetic ganglia (Laurikainen *et al.*, 2000a,b). Alterations in the GDNF family of neurotrophic substances or members of the GDNF family of receptors are well positioned to have a role in the pathogenesis or therapy of diabetes-induced dysfunction of parasympathetic ganglia, visceral sensory, or intramural enteric ganglion function in diabetes. Diabetes-induced depletion of mouse central spinal cord projections of nonpeptidergic nociceptive spinal afferent fibers is restored by the administration of exogenous GDNF (Akkina *et al.*, 2000).

D. Abnormalities of Regeneration and Synaptic Dysplasia

The association of neuroaxonal dystrophy with regenerative axonal sprouts in diabetic rat mesenteric nerves (Schmidt and Scharp, 1982) and diabetic rat and Chinese hamster SMG (Fig. 3C and Schmidt *et al.*, 1986, 1989b) and the ultrastructural resemblance of dystrophic axons to growth cones (Schmidt and Scharp, 1982), especially those whose regenerative progress had been intentionally frustrated (Schmidt and Plurad, 1985), suggests a possible interrelationship of dystrophic and regenerative processes (Jellinger, 1973). Previous studies of axotomized diabetic rat sciatic nerve have consistently demonstrated a defect in axonal regeneration (Longo *et al.*, 1986). However, regeneration of sympathetic axons following *in vivo* 6-hydroxydopamine terminal degeneration was actually faster in diabetic rats than controls (Vo and Tomlinson, 1999) and following *in vitro* explantation of diabetic rat sympathetic ganglia (Schmidt *et al.*, unpublished data), perhaps reflecting a conditioning effect of ongoing distal axonopathy. Regeneration of the parasympathetic innervation of the iris following oculomotor nerve crush was deficient in STZ-diabetic rats (Vandertop *et al.*, 1995) but was enhanced in diabetic mouse nodose ganglion *in vitro* (Saito *et al.*, 1999).

It has been proposed that neuroaxonal dystrophy, which develops in diabetic prevertebral ganglia, represents an abnormal outcome of ongoing cycles of synaptic turnover or remodeling, a normal process that may underlie synaptic plasticity (Schmidt, 1996b). This process may be exaggerated by postganglionic axotomy or distal diabetes-induced axonopathy similar to the process that produces synaptic detachment and reattachment following postganglionic axonal regeneration (Matthews and Nelson, 1975; Purvis, 1975; Ohara *et al.*, 1995). Failure of synaptic reattachment, irrespective of its origin by routine, day-to-day synaptic turnover or exaggerated by axonal injury, may eventually result in neuroaxonal dystrophy. Neuroaxonal dystrophy may, therefore, represent the morphological equivalent of the failure to complete the process of synaptic turnover or to eventually inhibit the regenerative/sprouting process (i.e., to institute a "stop" program).

E. GLYCATION AND ADVANCED GLYCOSYLATION END PRODUCTS (AGEs)

General influences of glycation products and the mechanisms of action of aminoguanidine are reviewed in the chapter by Thornalley. Hyperglycemia-induced glycosylation of various intracellular and extracellular proteins may be followed by chemical rearrangement into advanced glycosylation end products with subsequent cross-linking of a variety of proteins and subcellular organelles. Treatment of experimental diabetic animals with aminoguanidine, an agent that prevents AGE-dependent protein cross-linking but not its initial glycosylation step, has not demonstrated a salutary effect on neuroaxonal dystrophy in SMG after 7–10 months of treatment (Schmidt *et al.*, 1996) or experimental autonomic nervous system dysfunction (Way and Reid, 1994; Eika *et al.*, 1992). However, aminoguanidine does not interfere with the formation of early glucose adducts and their autooxidation to form superoxide anions, which may contribute to the pathogenesis of neuroaxonal dystrophy. Glycation of laminin, a constituent of the extracellular matrix, interfered with neurite outgrowth from neuroblastoma cells in culture (Federoff *et al.*, 1993). The AGE adduct carboxymethyllysine has been demonstrated in diabetic rat perineurial basal membranes, Schwann cells, the endoneurial microvasculature, and myelinated and unmyelinated axons (Sugimoto *et al.*, 1997).

F. Oxidative Stress

The reader is referred to chapters by Obrosova and Thornalley for details of mechanisms of diabetes-induced oxidative stress. Increased oxidative stress and reduction in antioxidant defenses have been proposed in the pathogenesis of diabetic neuropathy (reviewed in Low *et al.*, 1997). Increased indices of oxidative stress (lipid hydroperoxides, malondialdehyde, conjugated dienes, and decreased levels of GSH) have also been reported in the SCG of STZ-diabetic rats (Nagamatsu *et al.*, 1995). The diabetic rat SCG in some studies shows no deficiency or actually increased mRNA for antioxidant enzymes; catalase, for example, increases 20%, which may reflect compensatory upregulation in response to increased oxidative stress (Kishi *et al.*, 2000). Studies of diabetic sciatic nerve by Obrosova and colleagues (2000) have found that aldose reductase inhibitors diminish and sorbitol dehydrogenase inhibitors exaggerate biochemical parameters of oxidative damage, which may result in decreased or increased neuroaxonal dystrophy, respectively. Treatment of diabetic rats with DL-α-lipoic acid, which is metabolized to a potent antioxidant capable of scavenging superoxide, hydroxyl and peroxyl radical, and singlet oxygen, as well as regenerating reduced glutathione, has been reported to promote neuritogenesis (Dimpfel *et al.*, 1990) and to have salutary effects on diabetic autonomic dysfunction (Boulton and Malik, 1998). Dietary supplementation with the antioxidant vitamin E prevents structural changes in the noradrenergic innervation of the diabetic rat heart (Rösen *et al.*, 1995).

G. Autoimmune Mechanisms

An autoimmune pathogenesis has been proposed for diabetic autonomic neuropathy (Rabinowe *et al.*, 1990). Lymphocytic infiltrates in the sympathetic ganglia of symptomatic diabetics (Duchen *et al.*, 1980) represent a nonspecific finding (Schmidt, 1996) and are not compelling evidence for a selective diabetes-induced autoimmune attack. Antibodies directed against the adrenal gland, sympathetic ganglia, and vagus nerve may be associated with an increased incidence of autonomic neuropathy (Rabinowe *et al.*, 1990; Muhr *et al.*, 1997). Other studies have failed to find an association of autoantibodies and autonomic neuropathy (Husebye *et al.*, 1996; Stroud *et al.*, 1997) or have found an inverse correlation (Sundkvist *et al.*, 1994) or only small effects limited to IDDM patients (Rösen *et al.*, 1995). Although abnormal regional myocardial ^{123}I-labeled MIBG uptake (a measure of sympathetic innervation of the heart) correlated with the presence of anti-SCG autoantibodies in long-term IDDM patients, examination of newly diagnosed diabetics provided little evidence that autoantibodies produced diabetic autonomic neuropathy (Schnell *et al.*,1996). The serum of diabetic patients, especially those with neuropathy, has been reported to produce apoptosis of neurons *in vitro* (Pettinger *et al.*, 1997; Srinivasan *et al.*, 1998). An increased incidence of autoimmune iritis in diabetic patients with autonomic neuropathy has been proposed to reflect the development of insulin autoantibodies with cross-reactivity to NGF (Guy *et al.*, 1984), although more recent studies have not supported this hypothesis (Zanone *et al.*, 1994).

H. OTHER MECHANISMS

1. *Ischemia*

No single area has proven more contentious over the years than the determination of blood flow alterations in somatic nerves of diabetic rats, and the debate is reviewed by Zochodne elsewhere in this volume. Only two studies in which blood flow has been investigated in diabetic rat SCG reported a 40–52% decrease (Sasaki *et al.*, 1997; Cameron *et al.*, 2001). The preservation of principal sympathetic neurons and the selectivity of degenerative and dystrophic processes for subpopulations of nerve terminals complicate the interpretation of this possible pathogenetic mechanism. Certainly it is difficult to see how such a simplistic mechanism can account for disparate effects of diabetes on different neuronal phenotypes.

2. *Synaptic Degradation of Organelles*

Some materials undergo orthograde transport to the nerve terminal but are not returned and, instead, undergo degradation within the terminal axon. Neurofilaments (NF) are not typically seen in the normal nerve terminal, presumably having been degraded by proteolytic processes located in the preterminal axon. Dysfunction of the degradative process in diabetic ganglia, particularly in diabetic human subjects, may produce the large neurofilamentous skeins that accumulate in dystrophic terminals.

3. *Apoptosis*

Studies have reported that cultured rat sympathetic SCG neurons respond to a substantial increase of glucose concentration by developing a variety of neuritic abnormalities and, eventually, undergoing apoptosis (Russell and Feldman, 1999). However, determination of neuron numbers using unbiased counting methods in the SCG and SMG of STZ rats diabetic for 10 months have failed to show significant neuron loss (Schmidt, 2001).

V. Summary

A single unifying hypothesis has not yet emerged from clinical and experimental animal studies of diabetic autonomic neuropathy. However, a recurring theme is the involvement of distal portions of axons and nerve terminals by degenerative or dystrophic changes. The selectivity of the diabetic process for some nerve terminals while sparing others confounds simple global explanations, such as ganglionic hypoxia or diminished blood flow. It is likely that diabetic autonomic neuropathy will be shown to have multiple causative mechanisms, which will interact to result in the variety of presentations of autonomic injury in diabetes. Some of these mechanisms will be shared with aging changes in the autonomic nervous system. The basic neurobiologic processes underlying the development of neuroaxonal dystrophy, synaptic dysplasia, defective axonal regeneration, and alterations in neurotrophic substances may be mechanistically related. It is hoped that the ability to rigorously test several of the proposed pathogenetic mechanisms will contribute to the understanding of the pathogenetic mechanisms underlying diabetic autonomic neuropathy.

References

Abdelbaky, T. M., Brock, G. B., and Huynh, H. (1998). Improvement of erectile function in diabetic rats by insulin: Possible role of the insulin-like growth factor system. *Endocrinology* **139**, 3143–3147.

Akkina, S. K., Riekhof, J. T., Patterson, C. L., and Wright, D. E. (2000). The role of GDNF on unmyelinated primary afferents in diabetic neuropathy. *Soc. Neurosci. Abst.* **26**, 1871.

Apfel, S. C. (2001). Neurotrophic factor therapy-prospects and problems. *Clin. Chem. Lab. Med.* **39**, 351–355.

Appenzeller, O., and Ogin, G. (1974). Myelinated fibers in human paravertebral sympathetic chain: White rami communicantes in alcoholic and diabetic patients. *J. Neurol. Neurosurg. Psychiat.* **37**, 1155–1161.

Appenzeller, O., and Richardson, E. P., Jr. (1966). The sympathetic chain in patients with diabetic and alcoholic polyneuropathy. *Neurology* **16**, 1205–1209.

Ballman, M., and Conlon, J. M. (1985). Changes in the somatostatin, substance P and vasoactive intestinal polypeptide content of the gastrointestinal tract following streptozotocin induced diabetes in the rat. *Diabetologia* **28**, 355–358.

Battle, W. M., Snape, W. J., Alavi, A., Cohen, S., and Braunstein, S. (1980). Colonic dysfunction in diabetes mellitus. *Gastroenterology* **79**, 1217–1221.

Beggs, J., Johnson, P. C., Olafsen, A., and Watkins, C. J. (1994). Innervation of the vasa nervorum: changes in human diabetics. *J. Neuropathol. Exp. Neurol.* **51**, 612–629.

Belai, A., and Burnstock, G. (1987). Selective damage of intrinsic calcitonin gene-related peptide-like immunoreactive enteric nerve fibers in streptozotocin-induced diabetic rats. *Gastroenterology* **92**, 730–734.

Belai, A., Calcutt, N. A., Carrington, A. L., Diemel, L. T., Tomlinson, D. R., and Burnstock, G. (1996). Enteric neuropeptides in streptozotocin-diabetic rats; effects of insulin and aldose reductase inhibition. *J. Auton. Nerv. Syst.* **58**, 163–169.

Belai, A., Lincoln, J., and Burnstock, G. (1987). Lack of release of vasoactive intestinal peptide and calcitonin gene-related peptide during electrical stimulation of enteric nerves in streptozotocin-diabetic rats. *Gastroenterology* **93**, 1034–1040.

Belai, A., Lincoln, J., Milner, P., and Burnstock, G. (1991). Differential effect of streptozotocin-induced diabetes on the innervation of the ileum and distal colon. *Gastroenterology* **100**, 1024–1032.

Belai, A., Milner, P., Aberdeen, J., and Burnstock, G. (1996). Selective damage to sensorimotor perivascular nerves in the mesenteric vessels of diabetic rats. *Diabetes* **45**, 139–143.

Bellavere, F., Balzani, I., De Masi, G., Carraro, M., Carenza, P., Cobelli, C., and Thomaseth, K. (1992). Power spectral analysis of heart-rate variations improves assessment of diabetic cardiac autonomic neuropathy. *Diabetes* **41**, 633–640.

Bestetti, G., and Rossi, G. L. (1980). Hypothalamic lesions in rats with long-term streptozotocin-induced diabetes mellitus: A semi-quantitative light- and electron microscopic study. *Acta Neuropathol.* **52**, 119–127.

Bitar, M. S., Al-Bustan, M., Nehme, C. L., and Pilcher, C. W. T. (1996). Antinociceptive action of intrathecally administered IGF-1 and the expression of its receptor in rat spinal cord. *Brain Res.* **737**, 292–294.

Bitar, M. S., Pilchher, C. W. T., Khan, I., and Waldbillig, R. J. (1997). Diabetes-induced suppression of IGF-1 and its receptor mRNA levels in rat superior cervical ganglia. *Diabet. Res. Clin. Pract.* **38**, 73–80.

Boulton, A. J. M., and Malik, R. A. (1998). Diabetic neuropathy. *Med. Clin. North Am.* **82**, 909–929.

Brownlee, M., Cerami, A., and Vlassara, H. (1988). Advanced glycosylation endproducts in tissue and the biochemical basis of diabetic complications. *N. Engl. J. Med.* **318**, 1315–1321.

Calcutt, N. A., Chen, P., and Hua, X.-Y. (1998). Effect of diabetes on tissue content and evoked release of calcitonin gene-related peptide-like immunoreactivity from rat sensory nerves. *Neurosci. Lett.* **254**, 129–132.

Cameron, N. E., Jack, A. M., and Cotter, M. A. (2001). Effect of a-lipoic acid on vascular responses and nociception in diabetic rats. *Free Radic. Biol. Med.* **31**, 125–135.

Camilleri, M., and Malagelada, J.-R. (1984). Abnormal intestinal motility in diabetics with the gastroparesis syndrome. *Eur. J. Clin. Invest.* **14**, 420–427.

Cellek, S., Rodrigo, J., Lobos, E., Fernandez, P., Serrano, J., and Moncada, S. (1999). Selective nitrergic neurodegeneration in diabetes mellitus-a nitric oxide-dependent phenomenon. *Br. J. Pharmacol.* **128**, 1804–1812.

Chang, E. B., Bergenstal, R. M., and Field, M. (1985). Diarrhea in streptozocin-treated rats: Loss of adrenergic regulation of intestinal fluid and electrolyte transport. *J. Clin. Invest.* **75**, 1666–1670.

Chang, E. B., Fedorak, R. N., and Field, M. (1986). Experimental diabetic diarrhea in rats: Intestinal mucosal denervation hypersensitivity and treatment with clonidine. *Gastroenterology* **91**, 364–369.

Chang, F. Y., Lee, S. D., Yeh, G. H., and Wang, P. S. (1996). Influence of blood glucose levels on rat liquid gastric emptying. *Dig. Dis. Sci.* **41**, 528–532.

Cheng, H.-L., Randolph, A., Yee, D., Delafontaine, P., Tennekoon, G., and Feldman, E. L. (1996). Characterization of insulin-like growth factor-I and its receptor and binding proteins in transected nerves and cultured Schwann cells. *J. Neurochem.* **66**, 525–536.

Chesta, J., Debnam, E. S., Srai, S. K., and Epstein, O. (1990). Delayed stomach to caecum transit time in the diabetic rat. Possible role of hyperglucagonaemia. *Gut* **31**, 660–662.

Conti, A. M., Fischer, S. J., and Windebank, A. J. (1997). Inhibition of axonal growth from sensory neurons by excess nerve growth factor. *Ann. Neurol.* **42**, 838–846.

Cotman, C. W., and Gomez-Pinilla, F. (1991). Basic fibroblast growth factor in the mature brain and its possible role in Alzheimer's disease. *Ann. N.Y. Acad. Sci.* **638**, 221–231.

Crowe, R., and Burnstock, G. (1988). An increase of vasoactive intestinal polypeptide-, but not neuropeptide Y-, substance P-, or catecholamine-containing nerves in the iris of the streptozotocin-induced diabetic rat. *Exp. Eye Res.* **47**, 751–759.

Crowe, R., Lincoln, J., Blacklay, P. F., Pryor, J. P., Lumnley, J. S. P., and Burnstock, G. (1983). Vasoactive intestinal polypeptide-like immunoreactive nerves in diabetic penis. A comparison between streptozotocin-treated rats and man. *Diabetes* **32**, 1075–1077.

Crowe, R., Milner, P., Lincoln, J., and Burnstock, G. (1987). Histochemical and biochemical investigation of adrenergic, cholinergic and peptidergic innervation of the rat ventral prostate 8 weeks after streptozotocin-induced diabetes. *J. Auton. Nerv. Syst.* **20**, 103–112.

DePonti, F., Fealey, R. D., and Malagelada, J.-R. (1987). Gastrointestinal syndromes due to diabetes mellitus. *In* "Diabetic Neuropathy" (P. J. Dyck, P. K. Thomas, A. K. Asbury, A. I. Winegrad, and D. Porte, eds.), pp. 155–161. Saunders, Philadelphia.

Dheen, S. T., Tay, S. S. W., and Wong, W. C. (1994). Ultrastructural changes in the hypothalamic supraoptic nucleus of the streptozotocin-induced diabetic rat. *J. Anat.* **184**, 615–623.

Diani, A. R., Grogan, D. M., Yates, M. E., Risinger, D. L., and Gerritsen, G. C. (1979). Radiologic abnormalities and autonomic neuropathology in the digestive tract of the ketonuric diabetic Chinese hamster. *Diabetologia* **17**, 33–40.

Diani, A. R., Peterson, T., and Gilchrist, B. J. (1983). Islet innervation of non-diabetic and diabetic Chinese hamsters. I. Acetylcholinesterase histochemistry and norepinephrine fluorescence. *J. Neural. Transm.* **56**, 223–238.

DiGiulio, A. M., Tenconi, B., La Croix, R., Mantegazza, P., Cattabeni, F., and Gorio, A. (1989). Denervation and hyperinnervation in the nervous system of diabetic animals. I. The autonomic neuronal dystrophy of the gut. *J. Neurosci. Res.* **24**, 355–361.

Dimpfel, W., Spuler, M., Pierau, F.-K., and Ulrich, H. (1990). Thioctic acid induces dose-dependent sprouting of neurites in cultured neuroblastoma cells. *Dev. Pharmacol. Ther.* **14**, 193–199.

Duchen, L. W., Anjorin, A., Watkins, P. J., and MacKay, J. D. (1980). Pathology of autonomic neuropathy in diabetes. *Ann. Intern. Med.* **92**, 301–303.

Dyck, P. J., Kratz, K. M., Karnes, J. L., Litchy, W. J., Klein, R., Pach, J. M., Wilson, D. M., O'Brien, P. C., and Melton, L. J. (1993). The prevalence by staged severity of various types of diabetic neuropathy, retinopathy, and nephropathy in a population-based cohort. *Neurology* **43**, 817–824.

Eaker, E. Y., Sallustio, J. E., Marchand, S. D., Sahu, A., Kalra, S. P., and Sninsky, C. A. (1996). Differential increase in neuropeptide Y-like levels and myenteric neuronal staining in diabetic rat intestine. *Reg. Peptides* **61**, 77–84.

Eika, B., Levin, R. M., and Longhurst, P. A. (1992). Collagen and bladder function in streptozocin-diabetic rats: Effects of insulin and aminoguanidine. *J. Urol.* **148**, 167–172.

Ekstrom, P. A. R., Kanje, M., and Skottner, A. (1989). Nerve regeneration and serum levels of insulin-like growth factor-1 in rats with streptozotocin-induced deficiency. *Brain Res.* **496**, 141–147.

Engber, T. M., Bhat, R. V., Dennis, S. A., Miller, M., and Contreras, P. C. (1995). Effects of IGF-1 infusion on serum IGF binding proteins and functional recovery following sciatic nerve crush. *Soc. Neurosci. Abst.* **21**, 1549.

Enomoto, H., Heuckeroth, R. O., Golden, J. P., Johnson, E. M., Jr., and Milbrandt, J. (2000). Development of cranial parasympathetic ganglia requires sequential actions of GDNF and neurturin. *Development* **127**, 4877–4889.

Ewing, D. J., Campbell, I. W., and Clarke, B. F. (1980). The natural history of diabetic autonomic neuropathy. *Quart. J. Med.* **49**, 95–108.

Faerman, I., Glocer, L., Celener, D., Jadzinsky, M., Fox, D., Maler, M., and Alvarez, E. (1973). Autonomic nervous system and diabetes: Histological and histochemical study of the autonomic nerve fibers of the urinary bladder in diabetic patients. *Diabetes* **22**, 225–237.

Faerman, I., Glocer, L., Fox, D., Jadzinsky, M. N., and Rapaport, M. (1974). Impotence and diabetes: Histological studies of the autonomic nervous fibers of the corpora cavernosa in impotent diabetic males. *Diabetes* **23**, 971–976.

Faerman, I., Faccio, E., Calb, I., Razumny, J., Franco, N., Dominguez, A., and Podesta, H. A. (1982). Autonomic neuropathy in the skin: A histological study of the sympathetic nerve fibres in diabetic anhidrosis. *Diabetologia* **22**, 96–99.

Fazan, R., Jr., Dias, D. S., Ballejo, G., and Salgado, H. C. (1999). Power spectra of arterial pressure and heart rate in streptozotocin-induced diabetes in rats. *J. Hypertens.* **17**, 489–495.

Federoff, H. J., Lawrence, D., and Brownlee, M. (1993). Nonenzymatic glycosylation of laminin and laminin peptide CIKVAVS inhibits neurite outgrowth. *Diabetes* **42**, 509–513.

Feldman, E. L., Sullivan, K. A., Kim, B., and Russell, J. W. (1997). Insulin-like growth factors regulate neuronal differentiation and survival. *Neurobiol. Dis.* **4**, 201–214.

Fernyhough, P., Diemel, L. T., Brewster, W. J., and Tomlinson, D. R. (1994). Deficits in sciatic nerve neuropeptide content coincide with a reduction in target tissue nerve growth factor mRNA in streptozotocin-diabetic rats: Effects of insulin treatment. *Neuroscience* **62**, 337–344.

Fernyhough, P., Diemel, L. T., and Tomlinson, D. R. (1998). Target tissue production and axonal transport of neurotrophin-3 are reduced in streptozotocin-diabetic rats. *Diabetologia* **41**, 300–306.

Gorin, P. D., and Johnson, E. M., Jr. (1980). Effects of long-term nerve growth factor deprivation on the nervous system of the adult rat: An experimental autoimmune approach. *Brain Res.* **198**, 27–42.

Gorio, A., Di Giulio, A. M., Donadoni, L., Tenconi, B., Germani, E., Bertelli, A., and Mantegazza, P. (1992). Early neurochemical changes in the autonomic neuropathy of the gut in experimental diabetes. *Int. J. Clin. Pharmacol. Res.* **12**, 217–224.

Greene, D. A., Lattimer, S., Ulbrecht, J., and Carroll, P. (1985). Glucose-induced alterations in nerve metabolism: Current perspective on the pathogenesis of diabetic neuropathy and future directions for research and therapy. *Diabet. Care* **8**, 290–299.

Greene, D. A., and Mackway, A. M. (1986). Decreased myo-inositol content and Na^+-K^+-ATPase activity in superior cervical ganglion of STZ-diabetic rat and prevention by aldose reductase inhibition. *Diabetes* **35**, 1106–1108.

Grover-Johnson, N. M., Baumann, F. G., Imparato, A. M., Kim, G. E., and Thomas, P. K. (1981). Abnormal innervation of lower limb epineurial arterioles in human diabetes. *Diabetologia* **20**, 31–38.

Gu, J., Polak, J. M., Lazarides, M., Morgan, R., Pryor, J. P., Marangos, P. J., Blank, M. A., and Bloom, S. R. (1984). Decrease of vasoactive intestinal polypeptide (VIP) in the penises from impotent men. *Lancet* **2**, 315–317.

Guo, Y.-P., McLeod, J. G., and Baverstock, J. (1987). Pathologic changes in the vagus nerve in diabetes and chronic alcoholism. *J. Neurol. Neurosurg. Psychiat.* **50**, 1449–1453.

Guy, R. J. C., Dawson, J. L., Barrett, J. R., Laws, J. W., Thomas, P. K., Sharma, A. K., and Watkins, P. J. (1984). Diabetic gastroparesis from autonomic neuropathy: Surgical considerations and changes in vagus nerve morphology. *J. Neurol. Neurosurg. Psychiat.* **47**, 686–691.

Guy, R. J. C., Richards, F., Edmonds, M. E., and Watkins, P. J. (1984). Diabetic autonomic neuropathy and iritis: An association suggesting an immunological cause. *Br. Med. J.* **289**, 343–345.

Hellweg, R., and Hartung, H.-D. (1990). Endogenous levels of nerve growth factor (NGF) are altered in experimental diabetes mellitus: A possible role for NGF in the pathogenesis of diabetic neuropathy. *J. Neurosci. Res.* **26**, 258–267.

Hellweg, R., Wohrle, M., Hartung, H.-D., Stracke, H., Hock, C., and Federlin, K. (1991). Diabetes mellitus-associated decrease in nerve growth factor levels is reversed by allogeneic pancreatic islet transplantation. *Neurosci. Lett.* **125**, 1–4.

Hensley, G. T., and Soergel, K. H. (1968). Neuropathologic findings in diabetic diarrhea. *Arch. Pathol.* **85**, 587–597.

Heuckeroth, R. O., Enomoto, H., Grider, J. R., Golden, J. P., Hanke, J. A., Jackman, A., Molliver, D. C., Bardgett, M. E., Snider, W. D., Johnson, E. M., Jr., and Milbrandt, J. (1999). Gene targeting reveals a critical role for neurturin in the development and maintenance of enteric sensory, and parasympathetic neurons. *Neuron* **22**, 253–263.

Hirai, A., Yasuda, H., Joko, M., Maeda, T., and Kikkawa, R. (2000). Evaluation of diabetic neuropathy through the quantitation of cutaneous nerves. *J. Neurol. Sci.* **172**, 55–62.

Hosking, D. J., Bennett, T., and Hampton, J. R. (1978). Diabetic autonomic neuropathy. *Diabetes* **27**, 1043–1055.

Husebye, E. S., Winqvist, O., Sundkvist, G., Kampe, O., and Karlsson, F. A. (1996). Autoantibodies against adrenal medulla in type 1 and type 2 diabetes mellitus: No evidence for an association with autonomic neuropathy. *J. Int. Med.* **239**, 139–146.

Ishii, D. N. (1995). Implication of insulin-like growth factors in the pathogenesis of diabetic neuropathy. *Brain Res. Rev.* **20**, 47–67.

Ishii, D. N., Glazner, G. W., and Pu, S. F. (1994). Role of insulin-like growth factors in peripheral nerve regeneration. *Pharmacol. Ther.* **62**, 125–144.

Ishii, D. N., Glazner, G. W., and Whalen, L. R. (1993). Regulation of peripheral nerve regeneration by insulin-like growth factors. *Ann. N.Y. Acad. Sci.* **692**, 172–182.

Jellinger, K. (1973). Neuroaxonal dystrophy: Its natural history and related disorders. *Prog. Neuropathol.* **2**, 129–180.

Kamal, A. A. J., Tay, S. S. W., and Wong, W. C. (1991). The cardiac ganglia in streptozotocin-induced diabetic rats. *Arch. Histol. Cytol.* **54**, 41–49.

Kamijo, M., and Preston, J. (1993). Impaired nerve fiber regeneration in the BB/W rat is restored to supranormal values following ARI treatment. *Diabetes* **42**(Suppl. 1), 194. [Abstract].

Kennedy, W. R., and Wendelschafer-Crabb, G. (1997). Innervation of the skin. *In* "Clinical Autonomic Disorders: Evaluation and Management" (P. A. Low, ed.), pp. 109–115. Lippincott-Raven, Philadelphia.

Kinoshita, J. H., and Nishimura, C. (1988). The involvement of aldose reductase in diabetic complications. *Diabet. Metab. Rev.* **4**, 323–337.

Kishi, Y., Nickander, K. K., Schmelzer, J. D., and Low, P. A. (2000). Gene expression of antioxidant enzymes in experimental diabetic neuropathy. *J. Periph. Nerv. Syst.* **5**, 11–18.

Kreiner, G., Wolzt, M., Fasching, P., Leitha, T., Edlmayer, A., Korn, A., Waldhausl, W., and Dudczak, R. (1995). Myocardial m[^{123}I] iodobenzylguanidine scintigraphy for the assessment of adrenergic cardiac innervation in patients with IDDM. *Diabetes* **44**, 543–549.

Kristensson, K., Nordborg, C., Olsson, Y., and Sourander, P. (1971). Changes in the vagus nerve in diabetes mellitus. *Acta Pathol. Microbiol. Scand.* **79A**, 684–685.

Laurikainen, A., Hiltunen, J. O., Thomas-Crusells, J., Vanhatalo, S., Arumae, U., Airaksinen, M. S., Klinge, E., and Saarma, M. (2000a). Neurturin is a neurotrophic factor for penile parasympathetic neurons in adult rat. *J. Neurobiol.* **43**, 198–205.

Laurikainen, A., Hiltunen, J. O., Vanhatalo, S., Klinge, E., and Saarma, M. (2000b). Glial cell line-derived neurotrophic factor is expressed in penis of adult rat and retrogradely transported in penile parasympathetic and sensory nerves. *Cell Tissue Res.* **302**, 321–329.

Lee, P.-G., Hohman, T. C., Cai, F., Regalia, J., and Helke, C. J. (2001). Streptozotocin-induced diabetes causes metabolic changes and alterations in neurotrophin content and retrograde transport in the cervical vagus nerve. *Exp. Neurol.* **170**, 149–161.

Lee, S.-M., Schade, S. Z., and Doughty, C. C. (1985). Aldose reductase, NADPH and NADP$^+$ in normal, galactose-fed and diabetic rat lens. *Biochim. Biophys. Acta* **841**, 247–253.

Levi-Montalcini, R., Aloe, L., Mugnaini, E., Oesch, F., and Thoenen, H. (1975). Nerve growth factor induces volume increase and enhances tyrosine hydroxylase synthesis in chemically axotomized sympathetic ganglia of newborn rats. *Proc. Natl. Acad. Sci. USA* **72**, 595–599.

Liuzzi, F. J. (1990). Proteolysis is a critical step in the physiological stop pathway: Mechanisms involved in the blockade of axonal regeneration by mammalian astrocytes. *Brain Res.* **512**, 277–283.

Loesch, A., Belai, A., Lincoln, J., and Burnstock, G. (1986). Enteric nerves in diabetic rats: Electron microscopic evidence for neuropathy of vasoactive intestinal polypeptide-containing fibres. *Acta Neuropathol.* **70**, 161–168.

Longo, F. M., Powell, H. C., Lebeau, J., Gerrero, M. R., Heckman, H., and Myers, R. R. (1986). Delayed nerve regeneration in streptozotocin diabetic rats. *Muscle Nerve* **9**, 385–393.

Low, P. A., Nickander, K. K., and Tritschler, H. J. (1997). The roles of oxidative stress and antioxidant treatment in experimental diabetic neuropathy. *Diabetes* **46**(Suppl. 2), S38–S42.

Low, P. A., Walsh, J. C., Huang, C. Y., and McLeod, J. G. (1975). The sympathetic nervous system in diabetic neuropathy: A clinical and pathological study. *Brain* **98**, 341–356.

Lucas, P. D., and Sardar, A. M. (1991). Effects of diabetes on cholinergic transmission in two rat gut preparations. *Gastroenterology* **100**, 123–128.

Luiten, P. G. M., TerHorst, G. J., Buijs, R. M., and Steffens, A. B. (1986). Autonomic innervation of the pancreas in diabetic and non-diabetic rats: A new view on intraneural sympathetic structural organization. *J. Auton. Nerv. Syst.* **15**, 33–44.

Lysy, J., Karmeli, F., and Goldin, E. (1993). Substance P levels in the rectal mucosa of diabetic patients with normal bowel function and constipation. *Scand. J. Gastroenterol.* **28**, 49–52.

Mackway, A. M., Greene, D. A., Booth, A. M., and DeGroat, W. C. (1986). Impaired muscarinic transmission in diabetic rat superior cervical ganglia. *Diabetes* **35**(Suppl. 1), 44A. [Abstract]

Matthews, R. R., and Nelson, V. H. (1975). Detachment of structurally intact nerve endings from chromatolytic neurons of rat superior cervical ganglia during the depression of synaptic transmission induced by postganglionic axotomy. *J. Physiol.* **245**, 91–135.

McEwen, T. A. J., and Sima, A. A. F. (1987). Autonomic neuropathy in BB rat: Assessment by improved method for measuring heart rate variability. *Diabetes* **36**, 251–255.

McVary, K. T., Rathnau, C. H., and McKenna, K. E. (1997). Sexual dysfunction in the diabetic BB/WOR rat: A role of central neuropathy. *Am. J. Physiol.* **272**, R259–R267.

Melman, A., and Henry, D. (1979). The possible role of the catecholamines of the corpora in penile erection. *J. Urol.* **121**, 419–421.

Migdalis, I. N., Kalogeropoulou, K., Kalantzis, L., Nounopoulos, C., Bouloukos, A., and Samartzis, M. (1995). Insulin-like growth factor-1 and IGF-1 receptors in diabetic patients with neuropathy. *Diabet. Med.* **12**, 823–827.

Moscoso, G. J., Driver, M., and Guy, R. J. (1996). A form of necrobiosis and atrophy of smooth muscle in diabetic autonomic neuropathy. *Pathol. Res. Pract.* **181**, 188–194.

Moss, H. E., Crowe, R., and Burnstock, G. (1987). The seminal vesicle in eight and 16 week streptozotocin-induced diabetic rats: Adrenergic, cholinergic and peptidergic innervation. *J. Urol.* **138**, 1273–1278.

Muhr, D., Mollenhauer, U., Ziegler, A.-G., Maslbeck, M., Standl, E., and Schnell, O. (1997). Autoantibodies to sympathetic ganglia, GAD, or tyrosine phosphatase in long term IDDM with and without ECG-based cardiac autonomic neuropathy. *Diabet. Care* **20**, 1009–1012.

Nagamatsu, M., Nickander, K. K., Schmelzer, J. D., Raya, A., Wittrock, D. A., Tritschler, H., and Low, P. A. (1995). Lipoic acid improves nerve blood flow, reduces oxidative stress, and improves distal nerve conduction in experimental diabetic neuropathy. *Diabet. Care* **18**, 1160–1167.

Nowak, T. V., Castelaz, C., Ramaswamy, K., and Weisbruch, J. P. (1995). Impaired rodent vagal nerve sodium-potassium-ATPase activity in streptozotocin diabetes. *J. Lab. Clin. Med.* **125**, 182–186.

Nowak, T. V., Harrington, B., Kalbfleisch, J. A., and Anatruda, J. M. (1986). Evidence for abnormal cholinergic transmission in diabetic rat small intestine. *Gastroenterology* **91**, 124–132.

Obrosova, I. G., Greene, D. A., and Lang, H.-J. (2000). Antioxidative defense in diabetic peripheral nerve: Effects of DL-α-lipoic acid, aldose reductase inhibitor and sorbitol dehydrogenase inhibitor. *In* "Oxidative Stress and Diabetic Complications" (L. Packer, P. Rosen, H. J. Tritschler, G. L. King, and A. Azzi, eds.), pp. 93–110. Dekker, Frankfurt.

Ohara, S., Beaudet, L. N., and Schmidt, R. E. (1995). Transganglionic response of GAP-43 in the gracile nucleus to sciatic nerve injury in young and aged rats. *Brain Res.* **705**, 325–331.

Ohtani, N., Ohta, M., and Sugano, T. (1997). Microdialysis study of modification of hypothalamic neurotransmitters in streptozotocin-diabetic rats. *J. Neurochem.* **69**, 1622–1628.

Oja, S. S., and Kontro, P. (1990). Neuromodulatory and trophic actions of taurine. *Prog. Clin. Biol. Res.* **351**, 69–76.

Olsson, Y., and Sourander, P. (1968). Changes in the sympathetic nervous system in diabetes mellitus: A preliminary report. *J. Neurovisc. Relat.* **31**, 86–95.

Ordog, T., Takayama, I., Cheung, W. K. T., Ward, S. M., and Sanders, K. M. (2000). Remodeling of networks of interstitial cells of Cajal in a murine model of gastroparesis. *Diabetes* **49**, 1731–1739.

Park, S. H., Park, K. A., Kang, H. S., Baik, E. J., and Cho, W. J. (1997). The effects of NGF (nerve growth factor) on enteric nervous system in streptozotocin-diabetic rat. *Soc. Neurosci. Abstr.* **23**, 1429.

Paro, M., Prosdocimi, M., Fiori, M. G., and Sima, A. A. F. (1991). Autonomic innervation of the bladder in two models of experimental diabetes in the rat: Functional abnormalities, structural alterations and effects of ganglioside administration. *Urodinamica* **1**, 161–163.

Patel, M. J., and Rice, F. L. (1999). Selective impact of diabetes on subsets of unmyelinated cutaneous and vascular innervation in streptozotocin treated rats. *Soc. Neurosci. Abstr.* **25**, 1008.

Pittenger, G. L., Liu, D., and Vinik, A. I. (1997). The apoptotic death of neuroblastoma cells caused by serum from patients with insulin-dependent diabetes and neuropathy may be Fas-mediated. *J. Neuroimmunol.* **76**, 153–160.

Purvis, D. (1975). Functional and structural changes of mammalian sympathetic neurons following interruption of their axons. *J. Physiol.* **252**, 429–463.

Quigley, E. M. (1997). The pathophysiology of diabetic gastroenteropathy: More vague than vagal. *Gastroenterology* **113**, 1790–1794.

Rabinowe, S. L., Brown, F. M., Watts, M., and Smith, A. M. (1990). Complement-fixing antibodies to sympathetic and parasympathetic tissues in IDDM. *Diabet. Care* **13**, 1084–1088.

Reagan, L. P., Magarinos, A. M., and McEwen, B. S. (1999). Neurological changes induced by stress in streptozotocin diabetic rats. *Ann. N.Y. Acad. Sci.* **893**, 126–137.

Recio-Pinto, E., Rechler, M. M., and Ishii, D. N. (1986). Effects of insulin, insulin-like growth factor-II, and nerve growth factor in neurite formation and survival in cultured sympathetic and sensory neurons. *J. Neurosci.* **6**, 1211–1219.

Rösen, P., Ballhausen, T., Bloch, W., and Addicks, K. (1995). Endothelial relaxation is disturbed by oxidative stress in the diabetic rat heart: Influence of tocopherol as antioxidant. *Diabetologia* **38**, 1157–1168.

Robertson, D. M., and Sima, A. A. F. (1980). Diabetic neuropathy in the mutant mouse [C57/BL/ks (db/db)]: A morphometric study. *Diabetes* **25**, 60–67.

Rossi, J., Luukko, K., Poteryaev, D., Laurikainen, A., Sun, Y. F., Laakso, T., Eerikainen, S., Tuominen, R., Lakso, M., Rauvala, H., Arumae, U., Pasternack, M., Saarma, M., and Airaksinen, M. S. (1999). Retarded growth and deficits in the enteric and parasympathetic nervous system in mice lacking GFRa2, a functional neurturin receptor. *Neuron* **22**, 243–252.

Russell, J. W., and Feldman, E. L. (1999). Insulin-like growth factor-I prevents apoptosis in sympathetic neurons exposed to high glucose. *Horm. Metab. Res.* **31**, 90–96.

Russell, J. W., Sullivan, K. A., Windebank, A. J., Herrmann, D. N., and Feldman, E. L. (1999). Neurons undergo apoptosis in animal and cell culture models of diabetes. *Neurobiol. Dis.* **6**, 347–363.

Sahenk, Z., and Brown, A. (1991). Weak base amines inhibit the anterograde-retrograde conversion of axonally transported vesicles in nerve terminals. *J. Neurocytol.* **20**, 365–375.

Said, G., Goulon-Goeau, C., and Tchbroutsky, G. (1992). Severe early-onset polyneuropathy in insulin dependent diabetes mellitus: A clinical and pathological study. *N. Engl. J. Med.* **326**, 1257–1263.

Saito, H., Sango, K., Horie, H., Ikeda, H., Ishigatsubo, Y., Ishikawa, Y., and Inoue, S. (1999). Enhanced neural regeneration from transected vagus nerve terminal in diabetic mice *in vitro. Neuroreport* **10**, 1025–1028.

Sampson, M. J., Wilson, S., Karagiannis, P., Edmonds, M., and Watkins, P. J. (1990). Progression of diabetic autonomic neuropathy over a decade in insulin-dependent diabetics. *Quart. J. Med.* **278**, 635–646.

Sasaki, H., Schmelzer, J. D., Zollman, P. J., and Low, P. A. (1997) Neuropathology and blood flow of nerve, spinal roots and dorsal root ganglia in longstanding diabetic rats. *Acta Neuropathol.* **93**, 118–128.

Schmid, H., Forman, L. A., Cao, X., Sherman, P. S., and Stevens, M. J. (1999). Heterogenous cardiac sympathetic denervation and decreased myocardial nerve growth factor in streptozotocin-induced diabetic rats: Implications for cardiac sympathetic dysinnervation complicating diabetes. *Diabetes* **48**, 603–608.

Schmidt, H., Riemann, J. F., Schmid, A., and Sailer, D. (1984). Ultrastruktur der diabetischen autonomen neuropathie des gastrointestinal traktes. *Klin. Wochenschr.* **62**, 399–405.

Schmidt, R. E. (1996a). Synaptic dysplasia in sympathetic autonomic ganglia. *J. Neurocytol.* **25**, 777–791.

Schmidt, R. E. (1996b). The neuropathology of human sympathetic autonomic ganglia. *Microscop. Res. Tech.* **35**, 107–121.

Schmidt, R. E. (2001). Neuronal preservation in the sympathetic ganglia of rats with chronic streptozotocin-induced diabetes. *Brain Res.* **921**, 256–259.

Schmidt, R. E., and Cogswell, B. E. (1989). Tyrosine hydroxylase activity in sympathetic nervous system of rats with streptozocin-induced diabetes. *Diabetes* **38**, 959–968.

Schmidt, R. E., Dorsey, D. A., Beaudet, L. N., Parvin, C. A., and Escandon, E. (2001a). Effect of NGF and neurotrophin-3 treatment on experimental diabetic autonomic neuropathy. *J. Neuropathol. Exp. Neurol.* **60**, 263–273.

Schmidt, R. E., Dorsey, D. A., Beaudet, L. N., Plurad, S. B., Parvin, C. A., and Miller, M. A. (1999). Insulin-like growth factor I reverses experimental diabetic autonomic neuropathy. *Am. J. Pathol.* **155**, 1651–1660.

Schmidt, R. E., Dorsey, D. A., Beaudet, L. N., Plurad, S. B., Parvin, C. A., Yarasheski, K. E., Smith, S. R., Lang, H.-J., Williamson, J. R., and Ido, Y. (2001b). Inhibition of sorbitol dehydrogenase exacerbates autonomic neuropathy in rats with streptozotocin-induced diabetes. *J. Neuropathol. Exp. Neurol.* **60**, 1153–1169.

Schmidt, R. E., Dorsey, D. A., Beaudet, L. N., Plurad, S. B., Williamson, J. R., and Ido, Y. (1998). Effect of sorbitol dehydrogenase inhibition on experimental diabetic autonomic neuropathy. *J. Neuropathol. Exp. Neurol.* **57**, 1175–1189.

Schmidt, R. E., Dorsey, D. A., Beaudet, L. N., Reiser, K. M., Williamson, J. R., and Tilton, R. G. (1996). Effect of aminoguanidine on the frequency of neuroaxonal dystrophy in the superior mesenteric sympathetic autonomic ganglia of rats with streptozocin-induced diabetes. *Diabetes* **45**, 284–290.

Schmidt, R. E., Dorsey, D. A., Roth, K. A., Parvin, C. A., Hounsom, L., and Tomlinson, D. R. (2000). Effects of streptozotocin-induced diabetes on NGF, p75^{NTR} and TrkA content of prevertebral and paravertebral rat sympathetic ganglia. *Brain Res.* **867**, 149–156.

Schmidt, R. E., Modert, C. W., and Grabau, G. G. (1987). Orthograde and retrograde axonal transport of dopamine-β-hydroxylase in ileal mesenteric nerves of rats with chronic streptozotocin diabetes. *Brain Res.* **401**, 142–146.

Schmidt, R. E., Modert, C. W., Yip, H. K., and Johnson, E. M., Jr. (1983a). Retrograde axonal transport of intravenously administered ^{125}I-nerve growth factor in rats with streptozotocin-induced diabetes. *Diabetes* **32**, 654–663.

Schmidt, R. E., Johnson, E. M. Jr., and Nelson, J. S. (1981). Experimental diabetic autonomic neuropathy. *Am. J. Pathol.* **103**, 210–225.

Schmidt, R. E., and Plurad, S. B. (1985). Ultrastructural appearance of intentionally frustrated axonal regeneration in rat sciatic nerve. *J. Neuropathol. Exp. Neurol.* **44**, 130–146.

Schmidt, R. E., and Plurad, S. B. (1986). Ultrastructural and biochemical characterization of autonomic neuropathy in rats with chronic streptozotocin diabetes. *J. Neuropathol. Exp. Neurol.* **45**, 525–544.

Schmidt, R. E., Plurad, S. B., Coleman, B. D., Williamson, J. R., and Tilton, R. G. (1991). Effects of sorbinil, dietary myo-inositol supplementation and insulin on resolution of neuroaxonal dystrophy in mesenteric nerves of streptozocin-induced diabetic rats. *Diabetes* **40**, 574–582.

Schmidt, R. E., Plurad, S. B., and Modert, C. W. (1983b). Experimental diabetic autonomic neuropathy: Characterization in streptozotocin-diabetic Sprague–Dawley rats. *Lab. Invest.* **49**, 538–552.

Schmidt, R. E., Plurad, S. B., Olack, B. J., and Scharp, D. W. (1989a). The effect of pancreatic islet transplantation and insulin therapy on neuroaxonal dystrophy in sympathetic autonomic ganglia of chronic streptozocin-diabetic rats. *Brain Res.* **497**, 393–398.

Schmidt, R. E., Plurad, S. B., Parvin, C. A., and Roth, K. A. (1993). The effect of diabetes and aging on human sympathetic autonomic ganglia. *Am. J. Pathol.* **143**, 143–153.

Schmidt, R. E., Plurad, D. A., Plurad, S. B., Cogswell, B. E., Diani, A. R., and Roth, K. A. (1989b). Ultrastructural and immunohistochemical characterization of autonomic neuropathy in genetically diabetic Chinese hamsters. *Lab. Invest.* **61**, 77–92.

Schmidt, R. E., Plurad, S. B., Saffitz, J. E., Grabau, G. G., and Yip, H. K. (1985). Retrograde axonal transport of ^{125}I-nerve growth factor in rat ileal mesenteric nerves: Effect of streptozotocin diabetes. *Diabetes* **334**, 1230–1240.

Schmidt, R. E., Plurad, S. B., Sherman, W. R., Williamson, J. R., and Tilton, R. G. (1989c). Effects of aldose reductase inhibitor sorbinil on neuroaxonal dystrophy and levels of myo-inositol and sorbitol in sympathetic autonomic ganglia of streptozocin-induced diabetic rats. *Diabetes* **38**, 569–579.

Schmidt, R. E., and Scharp, D. W. (1982). Axonal dystrophy in experimental diabetic autonomic neuropathy. *Diabetes* **31**, 761–770.

Schnell, O., Muhr, D., Dresel, Tatsch, K., Ziegler, A. G., Haslbeck, M., and Standl, E. (1996). Autoantibodies against sympathetic ganglia and evidence of cardiac sympathetic dysinnervation in newly diagnosed and long-term IDDM patients. *Diabetologia* **39**, 970–975.

Schroer, J. A., Beaudet, L. N., and Schmidt, R. E. (1995). Effect of chronic autoimmune nerve growth factor deprivation on sympathetic neuroaxonal dystrophy in rats. *Synapse* **20**, 249–256.

Scott, L. D., and Ellis, T. M. (1980). Small intestinal transit and myoelectric activity in diabetic rats. *In* "Gastrointestinal Motility" (J. Christensen, ed.), pp. 395–399. Raven Press, New York.

Sima, A. A. F., Merry, A. C., and Levitan, I. (1997). Increased regeneration in ARI-treated diabetic nerve is associated with upregulation of IGF-I and NGF receptors. *Exp. Clin. Endocrinol. Diabet.* **105**, 60–62.

Sims, M. A., Hasler, W. L., Chey, W. D., Kim, M. S., and Owyang, C. (1995). Hyperglycemia inhibits mechanoreceptor-mediated gastrocolonic responses and colonic peristaltic reflexes in healthy humans. *Gastroenterology* **108**, 350–359.

Singhal, A., Cheng, C., Sun, H., and Zochodne, D. W. (1997). Near nerve local insulin prevents conduction slowing in experimental diabetes. *Brain Res.* **763**, 209–214.

Spangeus, A., Suhr, O., and El-Salhy, M. (2000). Diabetic state affects the innervation of gut in an animal model of human type 1 diabetes. *Histol. Histopathol.* **15**, 739–744.

Srinivasan, S., Stevens, M. J., Sheng, H., Hall, K. E., and Wiley, J. W. (1998). Serum from patients with type 2 diabetes with neuropathy induces complement-independent, calcium dependent apoptosis in culture neuronal cells. *J. Clin. Invest.* **102**, 1454–1462.

Steers, W. D., Mackway-Gerardi, A. M., Ciambotti, J., and de Groat, W. C. (1994). Alterations in neural pathways to the urinary bladder of the rat in response to streptozotocin-induced diabetes. *J. Auton. Nerv. Syst.* **47**, 83–94.

Stevens, M. J., Raffel, D. M., Allman, K. C., Schwaiger, M., and Wieland, D. M. (1999). Regression and progression of cardiac sympathetic dysinnervation complicating diabetes: An assessment by C-11 hydroxyephedrine and positron emission tomography. *Metabolism* **48**, 92–101.

Stroud, C. R., Heller, S. R., Ward, J. D., Hardisty, C. A., and Weetman, A. P. (1997). Analysis of antibodies against components of the autonomic nervous system in diabetes mellitus. *Quart. J. Med.* **90**, 577–585.

Sugimoto, K., Murakawa, Y., and Sima, A. A. F. (2000). Diabetic neuropathy: A continuing enigma. *Diabet. Metab. Res. Rev.* **16**, 408–433.

Sugimoto, K, Nishizawa, Y., Horiuchi, S., and Yagihashi, S. (1997). Localization in human diabetic peripheral nerve of N-carboxymethyllysine-protein adducts, one of advanced glycation end products. *Diabetologia* **40**, 1380–1387.

Sunagawa, R., Nagamine, F., F., Murakami, K., Mimura, G., and Sakanashi, M. (1989). Effects of dobutamine and acetylcholine on perfused hearts isolated from streptozocin-induced diabetic rats. *Arzn. Forsch.* **39**, 470–474.

Sundkvist, G., Velloso, L. A., Kampe, O., Rabinowe, S. L., Ivarsson, S. A., Lilja, B., and Karlsson, F. A. (1994). Glutamic acid decarboxylase antibodies, autonomic nerve antibodies and autonomic neuropathy in diabetic patients. *Diabetologia* **37**, 293–299.

Taborsky, G. J., Jr., Ahren, B., and Havel, P. J. (1998). Autonomic mediation of glucagon secretion during hypoglycemia: Implications for impaired α-cell responses in type I diabetes. *Diabetes* **47**, 995–1005.

Takahashi, T., Nakamura, K., Itoh, H., Sima, A. A. F., and Owyang, C. (1997). Impaired expression of nitric oxide synthase in the gastric myenteric plexus of spontaneously diabetic rats. *Gastroenterology* **113**, 1535–1544.

Taniuchi, M., Clark, H. B., and Johnson, E. M., Jr. (1986). Induction of nerve growth factor receptor in Schwann cells after axotomy. *Proc. Natl. Acad. Sci. USA* **83**, 4094–4098.

Tessari, F., Travagli, R. A., Zanoni, R., and Prosdocimi, M. (1988). Effects of long-term diabetes and treatment with gangliosides on cardiac sympathetic innervation: A biochemical and functional study in mice. *J. Diab. Complicat.* **2**, 34–37.

Tomlinson, D. R. (1983). Axonal transport of noradrenaline, protein, and glycoprotein in cat hypogastric nerves *in vitro* under conditions of high extracellular glucose. *Diabetologia* **24**, 172–178.

Turpeinen, A. K., Vanninen, E., Kuikka, J. T., and Uusitupa, M. I. J. (1996). Demonstration of regional sympathetic denervation of the heart in diabetes: Comparison between patients with NIDDM and IDDM. *Diabet. Care* **19**, 1083–1090.

Ustinova, E. E., Barrett, C. J., Sun, S.-Y., and Schultz, H. D. (2000). Oxidative stress impairs cardiac chemoreflexes in diabetic rats. *Am. J. Physiol.* **279**, H2176–H2187.

Vandertop, W. P., de Vries, W. B., Notermans, N. C., Tulleken, C. A., and Gispen, W. H. (1995). Neural influences on the iris of diabetic rats and effect of oculomotor nerve crush. *J. Auton. Nerv. Syst.* **55**, 112–114.

Vinik, A. I., Holland, M. T., LeBeau, J. M., Liuzzi, F. J., Stansberry, K. B., and Colen, L. B. (1992). Diabetic neuropathies. *Diabet. Care* **15**, 1926–1975.

Vinik, A. I., Newlon, P. G., Lautero, T. J., Liuzzi, F. J., Depto, A. S., Pittenger, G. L., and Richardson, D. W. (1995). Nerve survival and regeneration in diabetes. *Diabet. Rev.* **3**, 139–157.

Vo, P. A., and Tomlinson, D. R. (1999). The regeneration of peripheral noradrenergic nerves after chemical sympathectomy in diabetic rats: Effect of nerve growth factor. *Exp. Neurol.* **157**, 127–134.

Watkins, C. C., Sawa, A., Jaffrey, S., Blackshaw, S., Barrow, R. K., Snyder, S. H., and Ferris, C. D. (2000). Insulin restores neuronal nitric oxide synthase expression and function that is lost in diabetic gastropathy. *J. Clin. Invest.* **106**, 373–384.

Way, K. J., and Reid, J. J. (1994). Effect of aminoguanidine on the impaired nitric oxide-mediated neurotransmission in anococcygeus muscle from diabetic rats. *Neuropharmacology* **35**, 1315–1322.

Way, K. J., Young, H.M, and Reid, J. J. (1999). Diabetes does not alter the activity and localization of nitric oxide synthase in the rat anococcygeus muscle. *J. Auton. Nerv. Syst.* **76**, 35–44.

Wrzos, H. F., Cruz, A., Polavarapu, R., Shearer, D., and Ouyang, A. (1997). Nitric oxide synthase (NOS) expression in the myenteric plexus of streptozotocin-diabetic rats. *Dig. Dis. Sci.* **42**, 2106–2110.

Yagihashi, S., and Sima, A. A. F. (1985a). The distribution of structural changes in sympathetic nerves of the BB rat. *Am. J. Pathol.* **121**, 138–147.

Yagihashi, S., and Sima, A. A. F. (1985b). Diabetic autonomic neuropathy in the BB-rat: Ultrastructural and morphometric changes in sympathetic nerves. *Diabetes* **34**, 558–564.

Yagihashi, S., and Sima, A. A. F. (1986a). Neuroaxonal and dendritic dystrophy in diabetic autonomic neuropathy. *J. Neuropathol. Exp. Neurol.* **45**, 545–565.

Yagihashi, S., and Sima, A. A. F. (1986b). Diabetic autonomic neuropathy in BB rat. Ultrastructural and morphometric changes in parasympathetic nerves. *Diabetes* **35**, 733–743.

Yoshida, M. M., Schuffler, M. D., and Sumi, S. M. (1988). There are no morphological abnormalities of the gastric wall or abdominal vagus in patients with diabetic gastroparesis. *Gastroenterology* **94**, 907–914.

Zanone, M. M., Banga, J. P., Peakman, M., Edmonds, M., and Watkins, P. J. (1994). An investigation of antibodies to nerve growth factor in diabetic autonomic neuropathy. *Diabet. Med.* **11**, 378–383.

Zhang, W.-X., Chakrabarti, S., Greene, D. A., and Sima, A. A. F. (1990). Diabetic autonomic neuropathy in BB-rats: The effect of ARI treatment on heart-rate variability and vagus nerve structure. *Diabetes* **36**, 613–618.

Zhou, X.-F., Chie, E. T., Deng, Y.-S., and Rush, R. A. (1997). Rat mature sympathetic neurones derive neurotrophin-3 from peripheral effector tissues. *Eur. J. Neurosci.* **9**, 2753–2764.

Zhuang, H.-X., Snyder, C. K., Pu, S.-F., and Ishii, D. N. (1996). Insulin-like growth factors reverse or arrest diabetic neuropathy: Effects on hyperalgesia and impaired nerve regeneration in rats. *Exp. Neurol.* **140**, 198–205.

Zhuang, H.-X., Wuarin, L., Fei, Z.-J., and Ishii, D. N. (1997). Insulin-like growth factor (IGF) gene expression is reduced in neural tissues and liver from rats with non-insulin-dependent diabetes mellitus, and IGF treatment ameliorates diabetic neuropathy. *J. Pharmacol. Exp. Ther.* **283**, 366–374.

Zochodne, D. W. (1999). Diabetic neuropathies: Features and mechanisms. *Brain Pathol.* **9**, 369–391.

ROLE OF THE SCHWANN CELL IN DIABETIC NEUROPATHY

Luke Eckersley

Neuroscience Division, University of Manchester
School of Biological Sciences Manchester M13 9PT
United Kingdom

The relationships among Schwann cells, axons, and the perineurial barrier emphasize the key role Schwann cells play in normal function of the nerve. Schwann cells are responsible for action potential velocity through insulation of axons, maintenance of axonal caliber, and correct localization of Na^+ channels; immunological and functional integrity of the nerve through the perineurial blood–nerve–barrier; and effective nerve regeneration. In diabetic neuropathy, many of these facets of nerve function are defective. Hypoxia, hyperglycemia, and increased oxidative stress contribute directly and indirectly to Schwann cell dysfunction. The results include impaired paranodal barrier function, damaged myelin, reduced antioxidative capacity, and decreased neurotrophic support for axons. This chapter discusses the role of the Schwann cell in the normal or regenerating nerve and in the altered metabolic conditions of diabetes. © 2002, Elsevier Science (USA).

I. Peripheral Nervous System Development

During embryogenesis, the nervous system is the first organ to be generated, soon after denotation of the body axis. Motor neurons differentiate from a distinct region of the ventral neural tube, the floor plate, and extend processes into the surrounding tissue, contacting their target muscles. Highly migratory pluripotent cells of the trunk neural ectoderm (neural crest) form the sensory system and peripheral glia. The sensory neurons migrate laterally into the dorsal roots and colonize the ganglia. They subsequently send processes both into peripheral tissue to synapse with sensory receptors and into the central nervous system to synapse with commissar neurons of the spinal reflex circuit and ascending afferents to the brain.

In the rat, at embryonic day 9 to 10 (E9 to E10), prospective peripheral glial cells migrate laterally and ventrally into the dorsal and ventral roots and contact the axons of sensory and motor neurons, continuing to proliferate as they migrate along the axon toward the target tissue (Frank and Sanes, 1991). A major developmental switch occurs 2–3 days before birth (E16–E18), resulting in differentiation from Schwann cell precursors to early Schwann cells, dependent on signals from the apposing axon. The early Schwann cells invade the axon bundles, forming one-to-one relationships with axons to be heavily myelinated and relationships with clusters of other axons. High levels of the low-affinity neurotrophin receptor (p75^{NTR}), brain-derived neurotrophic factor (BDNF), and glial cell line-derived neurotrophic factor (GDNF) are expressed during Schwann cell maturation (Schecterson and Bothwell, 1992).

Neuregulins form the major survival factors for Schwann cell precursors. (Dong *et al.*, 1995). Neuregulins are a family of neuronally synthesized. (Marchionni *et al.*, 1993) factors, including neu differentiation factor, heregulin, acetylcholine-inducing activity (ARIA), and glial growth factors 1 and 2, derived from a single gene by alternative splicing. *In vitro*, postnatal day 3 (P3) rat Schwann cells undergo apoptosis if serum is removed, and neuregulin-β can prevent this. (Syroid *et al.*, 1996). Glial growth factor-2 (GGF-2) has also been demonstrated to be a Schwann cell survival factor following axotomy at the neonatal neuromuscular junction (Trachtenberg and Thompson, 1996). ErbB2 and erbB3, two transmembrane proteins forming high-affinity heterodimeric receptors for neuregulins, are expressed in Schwann cells *in vivo* and ErbB3 -/- mice lack Schwann cell precursors. An extra 79 and 82% of motor and sensory neurons, respectively, undergo embryonic cell death in these mice. (Riethmacher *et al.*, 1997), suggesting reciprocal survival signals between axons and Schwann cell precursors or early Schwann cells. Insulin-like growth factor-1 (IGF-1) along with GDNF and its family members neurturin and persephin are some candidates for Schwann cell-derived neuronal survival signals (Arce *et al.*, 1998; Buj-Bello *et al.*, 1995).

Wrapping of the axon with a process coincides with the downregulation of p75^{NTR} and upregulation of myelin protein expression in myelinating Schwann cells (Jessen *et al.*, 1990). Three transcription factors important in the differentiation of Schwann cells (Fig. 1) to the quiescent adult

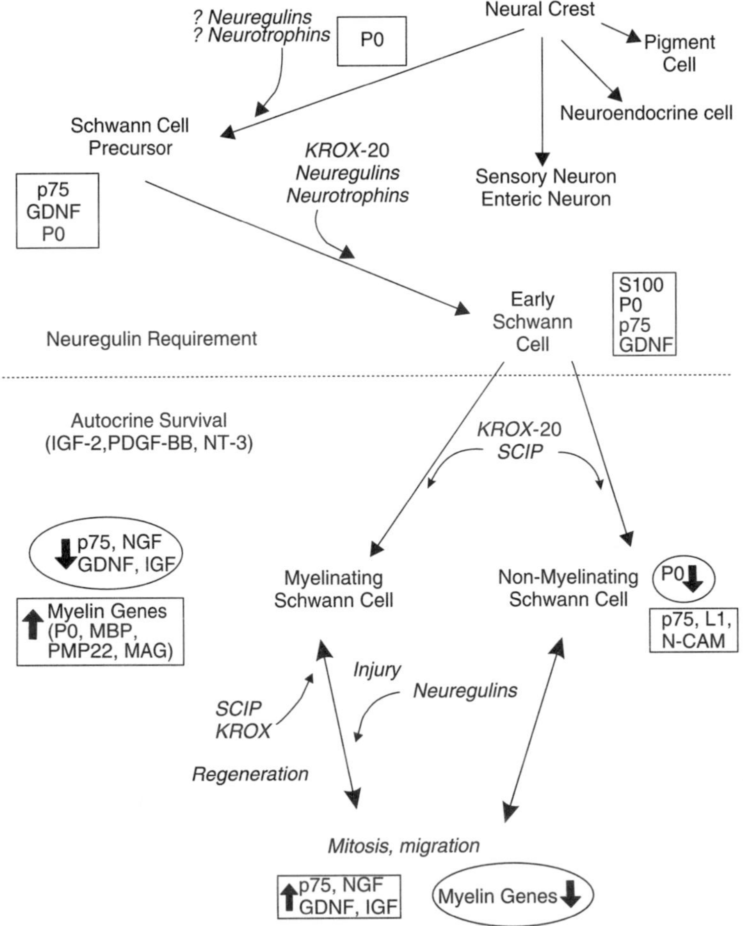

FIG. 1. Schwann cell development. Glial-committed P0-positive neural crest cells develop into Schwann cell precursors and early Schwann cells in response to unknown growth factors, including a survival requirement for neuregulin. They subsequently develop into mature myelinating or nonmyelinating Schwann cells. In response to injury, a phenotypic reversion to an early Schwann cell-like phenotype occurs, which is reversed upon reestablishment of axonal contact.

state are SCIP, Krox-20, and Krox-24. SCIP (Oct-6) is transiently expressed during maturation and is likely to be important in general differentiation; Krox-20 is expressed 24–36 h later in both development and regeneration and is maintained in the mature myelinating Schwann cell; and Krox-24 shows a reciprocal relationship to Krox-20, being downregulated in myelinating Schwann cells, and high in development, following nerve injury and in mature nonmyelinating Schwann cells. Krox-20 is postulated to mediate myelination (Zorick *et al.*, 1996), although it is probably not directly responsible for it (Mirsky and Jessen, 1999). The axonal molecular signals that control the myelinating and nonmyelinating phenotypes are still unidentified.

Schwann cell signaling is also important for formation of the sheaths surrounding and protecting the axon–Schwann cell units. Desert hedgehog (Dhh) is expressed in Schwann cell precursors and in Schwann cells until at least postnatal day 10. The receptor for hedgehog proteins, Patched (Ptch), is seen in mesenchymal cells surrounding the developing nerve at E15. Dhh -/- mice show almost no epineurium and a thin and diffuse perineurium. Perineurial-like cells invade the endoneurial space, forming minifascicles around small bundles of fibers. The diffusion and cellular infiltration barrier is compromised (Mirsky *et al.*, 1999). Dhh -/- mice have a decreased nerve conduction velocity, epineurial collagen is reduced, and the perineurium has patchy basal lamina, lacks connexin-43, and has abnormal tight junctions. Dhh upregulates mesenchymal Ptch. An exon 1 missense mutation in human Dhh leads to partial gonadal dysgenesis and polyneuropathy consisting of minifascicles and a decreased density of myelinated fibers (Umehara *et al.*, 2000).

II. Function of Schwann Cells

A. Role in Quiescent Nerve

In the intact nerve, the most obvious function of the Schwann cell is to act as an insulating barrier to ion movement across the internode region, thus allowing fast conduction to occur via depolarizing currents conducted longitudinally down the axon, crossing the membrane only at the nodes of Ranvier. Nerve conduction can therefore be described in terms of amplitude, which is dependent on the number and the size of fibers activated, and velocity. The velocity usually measured is that of the largest, fastest fibers. It depends on several factors: (1) the caliber of the largest fibers — greater fiber caliber allows greater current and greater current allows a greater distance between nodes; (2) the insulation barrier, which normal (Aguayo *et al.*, 1977). Figure 2 summarizes some of the known roles of Schwann cells in the nerve.

B. Response to Injury

Distal to a site of injury, axons and myelin degenerate and the resulting debris is phagocytosed by Schwann cells (Reichert *et al.*, 1994) and invading macrophages in a process termed Wallerian degeneration. Schwann cells "dedifferentiate" when they lose axonal contact; myelinating and nonmyelinating cells form a single population similar to that during late embryogenesis, where proliferation, migration, and neurotrophin production can occur (Grinspan *et al.*, 1996).

Distal Schwann cells remain within the endoneurial basal lamina that surround the axon–Schwann cell unit, proliferating and forming distinctive cordons called bands of Bungner. Endoneurial tubes can provide a low resistance pathway, allowing axons to extend branching neurites (10–20 mm). However, Schwann cells are necessary and sufficient for further extension (Gulati, 1996; Ansselin *et al.*, 1997). In Schwann cells distal to a rat sciatic nerve lesion (Fig. 1), p75^{NTR} (Taniuchi *et al.*, 1986), BDNF, neurotrophin-4 (NT-4), nerve growth factor (NGF), GDNF, GGF2, leukemia inhibitory factor (LIF), interleukin-6 (IL-6), and IGF expression increase markedly (Grinspan *et al.*, 1996; Bolin *et al.*, 1995; Curtis *et al.*,

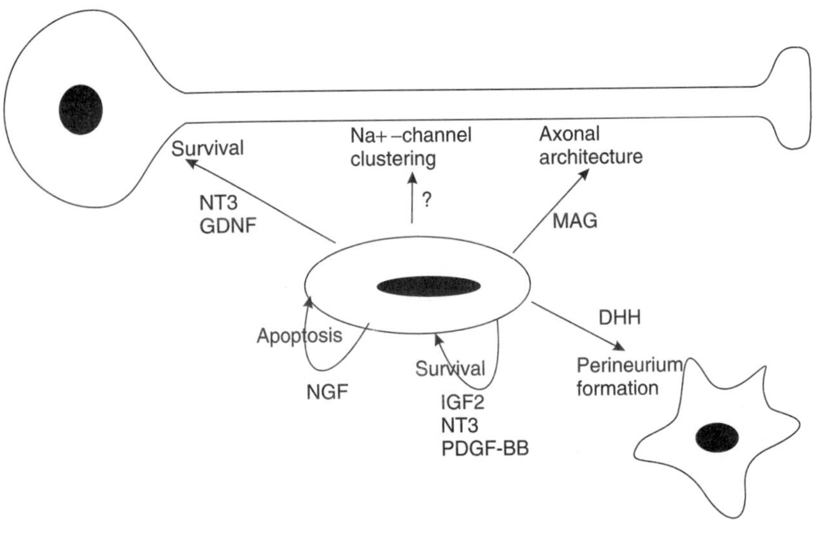

FIG. 2. Summary of Schwann cell roles in the intact nerve and the factors involved.

depends on the myelin thickness and its effectiveness in preventing
and (3) the currents generated at the nodes, which depend in turn
ionic gradients, generated by pumps, and the number and viabilit
channels.

Ensheathment of axons, either by myelin or the membrane of no
nating Schwann cells, is crucial to the function of the nerve. Of the
proteins, P0 is important to compaction of the extracellular and cyto
faces of the myelin membrane. P0+/− mice show late morpho
changes. By 4 months of age, myelin sheaths are too thin, and p
ating and supernumerary Schwann cells are present forming onion
indicative of myelin degeneration. Myelin-associated glycoprotein (
stabilizes the interaction between the axon and the inner Schwar
membrane. Knockout mice show very similar morphological chan
P0+/−. Knockout of a third myelin protein, of unknown function, p
eral myelin protein 22 (PMP22), also results in similar nerve phen
changes. The phenotypes in heterozygous mice suggest a concent
dependence on these myelin proteins. Interestingly, they appear to
quite similar changes to those seen in diabetic nerve (see later). Con
32, a gap junction protein involved in connecting the cytoplasm of
nodal loops, and therefore necessary for nuclear communication wit
axonal cytoplasm, is also important in myelin stability through unkr
mechanisms (Neuberg *et al.*, 1999).

Schwann cells show extensive and poorly understood interactions
the extracellular matrix and basal lamina through a range of integrins
other receptors. These are probably responsible for signaling from C
axons to nonmyelinating Schwann cells, migration during developr
or following injury (Milner *et al.*, 1997), and correct maintenance of
node of Ranvier, for example. The up or downregulation of severa
these adhesion molecules has been demonstrated to signal through pro
kinase C (PKC) or PKA-mediated pathways and is responsive to a rang
growth factors, e.g., transforming growth factor β (TGFβ) (Einheber *et*
1995).

Schwann cells also appear to control the caliber of the axon, match
it to the thickness of the myelin sheath through a mechanism that invol
MAG. This leads to an increased phosphorylation of neurofilament arr
The negatively charged phosphate groups are thought to repulse ea
other, increasing the spacing of the neurofilament chains, and the calib
of the axon (De Waegh *et al.*, 1992). This role is supported by ea
experiments where Schwann cells from Trembler mice, which have
genetic defect in PMP22 production, were grafted into normal anima
Trembler grafts showed reduced caliber and neurofilament spacing
regenerated axons, whereas proximal and distal to the graft, caliber w

1994; Funakoshi *et al.*, 1993; Matsuoka *et al.*, 1991; Meyer *et al.*, 1992; Pu *et al.*, 1995). Adhesion molecules, including L1, N-CAM, and N-cadherin, also increase. Myelin genes are reexpressed in Schwann cells apposed to a regenerated axon.

A model for a Schwann cell neurotrophic guidance system was put forward by Johnson and colleagues (1987). They proposed that the expression of p75^{NTR} on Schwann cells and NGF secretion led to high local concentrations of NGF. High-affinity NGF receptor TrkA receptors on neurites were proposed to interact either directly with these complexes or stimulate the axons to follow an NGF gradient (Fig. 3). The facts that BDNF

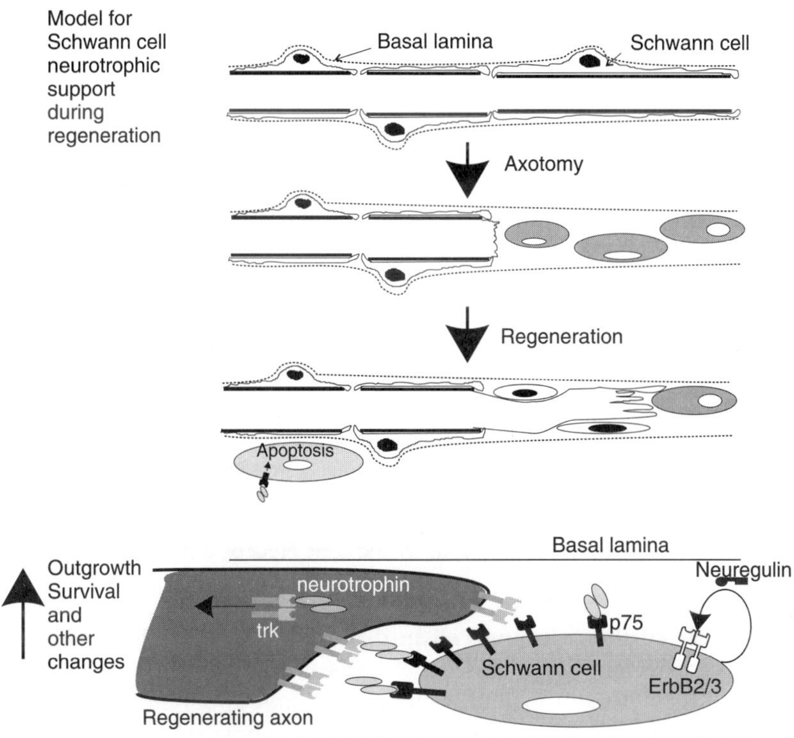

FIG. 3. In response to injury, Schwann cells dedifferentiate, migrate, and upregulate p75^{NTR} and a range of neurotrophic factors (see text). These are postulated to be concentrated on the Schwann cell surface, providing a neurotropic and neurotrophic attraction for regenerating axons, which have higher affinity receptors (trks) for the neurotrophins. The neurotrophins are endocytosed and retrogradely transported to the cell body to regulate cell survival, neurofilament production, and so on. Neuregulins are mitotic for Schwann cells and inhibit myelination. Once Schwann cell–axon contact is reestablished, supernumerary Schwann cells are thought to apoptose via signalling through p75^{NTR}.

and NT-3 bind to p75^{NTR} with greater stability than NGF (Rosenthal *et al.*, 1990) and that BDNF receptor TrkB truncated forms are downregulated distal and upregulated proximal to an injury have led to the revised hypothesis that BDNF and/or NT-3 may play a similar chemotactic role (Meyer *et al.*, 1992; Funakoshi *et al.*, 1993).

It has been shown that BDNF and NGF can stimulate transcription of Schwann cell myelin protein (SMP), but neurotrophin-3 (NT-3) can inhibit this; both effects are dependent on p75^{NTR} availability (Pruginin-Bluger *et al.*, 1997). NT-3 has been shown to be transported anterogradely from the dorsal root ganglia (DRG) (Altar and DiStefano, 1998; Fernyhough *et al.*, 1998). These findings raise the possibility of central/peripheral process transport selectivity and the release of neurotrophins with neurotransmitter peptides. Neurotrophins may therefore play a regulatory role in Schwann cells and other tissue differentiation and regenerative responses.

NGF-p75^{NTR} interaction may play a role in Schwann cell migration (Anton *et al.*, 1994), and Schwann cell apoptosis may also be controlled by NGF through p75^{NTR} (Syroid *et al.*, 2000). As there is no evidence of uncontacted Schwann cells or unsheathed axons in nerve bodies, a mechanism by which axon and Schwann cell numbers are closely matched has long been postulated to exist. p75^{NTR} -/- mice show reduced Schwann cell apoptosis following nerve injury (Ferri and Bisby, 1999), and NGF can stimulate Schwann cell apoptosis *in vitro* (Syroid *et al.*, 2000; Soilu-Hänninen *et al.*, 1999). Because p75^{NTR} and NGF are sharply downregulated following axonal contact, their continuing presence may signal apoptosis, thus removing supernumerary Schwann cells (Fig. 3).

III. Schwann Cells in Diabetic Neuropathy

The role of the Schwann cell in diabetic neuropathy has been investigated surprisingly little. This may be due to the early success of aldose reductase inhibitors in experimental models and apparent localization of this enzyme to Schwann cells. Thus, measurements derived from whole nerve were largely assumed to relate to Schwann cells. Although increasing knowledge of the complexity of the etiology of diabetic neuropathy has highlighted the need for *in vitro* studies, much of this work has yet to be done. This section (1) attempts to relate known etiological mechanisms to a potential Schwann cell location and (2) outlines the effects of these changes on nerve function.

IV. Neuropathy in Diabetic Patients

Symptomatic diabetic neuropathy is reported by 25% of the diabetic population; 50% have easily clinically detectable neuropathy, and 90% have a detectable nerve function defect (Pittenger *et al.*, 1999). The most common form of neuropathy in diabetes is known as distal symmetrical polyneuropathy: it can act on all types of neurons, but affects those with longest axons first. Distally innervating neurons must maintain a larger cellular volume (and hence higher metabolic efficiency). This also implies a greater dependence on Schwann cells and target tissues for mechanical and neurotrophic support. Therefore, these neurons are susceptible to defects in support cells. Although early symptoms are predominately sensory, this is likely to be a result of longer (and hence more susceptible) axons necessary to reach foot skin than muscles and more easily detected sensory deficits: postmortems show that all fiber types and sizes are affected.

V. Etiological Models for Nerve Dysfunction

Currently, changes in nerve function in diabetic neuropathy are postulated to arise from three main interacting and interrelated sources. These are reviewed in detail in other chapters of this book, but their special relations to Schwann cells require brief mention here.

1. Hypoxia — decreased oxygenation in nerve arises due to a decreased ratio of endothelial relaxing to constricting factors, alterations in basement membrane thickness, plasma viscosity, erythrocyte deformability, and hyperglycemia-related changes in endothelial cell ionic and energy balance (De Vriese *et al.*, 2000). Hypoxia may alter Schwann cell and axon energy and redox balance. Restoration of nerve conduction velocity (NCV) with vasodilators, essential fatty acids (precursors of prostanoids), and, more recently, angiotensin-converting enzyme (ACE) inhibitors and angitotensin II receptor antagonists support this etiological factor.
2. Oxidative damage — a sustained high level of glucose results in increases in mitochondrial-generated reactive oxygen species (ROS) (Nishikawa *et al.*, 2000), glucose autooxidation, and formation of advanced glycosylated end products (AGEs). Myelin insulation and membrane protein clustering function may suffer severely, and myelin may also become a target for phagocytosis. Other Schwann

cell proteins, including enzymes and neurotrophins, important to nerve function would also be affected. Treatment with aminoguandine, which is thought to decrease cross-linking of AGEs, and antioxidants such as probucol, vitamin E, and ascorbic acid confirm this is a cause of nerve dysfunction, at least in some animal models of diabetic neuropathy.

3. Metabolic disturbance — hyperglycemia is responsible for altered levels of a range of metabolites, enzymes, receptors, and growth factors. Many of these changes have been related to increased flux through the polyol pathway. The polyol pathway is an alternative pathway of glucose metabolism, perhaps functioning to prevent intracellular hyperosmolarity when the capacity of the glycolytic hexokinase-mediated pathway is saturated. Glucose is converted to sorbitol by aldose reductase (ALR2, or AR), oxidizing NADPH, and then to fructose by sorbitol dehydrogenase, producing NADH. Within the peripheral nerve, the main localization of aldose reductase is Schwann cells. The action of sorbitol as an intracellular osmolyte results in compensatory decreases in *myo*-inositol and taurine, which are also osmolytes (Stevens *et al.*, 1993). Utilization of NADPH by AR may reduce cellular availability for other enzymes, importantly nitric oxide synthase and glutathione reductase. Aldose reductase inhibitors have been tested extensively as a treatment, with success in animal models, although notably much less in human patients.

A. Early Changes: Hypoxia and Energy Deficits

In experimental models, the lack of gross morphological change seen at early time points, when conduction velocity, metabolic, and nerve blood flow deficits are all present, suggests a pathogenesis in which reversible metabolic and energy defects progress to irreversible oxidative, morphological, and possibly apoptotic damage.

Three weeks of STZ diabetes with or without a vasodilator, prazosin, reveals that simply restoring blood flow normalizes motor conduction velocity, cytosolic or mitochondrial NAD^+/NADH balance, and phosphocreatine:creatine in sciatic nerve, while not influencing the increased sorbitol or decreased *myo*-inositol, taurine, and Na^+-K^+-ATPase activity at this time point (Obrosova *et al.*, 2000). Furthermore, lipid peroxidation products and glutathione (GSH) were not affected by prazosin treatment. The authors argue that antioxidants that restore NCV act on vascular tissue in the nerve in early diabetes. This is supported by studies showing that the ameliorating effects of both aminoguanidine, an AGE inhibitor, and BM

15.0639, a free radical scavenger, are prevented by a nitric oxide synthase inhibitor, thus implying that nerve blood flow is the common feature restored.

Therefore, hypoxia and energy metabolism are most important to nerve conduction velocity early in diabetic neuropathy. Phosphocreatine, which was restored by prazosin, is the most direct metabolic regenerator of ATP levels so that restoration of an energy-dependent process may explain the restoration of NCV. Activity of ATP-dependent ion gradients or ATP-dependent phosphorylation are two possible candidates. Viewed another way, restoring nerve blood flow did not normalize nerve function; notably, many of the remaining changes are likely to be localized to Schwann cells, suggesting that early changes in Schwann cells are metabolic and oxidative, not hypoxic, in origin. A time course of the ability of prazosin treatment to restore NCV may provide a valuable method of evaluating if and when metabolic and oxidative changes become important to nerve conduction velocity. Treatment with *myo*-inositol or taurine has been effective in restoring NCV at later stages (Tomlinson and Mayer, 1985; Pop-Busui *et al.*, 2001).

B. Oxidative Damage

Advanced glycosylated end products have been implicated in diabetic neuropathy. In diabetic patients, AGE have been found in the perineurium, endothelial cells, and pericytes of endoneurial microvessels, as well as in myelinated and unmyelinated fibers. At the submicroscopic level, AGE deposition appeared focally as irregular aggregates in the cytoplasm of endothelial cells, pericytes, axoplasm, and Schwann cells of both myelinated and unmyelinated fibers. Interstitial collagens and basement membranes of the perineurium are also positive. (Sugimoto *et al.*, 1997). With such a widespread distribution, they are likely to affect the function of many components of the nerve.

Glycosylation of myelin occurs, which could affect nerve conduction velocity (Vlassara *et al.*, 1981) or provide a target for phagocytosis, therefore leading to segmental demyelination and the requirement for a regenerative response (see later). Altered myelin protein levels have also been identified in diabetic Schwann cells (Conti *et al.*, 1993).

C. Effects of Aldose Reductase Overactivity

Aldose reductase was initially localized to the Schwann cell cytoplasm in the rat sciatic nerve (Kern and Engerman, 1982; Ludvigson and Sorenson,

1980); importantly, these reports did not find any AR in nerve capillary endothelial cells, although endothelial cells of larger blood vessels did contain AR. However, a subsequent report in BB/W rats also found AR in sciatic nerve endothelial cells and pericytes (Chakrabarti *et al.*, 1987), and AR is induced in cultured microvascular endothelial cells by exposure to glycated bovine serum albumen or diabetic patients serum (Nakamura *et al.*, 2000).

D. OSMOTIC CHANGES OR METABOLIC FLUX?

myo-Inositol is an osmolyte and precursor of phosphoinositide species, which are important in many signaling pathways through the phospholipase formation of diacylglycerol (DAG), which activates PKC; and inositol triphosphate (IP$_3$), which increases intracellular calcium release. In nerve, depleted *myo*-inositol is proposed to link AR to decreased Na$^+$-K$^+$-ATPase activity through decreased DAG and PKC activity. In Schwann cell cultures grown in high glucose conditions, *myo*-inositol uptake and incorporation into phospholipids is decreased. An aldose reductase inhibitor or ascorbic acid (an antioxidant) partially reversed this (Karihaloo *et al.*, 1997). Neither had an affect in the presence of high sorbitol, suggesting that sorbitol is responsible in part for the *myo*-inositol decrease. However, a second study found that sorbitol did not accumulate in Schwann cells under high glucose conditions unless Na$^+$-induced hyperosmotic stress was applied (Suzuki *et al.*, 1999). The level of activity of AR may be the important factor in *myo*-inositol depletion under hyperglycemic conditions. In support, decreased glycerophospholipid arachidonoyl-containing molecular species (ACMS) occurring in tumor-derived human Schwann cell line NF1T in high glucose conditions has been shown to be due to increased aldose reductase activity, rather than an osmotic effect (Kuruvilla and Eichberg, 1998).

Therefore, increased aldose reductase activity may have physiologically relevant effects in the metabolically deranged Schwann cell. DAG and PKC levels increase in endothelial cells under high glucose conditions, whereas both these species decrease in nerve. This suggests that Schwann cells, rather than endothelial changes, predominate in diabetic nerve. Further studies in cultured Schwann cells under high glucose will therefore be important to identify real differences to the endothelium, and if so, what underlies them. Aldose reductase activity in endothelial cells is less than in Schwann cells of the nerve (Kern and Engerman, 1982). Levels of DAG and PKC activity could depend on the predominant pathway available for glucose metabolism.

Isoform-specific studies of PKC and Na$^+$-K$^+$-ATPase levels and activity suggest an even more complex picture. A study that found Schwann cell cultures express PKC α and β II isoforms, whereas axons probably contain mostly β I and ϵ, also detected no changes in compound activity or localization in 6- to 12-week STZ-diabetic rats (Borghini *et al.*, 1994). Subsequently, it has been reported in 6-week STZ-diabetic sciatic nerve that the PKC α subunit undergoes translocation from cytoplasm to membrane, consistent with activation. The Schwann cell β II subunit expression decrease is probably responsible for the overall activity decrease detectable (Roberts and McLean, 1997).

The decreased Na^+-K^+-ATPase $\alpha 2$ subunit is found at both the node of Ranvier and the myelin domain, and an upregulation of the $\beta 2$ subunit in the myelin domain only (Gerbi *et al.*, 1999). Little is known about the regulation of specific isoforms of PKC or Na^+-K^+-ATPase by glucose or DAG/PKC regulation of Na^+-K^+-ATPase in Schwann cells.

E. GALACTOSE INTOXICATION: PURE POLYOL?

Galactose intoxication is a model for the effect of increased polyol activity without the other complicating features in diabetic neuropathy. A primary Schwann cell defect has been strongly supported under these conditions. In this model, only 7 days of galactose feeding lead to increased water and galactitol (the sorbitol equivalent) content and depleted *myo*-inositol in sciatic nerve. Morphological changes are also seen at this time point (see later) (Mizisin and Powell, 1997). Notably, nerve or endoneurial Na^+-K^+-ATPase activity is increased in galactose-fed mice (Calcutt *et al.*, 1990), and the differences between diabetic and galactose neuropathy may be greater than anticipated; aldose reductase catalyzes lipid peroxidation product detoxification, and sorbitol, but not galactitol, is further metabolized to fructose, which creates further oxidative damage.

F. REDOX DISTURBANCES

Nitric oxide is a potent vasodilator and sympathetic inhibitor; a decrease in its production (in large vessel endothelium) may contribute to decreased nerve blood flow and hypoxia. However, chronic inhibition of aldose reductase in diabetic rats which normalized conduction velocity had little effect on resting sciatic nerve blood flow (Tomlinson *et al.*, 1998). This is consistent with a less significant role for hypoxia in later stages of experimental diabetes and more significant metabolic effects in Schwann cells.

G. ALDOSE REDUCTASE AS AN ANTIOXIDANT

An interesting alteration in the way in which AR is viewed is due to the observation that the lipid peroxidation product, 4-hydroxynonenal

(HNE), induces aldose reductase expression and is the best known substrate (Srivastava *et al.*, 1995). Other toxic aldehydes are also good substrates for AR. In giant cell arthritis, aldose reductase inhibitor treatment leads to increased HNE and increased apoptosis in the arterial wall (Rittner *et al.*, 1999). Conversion of HNE leads to the formation of lipid alcohols, which irreversibly inhibit AR. However, glutathione modification of AR prevents this, as does $NADP^+$ (Del Corso *et al.*, 1998).

The improvements seen in clinical trials of aldose reductase inhibitors were disappointing (Airey *et al.*, 2000). In the BB/W rat, the increased levels of sorbitol and decreases in *myo*-inositol ameliorate with time (Sima and Sugimoto, 1999). This suggests that other pathological mechanisms both independent of and secondary to aldose reductase overactivity have become more significant, and the irreversible changes caused by these secondary effects cannot be reversed by aldose reductase inhibition. The inactivation of AR by HNE may be the cause of normalization of sorbitol and *myo*-inositol levels. Tissue-specific mechanisms of AR induction are largely unexplored, and it would be unsurprising to find differences between endothelial and Schwann cells.

Initially, increased aldose reductase activity may reflect a response to high levels of toxic aldehydes due to ROS lipid peroxidation. Increased AR activity and glutatathione turnover may combine to deplete NADPH, thus reducing GSH and leaving AR open to irreversible inhibition by toxic aldehydes. It may be the failure of the AR and glutathione response to overcome the oxidative stress present, rather than the effect on cell metabolism or redox status, which leads to further deficits, such as impaired neurotrophism, morphological defects, and impaired regeneration, as outlined later. It will be important to discover if AR in Schwann cells has an antioxidative role as seen in endothelial cells.

H. NEUROTROPHIC ALTERATIONS IN DIABETIC NEUROPATHY

ROS, hypoxia, and glycation as outlined earlier may also cause the neurotrophic alterations in neuropathy, although the primary cause has not been established. Changes in end organ expression, neural receptor expression, and retrograde transport have all been implicated. The role of Schwann cell growth factor production in the intact nerve is not well established. However, in the STZ rat sciatic nerve, expression of NT-3 and NT-4 is decreased after 12 weeks of diabetes. $p75^{NTR}$ is decreased at both 6 and 12 weeks (Rodríguez-Pena *et al.*, 1995). Furthermore, NT-3 treatment of STZ-diabetic rats improves sensory, but not motor NCV (Fernyhough *et al.*, 1996), consistent with the presence of trkC, the NT-3 high-affinity receptor, on large proprioceptive neurons (Hory-Lee *et al.*, 1993).

The NGF content of sciatic nerve is decreased after 8 weeks of STZ diabetes, which is restored by an ARI, epalrestat. Furthermore, epalrestat stimulated NGF synthesis and secretion in cultured rat Schwann cells (Ohi *et al.*, 1998). CNTF, expressed in Schwann cells, has been reported to be reduced slightly (Calcutt *et al.*, 1992) or unchanged (Ohi *et al.*, 1998) after 8weeks of STZ diabetic neuropathy, but unchanged in human diabetic neuropathy patients (Lee *et al.*, 1996).

It has become clear that vascular endothelial growth factor (VEGF), characterized for its extensive role in angiogenesis, is also important in the regeneration of nervous tissue. In 12-week diabetic rats, VEGF is increased markedly in both DRG neurons and Schwann cells of the sciatic nerve, which is reversed by treatment with insulin or NGF. Interestingly, NGF increased VEGF staining in controls (Samii *et al.*, 1999).

I. Insulin, Insulin-like Growth Factors, and C Peptide

Although hyperglycemia is established as an important etiological factor in diabetic neuropathy, differences between type 1 and type 2 diabetic nerve in the progression of neuropathy argue in favor of other factors. Differences in glycemic control may explain some of the discrepancy, but the involvement of insulin and C peptide has been suggested by some workers. Insulin and C peptide are decreased markedly in type 1 diabetes, and although insulin is to some extent replaced, C peptide is not. In type 2 diabetes, both insulin and C peptide can be increased (early) or decreased (late). The only role of C peptide was thought to be in correctly linking the A and B chains of insulin during biosynthesis. This may have been due to maximal activation of its pathways at normal physiological concentrations; in experimental diabetes, it has been shown to restore NCV and increase nerve Na^+-K^+-ATPase activity (Ido *et al.*, 1997). Furthermore, it reduced impaired red blood cell deformability in type 1 diabetic patients through a Na^+-K^+-ATPase-dependent mechanism. Preliminary data suggest that a vasodilatory response to C peptide is via Ca^{2+}-mediated activation of eNOS (Johansson *et al.*, 2000; Kunt *et al.*, 1999). While it appears that C peptide acts via binding to a pertussis toxin-sensitive G-protein-coupled receptor (Wahren *et al.*, 2000), it also can increase activity of the insulin receptor.

In rat nerve, the high-affinity insulin receptor (IR) (lacking exon 11) is localized to paranodal terminal Schwann cell loops and microvilli, to the paranodal axolemma and Schmidt–Lantermann incisures. Endoneurial vessels show localization on plasma membranes and in endocytotic vesicles of endothelial cells and pericytes (Sugimoto *et al.*, 2000).

It has been suggested that decreased insulin levels may lead to decreased Schwann cell IGF-1 secretion. IGF-1 and IGF-1R mRNA levels and immunoreactivity in sural nerves from insulin-treated patients were higher than in noninsulin-treated subjects (Grandis *et al.*, 2001). IGF-1 is protective against hyperglycemia-induced apoptosis. Hyperglycemia induces caspase-3 activation and morphological changes in Schwann cells and neurons of dorsal root ganglia, consistent with apoptotic death, *in vivo* and *in vitro*. Overexpression of Bcl-xL, or IGF-I, signaling via PI3-kinase protects Schwann cells from glucose-mediated apoptosis *in vitro* (Delaney *et al.*, 1999, 2001). The addition of IGF-I at physiological concentrations prevents activation of caspase-3 and neuronal apoptosis *in vitro* (Russell *et al.*, 1999). Oxidative stress may promote mitochondrial changes in diabetic animals and lead to activation of programmed cell death caspase pathways.

Therefore, decreased insulin per se may alter Schwann cell and neuronal survival through altered Schwann cell production of IGF. Insulin itself and C peptide may have other direct effects on Schwann cells.

TABLE I

AXONAL CHANGES[a,b]

Changes observed	Sural biopsy	Dermis	Other primate	Galactose	BB rats	Alloxan rats	STZ rats
Degeneration/ reduction in myelinated fiber number	2,4,5[c]	15	No		27	20,21	23, 26
Microfibril deposits (type VI collagen)	6,7				6		
Neurofilament changes	8						24,28
Axonal glycogen accumulation	8				27	21	
Axonal shrinkage	8, 4		17	18	27		25[c],26, 28
Axonal swelling		15					23,24
Axonal vacuolization		15					24
Decreased axoplasmic organelles	8						

[a]Published morphological changes in nerve from diabetic patients and various models, References accompany the table in which the reference first appears.
[b]Sources: 2; Behse *et al.* (1977); 4; Dyck *et al.* (1986a); 5; Dyck *et al.* (1986b); 6; Muona *et al.* (1993); 7; Bradley *et al.* (2000); 8; Yagihashi *et al.* (1979); 15; Yasuda *et al.* (1985); 17; Birrell *et al.* (2000); 18; Sharma *et al.* (1976); 20; Powell and Myers (1984); 21; Powell *et al.* (1977); 23; Wattig *et al.* (1986); 24; Jirmanová (1993); 25; Sima and Robertson (1979); 26; Yagihashi *et al.* (1990); 27; Sima (1980); 28; Medori *et al.* (1988).
[c]Asserts that this is primary; and the sc or axonal changes also seen are secondary.

J. Morphological Changes

Ultrastructural findings in diabetic nerve are shown in Table I–III. Authors commonly comment that axon and Schwann cell deficits appear to proceed independently of each other. In the most common model, the diabetic rat, clear differences exist between (insulin-dependent type I) BB/W rats and (noninsulin-dependent type I/II) STZ rats. In STZ-treated rats, little consistent evidence remains for morphological changes in the nerve of diabetic rats at the time points that most studies exploit [up to 8 weeks after STZ treatment, no morphological changes are seen (Thomas and Tomlinson, 1992)] and at which nerve conduction, polyol,

TABLE II

Schwann Cell Changes[a,b]

Changes observed	Sural biopsy	Dermis	Primates	Galactose	BB rats	Alloxan rats	STZ rats
Axoglial dysjunction	(−) 9				1,3		
Schwann cell inclusion bodies	8,10,3			10		20	22
Schwann cell glycogen	8,10,(+) 9			10			22
Schwann cell basement membrane thickening	8	15				21	
Schwann cell proliferation	10	15		10			22
Schwann cell organelle degeneration	10			10			22
Schwann cell mitochondrial damage	10			10			22
Segmental remyelination	12,5					20, 21	
Decreased myelin thickness			17				23, 24
Myelin degeneration	8,10,12,5			10		20,21	22,24,25, 26

[a]Published morphological changes in nerve from diabetic patients and various models, References accompany the table in which the reference first appears.
[b]Sources: 1; Greene *et al.* (1987); 3; Sima *et al.* (1986); 9; Thomas *et al.* (1996); 10; Kalichman *et al.* (1998); 12; Thomas and Lascelles (1965; 1966); 22; Mizisin *et al.* (1998).

TABLE III
ENDOTHELIAL AND EXTRACELLULAR CHANGES[a,b]

Endoneurial edema			18,19	
Increased perineurial collagen/BM thickened	7,13	6		
Increased microvessel collagen/BM thickening	7			20,21
Damaged microvessel innervation	14			

[a]Published morphological changes in nerve from diabetic patients and various models. References accompany the table in which the reference first appears.
[b]Sources: 13; Johnson (1983); 14; Beggs *et al.* (1992); 19; Powell *et al.* (1981).

and neurotrophic deficits have been shown to be present. Specimens from diabetic patients in the early stages of neuropathy might reveal a similar etiology and progression in humans; however, the sporadic nature of hyperglycemia, the chronicity of the condition, and the intermittent administration of insulin all represent significant differences between the human condition and animal models.

Both axonal and myelin/Schwann cell deficits are usually observed; whether one precedes or causes the other is extremely difficult to judge. Many of the Schwann cell morphological defects resemble those seen in long-term experimental galactose neuropathy, a model of increased polyol pathway activity. In human and feline diabetic neuropathy, Mizisin and colleagues have found Schwann cell pathology in the absence of axonal changes. The same authors have shown that after only 7 days of galactose treatment, Schwann cells in myelinated fibers show increased cytoplasmic volume, lipid droplets, π granules of Reich, enlarged mito-chondria, accumulation of intermediate filaments in the inner glial loop, periaxonal swelling, enlarged mitochondria without recognizable cristae, lysis of Schwann cell cytoplasm, and demyelination. ARI treatment atten-uated these changes (Mizisin and Powell, 1997). Later, a shift toward smaller fibers is seen, with a decrease in axon diameter (Forcier *et al.*, 1991).

VI. Paranodal Structure and Axoglial Dysjunction

Sima and colleagues have suggested that axoglial dysjunction, loss of the attachment of the helical Schwann cell spiral to the axolemma at the nodes of Ranvier, is a feature of type I diabetes (Sima *et al.*, 1988) and the type I BB/W rat model (Sima *et al.*, 1986). They have proposed that this may be associated with the progressive decline in NCV seen in both human and BB/W rats due to the redistribution of nodal Na channels to internodes (Magnani *et al.*, 1996). Changes in the adhesion molecules, which constitute the mechanical paranodal barrier, can be detected by ELISA in 6-month BB/W rats. Immunohistochemistry did not detect any changes. Therefore, subtle changes in levels of the molecules that create the Schwann cell loop–paranodal axolemma barrier may be responsible for the gross physiological deficits in diabetic neuropathy (Merry *et al.*, 1998).

Thomas *et al.* (1996) did not find any axoglial dysjunction or changes in the paranodes of sural nerve from diabetic patients. They raise several issues in the morphometric analysis of paranodal apparatus. First, there is a sampling bias toward smaller fibers, and regenerating fibers, with smaller internodal distances, which is complemented by an increased artifactual change in larger paranodes. As larger fibers are responsible for the measured nerve conduction velocity, an analysis of small fiber paranodes is insufficient to explain nerve conduction velocity deficits. Second, transverse bands, the electron-dense myelin anchors into the axolemma, are difficult to quantitate adequately. Giannini and Dyck (1996) have cast doubt on the observed decrease in transverse bands due to insufficient magnification used, lack of serial sectioning, and the necessity of exclusion judgements as to which loops had previously been in contact with the axolemma. Furthermore, a developmental study revealed that a significant number of terminal loops do not contact the axolemma in normal adults (Bertram and Schroder, 1993), which would make adequate and unbiased sampling even more important.

VII. Is Regeneration Abnormal in Diabetic Neuropathy?

A theory put forward by Bathgate (1993) suggested that as the peripheral nervous system undergoes constant remodeling and regeneration on a small scale, as do most other tissues, abnormality in regenerative function in diabetic nerve could result in a chronic neuropathy through a gradual loss of signaling efficiency. As Schwann cells are essential to the successful regeneration of axons, abnormal Schwann cell function is a necessary implication of this hypothesis.

In human sural nerve biopsies, abnormal persistence and shape of the Schwann cell basal lamina have been seen associated with regenerating fibers (King *et al.*, 1989). This could indicate oxidative damage to this component of the extracellular matrix; thickening of the perineurial basal lamina and reduplication of basal laminae around endoneurial vessels have also been observed. In addition, the number of regenerating fibers was decreased in proportion to total myelinated fibers, suggesting a failure of regeneration with the progression of diabetic neuropathy, which was greater in type II than type I diabetics (Bradley *et al.*, 1995). Furthermore, the extent of a regenerative response is less in nonpainful diabetic neuropathy associated with foot ulcers than in painful diabetic neuropathy (Britland *et al.*, 1990).

Impaired sciatic nerve regeneration has been observed in rat diabetic neuropathy (Longo *et al.*, 1986; Ekström and Tomlinson, 1989). Interestingly, regeneration through silicone chambers after only 1 week of diabetes in rats showed an impairment in multiple aspects of the regenerative process, including "cable formation, Schwann cell migration, and axonal regeneration" (Tantuwaya *et al.*, 1997), suggesting that metabolic and/or hypoxic defects may play a role in impaired regeneration. Rats with galactose neuropathy also show impaired regeneration (Powell *et al.*, 1986).

An abnormal neurotrophic response to injury also occurs in nerves of animals with STZ diabetes. Following crush, the trkA transcript decrease is greater in diabetic than normal animals in the DRG, and in diabetic animals there is no change in p75; either in transcription or axonal transport. The NGF increase in distal Schwann cells following axotomy is greater than normal (Maeda *et al.*, 1996). The incapacity of the Schwann cell to increase IGF-I expression after severe nerve damage may also be important, and ARI-treated rats and humans show increased nerve regeneration capacity and increased IGF-1 mRNA (Sima *et al.*, 1996) NGF or IGF-1 treatment improves nerve regeneration (Whitworth *et al.*, 1995; Sjöberg and Kanje, 1989).

The neuregulin receptor ErbB2 increase is impaired in diabetic sciatic nerve following transection (Eckersley *et al.*, 2001). The level of neuregulins in the intact or lesioned nerve of diabetic patients or in animal models is unknown. Given the importance of neuregulin signaling pathways in Schwann cell function and the vital role Schwann cells play in regeneration, the ability of the neuregulin axis to function in diabetic animals is worthy of investigation.

Tellurium-induced demyelination is increased in diabetic rats, suggesting an increased sensitivity to insult. However, remyelination on tellurium withdrawal is not impaired (Jaffey and Gelman, 1996). Britland and colleagues (1990) have reported a failure of normal maturation of myelinated fibers, with the normal proximo-distal taper failing to develop. This may account for differences in "axonal swelling," which may represent failure of tapering, and would depend on where in the nerve the samples were taken. However, it seems unlikely this would be important in humans, where the majority of (type 2) diabetes develops at ages > 40.

VIII. Reversible or Irreversible Schwann Cell Injury?

The prevailing evidence suggests that both axons and Schwann cells are dysfunctional in diabetic neuropathy—unsurprising under the weight of such varied and substantial pathological attack! A key question for therapy is: are these changes reversible?

If Schwann cells are placed from an 8-week diabetic rat into a control nerve, the cells are able to sustain a regenerative response, which suggests signs of recovery within a 2-week period (Eckersley *et al.*, 2001). Additionally, Schwann cells cultured from diabetic patients show no difference from controls in mitotic capacity or morphological characteristics (Scarpini *et al.*, 1992). This suggests that defects in Schwann cells in diabetic neuropathy are reversible. Given the overarching role of the Schwann cell as a support cell for neuronal regeneration and function, they may be a natural target for the treatment of diabetic neuropathy.

IX. Conclusions

Early reversible hypoxic and energy deficiencies, which initially impair nerve function, give way to metabolic and redox defects accompanied by an increased production of ROS. Decreased NADPH due to an increased activity of Schwann cell antioxidant enzymes glutathione peroxidase and possibly aldose reductase may result in aldose reductase inactivation, increased AGEs and ROS, and decreased neurotrophins and finally lead to segmental demyelination, axonal degeneration, reduced regenerative capacity, and gross nerve dysfunction. Many of the enzymes and metabolites underlying these deficits localize either exclusively or inclusively to Schwann cells. This, along with the natural role of Schwann cells and their high recovery potential, suggests a need for further specific *in vitro* study and a target for therapeutic intervention.

References

Aguayo, A. J., Attiwell, M., Trecarten, J., Perkins, S., and Bray, G. M. (1977). Abnormal myelination in transplanted Trembler mouse Schwann cells. *Nature* **265**, 73–75.

Airey, M., Bennett, C., Nicolucci, A., and Williams, R. (2000). Aldose reductase inhibitors for the prevention and treatment of diabetic peripheral neuropathy. *Cochrane Database Syst. Rev.* CD002182.

Altar, C. A., and DiStefano, P. S. (1998). Neurotrophin trafficking by anterograde transport. *Trends Neurosci.* **21**, 433–437.

Ansselin, A. D., Fink, T., and Davey, D. F. (1997). Peripheral nerve regeneration through nerve guides seeded with adult Schwann cells. *Neuropathol. Appl. Neurobiol.* **23**, 387–398.

Anton, E. S., Weskamp, G., Reichardt, L. F., and Matthew, W. D. (1994). Nerve growth factor and its low-affinity receptor promote Schwann cell migration. *Proc. Natl. Acad. Sci. USA* **91**, 2795–2799.

Arce, V., Pollock, R. A., Philippe, J. M., Pennica, D., Henderson, C. E., and de Lapeyriere, O. (1998). Synergistic effects of schwann- and muscle-derived factors on motoneuron survival involve GDNF and cardiotrophin-1 (CT-1). *J. Neurosci.* **18**, 1440–1448.

Bathgate, R. H. (1993). A model of nerve regeneration in diabetic neuropathy. *Med. Hypoth.* **41**, 63–77.

Beggs, J., Johnson, P. C., Olafsen, A., and Watkins, C. J. (1992). Innervation of the vasa nervorum: Changes in human diabetics. *J. Neuropathol. Exp. Neurol.* **51**, 612–629.

Behse, F., Buchthal, F., and Carlsen, F. (1977). Nerve biopsy and conduction studies in diabetic neuropathy. *J. Neurol. Neurosurg. Psychiat.* **40**, 1072–1082.

Bertram, M., and Schroder, J. M. (1993). Developmental changes at the node and paranode in human sural nerves: Morphometric and fine-structural evaluation. *Cell Tissue Res.* **273**, 499–509.

Birrell, A. M., Heffernan, S. J., Ansselin, A. D., McLennan, S., Church, D. K., Gillin, A. G., and Yue, D. K. (2000). Functional and structural abnormalities in the nerves of type I diabetic baboons: Aminoguanidine treatment does not improve nerve function. *Diabetologia* **43**, 110–116.

Bolin, L. M., Verity, A. N., Silver, J. E., Shooter, E. M., and Abrams, J. S. (1995). Interleukin-6 production by Schwann cells and induction in sciatic nerve injury. *J. Neurochem.* **64**, 850–858.

Borghini, I., Ania-Lahuerta, A., Regazzi, R., Ferrari, G., Gjinovci, A., Wollheim, C. B., and Pralong, W.-F. (1994). $\alpha, \beta I, \beta II, \delta$, and ε protein kinase C isoforms and compound activity in the sciatic nerve of normal and diabetic rats. *J. Neurochem.* **62**, 686–696.

Bradley, J. L., King, R. H. M., Muddle, J. R., and Thomas, P. K. (2000). The extracellular matrix of peripheral nerve in diabetic polyneuropathy. *Acta Neuropathol.* **99**, 539–546.

Bradley, J. L., Thomas, P. K., King, R. H. M., Muddle, J. R., Ward, J. D., Tesfaye, S., Boulton, A. J. M., Tsigos, C., and Young, R. J. (1995). Myelinated nerve fibre regeneration in diabetic sensory polyneuropathy: correlation with type of diabetes. *Acta Neuropathol.* **90**, 403–410.

Britland, S. T., Young, R. J., Sharma, A. K., and Clarke, B. F. (1990). Association of painful and painless diabetic polyneuropathy with different patterns of nerve fiber degeneration and regeneration. *Diabetes* **39**, 898–908.

Buj-Bello, A., Buchman, V. L., Horton, A., Rosenthal, A., and Davies, A. M. (1995). GDNF is an age-specific survival factor for sensory and autonomic neurons. *Neuron* **15**, 821–828.

Calcutt, N. A., Muir, D., Powell, H. C., and Mizisin, A. P. (1992). Reduced ciliary neuronotrophic factor-like activity in nerves from diabetic or galactose-fed rats. *Brain Res.* **575**, 320–324.

Calcutt, N. A., Tomlinson, D. R., and Biswas, S. (1990). Coexistence of nerve conduction deficit with increased Na^+-K^+-ATPase activity in galactose-fed mice: Implications for polyol pathway and diabetic neuropathy. *Diabetes* **39**, 663–666.

Chakrabarti, S., Sima, A. A. F., Nakajima, T., Yagihashi, S., and Greene, D. A. (1987). Aldose reductase in the BB rat: Isolation, immunological identification and localization in the retina and peripheral nerve. *Diabetologia* **30**, 244–251.

Conti, A. M., Malosio, M. L., Scarpini, E., Di Giulio, A. M., Scarlato, G., Mantegazza, P., and Gorio, A. (1993). Myelin protein transcripts increase in experimental diabetic neuropathy. *Neurosci. Lett.* **161**, 203–206.

Curtis, R., Scherer, S. S., Somogyi, R., Adryan, K. M., Ip, N. Y., Zhu, Y., Lindsay, R. M., and DiStefano, P. S. (1994). Retrograde axonal transport of LIF is increased by peripheral nerve injury: Correlation with increased LIF expression in distal nerve. *Neuron* **12**, 191–204.

De Vriese, A. S., Verbeuren, T. J., Van de Voorde, J., Lameire, N. H., and Vanhoutte, P. M. (2000). Endothelial dysfunction in diabetes. *B. J. Pharmacol.* **130**, 963–974.

De Waegh, S. M., Lee, V. M. Y., and Brady, S. T. (1992). Local modulation of neurofilament phosphorylation, axonal caliber, and slow axonal transport by myelinating Schwann cells. *Cell* **68**, 451–463.

Del Corso, A., Dal Monte, M., Vilardo, P. G., Cecconi, I., Moschini, R., Banditelli, S., Cappiello, M., Tsai, L., and Mura, U. (1998). Site-specific inactivation of aldose reductase by 4-hydroxynonenal. *Arch. Biochemi. Biophys.* **350**, 245–248.

Delaney, C. L., Cheng, H.-L., and Feldman, E. L. (1999). Insulin-like growth factor-I prevents caspase-mediated apoptosis in Schwann cells. *J. Neurobiol.* **41**, 540–548.

Delaney, C. L., Russell, J. W., Cheng, H. L., and Feldman, E. L. (2001). Insulin-like growth factor-I and over-expression of Bcl-xL prevent glucose-mediated apoptosis in Schwann cells. *J. Neuropathol. Exp. Neurol.* **60**, 147–160.

Dong, Z., Brennan, A., Liu, N., Yarden, Y., Lefkowitz, G., Mirsky, R., and Jessen, K. R. (1995). Neu differentiation factor is a neuron-glia signal and regulates survival, proliferation, and maturation of rat schwann cell precursors. *Neuron* **15**, 585–596.

Dyck, P. J., Karnes, J. L., O'Brien, P. C., Okazaki, H., Lais, A., and Engelstad, J. K. (1986a). The spatial distribution of fiber loss in diabetic polyneuropathy suggests ischemia. *Ann. Neurol.* **19**, 440–449.

Dyck, P. J., Lais, A., Karnes, J. L., O'Brien, P. C., and Rizza, R. A. (1986b). Fiber loss is primary and multifocal in sural nerves in diabetic polyneuropathy. *Ann. Neurol.* **19**, 425–439.

Eckersley, L., Ansselin, A. D., and Tomlinson, D. R. (2001). Effects of experimental diabetes on axonal and Schwann cell changes in sciatic nerve isografts. *Brain Res. Mol. Brain Res.*, **92**, 128–137.

Einheber, S., Hannocks, M.-J., Metz, C. N., Rifkin, D. B., and Salzer, J. L. (1995). Transforming growth factor-β 1 regulates Axon/Schwann cell interactions. *J. Cell Biol.* **129**, 443–458.

Ekström, P. A. R., and Tomlinson, D. R. (1989). Impaired nerve regeneration in streptozotocin-diabetic rats: Effects of treatment with an aldose reductase inhibitor. *J. Neurol. Sci.* **93**, 231–237.

Fernyhough, P., Diemel, L. T., and Tomlinson, D. R. (1996). Neurotrophin-3 treatment of diabetic rats prevents reduced sensory nerve conduction velocity. *Diabet. Medi.* **13**(Suppl. 3), S5.[Abstract]

Fernyhough, P., Diemel, L. T., and Tomlinson, D. R. (1998). Target tissue production and axonal transport of neurotrophin-3 are reduced in streptozocin-diabetic rats. *Diabetologia* **41**, 300–306.

Ferri, C. C., and Bisby, M. A. (1999). Improved survival of injured sciatic nerve Schwann cells in mice lacking the p75 receptor. *Neurosci. Lett.*, **272**, 191–194.

Forcier, N. J., Mizisin, A. P., Rimmer, M. A., and Powell, H. C. (1991). Cellular pathology of the nerve microenvironment in galactose intoxication. *J. Neuropathol. Exp. Neurol.* **50**, 235–255.

Frank, E., and Sanes, J. R. (1991). Lineage of neurons and glia in chick dorsal root ganglia: Analysis *in vivo* with a recombinant retrovirus. *Development* **111**, 895–908.

Funakoshi, H., Frisén, J., Barbany, G., Timmusk, T., Zachrisson, O., Verge, V. M. K., and Persson, H. (1993). Differential expression of mRNAs for neurotrophins and their receptors after axotomy of the sciatic nerve. *J. Cell Biol.* **123**, 455–465.

Gerbi, A., Sennoune, S., Pierre, S., Sampol, J., Raccah, D., Vague, P., and Maixent, J. M. (1999). Localization of Na,K-ATPase α/β isoforms in rat sciatic nerves: Effect of diabetes and fish oil treatment. *J. Neurochem.* **73**, 719–726.

Giannini, C., and Dyck, P. J. (1996). Axoglial dysjunction: A critical appraisal of definition, techniques, and previous results. *Microsc. Res. Techni.* **34**, 436–444.

Grandis, M., Nobbio, L., Abbruzzese, M., Banchi, L., Minuto, F., Barreca, A., Garrone, S., Mancardi, G. L., and Schenone, A. (2001). Insulin treatment enhances expression of IGF-I in sural nerves of diabetic patients. *Muscle Nerve* **24**, 622–629.

Greene, D. A., Chakrabarti, S., Lattimer, S. A., and Sima, A. A. F. (1987). Role of sorbitol accumulation and *myo*-inositol depletion in paranodal swelling of large myelinated nerve fibres in the insulin dependant spontaneously diabetic BB rat: Reversal by insulin replacement, ARI and *myo*-inositol. *J. Clin. Invest.* **79**, 1479–1485.

Grinspan, J. B., Marchionni, M. A., Reeves, M., Coulaloglou, M., and Scherer, S. S. (1996). Axonal interactions regulate Schwann cell apoptosis in developing peripheral nerve: Neuregulation receptors and the role of neuregulins. *J. Neurosci.* **16**, 6107–6118.

Gulati, A. K. (1996). Peripheral nerve regeneration through short- and long-term degenerated nerve transplants. *Brain Res.* **742**, 265–270.

Hory-Lee, F., Russell, M., Lindsay, R. M., and Frank, E. (1993). Neurotrophin 3 supports the survival of developing muscle sensory neurons in culture. *Proc. Nat. Acad. Sci. USA* **90**, 2613–2617.

Ido, Y., Vindigni, A., Chang, K., Stramm, L., Chance, R., Heath, W. F., DiMarchi, R. D., Di Cera, E., and Williamson, J. R. (1997). Prevention of vascular and neural dysfunction in diabetic rats by C-peptide. *Science* **277**, 563–566.

Jaffey, P. B., and Gelman, B. B. (1996). Increased vulnerability to demyelination in streptozotocin diabetic rats. *J. Comp. Neurol.* **373**, 55–61.

Jessen, K. R., Morgan, L., Stewart, H. J., and Mirsky, R. (1990). Three markers of adult non-myelin-forming Schwann cells, 217c(Ran-1), A5E3 and GFAP: Development and regulation by neuron-Schwann cell interactions. *Development* **109**, 91–103.

Jirmanová, I. (1993). Giant axonopathy in streptozotocin diabetes of rats. *Acta Neuropathol.* **86**, 42–48.

Johansson, B. L., Borg, K., Fernqvist-Forbes, E., Kernell, A., Odergren, T., and Wahren, J. (2000). Beneficial effects of C-peptide on incipient nephropathy and neuropathy in patients with Type 1 diabetes mellitus. *Diabet. Med.* **17**, 181–189.

Johnson, E. M., Jr., Taniuchi, M., Clark, H. B., Springer, J. E., Koh, S., Tayrien, M. W., and Loy, R. (1987). Demonstration of the retrograde transport of nerve growth factor in the peripheral and central nervous system. *J. Neurosci.* **7**, 923–929.

Johnson, P. C. (1983). Thickening of the human dorsal root ganglion perineurial cell basement membrane in diabetes mellitus. *Muscle Nerve* **6**, 561–565.

Kalichman, M. W., Powell, H. C., and Mizisin, A. P. (1998). Reactive, degenerative, and proliferative Schwann cell responses in experimental galactose and human diabetic neuropathy. *Acta Neuropathol.* **95**, 47–56.

Karihaloo, A. K., Joshi, K., and Chopra, J. S. (1997). Effect of sorbinil and ascorbic acid on *myo*-inositol transport in cultured rat Schwann cells exposed to elevated extracellular glucose. *J. Neurochem.* **69**, 2011–2018.

Kern, T. S., and Engerman, R. L. (1982). Immunohistochemical distribution of aldose reductase. *Hist. J.* **14**, 507–515.

King, R. H., Llewelyn, J. G., Thomas, P. K., Gilbey, S. G., and Watkins, P. J. (1989). Diabetic neuropathy: Abnormalities of Schwann cell and perineurial basal laminae. *Neuropathol. Appl. Neurobiol.* **15**, 339–355.

Kunt, T., Schneider, S., Pfutzner, A., Goitum, K., Engelbach, M., Schauf, B., Beyer, J., and Forst, T. (1999). The effect of human proinsulin C-peptide on erythrocyte deformability in patients with type I diabetes mellitus. *Diabetologia* **42**, 465–471.

Kuruvilla, R., and Eichberg, J. (1998). Depletion of phospholipid arachidonoyl-containing molecular species in a human Schwann cell line grown in elevated glucose and their restoration by an aldose reductase inhibitor. *J. Neurochem.* **71**, 775–783.

Lee, D. A., Gross, L., Wittrock, D. A., and Windebank, A. J. (1996). Localization and expression of ciliary neurotrophic factor (CNTF) in postmortem sciatic nerve from patients with motor neuron disease and diabetic neuropathy. *J. Neuropathol. Exp. Neurol.* **55**, 915–923.

Longo, F. M., Powell, H. C., Lebeau, J., Gerrero, M. R., Heckman, H., and Myers, R. R. (1986). Delayed nerve regeneration in streptozotocin diabetic rats. *Muscle Nerve* **9**, 385–393.

Ludvigson, M. A., and Sorenson, R. L. (1980). Immunohistochemical localization of aldose reductase. I. Enzyme purification and antibody preparation: Localization in peripheral nerve, artery and testis. *Diabetes* **29**, 438–449.

Maeda, K., Fernyhough, P., and Tomlinson, D. R. (1996). Regenerating sensory neurones of diabetic rats express reduced levels of mRNA for GAP-43, gamma-preprotachykinin and the nerve growth factor receptors, trkA and p75^{NGFR}. *Mol. Brain Res.* **37**, 166–174.

Magnani, P., Cherian, P. V., Gould, G. W., Greene, D. A., Sima, A. A. F., and Brosius, F. C., III (1996). Glucose transporters in rat peripheral nerve: Paranodal expression of GLUT1 and GLUT3. *Metab. Clin. Exp.* **45**, 1466–1473.

Marchionni, M. A., Goodearl, A. D. J., Chen, M. S., Bermingham-McDonogh, O., Kirk, C., Hendricks, M., Danehy, F., Misumi, D., Sudhalter, J., Kobayashi, K., Wroblewski, D., Lynch, C., Baldassare, M., Hiles, I., Davis, J. B., Hsuan, J. J., Totty, N. F., Otsu, M., McBurney, R. N., Waterfield, M. D., Stroobant, P., and Gwynne, D. (1993). Glial growth factors are alternatively spliced erbB2 ligands expressed in the nervous system. *Nature* **362**, 312–318.

Matsuoka, I., Meyer, M., Hofer, M., and Thoenen, H. (1991). Differential regulation of nerve growth factor and brain-derived neurotrophic factor expression in the peripheral nervous system. *Ann. N. Y. Acad. Sci.* **633**, 550–552.

Medori, R., Autilio-Gambetti, L., Jenich, H., and Gambetti, P. (1988). Changes in axon size and slow axonal transport are related in experimental diabetic neuropathy. *Neurology* **38**, 597–601.

Merry, A. C., Yamamoto, K., and Sima, A. A. (1998). Imbalances in N-CAM, SAM and polysialic acid may underlie the paranodal ion channel barrier defect in diabetic neuropathy. *Diabet. Res. Clin. Pract.* **40**, 153–160.

Meyer, M., Matsuoka, I., Wetmore, C., Olson, L., and Thoenen, H. (1992). Enhanced synthesis of brain-derived neurotrophic factor in the lesioned peripheral nerve: Different mechanisms are responsible for the regulation of BDNF and NGF mRNA. *J. Cell Biol.* **119**, 45–54.

Milner, R., Wilby, M., Nishimura, S., Boylen, K., Edwards, G., Fawcett, J., Streuli, C., Pytela, R., and Ffrench-Constant, C. (1997). Division of labor of Schwann cell integrins during migration on peripheral nerve extracellular matrix ligands. *Dev. Biol.* **185**, 215–228.

Mirsky, R., and Jessen, K. R. (1999). The neurobiology of Schwann cells. *Brain Pathol.* **9**, 293–311.

Mirsky, R., Parmantier, E., McMahon, A. P., and Jessen, K. R. (1999). Schwann cell-derived desert hedgehog signals nerve sheath formation. *Ann. N. Y. Acad. Sci.* **883**, 196–202.

Mizisin, A. P., and Powell, H. C. (1997). Schwann cell changes induced as early as one week after galactose intoxication. *Acta Neuropathol.* **93**, 611–618.

Mizisin, A. P., Shelton, G. D., Wagner, S., Rusbridge, C., and Powell, H. C. (1998). Myelin splitting, Schwann cell injury and demyelination in feline diabetic neuropathy. *Acta Neuropathol.* **95**, 171–174.

Muona, P., Jaakkola, S., Salonen, V., and Peltonen, J. (1993). Expression of glucose transporter 1 in adult and developing human peripheral nerve. *Diabetologia* **36**, 133–140.

Nakamura, N., Obayashi, H., Fujii, M., Fukui, M., Yoshimori, K., Ogata, M., Hasegawa, G., Shigeta, H., Kitagawa, Y., Yoshikawa, T., Kondo, M., Ohta, M., Nishimura, M., Nishinaka, T., and Nishimura, C. Y. (2000). Induction of aldose reductase in cultured human microvascular endothelial cells by advanced glycation end products. *Free Radic. Biol. Med.* **29**, 17–25.

Neuberg, D. H., Sancho, S., and Suter, U. (1999). Altered molecular architecture of peripheral nerves in mice lacking the peripheral myelin protein 22 or connexin32. *J. Neurosci. Res.* **58**, 612–623.

Nishikawa, T., Edelstein, D., Du, X. L., Yamagishi, S., Matsumura, T., Kaneda, Y., Yorek, M. A., Beebe, D., Oates, P. J., Hammes, H. P., Giardino, I., and Brownlee, M. (2000). Normalizing mitochondrial superoxide production blocks three pathways of hyperglycaemic damage. *Nature* **404**, 787–790.

Obrosova, I. G., Van Huysen, C., Fathallah, L., Cao, X., Stevens, M. J., and Greene, D. A. (2000). Evaluation of α_1-adrenoceptor antagonist on diabetes-induced changes in peripheral nerve function, metabolism, and antioxidative defense. *FASEB J.* **14**, 1548–1558.

Ohi, T., Saita, K., Furukawa, S., Ohta, M., Hayashi, K., and Matsukura, S. (1998). Therapeutic effects of aldose reductase inhibitor on experimental diabetic neuropathy through synthesis/secretion of nerve growth factor. *Exp. Neurol.* **151**, 215–220.

Pittenger, G. L., Malik, R. A., Burcus, N., Boulton, A. J., and Vinik, A. I. (1999). Specific fiber deficits in sensorimotor diabetic polyneuropathy correspond to cytotoxicity against neuroblastoma cells of sera from patients with diabetes. *Diabet. Care* **22**, 1839–1844.

Pop-Busui, R., Sullivan, K. A., Van Huysen, C., Bayer, L., Cao, X., Towns, R., and Stevens, M. J. (2001). Depletion of taurine in experimental diabetic neuropathy: Implications for nerve metabolic, vascular, and functional deficits. *Exp. Neurol.* **168**, 259–272.

Powell, H., Knox, D., Lee, S., Charters, A. C., Orloff, M. J., Garrett, R. S., and Lampert, P. (1977). Alloxan diabetic neuropathy: Electron microscopic studies. *Neurology* **27**, 60–66.

Powell, H. C., Costello, M. L., and Myers, R. R. (1981). Galactose neuropathy: Permeability studies, mechanism of edema, and mast cell abnormalities. *Acta Neuropathol.* **55**, 89–95.

Powell, H. C., Longo, F. M., Lebeau, J. M., and Myers, R. R. (1986). Abnormal nerve regeneration in galactose neuropathy. *J. Neuropathol. Exp. Neurol.* **45**, 151–160.

Powell, H. C., and Myers, R. R. (1984). Axonopathy and microangiopathy in chronic alloxan diabetes. *Acta Neuropathol.* **65**, 128–137.

Pruginin-Bluger, M., Shelton, D. L., and Kalcheim, C. (1997). A paracrine effect for neuron-derived BDNF in development of dorsal root ganglia: Stimulation of Schwann cell myelin protein expression by glial cells. *Mech. Dev.* **61**, 99–111.

Pu, S. F., Zhuang, H. X., and Ishii, D. N. (1995). Differential spatio-temporal expression of the insulin-like growth factor genes in regenerating sciatic nerve. *Mol. Brain Res.* **34**, 18–28.

Reichert, F., Saada, A., and Rotshenker, S. (1994). Peripheral nerve injury induces Schwann cells to express two macrophage phenotypes: Phagocytosis and the galactose-specific lectin MAC-2. *J. Neurosci.* **14**, 3231–3245.

Riethmacher, D., Sonnenberg-Riethmacher, E., Brinkmann, V., Yamaai, T., Lewin, G. R., and Birchmeier, C. (1997). Severe neuropathies in mice with targeted mutations in the ErbB3 receptor. *Nature* **389**, 725–730.

Rittner, H. L., Hafner, V., Klimiuk, P. A., Szweda, L. I., Goronzy, J. J., and Weyand, C. M. (1999). Aldose reductase functions as a detoxification system for lipid peroxidation products in vasculitis. *J. Clin. Investi.* **103**, 1007–1013.

Roberts, R. E., and McLean, W. G. (1997). Protein kinase C isozyme expression in sciatic nerves and spinal cords of experimentally diabetic rats. *Brain Res.* **754**, 147–156.

Rodríguez-Pena, A., Botana, M., González, M., and Requejo, F. (1995). Expression of neurotrophins and their receptors in sciatic nerve of experimentally diabetic rats. *Neurosci. Lett.* **200**, 37–40.

Rosenthal, A., Goeddel, D. V., Nguyen, T., Lewis, M., Shih, A., Laramee, G. R., Nikolics, K., and Winslow, J. W. (1990). Primary structure and biological activity of a novel human neurotrophic factor. *Neuron* **4**, 767–773.

Russell, J. W., Sullivan, K. A., Windebank, A. J., Herrman, D. N., and Feldman, E. L. (1999). Neurons undergo apoptosis in animal and cell culture models of diabetes. *Neurobiol. Dis.* **6**, 347–363.

Samii, A., Unger, J., and Lange, W. (1999). Vascular endothelial growth factor expression in peripheral nerves and dorsal root ganglia in diabetic neuropathy in rats. *Neurosci. Lett.* **262**, 159–162.

Scarpini, E., Doronzo, R., Baron, P., Moggio, M., Basellini, A., and Scarlato, G. (1992). Phenotypic and proliferative properties of Schwann cells from nerves of diabetic patients. *Int. J. Clin. Pharmacol. Res.* **12**, 211–215.

Schecterson, L. C., and Bothwell, M. (1992). Novel roles for neurotrophins are suggested by BDNF and NT-3 mRNA expression in developing neurons. *Neuron* **9**, 449–463.

Sharma, A. K., Thomas, P. K., and Baker, R. W. R. (1976). Peripheral nerve abnormalities related to galactose administration in rats. *J. Neurol. Neurosurg. Psychiat.* **39**, 794–802.

Sima, A. A., Lattimer, S. A., Yagihashi, S., and Greene, D. A. (1986). Axo-glial dysjunction. A novel structural lesion that accounts for poorly reversible slowing of nerve conduction in the spontaneously diabetic bio-breeding rat. *J. Clin. Invest.* **77**, 474–484.

Sima, A. A., and Robertson, D. M. (1979). Peripheral neuropathy in the diabetic mutant mouse: An ultrastructural study. *Lab. Invest.* **40**, 627–632.

Sima, A. A. F. (1980). Peripheral neuropathy in the spontaneously diabetic BB-wistar- rat: An ultrastructural study. *Acta Neuropathol.* **51**, 223–227.

Sima, A. A. F., Lattimer, S. A., Yagihashi, S., and Greene, D. A. (1986). Axo-glial dysjunction: A novel structural lesion that accounts for poorly reversible slowing of nerve conduction in the spontaneously diabetic bio-breeding rat. *J. Clin. Invest.* **77**, 474–484.

Sima, A. A. F., Nathaniel, V., Bril, V., McEwen, T. A. J., and Greene, D. A. (1988). Histopathological heterogeneity of neuropathy in insulin-dependent and non-insulin-dependent diabetes, and demonstration of axo-glial dysjunction in human diabetic neuropathy. *J. Clin. Invest.* **81**, 349–364.

Sima, A. A. F., Ristic, H., Merry, A., Kamijo, M., Lattimer, S. A., Stevens, M. J., and Greene, D. A. (1996). Primary preventive and secondary interventionary effects of acetyl-L-carnitine on diabetic neuropathy in the bio-breeding Worcester rat. *J. Clin. Invest.* **97**, 1900–1907.

Sima, A. A. F., and Sugimoto, K. (1999). Experimental diabetic neuropathy: An update. *Diabetologia* **42**, 773–788.

Sjöberg, J., and Kanje, M. (1989). Insulin-like growth factor (IGF-1) as a stimulator of regeneration in the freeze-injured rat sciatic nerve. *Brain Res.* **485**, 102–108.

Soilu-Hänninen, M., Ekert, P., Bucci, T., Syroid, D., Bartlett, P. F., and Kilpatrick, T. J. (1999). Nerve growth factor signaling through p75 induces apoptosis in Schwann cells via a Bcl-2-independent pathway. *J. Neurosci.* **19**, 4828–4838.

Srivastava, S., Chandra, A., Bhatnagar, A., Srivastava, S. K., and Ansari, N. H. (1995). Lipid peroxidation product, 4-hydroxynonenal and its conjugate with GSH are excellent substrates of bovine lens aldose reductase. *Biochem. Biophys. Res. Commun.* **217**, 741–746.

Stevens, M. J., Henry, D. N., Thomas, T. P., Killen, P. D., and Greene, D. A. (1993). Aldose reductase gene expression and osmotic dysregulation in cultured human retinal pigment epithelial cells. *Am. J. Physiol.* **265**, E428–E438.

Sugimoto, K., Murakawa, Y., Zhang, W., Xu, G., and Sima, A. A. (2000). Insulin receptor in rat peripheral nerve: Its localization and alternatively spliced isoforms. *Diabet. Metab. Res. Rev.* **16**, 354–363.

Sugimoto, K., Nishizawa, Y., Horiuchi, S., and Yagihashi, S. (1997). Localization in human diabetic peripheral nerve of Nε- carboxymethyllysine-protein adducts, an advanced glycation endproduct. *Diabetologia* **40**, 1380–1387.

Suzuki, T., Mizuno, K., Yashima, S., Watanabe, K., Taniko, K., and Yabe-Nishimura, C. (1999). Characterization of polyol pathway in Schwann cells isolated from adult rat sciatic nerves. *J. Neurosci. Res.* **57**, 495–503.

Syroid, D. E., Maycox, P. J., Soilu-Hänninen, M., Petratos, S., Bucci, T., Burrola, P., Murray, S., Cheema, S., Lee, K.-F., Lemke, G., and Kilpatrick, T. J. (2000). Induction of postnatal Schwann cell death by the low-affinity neurotrophin receptor *in vitro* and after axotomy. *J. Neurosci.* **20**, 5741–5747.

Syroid, D. E., Maycox, P. R., Burrola, P. G., Liu, N., Wen, D., Lee, K. F., Lemke, G., and Kilpatrick, T. J. (1996). Cell death in the Schwann cell lineage and its regulation by neuregulin. *Proc. Natl. Acad. Sci. USA* **93**, 9229–9234.

Taniuchi, M., Clark, H. B., and Johnson, E. M., Jr. (1986). Induction of nerve growth factor receptor in Schwann cells after axotomy. *Proc. Natl. Acad. Sci. USA* **83**, 4094–4098.

Tantuwaya, V. S., Bailey, S. B., Schmidt, R. E., Villadiego, A., Tong, J. X. X., and Rich, K. M. (1997). Peripheral nerve regeneration through silicone chambers in streptozocin-induced diabetic rats. *Brain Res.* **759**, 58–66.

Thomas, P. K., Beamish, N. G., Small, J. R., King, R. H. M., Tesfaye, S., Ward, J. D., Tsigos, C., Young, R. J., and Boulton, A. J. M. (1996). Paranodal structure in diabetic sensory polyneuropathy. *Acta Neuropathol.* **92**, 614–620.

Thomas, P. K., and Lascelles, R. G. (1965). Schwann-cell abnormalities in diabetic neuropathy. *Lancet* **1**, 1355–1357.

Thomas, P. K., and Lascelles, R. G. (1966). The pathology of diabetic neuropathy. *Quart. J. Med.* **35**, 489–509.

Thomas, P. K., and Tomlinson, D. R. (1992). Diabetic and hypoglycaemic neuropathy. In (P. J. Dyck, P. K. Thomas, J. W. Griffin, P. A. Low, and J. F. Poduslo, eds.), "Peripheral Neuropathy" pp. 1219–1250. Saunders, Philadelphia.

Tomlinson, D. R., Dewhurst, M., Stevens, E. J., Omawari, N., Carrington, A. L., and Vo, P. A. (1998). Reduced nerve blood flow in diabetic rats: Relationship to nitric oxide production and inhibition of aldose reductase. *Diabet. Med.* **15**, 579–585.

Tomlinson, D. R., and Mayer, J. H. (1985). Reversal of deficits in axonal transport and nerve conduction velocity by treatment of streptozotocin-diabetic rats with myo-inositol. *Exp. Neurol.* **89**, 420–427.

Trachtenberg, J. T., and Thompson, W. J. (1996). Schwann cell apoptosis at developing neuromuscular junctions is regulated by glial growth factor. *Nature* **379**, 174–177.

Umehara, F., Tate, G., Itoh, K., Yamaguchi, N., Douchi, T., Mitsuya, T., and Osame, M. (2000). A novel mutation of *desert hedgehog* in a patient with 46,XY partial gonadal dysgenesis accompanied by minifascicular neuropathy. *Am. J. Hum. Genet.* **67**, 1302–1305.

Vlassara, H., Brownlee, M., and Cerami, A. (1981). Nonenzymatic glycosylation of peripheral nerve protein in diabetes mellitus. *Proc. Natl. Acad. Sci. USA* **78**, 5190–5192.

Wahren, J., Ekberg, K., Johansson, J., Henriksson, M., Pramanik, A., Johansson, B. L., Rigler, R., and Jornvall, H. (2000). Role of C-peptide in human physiology. *Am. J. Physiol. Endocrinol. Metab.* **278**, E759–E768.

Wattig, B., Warzok, R., and Thomas, P. K. (1986). Experimental diabetic neuropathy. Morphometric studies on the rat N. suralis in short-term streptozotocin-induced diabetes. *Zentralblatt Allgemeine Pathol. Pathologische Anat.* **131**, 451–458.

Whitworth, I. H., Terenghi, G., Green, C. J., Brown, R. A., Stevens, E., and Tomlinson, D. R. (1995). Targeted delivery of nerve growth factor via fibronectin conduits assists nerve regeneration in control and diabetic rats. *Eur. J. Neurosci.* **7**, 2220–2225.

Yagihashi, S., Kamijo, M., and Watanabe, K. (1990). Reduced myelinated fiber size correlates with loss of axonal neurofilaments in peripheral nerve of chronically streptozotocin diabetic rats. *Am. J. Pathol.* **136**, 1365–1373.

Yagihashi, S., Nishihira, M., and Baba, M. (1979). Morphometrical analysis of the peripheral nerve lesions in experimental diabetes rats. *Tohoku J. Exp. Med.* **129**, 139–149.

Yasuda, H., Kikkawa, R., Hatanaka, I., Kobayashi, N., Taniguchi, Y., and Shigeta, Y. (1985). Skin biopsy as a beneficial procedure for morphological evaluation of diabetic neuropathy. *Acta Pathol. Japon.* **35**, 1–8.

Zorick, T. S., Syroid, D. E., Arroyo, E., Scherer, S. S., and Lemke, G. (1996). The transcription factors SCIP and Krox-20 mark distinct stages and cell fates in Schwann cell differentiation. *Mol. Cell. Neurosci.* **8**, 129–145.

PART IV
POTENTIAL TREATMENT

POLYOL PATHWAY AND DIABETIC PERIPHERAL NEUROPATHY

Peter J. Oates

Department of Cardiovascular and Metabolic Diseases, Pfizer Global Research
and Development, Groton, Connecticut 06340

This chapter critically examines the concept of the polyol pathway and how it relates to the pathogenesis of diabetic peripheral neuropathy. The two enzymes of the polyol pathway, aldose reductase and sorbitol dehydrogenase, are reviewed. The structure, biochemistry, physiological role, tissue distribution, and localization in peripheral nerve of each enzyme are summarized, along with current information about the location and structure of their genes, their alleles, and the possible links of each enzyme and its alleles to diabetic neuropathy. Inhibitors of pathway enzymes and results obtained to date with pathway inhibitors in experimental models and human neuropathy trials are updated and discussed. Experimental and clinical data are analyzed in the context of a newly developed metabolic model of the *in vivo* relationship between nerve sorbitol concentration and metabolic flux through aldose reductase. Overall, the data will be interpreted as supporting the hypothesis that *metabolic flux* through the polyol pathway, rather than nerve concentration of sorbitol, is the predominant polyol pathway-linked pathogenic factor in diabetic peripheral nerve. Finally, key questions and future directions for basic and clinical research in this area are considered. It is concluded that robust inhibition of *metabolic flux* through the polyol pathway in peripheral nerve will likely result in substantial clinical benefit in treating and preventing the currently intractable condition of diabetic peripheral neuropathy. To accomplish this, it is imperative to develop and test a new generation of ''super-potent'' polyol pathway inhibitors. © 2002, Elsevier Science (USA).

"There is occasions and causes why and wherefore in all things."
Henry V, V.I, William Shakespeare

I. Introduction

A. DIABETIC PERIPHERAL NEUROPATHY

Diabetic neuropathy is a pervasive and growing worldwide medical problem (Feldman *et al.*, 1997; Boulton and Malik, 1998; Vinik *et al.*, 2000; Sugimoto *et al.*, 2000). Despite decades of intensive basic research and clinical testing of numerous potential therapies, there is still no effective treatment (Calcutt and Dunn, 1997; Fedele and Giugliano, 1997; Dejgaard, 1998; Tomlinson, 1998; Costantino *et al.*, 1999; Ward, 1999; Zochodne, 1999; Bril, 2001). However, intensified therapy to lower blood glucose has been proven to slow the progression of diabetic neuropathy (DCCT Research Group, 1995). In addition, the morphological severity of diabetic peripheral neuropathy has been shown to be strongly linked to glycemic control in both type 1 and type 2 diabetic patients (Perkins *et al.*, 2001). Such data point clearly to the pathogenicity for peripheral nerve of chronic hyperglycemia and/or diminished insulin action. At present there is only limited evidence to incriminate insulinopenia, C-peptide deficiency, or growth factor deficits *per se* in the development of diabetic neuropathy. In contrast, evidence is abundant and strong that chronic hyperglycemia *per se* most likely is a major etiologic factor in the pathogenesis of diabetic neuropathy. Data are also strengthening that implicate elevated free fatty acids as another contributory factor. Excessive metabolism of these energy-rich molecules results in biochemical disturbances very similar in some key aspects to those produced by hyperglycemia (Williamson *et al.*, 1993; Edelstein *et al.*, 2000; Inoguchi *et al.*, 2000).

B. GLUCOTOXICITY

1. *Excess Extracellular Glucose: Hyperglycemia*

Chronic hyperglycemia is widely believed to exert its tissue-damaging effects via both extracellular and intracellular routes (Tomlinson, 1999). Extracellularly, glucose participates in glycation reactions (Brownlee, 1997). Glycation products, glucosylated as well as those that chemically rearrange to more chemically complex species, "advanced glycation end products" (AGE), can bind to preexisting cell surface receptors for glucosylated (Cohen, 1996) and/or AGE (Schmidt *et al.*, 2000; Vlassara, 2001) molecular species. The binding and/or internalization processes involving such receptors generate superoxide via perturbation of membrane NADPH oxidase (Wautier *et al.*, 2001). If the process is chronic and exaggerated, it could potentially cause tissue-damaging levels of superoxide and related mediators (Cohen, 1997).

2. *Excess Intracellular Glucose: Hyperglysolia*

One of the hallmarks of tissues that are susceptible to diabetic microvascular complications is that they do not depend on insulin for the uptake of extracellular glucose. In the presence of chronic hyperglycemia, such tissues are subjected to abnormally high diffusion rates of extracellular glucose into the cytosol of their cells. Persistent elevation of cytosolic glucose concentration and/or metabolic flux of glucose, "hyperglysolia," has become increasingly linked to a cascade of metabolic disturbances that result in the generation of intracellular oxidative stress. Intracellular oxidative stress is now widely believed to play a central role in the pathogenesis of all diabetic microvascular complications e.g., Fig. 1, see also color insert (Biessels and VanDam, 1997; Cameron and Cotter, 1997; Williamson *et al.*, 1999; Ceriello, 2000; Nishikawa *et al.*, 2000; Rosen *et al.*, 2001).

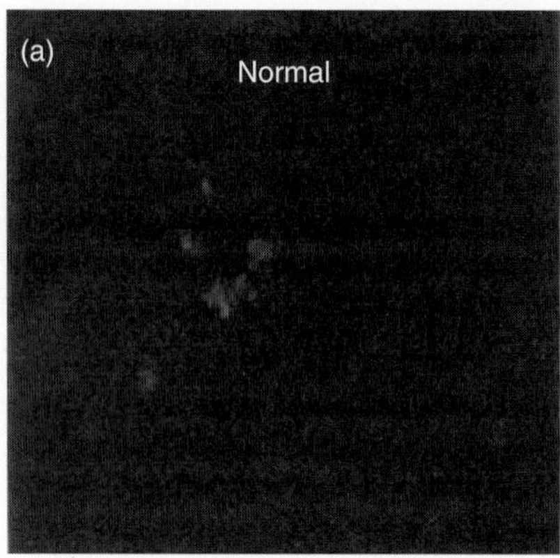

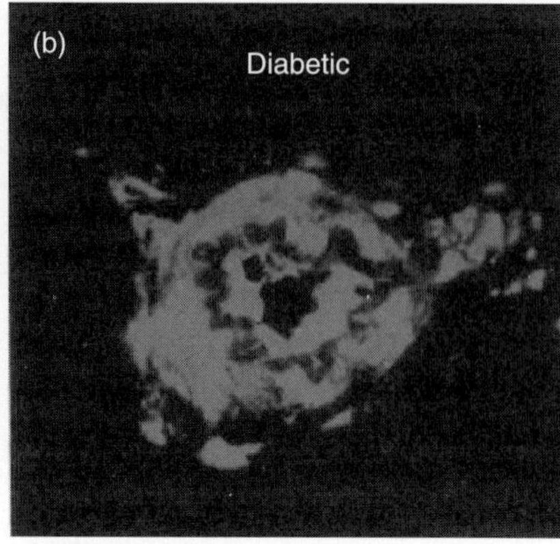

FIG. 1. Detection of superoxide levels in arterioles supplying sciatic nerve in normal and 3- to 4-week STZ-diabetic rats. Unfixed, frozen 0.5-μm-thick sections were stained with 2 μM hydroethidine and incubated for 30 min at 37°C. In the presence of superoxide, the dye is oxidized to highly fluorescent ethidium bromide, which intercalates into DNA. Fluorescence was detected with a Bio-Rad MRC-1024 laser-scanning confocal microscope with a krypton/argon laser using a 585-nm long-pass filter. Reproduced with permission from Coppey *et al.* (2001a). (See also color insert.)

C. HYPERGLYSOLIA, THE POLYOL PATHWAY, AND DIABETIC NEUROPATHY

1. *Raising Nerve Sorbitol Does Not Impair Nerve Function*

A widespread, but still poorly understood, pathway of intracellular glucose metabolism is the polyol pathway (Fig. 2, shaded area, see also color insert). Originally studied as an alternative pathway for producing fructose from glucose (Hers, 1956), the polyol pathway later came under suspicion of causing damage in diabetic peripheral nerve by producing pathogenic accumulations of sorbitol, a glucose metabolite (Gabbay, 1973; Kinoshita *et al.*, 1990). However, it thereafter became evident that sorbitol levels in diabetic nerves were in fact comparatively modest (Clements, 1986). Nevertheless, evidence also accrued that polyol pathway activity was somehow coupled to other potentially important metabolic activities in the diabetic nerve, such as maintenance of intracellular *myo*-inositol levels (Finegold *et al.*, 1983; Gillon *et al.*, 1983; Mayer and Tomlinson, 1983). Moreover, it has been observed that certain pharmacological agents robustly *raise* peripheral nerve sorbitol and can cause an increase in the sum of nerve sorbitol, fructose, and *myo*-inositol, yet do not cause detectable impairment, indeed can cause improvement, of peripheral nerve function (see Section VII,A,2,b). This observation has brought into strong question

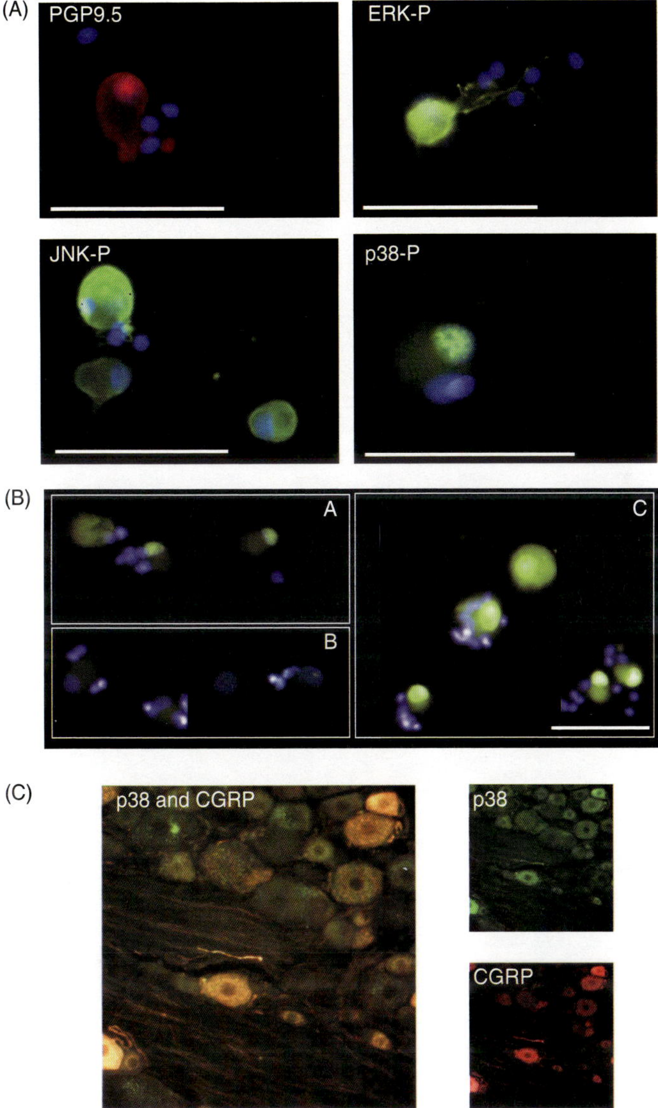

Fig. 4. (See chapter: Are MAP Kinases Glucose Tranducers?) Activation of MAP kinases in cultures of dorsal root ganglia (DRG) from adult rats shown by immunoreactivity (A and B) and colocalization of MAP kinase p38 with CGRP in sectioned intact DRG from rat (C). (A) Neuron-specific immunoreactivity to PGP 9.5 (red), with nonneurons PGP negative, but with DAPI-positive nuclei stained blue. Staining for phospho-MAP kinases, ERK, JNK, and p38, was confined to the neurons. (B) p38 activation by 50 mM glucose (subfigure C) as compared to 10 mM glucose (subfigure A). A negative control, without primary antibody, is shown in subfigure B. (C) Specific staining for p38 and CGRP in the small micrographs, with the combined picture in the larger micrograph.

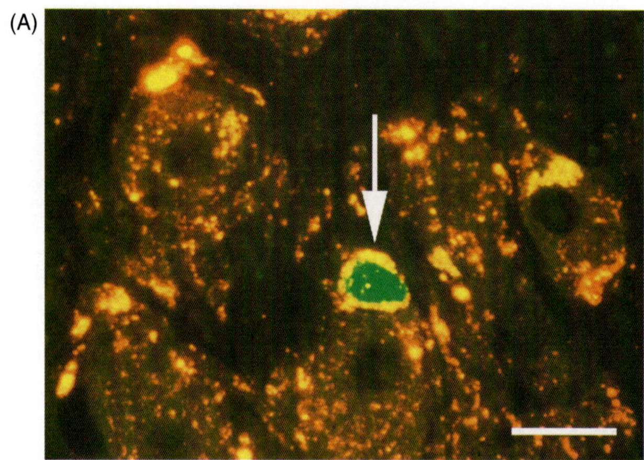

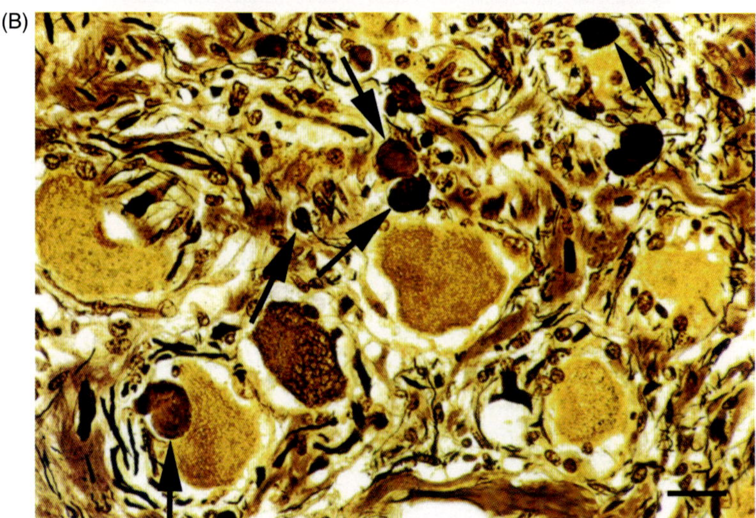

FIG. 3. (See chapter: Neurofilaments in Diabetic Neuropathy) Neuroaxonal dystrophy in diabetic human prevertebral sympathetic and dorsal root ganglia. (A) Simultaneous immunohistochemical detection of dopamine-β-hydroxylase (red) and highly phosphorylated NF-H (green) in a diabetic sympathetic ganglion shows their typical colocalization in a dystrophic axon (arrow), a pattern reflecting DβH containing neurotransmitter granules surrounding a neurofilamentous core (DβH and SMI-34 immunofluorescence) Bar: 30 μm. (B and C) Markedly enlarged neurofilament-laden dystrophic axons (arrows) are intimately apposed to principal dorsal root ganglion neurons (B: Bielschowsky silver stain; C: 1-μm plastic section) Bars: 20 μm.

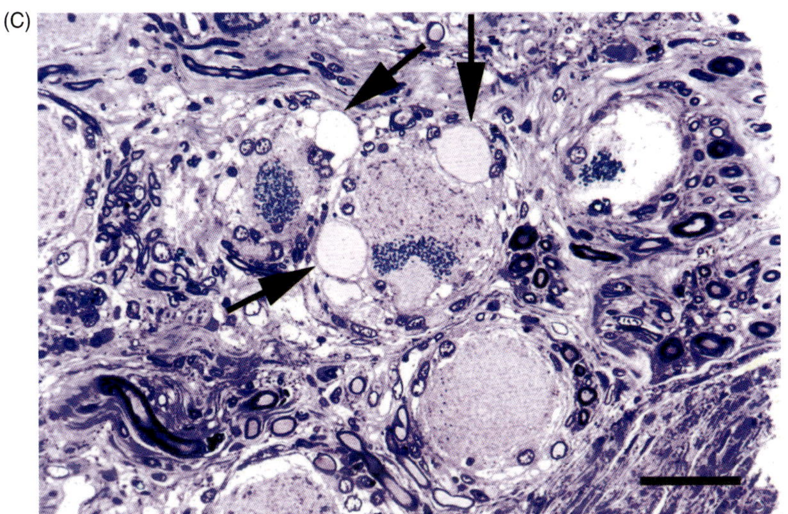

FIG. 3. (*continued*)

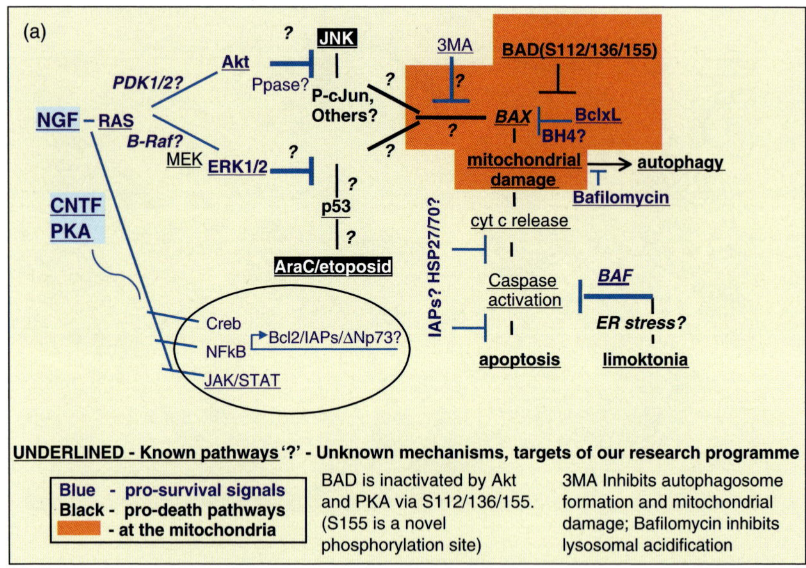

FIG. 1. (See chapter: Adoptosis in Diabetic Neuropathy) (a) Pro- and anti-apoptotic pathways in primary peripheral NGF-dependent neurons. NGF activates at least two survival signalling pathways via stimulation of Ras. The Akt (also known as PKB) pathway suppresses signalling via the pro-apoptotic JNK (stress kinase) pathway. The ERK (MAPK) pathway suppresses pro-apoptotic signals induced by p53. The precise targets of the survival kinases, and the mechanisms by which the pro-apoptotic signals induce apoptosis are still not well understood. One target is the protein BAD which is multiply phosphorylated by Akt and its downstream target RSK, and by protein kinase A. This phosphorylation keeps BAD from binding to anti-apoptotic Bcl2 family members. In addition, the transcription factors NFkB, CREB, and STATs, regulate the expression of several pro- and anti-apoptotic factors, illustrated by the examples of Bcl2 family members, IAPs, and a proposed dominant interfering inhibitor of the p53 pathways, the N-truncated splice variant of p73α. The orange box depicts mitochondria, where active Bax and Bak form pores that promote the release of pro-apoptotic factors, whose activity culminate in caspase activation and orderly demolition of the cells. Possible inhibitors are noted, for example, caspase inhibitors such as BAF (Boc.Asp(O-methyl).fluoromethylketone) and Heat shock proteins (HSP70/27). In the mitochondrial compartment, overexpression of Bcl2 will antagonise the actions of Bax/Bak and other BH3-only members of the Bcl-2 family. BH4 denotes peptides based on the protective regions of Bcl-2/Bcl-xL. The activation of autophagy, a bi-product of apoptotic signalling, is suppressed by agents such as 3MA (which is an inhibitor of autophagosome formation, but is quite nonspecific) and Bafilomycine A1 which is an inhibitor of the lysosomal H^+-ATPase. (b) Death receptor-mediated signalling pathways. The main motifs are the recruitment of caspase 8/10 to the receptor through a series of intermediate proteins recruited to the receptors upon their activation. A link into the mitochondrial pathway is depicted via cleavage of Bid, a BH3-only protein members of the Bcl-2 family. Another motif to note is the link to survival signals via JNK (which can be pro- or anit-apoptotic depending on cell context) and NFkB. (b) Courtesy of Malcolm I. Roberts.

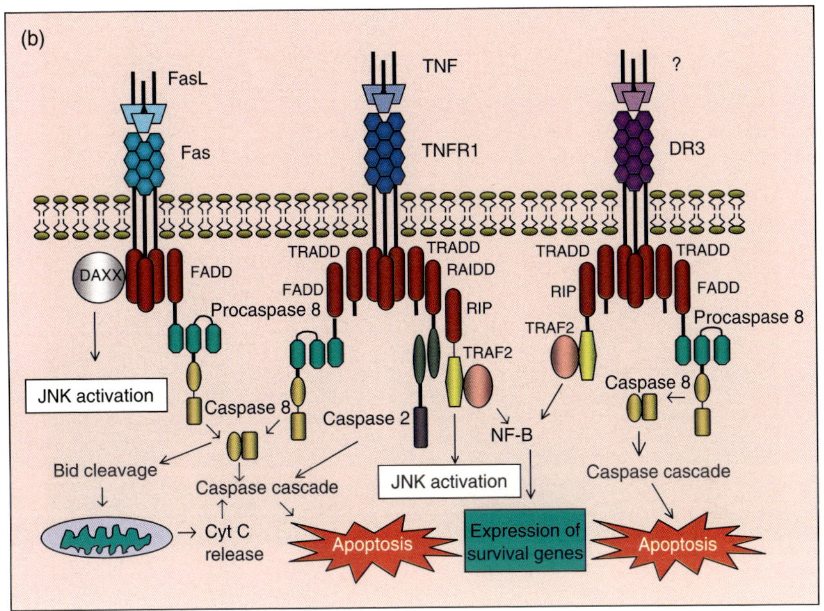

FIG. 1. (*continued*)

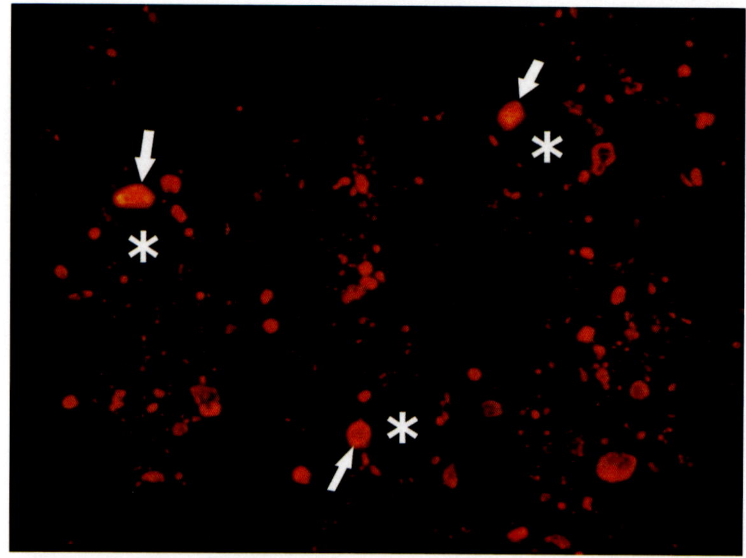

FIG. 2. (See chapter: Autonomic Neuropathy) DBH immunohistochemistry, diabetic human SMG. Swollen DBH immunoreactive dystrophic axons (arrows) cluster around relatively unlabeled perikarya (*). (Magnification: 300×)

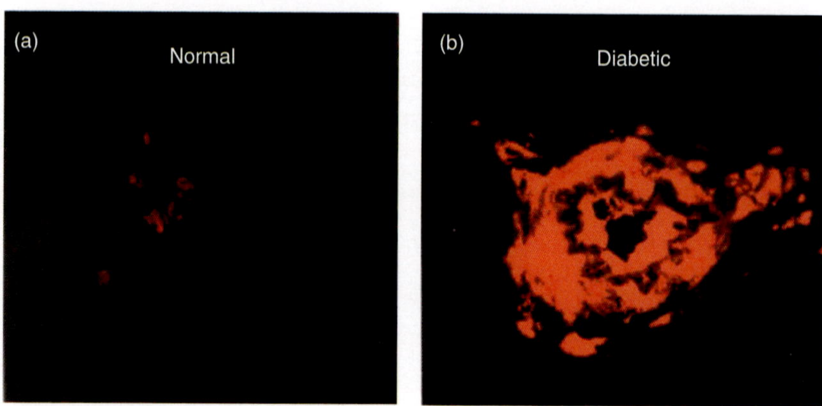

FIG. 1. (See chapter: Polyol Pathway and Diabetic Peripheral Neuropathy) Detection of superoxide levels in arterioles supplying sciatic nerve in normal and 3- to 4-week STZ-diabetic rats. Unfixed, frozen 0.5-μm-thick sections were stained with $2\ \mu M$ hydroethidine and incubated for 30 min at 37°C. In the presence of superoxide, the dye is oxidized to highly fluorescent ethidium bromide, which intercalates into DNA. Fluorescence was detected with a Bio-Rad MRC-1024 laser-scanning confocal microscope with a krypton/argon laser using a 585-nm long-pass filter. Reproduced with permission from Coppey *et al.* (2001a).

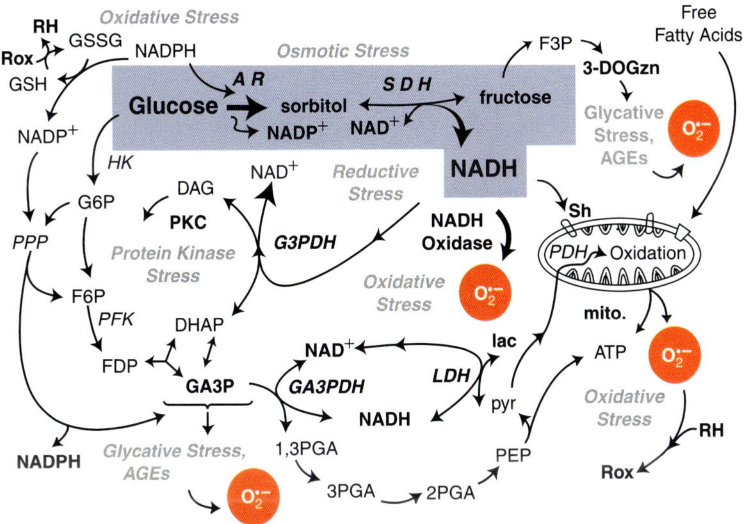

FIG. 2. (See chapter: Polyol Pathway and Diabetic Peripheral Neuropathy) Schematic of potential detrimental metabolic effects of hyperglysolia. Cytosolic glucose (upper left) is metabolized typically via hexokinase (HK) to glucose-6-phosphate (G6P) and to pyruvate (pyr) and ATP (lower right); the pathway for glycogen synthesis is omitted for simplicity. pyr enters cell mitochondria (mito.) and is further oxidized via pyruvate dehydrogenase (PDH), the citric acid cycle, electron transport, and oxidation phosphorylation processes (''oxidation'') to yield ATP (lower right). In some cell types, hyperglysolia stimulates glycolysis leading to increased turnover of NAD^+ via glyceraldehyde-3-phosphate dehydrogenase (*GA3PDH*) and production of pyruvate. pyr undergoes mitochondrial oxidation, although under normal conditions at rest this process is limited by the availability of ADP, and most pyr is converted in the cytoplasm to lactate (lac) by lactate dehydrogenase (LDH) with regeneration of NAD^+. In many cell types hyperglysolia stimulates metabolism through aldose reductase (AR) and sorbitol dehydrogenase (SDH) (shaded area) with a number of consequences, including (a) elevation of sorbitol and fructose metabolite pools (osmotic stress, upper center), (b) increased 3-deoxyglucosone (3-DOGzn), a highly reactive glycating agent (glycative stress and AGE formation, upper right), and (c) raised cytosolic $NADH/NAD^+$ ratio (reductive stress, center). Reductive stress can trigger excess production of reactive oxygen species, e.g., superoxide ($O_2^{·-}$) (oxidative stress), via (a) reaction of NADH with NADH oxidase (NADH Ox.) (center), and (b) overload of mitochondrial coenzyme shuttles (Sh) and matrix with NADH (lower right). In some cases, consumption of NADPH by AR can impair glutathione-based antioxidant defense (upper left, oxidative stress). Finally, plentiful substrate flux through HK concomitant with a high $NADH/NAD^+$ ratio can (a) cause a buildup of GA3P, a potent glycating agent (glycative stress, lower left), and (b) push metabolic flow of GA3P to α-glycerophosphate, a precursor of diacylglycerol (DAG), an activator of protein kinase C (PKC) (protein kinase stress, left center). See text for further details. 1,3PGA, 1,3-bisphosphoglyceric acid; 2PGA, 2-phosphoglyceric acid; 3PGA, 3-phosphoglyceric acid; DHAP, dihydroxyacetone phosphate; F3P, fructose-3-phosphate; F6P, fructose-6-phosphate; FDP, fructose-1,6-diphosphate; G3PDH, glycerol-3-phosphate dehydrogenase; GSH, reduced glutathione; GSSG, oxidized glutathione; $NADP^+$, oxidized NADPH; PEP, phosphoenolpyruvate; PFK, phosphofructokinase; PPP, pentose phosphate pathway; RH, reduced cellular molecule; Rox, oxidized form of RH. Modified from Oates and Mylari (1999).

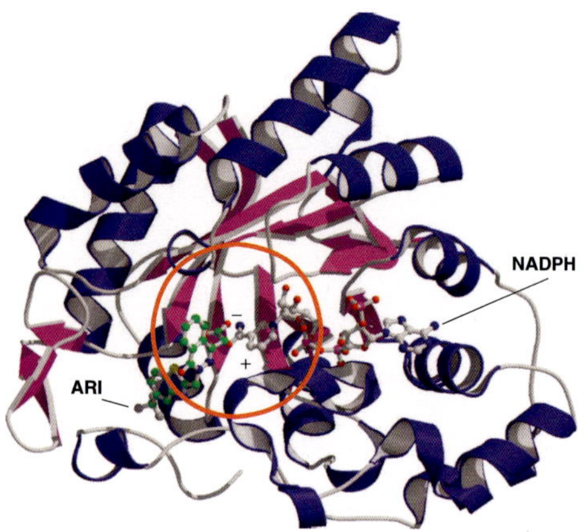

FIG. 3. (See chapter: Polyol Pathway and Diabetic Peripheral Neuropathy) X-ray structure of human aldose reductase with bound NADPH and ARI zopolrestat. The NADPH is in an extended conformation, leading from the adenine moiety along the ribose diphosphate backbone to the nicotinamide ring at the active site of AR, highlighted by the orange circle. Under conditions of steady-state turnover, the major species present is the enzyme-NADP$^+$ species in which the positively charged nicotinamide ring ("+") is positioned to interact with the negatively charged carboxylic acid moieties of ARIs such as zopolrestat ("−") (Harrison *et al.*, 1994; Bohren and Grimshaw, 2000). ARIs have strong hydrophobic interactions within the catalytic pocket as well. Ribbon diagram of human aldose reductase complexed with NADPH and zopolrestat [1MAR.pdb; Wilson *et al.* (1993)] prepared by V. L. Rath, using Molscript and Raster3D Image.

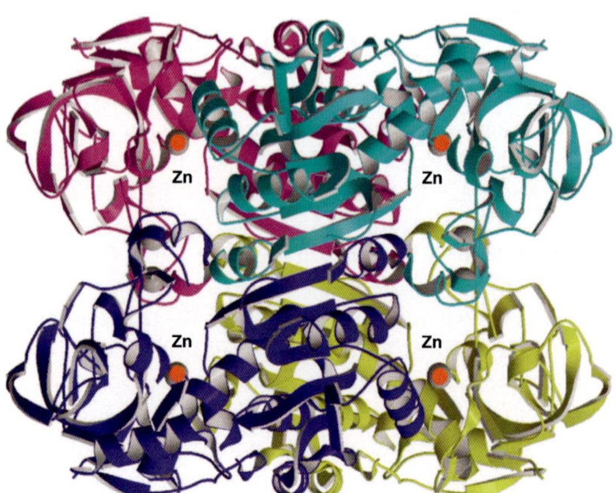

FIG. 6. (See chapter: Polyol Pathway and Diabetic Peripheral Neuropathy) X-ray structure of sorbitol dehydrogenase homotetramer from *Bemisia argentifolii*, the silverleaf whitefly, the first SDH structure published. Zn atoms at the catalytic sites are marked in orange. While the mammalian SDH has only the one catalytic Zn per monomer (Jeffery *et al.*, 1984b), the whitefly enzyme has a second structural Zn in each subunit, visible as gray balls near the equatorial axis of the image shown. Ribbon diagram of the homotetramer of silverleaf whitefly SDH with structural and catalytic zincs [1E3J.pdb; Banfield *et al.* (2001)] prepared by V. L. Rath, using Molscript and Raster3D.

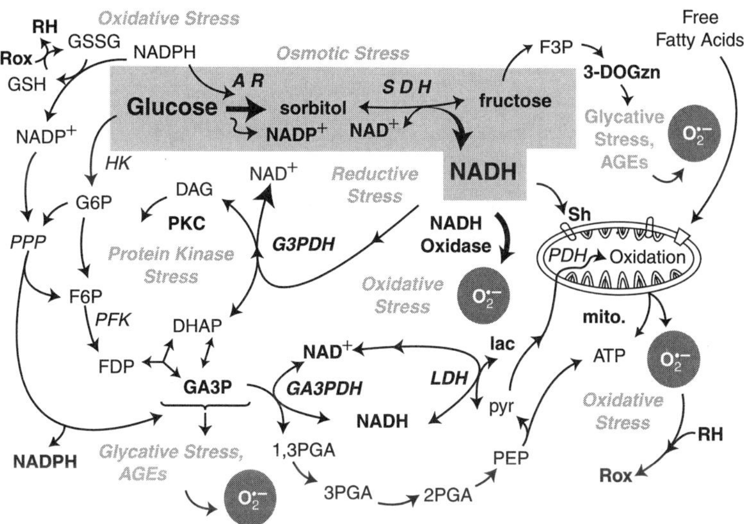

FIG. 2. Schematic of potential detrimental metabolic effects of hyperglysolia. Cytosolic glucose (upper left) is metabolized typically via hexokinase (HK) to glucose-6-phosphate (G6P) and to pyruvate (pyr) and ATP (lower right); the pathway for glycogen synthesis is omitted for simplicity. pyr enters cell mitochondria (mito.) and is further oxidized via pyruvate dehydrogenase (PDH), the citric acid cycle, electron transport, and oxidation phosphorylation processes ("oxidation") to yield ATP (lower right). In some cell types, hyperglysolia stimulates glycolysis leading to increased turnover of NAD^+ via glyceraldehyde-3-phosphate dehydrogenase (*GA3PDH*) and production of pyruvate. pyr undergoes mitochondrial oxidation, although under normal conditions at rest this process is limited by the availability of ADP, and most pyr is converted in the cytoplasm to lactate (lac) by lactate dehydrogenase (LDH) with regeneration of NAD^+. In many cell types hyperglysolia stimulates metabolism through aldose reductase (AR) and sorbitol dehydrogenase (SDH) (shaded area) with a number of consequences, including (a) elevation of sorbitol and fructose metabolite pools (osmotic stress, upper center), (b) increased 3-deoxyglucosone (3-DOGzn), a highly reactive glycating agent (glycative stress and AGE formation, upper right), and (c) raised cytosolic $NADH/NAD^+$ ratio (reductive stress, center). Reductive stress can trigger excess production of reactive oxygen species, e.g., superoxide ($O_2^{•-}$) (oxidative stress), via (a) reaction of NADH with NADH oxidase (NADH Ox.) (center), and (b) overload of mitochondrial coenzyme shuttles (Sh) and matrix with NADH (lower right). In some cases, consumption of NADPH by AR can impair glutathione-based antioxidant defense (upper left, oxidative stress). Finally, plentiful substrate flux through HK concomitant with a high $NADH/NAD^+$ ratio can (a) cause a buildup of GA3P, a potent glycating agent (glycative stress, lower left), and (b) push metabolic flow of GA3P to α-glycerophosphate, a precursor of diacylgycerol (DAG), an activator of protein kinase C (PKC) (protein kinase stress, left center). See text for further details. 1,3PGA, 1,3-bisphosphoglyceric acid; 2PGA, 2-phosphoglyceric acid; 3PGA, 3-phosphoglyceric acid; DHAP, dihydroxyacetone phosphate; F3P, fructose-3-phosphate; F6P, fructose-6-phosphate; FDP, fructose-1,6-diphosphate; G3PDH, glycerol-3-phosphate dehydrogenase; GSH, reduced glutathione; GSSG, oxidized glutathione; $NADP^+$, oxidized NADPH; PEP, phosphoenolpyruvate; PFK, phosphofructokinase; PPP, pentose phosphate pathway; RH, reduced cellular molecule; Rox, oxidized form of RH. Modified from Oates and Mylari (1999). (See also color insert.)

the hypothesis that diabetes-induced elevation of nerve sorbitol per se underlies dysfunction and pathology in the diabetic nerve.

2. *Lowering Nerve Sorbitol Correlates "Poorly" with Improved Nerve Function*

At the same time, data have emerged from several sources that make it evident that at least in rat peripheral nerve, there is a "poor," i.e., strongly nonlinear, correlation between nerve polyol levels and nerve function. Indeed, dose–response studies with several polyol pathway inhibitors show clear, graded improvements in nerve function among doses of the same inhibitor that correspond to extremely high degrees of polyol pathway metabolite suppression, i.e., where there are virtually no measurable differences between nerve sorbitol levels. These observations are difficult to reconcile with a simple osmotic hypothesis that attributes neural dysfunction to alterations in the pool size of polyol pathway metabolites, e.g., sorbitol.

3. *Evidence for a Pathogenic Role of the Polyol Pathway Continues to Mount*

In the face of the aforementioned observations, experimental genetic, biochemical, and pharmacological evidence nevertheless has continued to strengthen the link between elevated metabolism through the polyol pathway and diabetes-induced peripheral nerve dysfunction. Moreover, since 1995 there have been almost 30 reports on the possible association between certain allelic forms of the human gene for aldose reductase, the first enzyme of the polyol pathway, and clinical susceptibility to diabetic microvascular complications, including diabetic neuropathy. Many, but not all, investigators find that certain "high-expression" allelic forms of the human aldose reductase gene and/or biochemically elevated levels of aldose reductase are positively associated with the prevalence or rate of progression of diabetic complications, particularly with nephropathy and retinopathy, but also with neuropathy (reviewed in Oates and Mylari, 1999).

D. CHAPTER AIMS AND OUTLINE

Against the background of these exciting new developments, this chapter critically examines the concept of the polyol pathway and how it relates to the pathogenesis of diabetic peripheral neuropathy. The two enzymes of the polyol pathway, aldose reductase and sorbitol dehydrogenase, are reviewed. The structure, biochemistry, physiological role, tissue distribution, and localization in peripheral nerve of each enzyme are summarized, along with current information about the location and structure of their genes, their alleles, and the possible links of each enzyme and its alleles to diabetic neuropathy. Inhibitors of pathway enzymes and results obtained to date with pathway inhibitors in experimental models and human neuropathy trials are updated and discussed. Recent experimental and clinical data are analyzed in the context of a newly developed metabolic model of the *in vivo* relationship between nerve sorbitol concentration and metabolic flux through aldose reductase. Overall, the data are interpreted as supporting the hypothesis that *metabolic flux* through the polyol pathway, rather than nerve concentration of sorbitol, is the predominant polyol pathway-linked pathogenic factor in diabetic peripheral nerve. Finally, key questions and future directions for basic and clinical research in this area are considered. It is concluded that robust inhibition of *metabolic flux* through the polyol pathway in peripheral nerve will likely result in substantial clinical benefit in treating and preventing the currently intractable condition of diabetic peripheral neuropathy. To accomplish this, it is imperative to develop and test a new generation of "super-potent" polyol pathway inhibitors.

II. Polyol Pathway

A. Osmotic Hypothesis

The polyol pathway, first described in 1956 by Hers, consists of two oxidoreductases: aldose reductase (AR) and sorbitol dehydrogenase (SDH) (Fig. 2, shaded area). Acting in concert with the appropriate coenzyme, AR can transform glucose into sorbitol, and SDH can convert sorbitol into fructose. The presence of metabolites of the polyol pathway in rat peripheral nerve and their elevation in the diabetic state were first demonstrated by Gabbay *et al.* (1966). During the same period, the now classic studies of Kinoshita and colleagues on cataractogenesis in the diabetic and galactosemic rat lens gave rise to "the osmotic hypothesis." This concept emphasizes the pathogenic centrality of osmotic stress resulting from accumulation of polyol pathway intermediates, particularly sorbitol. In this paradigm, an abnormally high level of glucose is metabolized via the polyol pathway to create elevated intratissue levels of poorly membrane-permeable sorbitol and fructose. The accumulation of polyols results in osmotic imbalance, lens swelling, and ion and metabolite alterations and culminates in "sugar" cataract formation (Kinoshita, 1990). In regard to glucose-induced cataractogenesis, the osmotic hypothesis has received compelling confirmation from studies with transgenic mice that overexpress AR (Lee *et al.*, 1995a). When genetically combined with a mouse strain that has an

SDH-deficient phenotype (Holmes *et al.*, 1982), high AR-expressing hyper-glycemic mice exhibited even faster rates of lens sorbitol accumulation and cataract formation (Lee *et al.*, 1995a). These data provide an important *in vivo* genetic confirmation of the osmotic hypothesis as the central mechanism of sugar-induced cataractogenesis in the diabetic rat lens.

B. METABOLIC FLUX HYPOTHESIS

1. *Evolution of the Hypothesis*

The possible pathogenic role of abnormally high metabolic flux through the polyol pathway has long been recognized (e.g., Winegrad, 1973). However, a combination of factors has hindered the development of a pathogenic hypothesis based primarily on metabolic flux rather than on polyol accumulation. These factors include the striking success of the osmotic hypothesis in the lens (Kinoshita, 1990), the numerous observations in the diabetic nerve that are in fact compatible with the osmotic hypothesis (e.g., Gabbay, 1975), and the comparatively much greater technical difficulty of experimentally measuring metabolic flux (e.g., Cheng and Gonzalez, 1986).

Nevertheless, several decades of research in this area have led to a growing realization that in many tissues the polyol pathway is integrally linked via its coenzymes to a variety of other pathways (Fig. 2). For example, Cheng and Gonzalez (1986) measured polyol pathway flux with [^{13}C]NMR in rat lens and found that hyperglycemia caused a 3000% per hour turnover of NADPH and that aldose reductase competed with glutathione reductase for NADPH. A second example is the data of Williamson and colleagues that emphasize the importance of reductive stress (Fig. 2), in large part a result of linkage of polyol pathway flux to the rate of production of cytoplasmic NADH, a coenzyme fundamental to many metabolic processes (Williamson *et al.*, 1993; Ido *et al.*, 2001).

New and related pathogenic concepts continue to proliferate, evolve, and intermingle. They currently include elevated chronic oxidative stress, carbonyl stress, excess glycolytic metabolism, altered lipid metabolism, glycative stress, and abnormal protein kinase C activity, as well as deficiencies in detoxification, neurovascular supply, growth factors, and specific lipid species (Stevens *et al.*, 1995; King *et al.*, 1997; VanDam and Bravenboer, 1997; Tomlinson, 1998; Baynes and Thorpe, 1999; Cameron and Cotter, 2000; Nishikawa *et al.*, 2000). Throughout this evolution of hypotheses, the notion of the pathogenic role of elevated flux through the polyol pathway has persevered and is still thought by many to play a fundamental role in the genesis of the biochemical disturbances caused by hyperglysolia (Fig. 2), e.g., reviewed by Oates and Mylari (1999).

2. *Metabolic Flux Hypothesis vs Osmotic and Other Hypotheses*

With respect to the involvement of the polyol pathway in the pathogenesis of diabetic peripheral neuropathy, a central question has been to define the extent to which the mechanism underlying dysfunction and pathology in the diabetic nerve resembles the prototypic osmotic disturbance in the rat lens. That is, does neural dysfunction and damage result primarily from osmotic stress—even if this is in a small, confined tissue subcompartment—or do coenzyme imbalances linked to metabolic flux (Fig. 2) play a more prominent role? Although it seems likely that both osmotic and metabolic flux pathogenic mechanisms are simultaneously operative, evidence is examined and presented herein that gives primacy to the latter possibility, i.e., that in diabetic peripheral nerve, metabolic flux through the polyol pathway plays a more critical role in causing neural dysfunction than does accumulation of nerve sorbitol.

This discussion is not meant to exclude the potential pathogenic importance of related hypotheses such as elevated *glycolytic* flux (Nishikawa, 2000) or protein kinase C overactivation (King, 1997). The relationship of polyol pathway flux to other such mechanisms constitutes an additional key question beyond the scope of this chapter. The reader is referred to discussions by other authors in this volume, who will address in detail the fundamental and complex question of the relative impact of different pathogenic mechanisms in the overall pathogenesis of diabetic peripheral neuropathy. The reader can note, however, that the polyol pathway sits quite high in the cascade of events that ensue from hyperglysolia and that the polyol pathway is linked via its coenzymes and products to a variety of other proposed pathogenic mechanisms (Fig. 2).

III. Enzymes of the Polyol Pathway

A. ALDOSE REDUCTASE (AR)

1. *Aldose Reductase, the Enzyme*

a. General Characteristics. Aldose reductase (EC 1.1.1.21; ALD2, AKR1B1, or, less formally, AR) was first studied by Hers (1960) and has since been well characterized. A member of the aldo-keto reductase superfamily (Jez and Penning, 2001), AR is a cytoplasmic (Clements *et al.*, 1969; Ludvigson and Sorenson, 1980), monomeric enzyme of ~35,900 Da with a triose phosphate isomerase structural motif (Rondeau *et al.*, 1992) (Fig. 3, see also color insert). AR contains no metal ion or carbohydrate. The

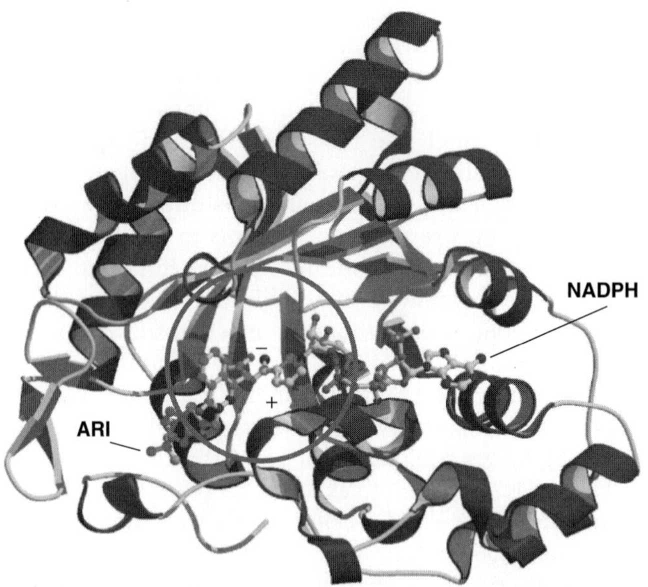

FIG. 3. X-ray structure of human aldose reductase with bound NADPH and ARI zopolrestat. The NADPH is in an extended conformation, leading from the adenine moiety along the ribose diphosphate backbone to the nicotinamide ring at the active site of AR, highlighted by the orange circle. Under conditions of steady-state turnover, the major species present is the enzyme-NADP$^+$ species in which the positively charged nicotinamide ring ("+") is positioned to interact with the negatively charged carboxylic acid moieties of ARIs such as zopolrestat ("−") (Harrison *et al.*, 1994; Bohren and Grimshaw, 2000). ARIs have strong hydrophobic interactions within the catalytic pocket as well. Ribbon diagram of human aldose reductase complexed with NADPH and zopolrestat [1MAR.pdb; Wilson *et al.* (1993)] prepared by V. L. Rath, using Molscript and Raster3D Image. (See also color insert.)

enzyme preferentially uses NADPH as a hydride donor in the "forward" direction to reduce an aldehyde substrate to the corresponding alcohol, e.g., glucose to sorbitol (Fig. 2, shaded area).

AR has broad substrate specificity, allowing it to reduce a wide variety of aldehydes (Hers, 1960). It reduces galactose with greater avidity than glucose (Kinoshita *et al.*, 1963), but its "natural" substrate is unknown (see Section IV,A,2,a). The catalytic center of AR contains a cysteine residue, Cys 298, which, when oxidized or mutated, causes AR to exhibit altered catalytic properties and insensitivity to inhibitors (Petrash *et al.*, 1992). Interestingly, although there is wide interindividual variability in the levels of AR in a particular tissue (see Section III,A,2,d), AR appears to occur primarily as a single, reduced enzyme form in human tissues, and only the reduced form of AR was found in nondiabetic and diabetic human kidneys (Robinson *et al.*, 1993). However, the enzyme is sensitive to oxidation during isolation (Jedziniak and Kinoshita, 1971; Vander Jagt *et al.*, 1990), and up to 20% of AR freshly isolated from nondiabetic human placenta was the oxidized or "activated" form (Grimshaw and Lai, 1996).

b. Kinetic Mechanism of AR and the Form of Its Glucose Substrate. The catalytic mechanism of AR is classified as ordered bi–bi with coenzyme binding first and leaving last (Grimshaw *et al.*, 1995). The enzyme binds both NADPH and NADP$^+$ tightly and with almost equal affinity (Grimshaw, 1992), but has much greater affinity for aldehydes than alcohols. The preference of AR for aldehyde substrates, coupled with a greater than 200-fold faster rate of hydride transfer in the forward vs backward direction, makes the forward reduction reaction essentially "one way" (Grimshaw, 1992; Grimshaw *et al.*, 1995).

Although the Michaelis constant (K_m) for glucose is typically reported in the 100–400 mM range (Iwata *et al.*, 1990; Sato, 1992), aldose reductase acts only on the aldehydic, straight-chain form of glucose, as opposed to the much more prevalent cyclic anomeric "boat" and "chair" forms of glucose present in solution (Inagaki *et al.*, 1982; Grimshaw, 1986). Because the straight-chain species constitutes only 0.0023% of the bulk glucose in solution (Los *et al.*, 1956; Hayward and Angyal, 1977), the K_m for straight-chain glucose is in the low micromolar range, as illustrated for recombinant rat AR in Fig. 4. For further details on the enzymology and structure of aldose reductase, the reader is referred to excellent published reviews and studies (Grimshaw, 1992; Petrash *et al.*, 1994; Grimshaw *et al.*, 1995).

2. *The Aldose Reductase Gene, ALD2*

a. ALD2 Localization and Structure. The functional gene coding for aldose reductase (*ALD2* or *AKR1B1*) resides on human chromosome 7 at locus q35 (Graham *et al.*, 1991b), a site that has also been linked to diabetic complications (Patel *et al.*, 1996; Imperatore *et al.*, 1998). The *ALD2* gene is composed of 10 exons distributed over ~18 kb that code for 316 amino acids (Chung and LaMendola, 1989; Graham *et al.*, 1991a). The basal promoter has consensus sequences for a TATA box (at −37) and a CCAAT box (at −104), and there is an androgen-like response element at −396 to −382 (Graham *et al.*, 1991a; Wang *et al.*, 1993). Three osmotic response elements, OreA, OreB, and OreC, span a 132-bp region ~1200 bp upstream of the transcription start site (Ko *et al.*, 1997). Pseudogenes of *ALD2* have also been reported (Bateman *et al.*, 1993).

b. Polymorphisms of ALD2. Three types of genetic polymorphisms associated with the *ALD2* gene have been described. One is a microsatellite (AC)$_n$ repeat region 2.1 kb upstream of the transcription start site (Ko *et al.*, 1995).

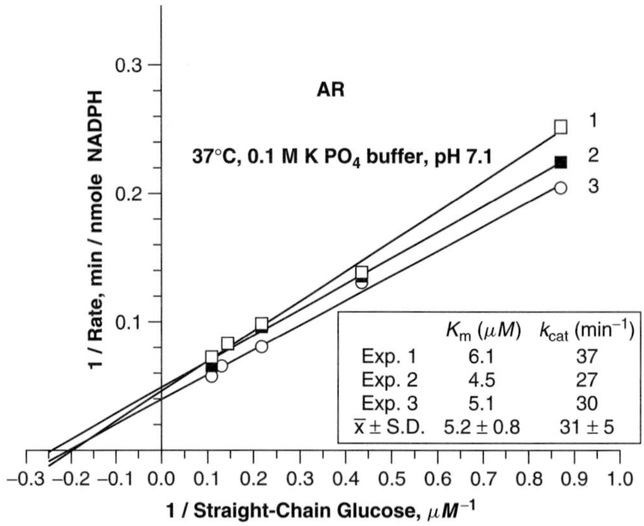

	K_m (μM)	k_{cat} (min^{-1})
Exp. 1	6.1	37
Exp. 2	4.5	27
Exp. 3	5.1	30
$\bar{x} \pm$ S.D.	5.2 ± 0.8	31 ± 5

FIG. 4. Kinetic constants of recombinant rat aldose reductase determined at pH 7.1 in 100 mM potassium phosphate buffer at 37°C. Rat AR (0.22 μM) was preincubated at 37°C with 300 μM NADPH for 10 min. The reaction was initiated by addition of the indicated amounts of glucose, and the decrease in optical density at 340 nm was monitored for 3 min at 37°C. Straight-chain percentage of glucose was calculated at 0.0023% of the nominal glucose concentration (Los *et al.*, 1956; Hayward and Angyal, 1977). Parameters were estimated from nonlinear regressions of data using GraphPad Prism software, v.3.02, GraphPad Prism Inc., but are displayed as Lineweaver–Burke plots. From Oates *et al.* (2001).

The most frequent of these variable length short-repeat alleles, (AC)$_{24}$, is arbitrarily designated "Z". The second one is a C(-106)T single nucleotide polymorphism (SNP) in the basal promoter region (Kao *et al.*, 1999a). The third type is a *Bam*HI site consisting of a single A-to-C substitution at the 95th nucleotide of intron 8 (Kao *et al.*, 1999b). The (AC)$_n$ and C($-$106)T polymorphisms are closely linked (Kao *et al.*, 1999a; Demaine *et al.*, 2000; Moczulski *et al.*, 2000).

c. ALD2 Polymorphisms and Complications Risk. In many, but not all, studies to date, certain allelic forms of these polymorphisms in the *ALD2* gene have been linked to increased risk for rapid onset or increased prevalence of diabetic complications. For example, the "Z − 2" (AC)$_n$ microsatellite polymorphism, i.e., (AC)$_{23}$, has been reported to be associated with rapid progression or increased prevalence of (a) diabetic retinopathy (Ko *et al.*, 1995; Demaine *et al.*, 2000; Chistyakov *et al.*, 2000; Olmos *et al.*, 2000); (b) diabetic nephropathy (Heesom *et al.*, 1997; Shah *et al.*, 1998; Moczulski *et al.*, 2000); and, less strongly, (c) diabetic neuropathy (Heesom *et al.*, 1998). In the latter case, a strong association was evident between overt diabetic neuropathy and a decrease in the "protective" "Z + 2" allele, i.e., $(AC)_{25}$ (linked to low AR expression; see Section III,A,3). A similar observation was made regarding diabetic retinopathy, i.e., the presence of the Z + 2 allele was associated with a lack of retinopathy (Ikegishi *et al.*, 1999).

However, several reports have not detected an association between *ALD2* alleles and complications risk (Moczulski *et al.*, 1999; Dyer *et al.*, 1999; Ng *et al.*, 2001). One Japanese study of type II diabetic (T2DM) patients, while not confirming an association of Z − 2 with proteinuria, nevertheless demonstrated a significant association of elevated erythrocyte AR content with proteinuria (Maeda *et al.*, 1999) (see also Section III,A,3). The reasons for negative findings in some genetic association studies are unclear at this time, but might result from issues of patient classification and heterogeneity, study design, or sample size (Oates and Mylari, 1999). For example, one important question in studies on nephropathy is whether lack of proteinuria (in the "nonnephropathic" group) in fact corresponds to a lack of renal lesions; poor correspondence of these two parameters has been shown previously (Chavers *et al.*, 1989). On this point, the severity of structural changes in glomeruli from nephropathic diabetic patients has been reported to correlate strongly with the intensity of immunoreactive AR ($r = 0.79$, $p < 0.01$) (Kasajima *et al.*, 2001).

d. ALD2 Polymorphisms and AR Expression. Importantly, genetic variations that confer increased risk for diabetic microvascular complications are the same alleles that result in high expression levels of AR. For example, the "CC" SNP of C(-106)T in the basal promoter of the *ALD2* gene, which is associated with the rapid progression of retinopathy in adolescent type I diabetics (Kao *et al.*, 1999a), has also been reported in cultured human retinal pigmented epithelial cells to be associated with elevated levels of *ALD2* transcription, AR mRNA, AR protein, and AR activity (Stevens *et al.*, 2000a). This "CC" SNP presumably enhances the efficiency of binding of transcription factors, as the -106 location is adjacent to the CCAAT box at -104 (Graham *et al.*, 1991b; Wang *et al.*, 1993). A CCAAT box is a widely employed promoter element in eukaryotic genes that influences the efficiency with which RNA polymerase transcribes the associated gene (Karp, 1999).

The Z − 2 microsatellite allele of *ALD2* is associated with two- to threefold elevated AR mRNA in human peripheral white blood cells (Shah *et al.*, 1998; Hodgkinson *et al.*, 2001a) and with elevated erythrocyte AR activity in diabetic and nondiabetic patients (Zou *et al.*, 2000) (Fig. 5a). "Short" (*per se*) microsatellite alleles of AR, including Z − 2 and Z − 4, i.e., $(AC)_{23}$ and $(AC)_{22}$, have also been linked with retinopathy (Fujisawa *et al.*, 1999), and the Z − 4 allele has also been associated with elevated erythrocyte AR activity and reporter gene expression (Ikegishi *et al.*, 1999). The molecular mechanism by which Z − 2 or Z − 4, located 2.1 kb upstream of the *ALD2* basal promoter region, influences the rate of transcription and expression at the *ALD2* gene is under active investigation. $(AC)_n$ elements are suggested to possibly play a role in chromatin organization and/or three-dimensional stabilization or enhancement of binding of transcription factor complexes (Brahmachari *et al.*, 1995).

3. *AR Activity and Diabetic Neuropathy*

The association of short *ALD2* alleles, e.g., Z − 2, with increased erythrocyte AR activity (Fig. 5a) is particularly relevant to diabetic neuropathy, as T2DM patients with duration of diabetes less than 10 years show a

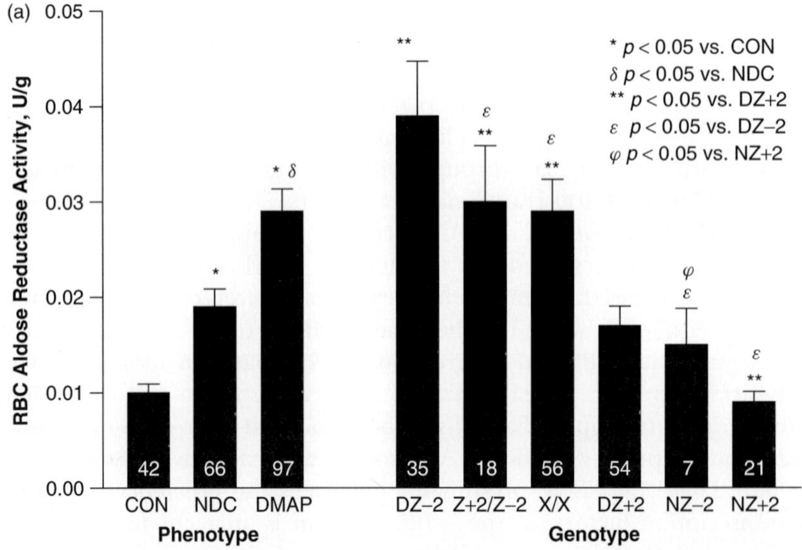

FIG. 5. Relationship among $(AC)_n$ microsatellite allele type of the human aldose reductase (AR) gene, enzymatic activity levels of AR in human erythrocytes, and diabetic neuropathy. (a) Erythrocyte (RBC) AR activity is seen to be elevated in noncomplicated diabetics (NDC) versus nondiabetics (CON) and is even higher in those patients that have diabetic microangiopathy (DMAP). In both diabetics and nondiabetics, the Z − 2 allele, $(AC)_{23}$, is associated with relatively high activity levels of AR. However, the Z + 2 allele, $(AC)_{25}$, is associated with relatively low AR activity. DZ − 2, diabetic patients homozygous for Z − 2; X, non Z − 2, non Z + 2; DZ + 2, diabetics homozygous for Z + 2; NZ − 2, nondiabetics homozygous for Z − 2; NZ + 2, nondiabetics homozygous for Z + 2. Data plotted from Zou *et al.* (2000). (b) In patients with a known duration of diabetes less than 10 years, the increasing immunoreactive content of RBC AR is associated with the increasing prevalence of diabetic neuropathy. From Tanimoto *et al.* (1998).

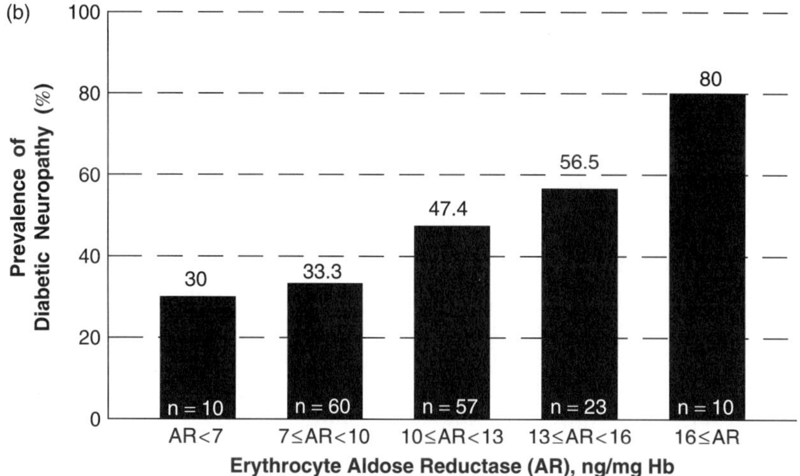

FIG. 5. (*continued*)

prevalence of diabetic neuropathy that is significantly increased in proportion to erythrocyte AR protein levels ($p < 0.05$) (Tanimoto *et al.*, 1998) (Fig. 5b). In the same study, the AR content of canine erythrocytes was found to correlate with that of sciatic nerve ($r = 0.62$, $p < 0.05$). Note in Fig. 5a that the $Z + 2$ allele is associated with comparatively low AR activity in both diabetic and nondiabetic individuals. In the as yet limited data available for diabetic neuropathy, while there was an increased frequency of $Z - 2$ alleles in diabetic patients with neuropathy in Plymouth, United Kingdom, there was an even stronger statistical association of neuropathy with a decreased frequency of $Z + 2$, $p < 0.00001$ (Heesom *et al.*, 1998). It can be concluded that while additional, conclusive data are still needed, high expression levels of AR in human diabetics (e.g., high frequency of $Z - 2$ alleles) or the lack of low expression levels of AR (e.g., low frequency of $Z + 2$ AR alleles) appear in many, but not all, studies to enhance susceptibility to and/or accelerate the progression rate of diabetic complications, including peripheral diabetic neuropathy.

B. SORBITOL DEHYDROGENASE (SDH)

1. Sorbitol Dehydrogenase, the Enzyme

a. General Characteristics. Sorbitol dehydrogenase (Blakley, 1951) (EC 1.1.1.14) is an enzyme of many names, e.g., "polyol dehydrogenase" (McCorkindale and Edson, 1954) and "L-iditol:NAD$^+$ oxidoreductase"

(Smith, 1962). Its official name is "L-iditol 2-dehydrogenase" (U.S. National Library of Medicine), while its systematic name is "L-iditol:NAD$^+$ 2-oxidoreductase" as given in the BRENDA system, which also lists eight other names in addition to those mentioned (accessible via ENZYME, Bairoch, 2000). The amino acid sequence of SDH has been determined for sheep, rat, human, silk worm *Bombyx mori*, yeast *Saccharomyces cerevisiae*, and bacteria *B. subtilis*; see Carr and Markham (1995). SDH is a member of the superfamily of medium-chain dehydrogenase/reductases (Persson *et al.*, 1994), and the X-ray structures of whitefly and rat SDH have been determined (Fig. 6, see also color insert) (Banfield *et al.*, 2001; Johansson *et al.*, 2001). The native enzyme is tetrameric (Moriyama *et al.*, 1973; Jeffery *et al.*, 1981; Banfield *et al.*, 2001; Johansson *et al.*, 2001), zinc containing (Jeffery *et al.*, 1984), uses NAD(H) as a cofactor (Jeffery and Jörnvall, 1988), and has 354–357 amino acids per monomer.

Although there is some variation in early reports, the subunit molecular masses of the enzyme from livers of human, rat, sheep, and cow

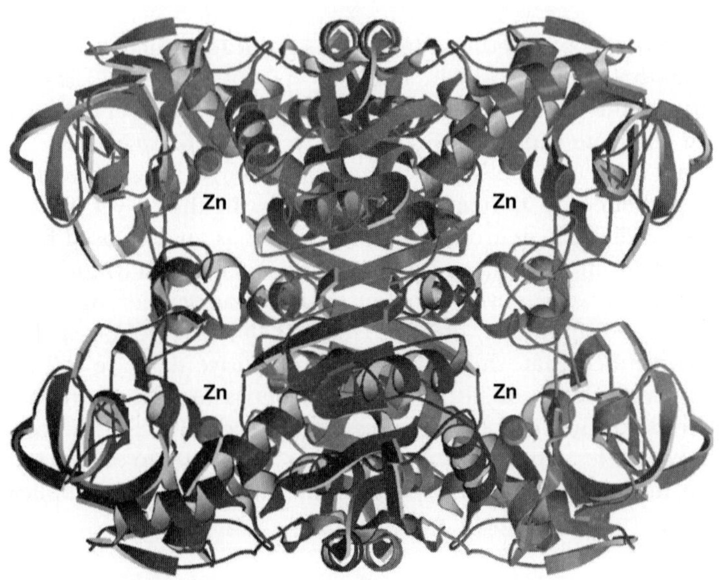

FIG. 6. X-ray structure of sorbitol dehydrogenase homotetramer from *Bemisia argentifolii*, the silverleaf whitefly, the first SDH structure published. Zn atoms at the catalytic sites are marked in orange. While the mammalian SDH has only the one catalytic Zn per monomer (Jeffery *et al.*, 1984b), the whitefly enzyme has a second structural Zn in each subunit, visible as gray balls near the equatorial axis of the image shown. Ribbon diagram of the homotetramer of silverleaf whitefly SDH with structural and catalytic zincs [1E3J.pdb; Banfield *et al.* (2001)] prepared by V. L. Rath, using Molscript and Raster3D. (See also color insert.)

are approximately 38,000–40,000 Da, with each tetramer possessing a molecular mass of about 140,000–160,000 Da (Walsall *et al.*, 1978; Karlsson, 1994). Ligand-binding studies in our laboratory with [^{14}C]NAD$^+$ and [^{3}H]sorbitol show a ratio of one substrate site and one coenzyme site per 38,000 Da of SDH, consistent with four coenzyme and four substrate sites per mole of 152,000 Da tetramer sheep SDH (Fig. 7). Each subunit of the mammalian SDH contains one catalytic zinc ion (Jeffery *et al.*, 1984b; Karlsson and Höög, 1993). The N-terminal amino acid of human and sheep SDH is acetylated (Fairwell *et al.*, 1984; Jeffery *et al.*, 1984a; Karlsson *et al.*, 1989).

b. Subcellular Localization and Isoforms. Histochemistry shows that SDH is primarily a cytoplasmic enzyme (Cohen, 1961; Chida *et al.*, 1975). Subcellular fractionation of human liver shows that a minor portion of SDH may be associated with the microsomal fraction, possibly linked to the

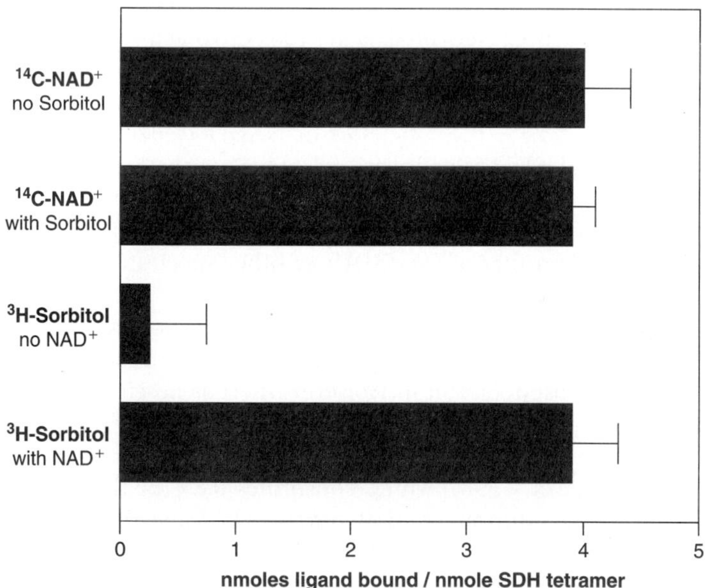

FIG. 7. NAD$^+$ and sorbitol ligand binding to sheep SDH. [^{14}C]NAD$^+$, [^{3}H]sorbitol, or both were mixed at 4°C with 10 nm sheep liver SDH in 100 m*M* potassium phosphate buffer, pH 7.0, and centrifuged at ~100 g for 2 min at 4°C in Microcon microseparators (Amicon, Beverly, MA, 20 kDa cutoff). Aliquots of eluate were taken for liquid scintillation counting. The microseparator membrane unit was inverted, cold potassium phosphate buffer was added to the upper chamber, and the unit was centrifuged as before. Aliquots of this second eluate were taken for counting to determine SDH-bound NAD$^+$ and/or sorbitol. Standard dual-label techniques for liquid scintillation counting were used. Overall recoveries were 85–110%. From Oates *et al.* (1997).

pentose phosphate pathway of the endoplasmic reticulum (Maret, 1991). A membrane-bound form of SDH was also reported in human erythrocytes (Alvarez *et al.*, 1993). There is also a particulate form of SDH that can be extracted from guinea pig liver mitochondria (Hollman and Touster, 1957). The rat liver enzyme was found to occur in multiple, similar molecular forms that are identifiable on starch-gel electrophoresis (Murray *et al.*, 1969), and the human liver enzyme shows three forms with p*I* values of 6.4, 7.5, and 8.8 (Maret and Auld, 1988). The basis for these microheterogeneities has not been identified.

   c. Substrate Specificity of SDH and the Form of Its Fructose Substrate. Although SDH shares many common structural features with alcohol dehydrogenase (Eklund *et al.*, 1985), it does not oxidize ethanol or other primary alcohols, but does react with a variety of secondary alcohols (Lindstad *et al.*, 1998). For example, in the "forward" direction, e.g., see shaded area in Fig. 2, the enzyme uses NAD^+ to oxidize the C2 carbon of sorbitol to a keto moiety, yielding fructose, NADH, and H^+. NADH can compete with NAD^+ in the forward reaction (Jedziniak *et al.*, 1981). No activity was found with $NADP^+$ or NADPH for rat or mouse SDH (Leissing and McGuinness, 1982; Burnell and Holmes, 1983), but NADPH cocatalyzed the backward reaction of human SDH at 25% the efficiency of NADH (Jedziniak *et al.*, 1981). Relative to sorbitol, SDH metabolizes galactitol poorly (Smith, 1962; Maret and Auld, 1988) or not at all (Kinoshita *et al.*, 1963; Murray *et al.*, 1969; Leissing and McGuinness, 1982). The substrate specificity of human liver SDH is broader than that of liver SDHs of other species (Maret and Auld, 1988).

   In contrast to the strong unidirectional catalytic preference of AR (Section III,A,1,b), SDH readily catalyzes its "reverse reaction," e.g., the reduction of fructose to sorbitol. However, there is no clear evidence that SDH (with AR) can generate glucose from fructose *in vivo* (Kinoshita *et al.*, 1963). The form of fructose SDH binds or releases is likely the straight-chain species (Maret, 1996), which constitutes 0.8% of the total fructose in an equilibrated solution (Angyal, 1984). The catalytic center of SDH evidences stereospecificity: the enzyme catalyzes the oxidation of the $R(-)$-enantiomer only of 1,2-propanediol (Lindstad and McKinley-McKee, 1993).

   d. Kinetic Mechanism of SDH. The kinetic mechanism of SDH is compulsory ordered (Theorell-Chance) bi–bi with coenzyme binding first and leaving last (Lindstad *et al.*, 1992; Lindstad and McKinley-McKee, 1995; Marini *et al.*, 1997). Although there were some early proposals of alternative mechanisms, such as rapid equilibrium random (O'Brien *et al.*, 1983; Leissing and McGuinness, 1983), the compulsory ordered nature of the mechanism is supported by later kinetic studies with pure enzyme and by the demonstrable sorbitol-independent binding of $[^{14}C]NAD^+$ to SDH,

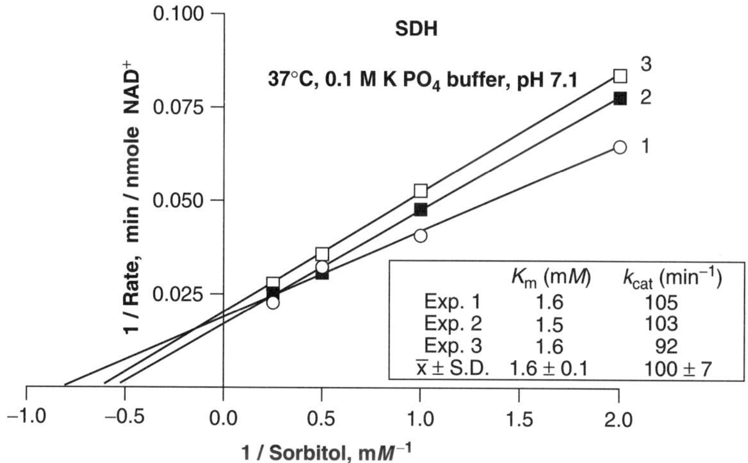

	K_m (mM)	k_{cat} (min⁻¹)
Exp. 1	1.6	105
Exp. 2	1.5	103
Exp. 3	1.6	92
$\bar{x} \pm$ S.D.	1.6 ± 0.1	100 ± 7

FIG. 8. Kinetic constants of recombinant rat sorbitol dehydrogenase at pH 7.1 in 100 mM potassium phosphate buffer at 37°C. Rat SDH (0.02 μM) was preincubated with 2 mM NAD$^+$ for 10 min at 37°C. The reaction was initiated by addition of the indicated amounts of sorbitol, and the decrease in the optical density at 340 nm was monitored for 3 min. Parameters were estimated from nonlinear fits of data using GraphPad Prism software, v. 3.02, GraphPad Prism Inc., but are displayed as Lineweaver–Burke plots. From Oates *et al.* (2001).

coupled with NAD$^+$-dependent binding of [^{3}H]sorbitol to SDH (Fig. 7). Kinetic constants for rat SDH at 37°C in 100 mM potassium phosphate buffer, pH 7.1, are given in Fig. 8. The K_m for sorbitol for the recombinant rat enzyme under these conditions is 1.6 mM, a value consistent with literature data (Smith, 1962; Burnell and Holmes, 1983; Karlsson, 1994). Note that under the approximately physiological conditions of these kinetic measurements, the catalytic rate constant, k_{cat}, is approximately three times larger for SDH than for AR (Figs. 4 and 8). This constant, which reflects the maximum rate of catalytic conversion of substrate to product per active site of enzyme under saturating substrate conditions, will be one of several important factors that determines the theoretical level of nerve sorbitol in the peripheral nerve (see Section IX).

2. Sorbitol Dehydrogenase Gene, SORD

a. SORD Localization. The SDH gene, *SORD*, has been mapped to human chromosome 15 (Donald *et al.*, 1980), but there is not a consensus on its precise position on this acrocentric chromosome. Based on fluorescence *in situ* hybridization (FISH), one assignment is 15q15 (Lee *et al.*, 1994; Carr and Markham, 1995). Another assignment, based on the same method, differs by one band, 15q21.1 (Iwata *et al.*, 1995). Two very closely homologous *SORD* sequences, *SORD1* and *SORD2*, are reported to lie within 0.5 megabases of each other at 15q15.3 (Carr *et al.*, 1998). Differing by only six amino acids, the two *SORD* sequences apparently result from a gene duplication. However, one of these copies, *SORD1*, is nonfunctional (Carr *et al.*, 1997).

b. SORD Structure. The structure of the human *SORD* gene spans approximately 30 kb, divided into nine exons and eight introns (Iwata *et al.*, 1995). In the 5′ noncoding region, classical TATAA or CCAAT elements were not found, but three Sp1 sites (CCCGCCCC) and a CACCC box were found, along with a unique repetitive $(CAAA)_5$ sequence. In all tissues analyzed, two transcriptional initiation signals were found 16 and 89 bp upstream of the translation initiation site. Human SDH mRNA includes an open reading frame that codes for 356 amino acids (Iwata *et al.*, 1995), as found for rat SDH mRNA (Karlsson *et al.*, 1991). However, sequencing rat testis SDH cDNA showed a second ATG translation start site codon 126 bp upstream from the first start site that could code for an additional 42 amino acid N-terminal peptide in a pre-SDH (Wen and Bekhor, 1993). Because pre-SDH has not yet been detected experimentally, this peptide is likely removed in posttranslational processing of the enzyme. Possibly the pre-SDH form is relevant to the mitochondrial form of SDH detected in guinea pig liver (Hollman and Touster, 1957). However, because only one ATG translation start site was reported for human *SORD* (Iwata *et al.*, 1995), it is unlikely that the reported human membrane-bound form of SDH (Alvarez *et al.*, 1993) is a pre-SDH.

c. SORD Polymorphisms. A number of electrophoretic or sequence-deduced SDH variants have been detected, consistent with possible allelic forms of the enzyme (Ibarra *et al.*, 1982, 1989; Karlsson *et al.*, 1989; Iwata *et al.*, 1995; Carr and Markham, 1995; Carr *et al.*, 1998). However, the SDH structural variants detected thus far have not been associated with diabetic neuropathy or other microvascular complications of diabetes. However, a wide variability in erythrocyte SDH activity is observed in nondiabetic humans and diabetic patients, and reduced SDH activity levels in red blood cells of certain members of particular families suggest transmission of an as yet unidentified polymorphism affecting SORD activity (Vaca *et al.*, 1982; Shin *et al.*, 1984). However, significant differences in erythrocyte SDH activity have not been seen between diabetic patients with and without cataracts, nor between patients with diabetes versus patients with idiopathic cataracts (Medina *et al.*, 1987). In contrast, plasma SDH levels were twofold higher in human diabetics vs nondiabetic controls ($p < 0.001$), and were higher still (18%, $p < 0.05$) in diabetics with vascular complications (Sgambato *et al.*, 1979).

IV. Physiological Role of Polyol Pathway Enzymes

A. Physiological Functions of AR and SDH

1. *Specialized Tissues: Fructose Production, Osmotic Regulation*

The polyol pathway was originally described by Hers (1956) as an enzymatic route for fructose formation in seminal fluid and fetal sheep liver, and the possible metabolic significance of polyols was first reviewed by Touster and Shaw (1962). Despite continuing investigation during the ensuing decades, there has been only limited progress in uncovering a fundamental physiological role for the polyol pathway. A partial exception is the specific case of the AR-rich inner medulla of the kidney. Here AR plays a role in producing intracellular sorbitol to help balance locally high osmotic interstitial pressures associated with urine concentration (Bagnasco *et al.*, 1987; Oates and Goddu, 1987; Burg *et al.*, 1997; Burger-Kentischer *et al.*, 1999). The inner medullary osmolyte system has coordinate reciprocal regulation such that pharmacological inhibition of AR results in the upregulation of other components of the system, e.g., the betaine transporter, which largely compensates for sorbitol reduction (Burg, 1996). However, in contrast to inner medullary AR mRNA, SDH mRNA levels are low and unresponsive to changes in hypertonicity (Burger-Kentischer *et al.*, 1999), suggesting that this overall body of evidence defines a physiological role more for AR than for the polyol pathway *per se*.

2. *"Typical" Tissues: AR and SDH Function Still Unknown*

a. Other Possible Metabolic Roles. Despite discovery and elucidation of the osmoregulatory role of AR in the renal inner medulla, a general physiological role for AR and SDH working together as a pathway remains obscure (Yabe-Nishimura, 1998). Because most tissues are not subjected to marked osmotic stress, one wonders if AR and SDH constitute part of a cellular system for "fine tuning" cytosolic osmolarity in response to minor osmotic variations in the cellular environment. Alternatively, is it possible that the polyol pathway plays a more fundamental role as a metabolic "fuel-sensing switch" that diverts glucose to uses other than glycolysis when fuel, i.e. glucose, is plentiful? In this regard, potential roles of sorbitol-6-phosphate (Srivastava *et al.*, 1982), sorbitol-3-phosphate (Szwergold *et al.*, 1989), and fructose-3-phosphate (Szwergold *et al.*, 1990) remain unidentified. Likewise, although it has been speculated that SDH might play a regulatory role in the formation of glycogen, the metabolic role of SDH remains "largely enigmatic" (Maret and Auld, 1981; Maret, 1996).

Other possible functions suggested for AR include participation in the metabolism of steroid (Petrash *et al.*, 1997) or norepinephrine (Kawamura *et al.*, 1997) intermediates, or detoxification of reactive aldehydes, e.g., (Grimshaw, 1992), or of glutathionylated derivatives of reactive aldehydes (Dixit *et al.*, 2000). However, it is as yet still difficult to fully embrace such proposals based on particular AR substrates because the broad substrate specificity of AR overlaps that of abundant, closely related enzymes like aldehyde reductase (Rees-Milton *et al.*, 1998) and aldehyde dehydrogenase (Vasiliou *et al.*, 2000). For example, although there may well be tissue-specific differences, the enzyme primarily responsible for reductive detoxification of 3-deoxyglucosone in rat liver has been clearly identified as aldehyde reductase (Takahashi *et al.*, 1993). Moreover, at least in diabetic and galactosemic peripheral nerve, pharmacological inhibition of AR (see Section VII,A,1,c) suppresses and reverses, rather than accentuates, markers of oxidative and aldehydic stress (Lowitt *et al.*, 1995; Hohman *et al.*, 1997a; Obrosova *et al.*, 2000).

b. AR$^{-/-}$ Mouse Is Essentially Normal.   We can now say that AR per se is not essential for life or reproduction based on the virtually normal histology and physiology of the AR$^{-/-}$ genetic "knockout" mouse (Ho *et al.*, 2000). Despite the absence of AR in all tissues, this genetically engineered mouse exhibits normal structures in all tissues and has normal biochemical parameters, reproductive behavior, and physiology. The only sign noted, which is consistent with the known osmoregulatory function of AR (Section IV,A,1), was mild polyuria, compensated by an increase in water intake (Ho *et al.*, 2000). An alteration in urine and blood divalent cation concentration has also been detected in this mouse, but it is currently unclear if the alteration is secondary to mild chronic diuresis or whether AR plays a more fundamental role in divalent cation homeostasis (Aida *et al.*, 2000). The absence of AR has no impact on nerve conduction velocity (NCV); however, in the diabetic state the absence of AR is strikingly protective of NCV (Ho *et al.*, 2001) (Section VII,B,2,a), as is pharmacological inhibition of AR in the normal and diabetic ischemic heart (Ramasamy *et al.*, 1997).

c. SDH-Deficient Sdh-1^c Mouse Is Virtually Normal.   SDH per se also does not seem to be critical for life or reproduction based on the apparent physiological normalcy of the severely SDH-deficient *Sdh-1^c* mouse (Holmes *et al.*, 1982). *Sdh-1^c* or C57Bl/LiA "null" SDH mice have 0–7% of normal SDH tissue protein and activity levels (Holmes *et al.*, 1982; Ng *et al.*, 1998). This deficiency results from a spontaneously arisen point mutation in the SDH gene at the splice junction of exon 8 and intron 8 that causes aberrant splicing of the mRNA (Lee *et al.*, 1997). Tissues of the SDH-1^c mouse that have little or no evidence of SDH activity by electrophoresis or by enzyme assay include sciatic nerve, liver, kidney, testis, heart, uterus, ovary, brain, seminal vesicles, epididymis, adrenals, intestine, and stomach (Holmes *et al.*, 1982). The "apparently healthy" SDH-deficient mice showed no obvious abnormalities in kidney, liver, lens, ileum, testis, coagulating gland, or seminal vesicle and showed normal blood glucose, weight gain pattern, reproductive pattern, survival rates, lens, and nerve function (Ng *et al.*, 1998).

B. Tissue Distributions of AR and SDH

1. *AR Tissue Distribution*

Insight into the physiological function of a protein can sometimes be gained through examination of its tissue distribution, e.g., the osmolyte function of AR (Section IV,A,1). However, but for the exception just mentioned, tissue distribution information for AR has been unhelpful because AR is present in all tissues examined (Markus *et al.*, 1983). In a survey of human tissues, immunoreactive AR was most abundant in the inner medulla of kidney (29 μg/mg protein) (Tanimoto *et al.*, 1998), consistent with the known role of AR in renal osmoregulation. The human tissue second richest in AR is sciatic nerve (5 μg/mg protein). This was followed by lens (3 μg/mg), testis and heart (2 μg/mg each), cornea (1 μg/mg), and liver, renal cortex, stomach, spleen, lung, small intestine, and colon ranging downward from 0.8 to 0.4 μg/mg. Of blood cells, monocytes were richest on a per cell basis at 44 ng/10^6 cells. Neutrophils were a distant second at 2 ng/10^6 cells, some sevenfold higher still than erythrocytes, at 0.3 ng/10^6 cells (Tanimoto *et al.*, 1998). AR was also immunochemically detected in human adrenal, aorta, brain, muscle, and placenta (Markus *et al.*, 1983; Srivastava *et al.*, 1984; Grimshaw and Mathur, 1989). A generally similar distribution pattern is seen in other species, e.g., rat, although there are some notable quantitative exceptions, e.g., the rat lens has much more AR activity than the human lens (Markus *et al.*, 1983).

2. *SDH Tissue Distribution*

SDH is also distributed in virtually all tissues and shows wide interspecies variability (Vaca *et al.*, 1984). In rat, the 1.8-kb SDH mRNA was detected in all tissues examined except small intestine (Estonius *et al.*, 1993). SDH mRNA was most abundant in rat testis, with high levels in liver, kidney, and lung. Muscle, duodenum, and eye were lowest, some 30-fold less than testis. Enzymatic SDH activity was detected in rabbit sciatic nerve (Gabbay and O'Sullivan, 1968), but not in rat skeletal muscle (Kida, 1974) nor in guinea pig epidermis (Roelfzema *et al.*, 1976). Mice have highest SDH activity in liver, kidney, seminal vesicle, and brain (Ng *et al.*, 1998) and reportedly lack SDH mRNA in the inner medulla of the kidney (Lee *et al.*, 1995b). In human tissues, high levels of *SORD* transcripts were observed by Northern blot analysis in lens and kidney (Iwata *et al.*, 1995). In contrast, pigs reportedly have virtually undetectable SDH activity in lenses and erythrocytes (Vaca *et al.*, 1984).

V. Localization of AR and SDH in Peripheral Nerve

A. Does the Polyol Pathway Exist in Peripheral Nerve?

1. *Issue of Tissue Subcompartments*

AR and SDH are two predominantly cytoplasmic enzymes that catalyze sequentially linked metabolic steps. When both enzyme activities or metabolic products are present in tissue extracts, it is often assumed that the enzymes will be in communication *in vivo* via their intermediate metabolite, giving a biochemical conduit through the two enzymes. However, particularly in the case of a highly compartmentalized tissue such as peripheral nerve, the possibility of distinct and separate enzyme and metabolite locations within the tissue must be kept in mind, especially when there are no published data showing *in vivo* colocalization of the two enzymes in the same neural cell type.

2. *Inhibitor Effects*

The metabolites of the polyol pathway were first demonstrated in homogenates of rat peripheral nerve by Gabbay *et al.* (1966). Subsequently, selective inhibitors of aldose reductase were given to diabetic rats and found to reduce the products of both the first step and the second step, i.e., sorbitol and fructose (e.g., Yue *et al.*, 1982; Whiting and Ross, 1988; Ao *et al.*, 1991; Cameron *et al.*, 1994). Assuming reasonable specificity for AR of the inhibitors, these data imply substrate-level communication between the two enzymes, i.e., the presence of the polyol pathway in rat nerve. More recently, treatment of rats with inhibitors of SDH coordinately elevated nerve sorbitol and suppressed nerve fructose, independently confirming the presence of the polyol pathway in rat nerve (Geisen *et al.*, 1994; Tilton *et al.*, 1995; Cameron *et al.*, 1997; Obrosova *et al.*, 1999). The same pattern, elevated nerve sorbitol and lowered nerve fructose, in the SDH-deficient mouse (Ng *et al.*, 1998) is also consistent with a functional polyol pathway in the mouse nerve. In the human nerve, ARI treatment of human diabetics was shown to reduce both sural nerve sorbitol (Dyck *et al.*, 1988; Sima *et al.*, 1988; Hohman *et al.*, 1997b; Greene *et al.*, 1999) and fructose (Hohman *et al.*, 1997b), confirming the existence of the polyol pathway in human peripheral nerve. Thus, the fundamental answer to the question of whether the pathway exists in peripheral nerve is a clear affirmative. This does not necessarily imply, however, that the absolute amounts or the ratios of the two enzymes are the same in all peripheral nerves (Cameron *et al.*, 1994) or in all cell types and subcompartments within a specific peripheral nerve.

B. AR AND SDH LOCALIZATION IN PERIPHERAL NERVE, IMMUNOCHEMICAL DATA

1. *AR Immunoreactivity and Gene Probe Hybridization*

Histochemical studies of AR and SDH in peripheral nerve have yielded little useful information, likely because of technical issues (Johnson, 1965; Orosz *et al.*, 1981). However, immunohistochemical data have provided significant insight. In postmortem human sciatic nerve, AR immunoreactivity was observed in Schwann cells and in pericytes and smooth muscle cells of endo- and epineurial microvessels (Kasajima *et al.*, 2001). Notably, immunoassayed AR content was 1.8-fold higher in sciatic nerve from diabetic patients versus that from nondiabetic patients ($p < 0.05$). These important new data are consistent with previous work done in rat, mouse, and rabbit peripheral nerve that localized AR primarily to the Schwann cell (Gabbay and O'Sullivan, 1968; Ludvigson and Sorenson, 1980; Chakrabarti *et al.*, 1987; Powell *et al.*, 1991; Fu *et al.*, 1996). In rat nerve, AR immunoreaction was particularly strong in the paranodal cytoplasm of Schwann cells of myelinated fibers, but the reaction product was also detected in pericytes and endothelial cells of endoneurial capillaries (Chakrabarti *et al.*, 1987). In another study of rat nerve, AR immunoreaction was noted to be particularly intense in the paranodal cytoplasmic region of Schwann cells and in Schmidt–Lanterman clefts, as well as in terminal expansions of paranodal myelin lamellae and in the nodal microvilli (Powell *et al.*, 1991). In that study, Schwann cell cytoplasm in unmyelinated fibers reacted weakly, whereas endoneurial endothelia, pericytes, and perineurium did not react appreciably with the AR antibody. In mouse sciatic nerve, *in situ* hybridization with ^{35}S-labeled gene-specific probes revealed that AR mRNA was strongly expressed in Schwann cells and was also detected in endothelial cells (Fu *et al.*, 1996).

2. *SDH Immunoreactivity and Gene Probe Hybridization*

In postmortem human sciatic nerve, immunostaining for SDH was described as "faint," although SDH immunoreactivity was comparatively strong in epineurial arterioles (Kasajima *et al.*, 2001). Mizisin *et al.* (1997) wrote that "Schwann cells appear to lack SDH, as this is found in axons (unpublished observation)." Presumably some, if only a relatively small amount, of SDH exists in Schwann cells, consistent with the comment of Ng *et al.* (1998) that their unpublished data "on the mRNA localization of AR and SDH in the neurons and Schwann cells in the sciatic nerve of untreated mice confirm our present finding of polyol pathway flux in normal sciatic nerve." Conversely, the relative insensitivity of immunohistochemical methodology does not preclude a comparatively small amount of AR from existing in the axon. Overall, it appears from the very limited immunochemical and hybridization data available that a significant amount of SDH occurs in the epineurial blood vessels, but in terms of the distribution of SDH between axon and Schwann cell, most of the SDH is axonal.

C. AR AND SDH LOCALIZATION, BIOCHEMICAL DATA

1. *Rabbit Nerve, Enzyme Activity Measurements*

Localization of AR and SDH in rabbit peripheral nerve has been examined using enzyme activity assays coupled with a surgical procedure to induce Wallerian degeneration (Gabbay and O'Sullivan, 1968). The paradigm is that the nerve is sectioned *in situ* at a particular location, causing the distal segment of the axon to degenerate, while the distal axon-ensheathing Schwann cells, being independent cellular units, survive. It was observed that 4 days after sectioning rabbit sciatic nerve, AR activity in the distal portion did not drop, whereas 84% of SDH activity was lost (Gabbay and O'Sullivan, 1968). This was interpreted to indicate that AR was localized primarily in Schwann cells, and SDH primarily in axons, a result consistent with later immunohistochemical data (Section V,B). However, a caveat with this approach is that Schwann cells theoretically could alter their AR or SDH contents during their dedifferentiation process that follows nerve section (Poduslo *et al.*, 1985).

2. *Rat Nerve, Sorbitol Measurements*

In a protocol with a rationale and design similar to that described in the preceding paragraph, an 82% loss in (net) sorbitol-accumulating capacity was observed in homogenates prepared from distal segments of ligated rat sciatic nerve (Chandler and Miller, 1986). Although this could be taken to imply a species difference versus the rabbit (where AR activity was not lost in a similar nerve segment), the consistency of immunohistochemical and mRNA hybridization data (Section V,B,1) suggests another explanation be sought. Speculations include the possibilities that, in contrast to the rabbit nerve, the rat nerve Schwann cells in the distal segment did not remain viable, that the AR content was lost rapidly in rat Schwann cells upon dededifferentiation, or that there was an induction of SDH in the distal segment of the dedifferentiating Schwann cells. More data are needed to distinguish among the various possibilities.

A potentially related observation is that Schwann cells isolated from rat sciatic nerves and cultured in hyperglycemic media did not accumulate sorbitol, despite the presence of ARI-inhibitable AR activity (Suzuki *et al.*, 1999). Glucose likely penetrated into the cells, as cultured Schwann cells predominantly express mRNA for the high-affinity, insulin-insensitive, facilitative glucose transporter GLUT1 (Magnani *et al.*, 1998). Moreover, like renal mesangial cells (Henry *et al.*, 1999), Schwann cells fail to down-regulate glucose transport under hyperglycemic conditions (Magnani *et al.*, 1998), implying that under hyperglycemic conditions, Schwann cell cytosol likely experiences hyperglysolia and elevated flux through AR, even if sorbitol does not accumulate. Consistent with this notion, in the study of Suzuki *et al.* (1999), extracellular sorbitol did increase significantly, suggesting that the cultured Schwann cells predominantly excreted the sorbitol produced. These data serve to draw attention to the fundamentally important distinction between sorbitol accumulation, which may be undetectable, and metabolic flux through AR, which may nevertheless be substantial (see Section IX).

VI. Inhibitors of Polyol Pathway Enzymes

A. Aldose Reductase Inhibitors (ARIs)

1. *Structures and Development Status*

Aldose reductase inhibitors have been reviewed adequately in the literature, (Hotta, 1997; Yabe-Nishimura, 1998; Oates and Mylari, 1999; Costantino *et al.*, 1999). Structures of selected ARIs are shown in Fig. 9. At the moment, epalrestat (Kinedak) is the only ARI commercialized successfully, available in Japan only. Fidarestat (SNK-860), a more potent analog of spirohydantoin sorbinil (Mizuno *et al.*, 1992), is presently in phase III in Japan and is in phase II neuropathy studies in the United States. Another potent ARI, minalrestat, has successfully completed a phase II study in neuropathic diabetic patients (see Section VIII,A).

Carboxylic acids zopolrestat and zenarestat showed enough promise in phase II trials to trigger phase III studies, but these agents encountered dose-limiting hepatic and renal toleration problems, respectively, which dictated discontinuation of both development programs. A potent carboxylic acid ARI tentatively named "lindolrestat" (IDD-676) is currently in phase II U.S. trials.

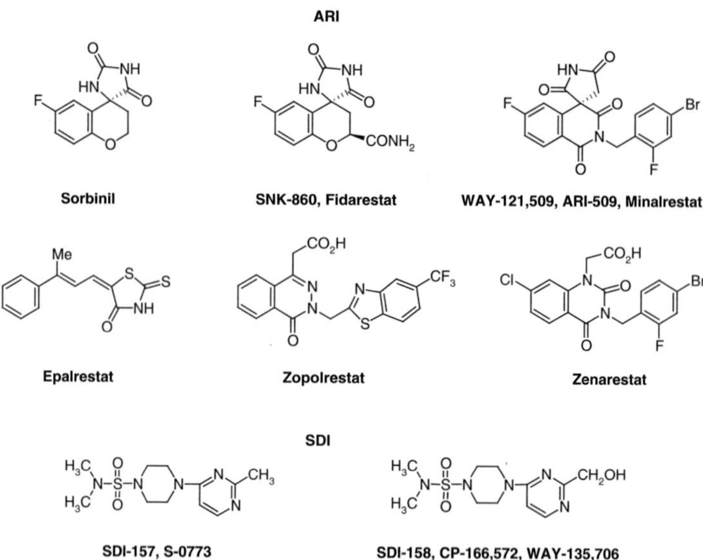

FIG. 9. Structures of selected aldose reductase inhibitors (ARIs) and sorbitol dehydrogenase inhibitors (SDIs).

2. Mechanism of Inhibition of AR by ARIs

After some initial uncertainty derived primarily from *in vitro* enzyme kinetic studies, it is now clear that ARIs bind at the active site of AR. This was first demonstrated by the X-ray structure of human AR complexed with ARI zopolrestat (Wilson *et al.*, 1993) (Fig. 3). Approximately nine other ARIs have now been seen by X-ray to be present in the active site of AR, some in slightly different binding orientations (Urzhumtsev *et al.*, 1997; El-Kabbani *et al.*, 2000; Oka *et al.*, 2000). The X-ray data prompted a reexamination of previous interpretations of enzyme kinetic data as to the mode of inhibition of ARIs. The elegant stopped flow and modeling studies of Bohren and Grimshaw (2000) have conclusively demonstrated that sorbinil and zopolrestat, and probably all ARIs capable of assuming a negative charge, bind at the active site of AR, primarily to the $NADP^+$ form of AR. This species of enzyme constitutes some 99% of the total AR enzyme species present at steady state in the aldehyde-reducing direction (Grimshaw *et al.*, 1995). Some 15-fold weaker inhibition also comes from binding of sorbinil to the NADPH form of the enzyme (Nakano and Petrash, 1996). However, the primary inhibitory effect occurs from binding of ARIs to the $AR–NADP^+$ complex, "the sorbinil trap" (Bohren and Grimshaw, 2000).

B. Sorbitol Dehydrogenase Inhibitors (SDIs)

1. *Structures and Nomenclature*

There is only one known class of sorbitol dehydrogenase inhibitors (SDIs), namely, piperazine pryimidines, discovered at Hoechst (Geisen, 1992). SDI-157 (S-0773) (Fig. 9) was initially described as a sorbitol-raising compound of unknown mechanism, but was soon found by Geisen *et al.* (1994) and others to be a prodrug, which gave rise to an active SDI, variously called SDI-158, CP-166,572, and WAY-135,706 (Fig. 9). SDI-157 and SDI-158 constitute a striking example of the pharmacological power of a small chemical change: *in vivo* hydroxylation of the methyl group of the pyrimidine ring of SDI-157 results in SDI-158, the latter an SDI with an IC_{50} of 0.25 μM, i.e., $\sim$1800 times more potent than SDI-157 (Oates *et al.*, 1994; Mylari *et al.*, 2001).

2. *Mechanism of Inhibition of SDH by SDI-158*

Inhibition of SDH by SDI-158 with respect to fructose has been reported to be competitive (Geisen *et al.*, 1994) or mixed noncompetitive (Lindstad and McKinley-McKee, 1997). We have observed both patterns and note that the slight discrepancy appears to be a function of NADH concentration used, with higher NADH giving the competitive pattern. Concordant with similar findings for ARIs and AR (Bohren and Grimshaw, 2000), kinetic data for SDI-158 and SDH suggest that SDI-158 binds primarily at the active site to the fructose-releasing NADH species of SDH. Based on the close structural similarity (Fig. 9) and on the striking $\sim$1800-fold increase in inhibitor potency of SDI-158 vis à vis SDI-157, the hydroxyl group of SDI-158 likely mimics a critical substate hydroxyl group by coordinating with the catalytic zinc ion at the active site. This hypothesis is consistent with available kinetic (Geisen *et al.*, 1994; Lindstad and McKinley-McKee, 1997), X-ray (Darmanin and El-Kabbani, 2001; Johansson *et al.*, 2001), and structure–activity data obtained with other more potent congeners of SDI-158 (Mylari *et al.*, 2001).

VII. Polyol Pathway Inhibition in Models of Diabetic Neuropathy

A. Rat Models of Diabetic Neuropathy

1. *Effects of ARIs*

a. Protective Effects of ARIs on Nerve Function and Structure.   Since the first demonstration of a positive effect on nerve conduction velocity (NCV) with

ARI sorbinil (Yue *et al.*, 1982), many laboratories have confirmed that AR inhibitors of various chemical structures suppress nerve polyols, protect and correct NCV, and, partially or completely, protect or correct other neuronal functional abnormalities, e.g., altered neural blood flow and neurovascular albumin leakage (Tilton *et al.*, 1989; Cameron *et al.*, 1994). The interested reader is also referred to numerous reviews and reports of complete or partial prevention and stabilization of nerve structural alterations (Schmidt *et al.*, 1991; Tomlinson *et al.*, 1992; Sarges and Oates, 1993; Yagihashi, 1997; Kato *et al.*, 2000; Sugimoto *et al.*, 2000). For reasons of brevity, the present discussion emphasizes primarily the *in vivo* dose–response relationship between nerve polyol pathway metabolites and NCV data in diabetic rats.

b. Dose Dependence of ARI Effects.   The *in vivo* dose–response relationship between nerve polyols and NCV in diabetic rats has now been examined for a number of ARIs (Ao *et al.*, 1991; Cameron and Cotter, 1992; Cameron *et al.*, 1994, 1996; Mizuno *et al.*, 1999; Shimoshige *et al.*, 2000). To date, all show a very similar pattern: a very high degree of suppression of nerve sorbitol (>85%) is associated with recovery (Fig. 10) or protection (Fig. 11) of NCV. This holds true whether the ARI is a noncarboxylic acid (Fig. 10) or a carboxylic acid (Fig. 11) and whether the nerve tract under study is sciatic (upper leg) (Fig. 10), distal tibial (lower leg) (Fig. 11), or caudal (Fig. 12). In the diabetic rat model, suppression of nerve fructose lags behind that of sorbitol by a factor of 3- to 20-fold (e.g., Fig. 10). Correction

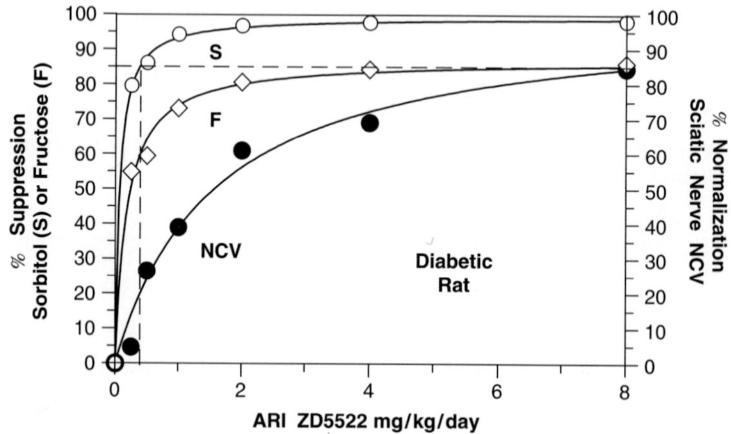

FIG. 10.   Dose–response relationship in diabetic rats between dose of sulfonylnitromethane ARI ZD5522 and suppression of sciatic nerve sorbitol, fructose, and recovery of nerve function as evidenced by improvement in sciatic nerve conduction velocity (NCV). Rats were diabetic for 2 months and were then treated daily for 1 month by oral gavage at the doses indicated. Data plotted from Cameron *et al.* (1994).

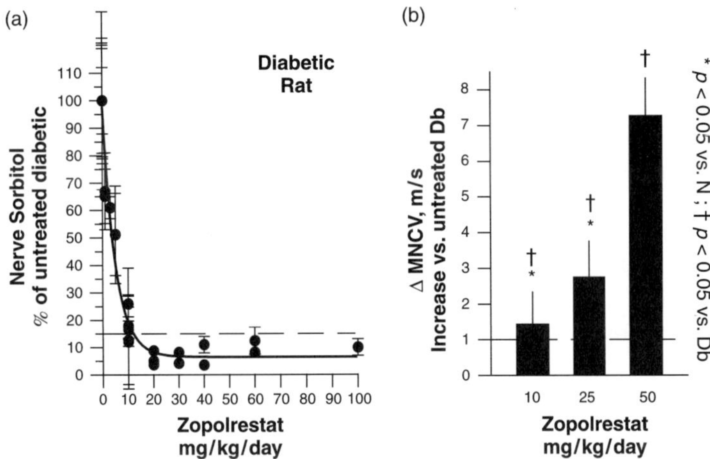

FIG. 11. Dose–response relationships in experimental diabetes with carboxylic acid ARI zopolrestat. (a) Nerve sorbitol content was measured in sciatic nerve after 5 days of once-daily dosing in rats that had been streptozocin (STZ) diabetic for 7 days when dosing was initiated. The horizontal dashed line at 15% (85% suppression) is for reference. (b) Tibial nerve NCV was measured after 4 weeks of preventative treatment with the indicated doses of zopolrestat. The horizontal dashed line at 1 m/s is for reference. Db, diabetic; N, nondiabetic. Data plotted from Oates *et al.* (1998).

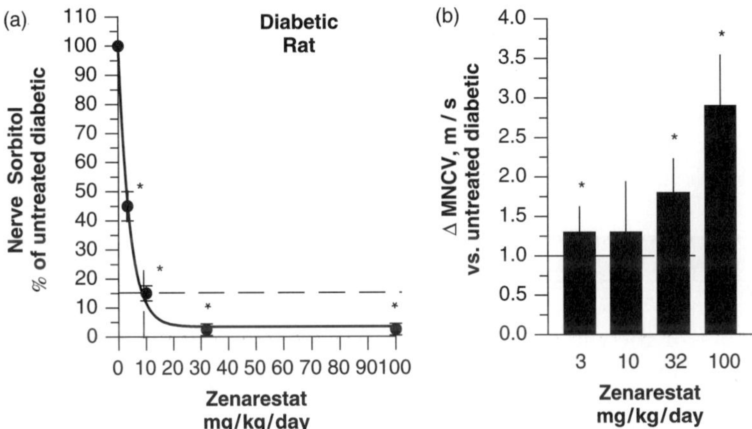

FIG. 12. Dose–response relationships in experimental diabetes with carboxylic acid ARI zenarestat. (a) Sorbitol content was measured in sciatic nerve after 2 weeks of zenarestat treatment of diabetic rats that had already been STZ diabetic for 2 weeks. (b) Improvement in caudal motor NCV was measured after 2 weeks of zenarestat treatment of rats that had already been STZ diabetic for 2 weeks. Data plotted from Ao *et al.* (1991).

of abnormal nerve blood flow requires still higher ARI doses, doses that match closely those doses necessary to correct NCV (Cameron *et al.*, 1994). Thus, as has been pointed out before (Cameron and Cotter, 1992; Cameron *et al.*, 1994), there is a "poor," i.e., seemingly very flat-sloped, correlation between nerve polyol levels and functional effects in these models. One interpretation of these data is that they are consistent with the need to normalize or overnormalize metabolic *flux* through neural AR, rather than to normalize polyol levels *per se* (see Section IX).

c. Do ARIs Have Direct Antioxidant Effects?. Another suggestion that hypothetically could underlie the reason why high ARI doses are needed to achieve robust functional effects in peripheral nerve models is that ARIs theoretically might have direct antioxidant effects, unrelated to inhibition of AR (Jiang *et al.*, 1991). This would be consistent with observations that ARIs protect reduced glutathione (GSH) levels in nerve (Hohman *et al.*, 1997a; Obrosova *et al.*, 1998) and normalize markers of oxidative stress in diabetic nerve, such as malondialdehyde (Lowitt *et al.*, 1995) and 4-hydroxynonenal (Obrosova *et al.*, 1998). Based on *in vitro* data, it has been speculated that ARIs of a particular chemical class, hydantoin, might cause some of their benefit through an antioxidant effect based on metal chelation (Jiang *et al.*, 1991). While this proposal is very difficult to evaluate experimentally in an *in vivo* setting, it can be observed that all potent ARIs examined, regardless of chemical class, exert an antioxidant effect in biological tissues and cells and that the primary basis of this antioxidant effect now appears indirect and most likely linked to inhibition of metabolic flux through aldose reductase (Fig. 2).

d. ARI Effects Likely Result from Inhibition of AR. For example, ARI zopolrestat is a member of the carboxylic acid class of ARIs, a class that does not have significant direct antioxidant properties (Jiang *et al.*, 1991). Nevertheless, despite its absence of direct antioxidant activity, zopolrestat dose-dependently suppresses nerve sorbitol (Fig. 11a), suppresses hyperglycemia-induced production of superoxide anion from vascular tissue (Gupta *et al.*, 2001), and dose dependently protects nerve function as reflected by NCV (Fig. 11b). Similar results are obtained with another member of the carboxylic acid class, zenarestat, used in an intervention protocol on a different nerve in diabetic rats (Fig. 12). A third carboxylic acid ARI, tolrestat, reportedly lowers oxygen-derived free radicals in the plasma of diabetic patients (Fondelli *et al.*, 1993) and improves impaired GSH and NADPH levels in erythrocytes of T2DM patients (Bravi *et al.*, 1997).

Moreover, results with the $AR^{-/-}$ mouse strongly support the concept that blockade of flux through the polyol pathway, rather than direct chemical antioxidant effects, underlies the antioxidant effect of ARIs: genetically deleting AR from the mouse genome results in complete protection of nerve GSH and NCV in the diabetic state (Ho *et al.*, 2001). Furthermore, in transgenic mice that overexpress aldose reductase in their lenses, "the flux of glucose through the polyol pathway was found to be the major cause of hyperglycemic oxidative stress in that tissue" (Lee and Chung, 1999). Thus, the beneficial effects of ARIs on NCV and oxidative stress end points in the diabetic rat nerve most likely result from inhibition of AR, not from putative chemical antioxidant effects. The indirect antioxidant effect of ARIs probably results from normalization of NADPH and NAD^+ coenzyme fluxes (Fig. 2). Although there is no evidence to suggest any direct effects of ARIs on mitochondrial electron transport (Nishikawa *et al.*, 2000) or NAD(P)H oxidases or other such potential sources of reactive oxygen species (Wolin, 2000), there are no data as yet to rule out these theoretically possible additional sites of action.

2. *Effect of SDIs on NCV*

a. Observations in Nondiabetic Rats. Treatment of normal, nondiabetic rats with high doses (>100 mg/kg/day) of SDI-158 caused nerve sorbitol to be elevated to levels indistinguishable from those in diabetic rats (e.g., Fig. 13a). Contrary to some early notions that the polyol pathway was largely dormant in the nondiabetic state, this observation indicated that there was significant metabolic flux through SDH even in the nondiabetic state. Consistent with the expected effect of an SDI (Fig. 2), fructose was lowered at the same time sorbitol was raised. After 5 weeks of SDI-158 treatment, the

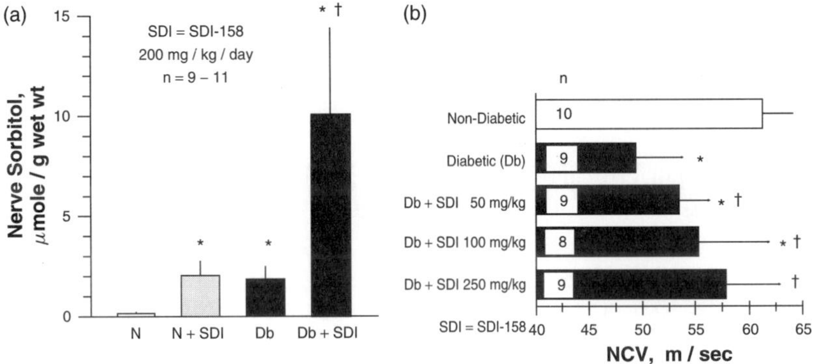

Fig. 13. Effects of SDI-158 on nerve sorbitol and on NCV in diabetic rats. (a) Normal and diabetic rats were treated shortly after STZ induction of diabetes with ~200 mg/kg/day SDI-158 in the drinking water for 5 weeks. Data plotted from Ido *et al.* (1995). (b) Diabetic rats were treated shortly after STZ induction of diabetes with ~50- to 250-mg/kg/day doses of SDI-158 in the drinking water for a period of 4 weeks, at which time tibial motor NCV was measured by standard electrophysiological methods. Data plotted from Oates *et al.* (1998).

sum of nerve sorbitol, fructose, and *myo*-inositol in nondiabetic rat nerve was increased 33% (Ido *et al.*, 1995). However, as originally observed by Geisen *et al.* (1994), SDI treatment "surprisingly" did not have an adverse impact on NCV in normal rats, an observation soon confirmed in other laboratories (Ido *et al.*, 1996; Cameron *et al.*, 1997). A caveat is that the dose of the short-acting SDI prodrug SDI-157 be greater than 100 mg/kg/day (Tilton *et al.*, 1995) (see Section VII,A,3,a).

   b. Observations in Diabetic Rats. Treatment of streptozocin (STZ)-diabetic rats with SDI-158 caused nerve sorbitol levels, elevated by diabetes, to rise even more markedly, e.g., some 500% in Fig. 13a. Nerve sorbitol builds up rapidly, within hours, in response to SDI-158, and the effect is persistent (Geisen *et al.*, 1994). After 5 weeks of SDI-158 treatment, the sum of neural sorbitol, fructose, and *myo*-inositol in STZ-diabetic rat nerve was increased 31% (Ido *et al.*, 1995). However, several laboratories reported that preventative treatment of diabetic rats with SDI-157 or SDI-158 had a protective, rather than exacerbative, effect on NCV in diabetic rats (Geisen *et al.*, 1994; Tilton *et al.*, 1995; Ido *et al.*, 1996), an effect that was observed to be dose dependent (Fig. 13b) (Oates *et al.*, 1998). Reversal of the NCV deficit in STZ rats with SDI-158 was also seen after 4 weeks of treatment (Ido *et al.*, 1999), but not after 2 weeks of treatment (Cameron *et al.*, 1997; Ido *et al.*, 1999).

   While it is always difficult to rule out completely the possibility of a secondary effect of any particular drug, taken together these observations appear inconsistent with the concept that the diabetes-induced NCV deficit results from sorbitol-induced osmotic stress. That is, if diabetes-induced nerve sorbitol accumulation was a key pathogenic driver of the NCV deficit, an agent that increases nerve sorbitol levels severalfold would be expected to exacerbate or at least accelerate, but not ameliorate, the diabetes-induced NCV deficit. Because SDI-158 has excellent cell and tissue penetration properties, it seems unlikely that it would be excluded from any potentially small, critical tissue compartment within the nerve, wherein local osmotic stress might hypothetically cause a NCV deficit in the diabetic state. Therefore, assuming SDI-158 exerts its protective effect on NCV (Fig. 13b) via inhibition of SDH, which is associated with a distinct elevation of nerve sorbitol (Fig. 13a), it is difficult to attribute a primary role to the elevation of nerve sorbitol in causing the NCV deficit. Rather these data support the notion that there is a significant role for metabolic flux through the polyol pathway, in this instance, through the second step of the pathway, in the development and maintenance of the NCV deficit in the STZ-diabetic rat.

   c. Further SDI Studies Needed in NCV Reversal Model. Normalization or near normalization of NCV in 6-week STZ-diabetic rats occurred after 2 weeks of treatment with potent ARIs, but only after 4 weeks of treatment with

SDI-158 (Cameron *et al.*, 1997; Ido *et al.*, 1999). The reason for the relative slowness of NCV recovery with SDI intervention vis à vis ARI intervention, although not understood, is intriguing. It could possibly be related to the comparatively modest potency of the SDI vs highly potent ARIs and the fact that SDI-158 is competitive with fructose (Section VI,B,2) and/or that the enzyme substrate builds up to such high levels for SDI (e.g., Fig. 13a) but not with ARI treatment (Ido *et al.*, 1999). Although the rates of reduction of nerve fructose by ARI vs SDI over the first 2 weeks were not reported in either study, nerve fructose was strongly suppressed at the 2-week time point of SDI treatment (Cameron *et al.*, 1997) and was not any lower after 4 weeks of intervention (Cameron *et al.*, 1997; Ido *et al.*, 1999). If the initial rate of normalization of polyol metabolites was in fact faster with ARI treatment than with SDI treatment, this might be taken to imply that the rate of return toward normalcy of flux through the polyol pathway could exert an important, if transient, influence on the rate of NCV recovery. It could possibly be related to a slow reversal of fructosylation or AGE disposal (Cameron *et al.*, 1997). Alternatively, it seems possible that the increased osmotic stress associated with the SDI, but not the ARI, might transiently slow down the rate of NCV recovery.

Another possibility arises from the very short half-life of SDI-158 (see next paragraph) and another facet of the demonstrated poor correlation between nerve sorbitol and fructose and nerve function. Withdrawal of treatment with ARI WAY-121,509 from diabetic rats caused relapse of GSH and NCV to pretreatment levels within 2 and 3–4 days, respectively, whereas nerve sorbitol and fructose remained below normal for 5 days (Basso *et al.*, 1998). One interpretation of these potentially important preliminary data is that oxidative stress is closely coupled to polyol pathway flux (via coenzyme turnover) and that NCV change is coupled more closely to oxidative stress and polyol pathway flux than to sorbitol and fructose levels *per se*. Available data clearly imply that normalization or even overnormalization of nerve fructose does not necessarily correspond to a condition of full blockade of neural polyol pathway flux. Therefore, although nerve fructose was overnormalized after 2 weeks of treatment with short-acting SDI-158 in the same model (Cameron *et al.*, 1997), this fact does not necessarily demonstrate that polyol pathway flux was sufficiently normalized to prevent an NCV deficit. Additional studies, with more detailed time courses and perhaps with more potent SDIs (Mylari *et al.*, 2001), are needed to clarify this potentially important point.

3. SDI Effects on Other Nerve Parameters, Biochemistry, and Structure

a. Potential Importance of SDI Half-Life. Blockade of SDH with an SDI can cause the buildup of substantial levels of sorbitol (Fig. 13a). As the drug effect wears off, abnormally high flux through SDH will ensue and create diabetes-like biochemical changes. Unfortunately, SDI-158 and its prodrug SDI-157 have a very short half-life ($t_{1/2} < 30$ min), both in serum and in nerve (Ballinger *et al.*, 1996). This property makes these compounds less than optimal pharmacological tools, necessitating that high doses be used to attempt to maintain constant SDH blockade. This pharmacokinetic property likely accounts at least in part for the finding that treatment with 100 mg/kg/day of prodrug SDI-157 caused deterioration of NCV in control, but not diabetic rats; nondiabetic rats discontinue drinking (in this case, SDI-containing) water during daylight hours, whereas diabetic rats tend not to (see Tilton *et al.*, 1995). Consistent with this interpretation, deterioration in nondiabetic control NCV was not seen in later studies at higher SDI doses of SDI-158, i.e., ≥ 200 mg/kg/day, in the same (Ido *et al.*, 1995) or other (Cameron *et al.*, 1997) laboratories.

b. Effects of SDIs on Nerve Biochemical End Points. No effects of prodrug SDI-157 were observed on nerve biochemical parameters in normal rats; however, a 100-mg/kg/day dose failed to improve or exacerbated several biochemical end points in the nerves of STZ-diabetic rats (Obrosova *et al.*, 1999). It is possible that there could be an SDI-mediated adverse impact on biochemical but not electrophysiological parameters in diabetic rats (Geisen *et al.*, 1994; Tilton *et al.*, 1995; Cameron *et al.*, 1997) (Fig. 13b). If so, this may be relevant to the comparative slowness of NCV recovery in SDI intervention vs ARI intervention (Section VII,A,2,c). However, in view of the comments in the preceding paragraph, this result will be best interpreted in light of additional data where both biochemical and electrophysiological parameters are measured in the same experiment, at a dose of active SDI-158 greater than 100 mg/kg/day, or perhaps with a more potent SDI.

c. Effects of SDIs on Nerve Structure. Morphometry of sciatic nerves from rats treated with $\sim$250 mg/kg/day of SDI-158 (CP-166,572) for 26 weeks revealed no alterations in axonal area or diameter (Ido *et al.*, 1996). However, an interaction of SDI-enhanced osmotic stress (Geisen *et al.*, 1994; Kador *et al.*, 1998) with an altered redox state was proposed to explain the acceleration of experimental neuroaxonal dystrophy observed in STZ-diabetic rats with SDI-158 treatment (Schmidt *et al.*, 1998). The morphological abnormality was not precipitated in nondiabetic rats, even though sorbitol levels in the sciatic nerve were elevated to as high a level as in diabetic rats. In contrast, ARI treatment was confirmed to inhibit this diabetes-linked autonomic morphological change in the absence or presence of SDI-158 (Schmidt *et al.*, 2001). Although the functional significance of this morphological change remains to be defined, it should be noted that this potentially adverse effect of SDI treatment was detected in the very same animals in which SDI treatment caused protection of peripheral NCV.

This highlights the need to define the possible heterogeneity of polyol pathway enzymes in different types of nerves (Cameron *et al.*, 1994). It is hoped that studies with structurally distinct SDIs and ARIs will help clarify the mechanism of these effects.

B. Mouse Models of Diabetic Neuropathy

1. *Effects of ARIs in Diabetic and Galactosemic Mice*

Mouse models have been comparatively little studied, in part because this species has low AR and is resistant to polyol accumulation. Histological sections of all nerves of the db/db mouse that were immunostained for AR were consistently negative, suggesting to the authors that the nerves of these mice lack the enzyme (Bianchi *et al.*, 1990). Another possibility is that the histological method was relatively insensitive to low levels of AR. Consistent with this latter notion, while the sciatic nerve of diabetic C57BL/Ks mice did not accumulate sorbitol or fructose, galactose feeding for 5 days resulted in marked accumulations of dulcitol, which was blocked by ARI ponalrestat (Calcutt *et al.*, 1988). The ARI also blocked a galactose-induced 22% reduction in NCV and an increase in Na^+K^+-ATPase activity (Calcutt *et al.*, 1990). A structurally distinct ARI improved NCV in the diabetic C57BL/Ks mouse (Miwa *et al.*, 1989), also indicating the likely presence of AR in the nerve of this particular mouse strain.

2. *Genetic Alterations in Polyol Pathway Enzymes*

a. Mice with Genetic Changes in Neural AR Levels. Results from studies with the $AR^{-/-}$ mouse, a strain genetically engineered to lack AR (Section IV,A,2,b), provide strong support for a critical role of AR and the polyol pathway in the development of diabetic neuropathy. Whereas STZ-diabetic wild-type mice showed a substantial deterioration in motor NCV after 4 weeks, NCV and nerve GSH were unaltered in STZ-diabetic $AR^{-/-}$ mice (Ho *et al.*, 2001). Complimentary data come from transgenic mice overexpressing human AR: when fed 30% galactose for 16 weeks, these mice had an accentuated decrease in motor NCV to 80% of the level of galactose-fed nontransgenic littermate mice (Yagihashi *et al.*, 1996). In either case, data imply that a high level of flux through AR adversely impacts nerve function as evidenced by NCV.

b. Sdh-1, the SDH-Deficient Mouse. $Sdh-1^c$ or C57BL/LiA mice are genetically deficient in SDH (Holmes *et al.*, 1982; Ng *et al.*, 1998), but have no loss of NCV in the nondiabetic state. However, in contrast to the $STZ-AR^{-/-}$ mouse, where no loss of NCV was seen (Ho *et al.*, 2001), SDH deficiency appeared to afford no protection compared to a similar fall in NCV in the diabetic C57BL/10N mouse. However, a positive control is lacking from this study, e.g., ARI treatment, to show that the NCV defect in this case is indeed responsive to polyol pathway blockade (Malone *et al.*, 1996). It will also be of great interest to examine the effects of various genetic combinations of mice with AR overexpression and SDH deficiency (Lee *et al.*, 1995a) on nerve function.

C. Canine Models of Diabetic Neuropathy

In a landmark 5-year diabetic dog study, NCV deterioration was completely prevented with ARI sorbinil (Engerman *et al.*, 1994), implying flux through the polyol pathway, or at least through AR, was critical to the development of the NCV deficit in the diabetic dog. Ulnar nerve sorbitol content was suppressed 87%; fructose was not reported. However, in galactose-fed dogs, although neural galactitol accumulated to levels some fourfold higher than in diabetic dogs, there was no NCV deficit at 44 months (Sugimoto *et al.*, 1999) or 60 months (Engerman *et al.*, 1994). One interpretation of this observation is that because galactitol is a very poor substrate for SDH (Section III,B,1,c), polyol accumulation *per se* is insufficient to induce a NCV deficit in the canine nerve, whereas flux through the second step, SDH, is critical for the development of the NCV deficit, as would occur in the diabetic canine nerve. Alternatively, because galactose feeding results in only intermittent hyperhexosemia relative to the diabetic state, it can be suggested that *sustained* flux through at least the first step of the polyol pathway is necessary for the development of the NCV deficit in this model.

VIII. Effects of Polyol Pathway Inhibitors in Human Diabetic Neuropathy

A. ARI Clinical Efficacy Data

It has long been known that ARIs have a positive effect on human NCV (Judzewitsch *et al.*, 1983). However, despite years of research with a variety of ARIs (Dvornik, 1987; Pfeifer *et al.*, 1997), it was only in 1999 that an ARI, zenarestat, was clearly shown to cause positive dose-dependent effects in human nerve (Greene *et al.*, 1999). As depicted in Fig. 14a, increasing doses of zenarestat caused increasing suppression of nerve sorbitol. This double-blind, placebo-controlled 12-month study also showed graded improvement in small myelinated nerve fiber density (not shown) and in NCV (Fig. 14b). Note that, plotted on the same scale as rat data, the ARI dose−response curve for sorbitol suppression by zenarestat in human nerve (Fig. 14a) is similar to that in the rat nerve (Fig. 12a), as well as to that with zopolrestat

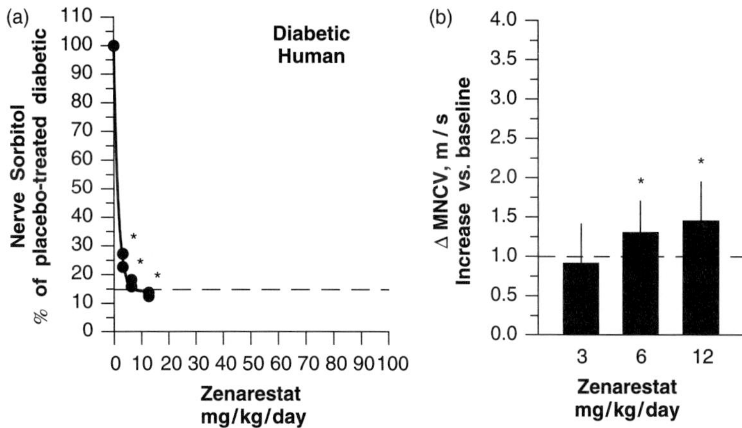

FIG. 14. Dose–response relationships in human diabetes with zenarestat. (a) Sorbitol content was measured in sural nerve biopsies after 6 and 52 weeks of each dose (drug doses are expressed per kilogram assuming 100 kg patient body weight). (b) Peroneal motor NCV was measured after 52 weeks of double-blind, placebo-controlled zenarestat treatment at the indicated doses. Compare with Fig. 12. Data plotted from Greene *et al.* (1999).

in rat nerve (Fig. 11a). Moreover, comparison of the extent of sorbitol lowering vs NCV improvement in the human nerve (Fig. 14) with similar end points for the same drug in the rat model (Fig. 12) indicates that the magnitude of the NCV response observed in the human nerve is consistent with the response noted in rat nerve.

Assuming that human nerve is similar to rat nerve in regard to linkage between polyol pathway flux and NCV response, these data further suggest that maximum possible pharmacologic inhibition of human nerve AR was not achieved with zenarestat treatment. On this point, in a 2-week phase II study, the potent ARI minalrestat (WAY-121,509) dosed at 20 mg/day showed robust lowering of both sorbitol (−87%) and fructose (−89%) in human nerve (Hohman *et al.*, 1996). This double-blind placebo-controlled study reported a 2.8 ± 0.7 m/s improvement in NCV ($p < 0.05$), although a baseline imbalance in the starting NCV values clouds this particular outcome somewhat (Hohman *et al.*, 1997b). Recall that in the rat nerve it takes 3- to 20-fold more of an ARI to reduce fructose vs sorbitol (Section VII,A,1,b) and that neither zenarestat nor any ARI, other than WAY-121,509, has been reported to show strong suppression of nerve fructose vs baseline value. One interpretation of these data is that near-maximal reduction of human nerve sorbitol may have occurred with zenarestat, but as evidenced by the presumed failure to substantially lower nerve fructose, the polyol pathway flux was still not maximally inhibited (see Sections VIII,B,4 and IX).

In a 12-month double-blind, placebo-controlled phase III trial, sorbinil analog fidarestat (SNK-860) at only 1 mg per day showed statistically significantly improved NCV parameters and, importantly, improvement in symptoms associated with diabetic neuropathy, including pain (Hotta *et al.*, 2001). If these data are confirmed in currently ongoing trials with this agent, it could represent a major step forward in therapy for diabetic neuropathy. In view of the relatively low dose used, the mechanism of such a positive result will be of interest for further study.

B. PERSPECTIVE ON ARI DIABETIC NEUROPATHY CLINICAL TRIALS TO DATE

1. *NCV End Point and Clinical Outcome*

High doses of ARIs have been used in diabetic animal models to establish a firm link between nerve function and polyol pathway activity (e.g., Figs. 10–12). However, ARIs tested thus far in double-blind, placebo-controlled trials in human diabetic neuropathy have shown encouraging, but seemingly weak, activity (e.g., Fig. 14). In addition to obvious dose–response considerations (Sections VIII,A and VIII,B,3-4), the 1- to 2-m/s improvement in NCV obtained with ARIs thus far should be viewed in context of the biological linkage between NCV in diabetic human nerve and clinical outcomes for diabetic peripheral neuropathy (Oates, 1997). Unfortunately, there are extremely little data available on this point. One research group studied the effect of intensified glycemic control for 4 years on nerve function in 55 poorly controlled type I diabetic (T1DM) patients, more than one-third of whom had chronic symptomatic lower limb sensory neuropathy at baseline (Ziegler *et al.*, 1992). Patients were grouped as a function of the degree of glycemic control achieved over months 3–48. The near-normoglycemic group (HbA$_1$ < 7.8%, $n = 19$) showed a 56% reduction in symptoms at 4 years ($p < 0.05$), but exhibited *no improvement in sural or peroneal NCV at 6 months or at any time during the 48 months*. Another even smaller study of the effect of improved glycemic control on nerve end points in neuropathic patients showed no effect on NCV at 4 months, but a 1.3-m/s improvement in NCV at 8 months and a significant improvement in vibratory sensation threshold, but with no relief of symptoms (Service *et al.*, 1985). Larger scale studies of this nature are sorely needed to define the relationship between improved glycemic control and neurological end points. Nevertheless, comparison of ARI results with these limited data offer a framework for assessment of what might be expected of a treatment designed to mimic improved glycemic control. An improvement of 1 m/s in a matter of 9 weeks (Judzewitsch *et al.*, 1983) or in 12 weeks (Arezzo *et al.*, 1996) with ARI treatment is a relatively robust effect by comparison.

2. NCV and Neuropathy End Points in the DCCT

A large trial, such as the landmark Diabetic Complications and Control Trial (DCCT), although initially selecting uncomplicated or retinopathic rather than neuropathic, T1DM patients, is nevertheless instructive, as NCV and clinical neuropathy end points were evaluated yearly in subsets of several hundred patients. In the intensified glycemic control group, a persistent rise of 1 m/s per year was observed during the first 2 years only. Importantly, however, at 5 years this group had a highly significant 64% reduction in the prevalence of diabetic peripheral neuropathy (DCCT Research Group, 1995). Such data, while limited, provide a more realistic backdrop for therapeutic expectations for ARI therapy, and again offer encouragement as to the therapeutic potential of a positive 1- to 2-m/s NCV effect maintained over the longer term. However, from a practical point of view, in order to gain commitment for marshaling the considerable resources necessary to execute any such longer term trial (Pfeifer *et al.*, 1997), a candidate ARI will have to demonstrate clear-cut efficacy on more than one of the available shorter term end points, such as NCV, quantitative sensory testing, and/or signs or symptoms.

3. ARI Clinical Doses Have Been Limited by Various Toxicities

Thus far it has not been possible to define maximal pharmacological effects of ARIs in humans because compound toxicity have invariably limited ARI dose selection. To date, tolerance problems with ARIs in humans typically have been distinct for different ARIs, implying that these problems are unrelated to AR inhibition *per se*, but rather to nonspecific or compound-specific effects. The virtual normality of the $AR^{-/-}$ mouse (Ho *et al.*, 2000) (Section IV,A,2,b) supports the concept that inhibition of AR *per se* can be well tolerated physiologically. The complete protection of NCV afforded in animal models by the AR-deficient state, whether it be genetic (Ho *et al.*, 2001) or pharmacological (Engerman *et al.*, 1994), argues strongly for the fundamental therapeutic efficacy of the ARI mechanism in animal models. The dose-dependent improvement seen in the clinical study with zenarestat (Fig. 14) and its reasonable correspondence with animal model data (Fig. 12) support the fundamental therapeutic efficacy of the ARI mechanism in human diabetics as well.

4. Very Strong AR Inhibition Likely Needed in Human, as in Rat, Nerve

The extent of impact on NCV in human diabetics has been disappointing to many. It should not be overlooked that ARIs taken into human trials were selected on the basis of their ability to lower sorbitol in rat nerve, an end point that may cause underestimation of the potency required (see Figs. 10–12 and Section IX). In addition, the issue of nerve "penetration," on which there are scant data, needs clearer definition. For example, the "free" fraction of a carboxylic acid ARI in the rat nerve (the amount of unbound drug available to inhibit AR) is reported to be fourfold lower than the free fraction of the same ARI in the serum (Dvornik *et al.*, 1994). Further, it was only relatively recently that it has been fully appreciated that, as in rat models, the salutary effects of ARIs on human nerve function and structure are strongly dose dependent, and a high degree of inhibition of polyol pathway flux, not simply lowering nerve sorbitol, will probably be required for maximal therapeutic effects (Figs. 12 and 14).

IX. Inhibition of Nerve Sorbitol versus AR Metabolic Flux

A. MODEL OF RAT SCIATIC NERVE SORBITOL VERSUS AR FLUX

1. *Pool Size vs Metabolic Flux*

The pharmacological activity of ARIs has long been assessed by the suppression of sorbitol in erythrocytes and in peripheral nerve (Malone *et al.*, 1984; Sima *et al.*, 1988; Greene *et al.*, 1999; Sobajima *et al.*, 2001) (Fig. 14a). Driven by the success of the "osmotic hypothesis" (Section II,A), the concentration or "pool size" of sorbitol has received major attention in most tissues, whereas the metabolic flux through AR and the sorbitol pool has been essentially ignored. In the rare case where the turnover of sorbitol was examined, e.g., in rat erythrocytes and aorta, it was found to be "very rapid" (Winegrad, 1973). In view of the preceding discussion (e.g., Section VIII,B,4), an attempt was made to define the relationship in the peripheral nerve between the steady-state level of sorbitol and flux through the polyol pathway. For this purpose, metabolic flux through the polyol pathway is considered equivalent to substrate flux through AR. Stating the question in more detail then: in human trials, ARIs have shown dose-dependent NCV improvements when nerve sorbitol was suppressed ~75–85% (Fig. 14); to how much inhibition of peripheral nerve AR does this correspond?

2. *One-Compartment Model*

a. Theoretical Assumptions. The question posed was first approached by exploring a simple model (Fig. 15a), predicated on the following assumptions: one compartment; unidirectional flux from glucose to fructose; AR

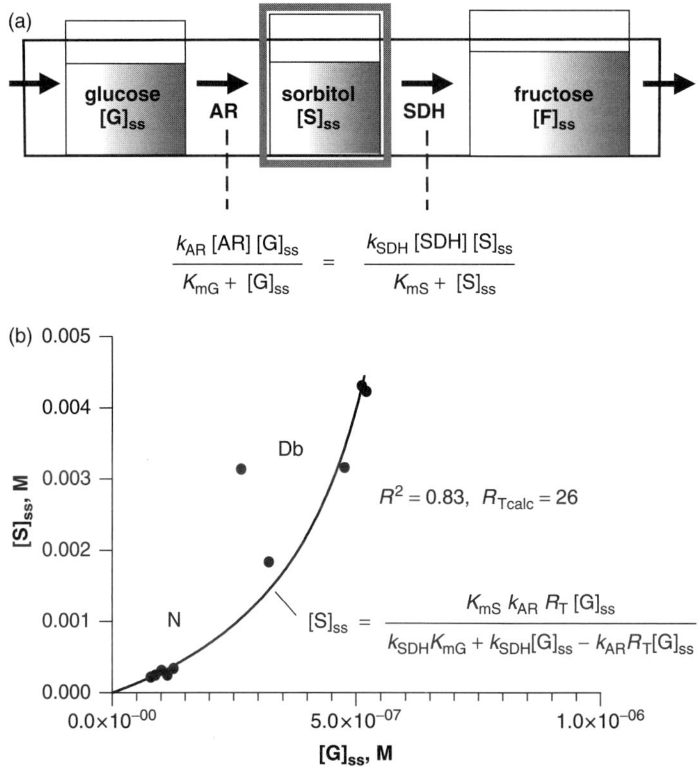

$$\frac{k_{AR}\,[AR]\,[G]_{ss}}{K_{mG} + [G]_{ss}} = \frac{k_{SDH}\,[SDH]\,[S]_{ss}}{K_{mS} + [S]_{ss}}$$

$$[S]_{ss} = \frac{K_{mS}\,k_{AR}\,R_T\,[G]_{ss}}{k_{SDH}K_{mG} + k_{SDH}[G]_{ss} - k_{AR}R_T[G]_{ss}}$$

FIG. 15. One-compartment model of the polyol pathway in peripheral nerve. (a) Nerve sorbitol is assumed to occur in one compartment and to be created enzymatically only by aldose reductase and transformed enzymatically only by sorbitol dehydrogenase in a unidirectional manner under steady-state conditions by enzymes that follow classical Michaelis–Menten kinetic behavior. See text for other assumptions and definitions of variables. (b) Results obtained with the one-compartment model. Kinetic constants for the two enzymes were taken from Figs. 4 and 8 and used in Eq. (4) to calculate the best fit value for R_T, the ratio of [AR] to [SDH], against experimental nerve sorbitol and glucose values (filled circles) from five literature reports. The best fit ($R^2 = 0.83$) was obtained with the computer-selected value of $R_T = 26$. See text for further details. From Oates *et al.* (2001).

and SDH are the sole participating enzymes in sorbitol metabolism; no dietary source or diffusional loss of sorbitol; classical hyperbolic kinetics for each enzyme; steady-state conditions with saturating coenzymes; and no depletion of glucose or coenzymes during steady-state metabolism. Given these assumptions, the rate of synthesis of sorbitol, is expressed on the left side of Eq. (1) by the classical Michaelis–Menten–Henri equation (e.g., Segel, 1975). At steady state, the rate of synthesis of sorbitol from glucose by AR will be equal to the rate of metabolic transformation of sorbitol to fructose by SDH, where the latter is expressed by a similar Michaelis–Menten–Henri equation on the right side of Eq. (1):

$$\frac{k_{AR}[AR][G]_{ss}}{K_{mG} + [G]_{ss}} = \frac{k_{SDH}[SDH][S]_{ss}}{K_{mS} + [S]_{ss}} \tag{1}$$

where, respectively, k_{AR} and k_{SDH} are catalytic rate constants of AR and SDH; [AR], $[G]_{ss}$, [SDH], and $[S]_{ss}$ are (intracellular) tissue concentrations of AR, straight-chain glucose (Section III,A,1,b), SDH, and sorbitol; and K_{mG} and K_{mS} are Michaelis constants of AR for glucose and of SDH for sorbitol.

  b. Equation for Sorbitol Pool Size. Solving Eq. (1) for steady-state sorbitol, $[S]_{ss}$, gives an equation for the steady-state concentration of this metabolic intermediate (Reiner, 1969; Varfolomeev, 1977):

$$[S]_{ss} = \frac{k_{AR}K_{mS}[AR][G]_{ss}}{k_{SDH}[SDH]K_{mG} + k_{SDH}[SDH][G]_{ss} - k_{AR}[AR][G]_{ss}}. \tag{2}$$

The ratio, R, of AR to SDH in the compartment is defined as

$$R = \frac{[AR]}{[SDH]}. \tag{3}$$

Rearranging Eq. (3) as $[AR] = R[SDH]$, substituting into Eq. (2) and simplifying:

$$[S]_{ss} = \frac{k_{AR}K_{mS}R[G]_{ss}}{k_{SDH}K_{mG} + k_{SDH}[G]_{ss} - k_{AR}R[G]_{ss}}. \tag{4}$$

This equation contains an important statement. Namely, in such a model, the steady-state level of tissue sorbitol depends only on the four kinetic constants of the two enzymes, the prevailing glysolia, $[G]_{ss}$, and the ratio of the two enzymes, R. In contrast, given steady state and other assumptions, the flux through AR, equal to the flux through SDH, will be directly proportional to the concentration of AR, [AR]. That is, because AR affects virtually "one-way" catalysis in the direction of aldehyde reduction (Grimshaw, 1992) (Section III,A,1,b) and because negative metabolite feedback has not been described, then to a first approximation, at steady state the metabolic *flux* through AR (and the system) will depend only on the left side of Eq. (1), i.e., will be a function only of the two kinetic constants of AR, the prevailing glysolia, $[G]_{ss}$, and the absolute amount of AR, [AR]. Thus, given the assumptions cited, to a first approximation, steady-state metabolic *flux* through AR (the polyol pathway) will depend on the *absolute amount (concentration) of AR* present in the compartment, [AR], whereas the size of the *sorbitol pool* will depend on R, the *ratio of [AR] to [SDH]*. This concept, that the size of the sorbitol pool will be proportional to the AR/SDH ratio, is consistent with the earlier suggestion and data of Griffin *et al.* (1987) and is implicit in Judziniak *et al.* (1981).

c. Application of Sorbitol Pool Size Equation to Rat Sciatic Nerve. Equation (4) was evaluated for its applicability to rat nerve as a one-compartment system in the following way. Kinetic constants determined for rat AR and SDH (Figs. 4 and 8) were combined with experimentally measured values of $[G]_{ss}$ and $[S]_{ss}$ from five literature papers that reported both normal and diabetic rat nerve glucose and sorbitol values (Fig. 15b, data points, filled circles) (Stewart *et al.*, 1966; Anand *et al.*, 1988; Thurston, 1990; Cameron *et al.*, 1994; Stevens *et al.*, 2000b). R_T, the tissue ratio of AR/SDH is defined as the average tissue concentration of AR, $[AR]_T$, divided by the average tissue concentration of SDH, $[SDH]_T$. In a one-compartment system, R_T is equal to R in Eqs. (3) and (4). Nonlinear regression based on Eq. (4) with $[G]_{ss}$ as the independent (x) variable was then used to pick the R_T value that gave the best fit of these experimental data from the literature. As can be seen in Fig. 15b, the best-fit ($R^2 = 0.83$) value generated was $R_{Tcalc} = 26$. That is, in the one-compartment model with the assumptions given and kinetic constants used, a theoretical value for the ratio of nerve AR to SDH, R_T, if it were equal to 26, would give the curve shown in Fig. 15b, i.e., would allow a reasonably good prediction of experimental nerve sorbitol values from experimental nerve glucose values by means of Eq. (4), but how reasonable is such a value for R_T?

d. Experimental Estimation of R_T in Rat Sciatic Nerve. An experimental value for rat sciatic nerve R_T [calculated from its definition in the preceding paragraph, $[AR]_T/[SDH]_T$] was estimated by measuring the average concentrations of immunoreactive AR and SDH using Western blots of sciatic nerve homogenates prepared according to Nishimura *et al.* (1993). The measured amounts of enzymes were assumed to be distributed uniformly in nerve water. Nerve water was found after freeze-drying nerve samples to constant weight to be 73% (w/w) for both normal and 12-day STZ-diabetic rat nerves. Although the sample sizes did not permit a definitive conclusion, immunoreactive enzyme contents for either enzyme were not statistically significantly different in normal vs diabetic rat nerves. Data were therefore pooled, yielding average values of 9.1 ± 2.7 ($n = 16$) and $1.5 \pm 0.8 \ \mu M$ ($n = 13$), respectively, for $[AR]_T$ and $[SDH]_T$. These results provide an experimental estimate of $R_{Texp} = 6.1$ for rat sciatic nerve, which in Eq. (4) gave a much poorer fit ($R^2 = -0.86$) vs the calculated best fit value of $R_{Tcalc} = 26$ ($R^2 = 0.83$). Therefore, a two-compartment model was next explored.

3. *Two-Compartment Model*

a. Theoretical Assumptions. A marked improvement in overall fit of model parameters vs experimental data was obtained by considering a simple two-compartment model (Fig. 16a). Such a model is more reflective of actual nerve structure, which is well known to be distinctly compartmentalized. For the purpose of the model, a periaxonal compartment (compartment 1) and an axonal compartment (compartment 2) were defined and assumed to occupy equal volumes. The volume occupied by the hydrophobic myelin sheaths was ignored. The periaxonal compartment was assumed to contain Schwann cells and surrounding endoneurial space, but not epineurium. Based on the known localization of AR in Schwann cells and vascular elements (see Section V,B,1), AR was assumed as a first approximation to be 100% localized to the periaxonal compartment, as was sorbitol. Glucose concentrations, $[G]_{ss_1}$ and $[G]_{ss_2}$, were assumed to be equal in the two compartments (Fig. 16a). Given these assumptions, there follow some simple definitions that take into account the presence of the two compartments (see Fig. 16a):

$$R_1 = \frac{[AR]_1}{[SDH]_1} \tag{5}$$

and, from Eq. (4),

$$[S]_{ss1} = \frac{k_{AR}K_{mS}R_1[G]_{ss1}}{k_{SDH}K_{mG} + k_{SDH}[G]_{ss1} - k_{AR}R_1[G]_{ss1}} \tag{6}$$

where

$$[AR]_1 = 2[AR]_T \tag{7}$$

and

$$[SDH]_2 = 2[SDH]_T - [SDH]_1 \tag{8}$$

and

$$R_T = \frac{[AR]_T}{[SDH]_T} = \frac{[AR]_1/2}{([SDH]_1 + [SDH]_2)/2}. \tag{9}$$

b. Application of Two-Compartment Sorbitol Equation to Rat Sciatic Nerve. Equation (6) was tested as before for Eq. (4) using the kinetic constants from Figs. 4 and 8 and the best fit R_1 value from nonlinear regression on the literature values of $[G]_{ss}$ and $[S]_{ss}$. As shown in Fig. 16b, the best fit, still quite strong ($R^2 = 0.74$), was obtained with $R_1 = 30.1$. Note that in the case of the two-compartment model, the best-fit value of R_1 forces a distribution of SDH between compartments 1 and 2 by virtue of Eqs. (5), (7), and (8). From the experimentally determined value of $[AR]_T$ (9.1 μM; Section IX,A,2,d)

(a)

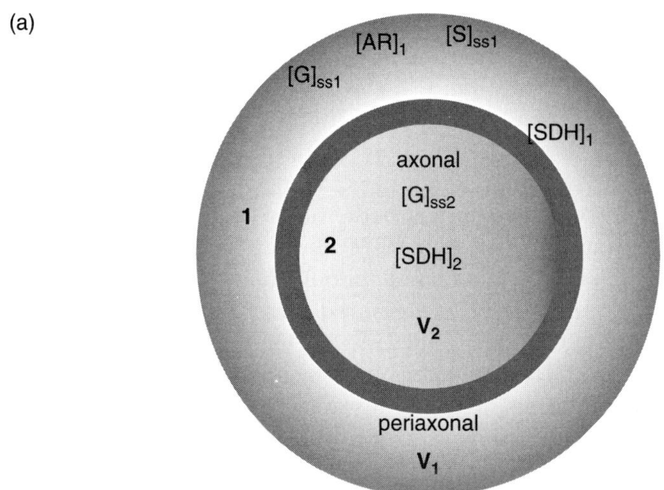

(b)

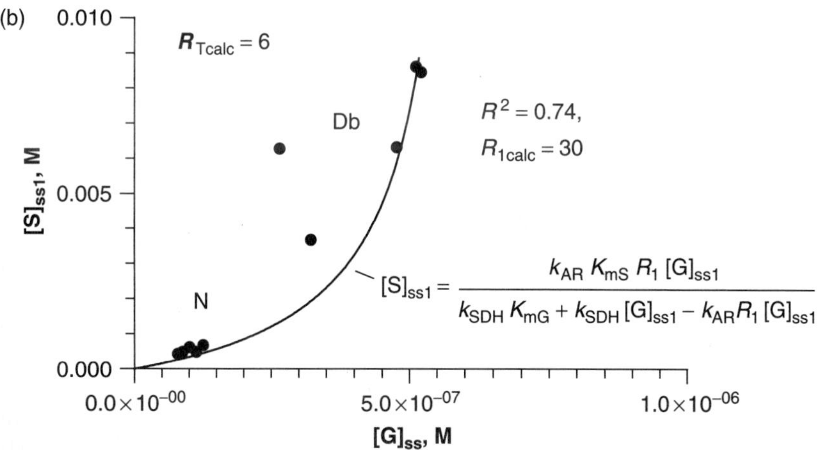

Fig. 16. Two-compartment model of the polyol pathway in peripheral nerve. (a) The nerve is considered to be composed of two compartments: a periaxonal compartment (compartment 1) and an axonal compartment (compartment 2). For simplicity, the volumes are assumed equal, and AR and sorbitol are confined to the Schwann cell and vascular element-containing periaxonal compartment. SDH is distributed between the two compartments according to the computer-calculated best-fit value for R_1, the ratio between [AR]$_1$ and [SDH]$_1$ in the periaxonal compartment. See text for further details. (b) Results obtained with the two-compartment model. The value for R_T was constrained to match the experimental value of 6. A strong fit ($R^2 = 0.74$) of experimental metabolite data (•) was obtained with Eq. (6) using the kinetic constants from Figs. 4 and 8 and the computer-selected value of $R_1 = 30$. See text for further details. From Oates *et al.* (2001).

and Eq. (7), $[AR]_1 = 18.2 \ \mu M$, and with $R_1 = 30.1$ (Fig. 16b), Eq. (5) gives $[SDH]_1 = 0.6 \ \mu M$. From the experimentally determined value for $[SDH]_T$ (1.5 μM; Section IX,A,2,d) and Eq. (8), $[SDH]_2$ is 2.4 μM. Therefore, this two-compartment model gives a good fit to literature metabolite data and distributes SDH 20% to the periaxonal compartment and 80% to the axonal compartment, a distribution very similar to the 16% periaxonal and 84% axonal observed for (desheathed) rabbit sciatic nerve subjected to Wallerian degeneration (Gabbay and O'Sullivan, 1968) (Section V,C,1). Note also that in this two-compartment model, R_T as defined by Eq. (9) is maintained at the experimentally measured value of 6.1. Thus, the two-compartment model (a) is consistent with the experimentally measured tissue ratio of $[AR]_T / [SDH]_T$ (as determined by immunoelectrophoresis), (b) fits the experimentally measured literature values of steady-state sorbitol and glucose in normal and diabetic nerve with an $R^2 = 0.74$, and (c) is consistent with existing data on the immunochemical and biochemical localizations of AR and SDH (Sections V,B and V,C).

4. Relationship between Inhibition of Nerve Sorbitol versus AR Flux

a. Equation Relating Change in Sorbitol Pool Size, p, to Change in AR Flux, i. At constant glucose, $[G]_{ss1}$, and constant $[SDH]_1$, Eqs. (5) and (6) show that sorbitol, $[S]_{ss1}$, is a function only of $[AR]_1$; all other variables in Eq. (6) are constants. Under the conditions of the model, flux through AR is directly proportional to the amount of AR present in the compartment, i.e., to $[AR]_1$ (see Section IX,A,2,b). We can therefore calculate the impact of reducing flux through $[AR]_1$ by an arbitrary percentage, i, by reducing the concentration of $[AR]_1$ by that percentage. This will then dictate some percentage fall, p, in the steady-state sorbitol pool size according to Eqs. (5) and (6). Thus, by designating initial values of $[S]_{ss1}$, $[AR]_1$, and $[SDH]_1$ with the superscript "°" and by substituting Eq. (5) into Eq. (6) and multiplying numerator and denominator by the initial value of $[SDH]_1$, we can define the dependence in the two-compartment model of p, the percentage reduction of initial steady-state nerve sorbitol (left side), on i, the percentage inhibition of initial nerve $[AR]_1$, (right side):

$$(1 - p)[S]^o_{ss1}$$

$$= \frac{k_{AR} K_{mS} [G]_{ss1} (1 - i)[AR]^o_1}{k_{SDH} K_{mG} [SDH]^o_1 + k_{SDH} [G]_{ss1} [SDH]^o_1 - k_{AR} [G]_{ss1} (1 - i)[AR]^o_1}. \tag{10}$$

Then by substituting into Eq. (10) the initial values of $[AR]_1$ from Eq. (5),

$$[AR]^o_1 = R^o_1 [SDH]^o_1, \tag{11}$$

and using Eq. (6) to define the initial value of $[S]_{ss1}$, p can be solved as

$$p = \left\{ 1 - \left[\left(\frac{k_{SDH}K_{mG} + k_{SDH}G_{ss1} - k_{AR}G_{ss1}R_1^o}{k_{AR}K_{ms}G_{ss1}R_1^o} \right) \right. \right.$$

$$\left. \left. \left(\frac{k_{AR}K_{ms}G_{ss1}R_1^o(1 - i)}{k_{SDH}K_{mG} + k_{SDH}G_{ss1} - k_{AR}G_{ss1}R_1^o(1 - i)} \right) \right] \right\}. \tag{12}$$

Note in Eq. (12) that at constant hyperglysolia $[G]_{ss1}$ = constant, and all other parameters are constant except variables p and i. The exact shape of the relationship will depend on the initial value of R_1, the initial ratio of $[AR]_1/[SDH]_1$, in any particular nerve. It can be seen by inspection of Eq. (12) that when $i = 0$, $p = 0$, and when $i = 1$, $p = 1$, as expected. When Eq. (12) is evaluated using the respective kinetic constants in Figs. 4 and 8 and at constant $[G]_{ss1} = 0.42\ \mu M$ [the average of the five published experimental values for diabetic nerve ("Db" in Fig. 16b)], the curves shown in Fig. 17 result.

b. Linear Region of the p versus i Relationship: $R_1 \leq 10$. As is evident in Fig. 17, the relationship between p, the percentage suppression of sorbitol, and i, the percentage inhibition of AR, depends on the initial ratio between [AR] and [SDH] in compartment 1. At an initial ratio of approximately 10:1 or lower, the model predicts that the relationship is essentially linear, with a slope approaching 1.0. That is, given the catalytic constants and the assumptions of the model, when initial $[AR]_1$ is not much greater than about 10-fold $[SDH]_1$, the steady-state sorbitol level is an essentially accurate reflection of the degree of activity of AR. For example, an 85% decrease in AR activity results in an approximately 83% drop in sorbitol (Fig. 17, right-hand vertical dotted line).

Stated differently, under these conditions, there is enough SDH present relative to AR such that the pool size of sorbitol reflects the metabolic flux through AR, not the catalytic inability of SDH to remove sorbitol in a timely manner. The approximately 10-fold ratio reflects the combined difference in (a) catalytic rate constants, with rat SDH capable of catalyzing substrate turnover some 3-fold faster than AR ($100\ min^{-1}$ vs $31\ min^{-1}$, Figs. 8 and 4, respectively); and (b) SDH is nominally operating about four times above its 1.6 mM K_m (Fig. 8) at $\sim$7 mM sorbitol (Fig. 16b, y axis), whereas AR is nominally operating some 10-fold below its $\sim$5 μM K_m for straight-chain glucose (Fig. 4), even under diabetic conditions, at $\sim$0.4 μM glucose (Fig. 16b, x axis).

c. Nonlinear Region of the p versus i Relationship, $R_1 > 10$; Rat Sciatic Nerve. At increasingly higher initial ratios of $[AR]_1$ to $[SDH]_1$ above $\sim$10:1, the relationship between p and i becomes increasing hyperbolic (Fig. 17). This ensues because the capacity of the available SDH to remove newly

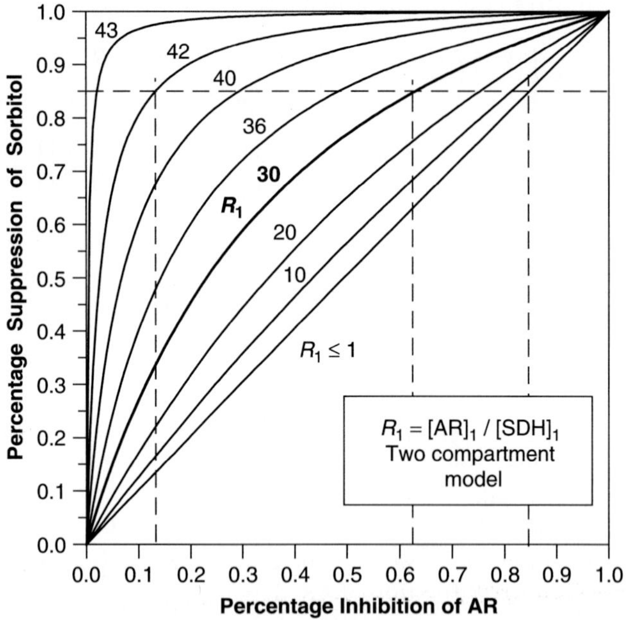

FIG. 17. Relationship between percentage suppression of nerve sorbitol pool size, p, and percentage inhibition of nerve aldose reductase, i, derived from the two-compartment model. Curves are calculated from Eq. (12) using kinetic constants from Figs. 4 and 8, and the average measured $[G]_{ss}$ value for diabetic nerve (0.42 μM), while arbitrarily varying the initial values of R_1, the ratio between $[AR]_1$ and $[SDH]_1$ in the periaxonal compartment. From Fig. 16b, the best fit of experimental data on nerve sorbitol and glucose was found with $R_1 = 30$, the bold line in Fig. 17. As indicated by dotted lines, at $R_1 = 30$, suppression of nerve sorbitol by 85% corresponds to an inhibition of nerve AR of 62%. From Oates *et al.* (2001).

created sorbitol is increasing exceeded, and sorbitol begins accumulating in a nonlinear fashion that becomes increasingly sensitive to the degree of initial relative excess of AR. For example, at an "extreme" R_1 value of say, 42, a collapse of the sorbitol pool of 85% corresponds to only a 13% reduction in AR flux (Fig. 17, left-hand vertical dotted line)! Thus, because the experimentally determined best-fit value of initial R_1 in the two-compartment model is 30 (Fig. 16b), the relationship in the rat nerve appears to be in the distinctly nonlinear range. Therefore, as shown in Fig. 17 (middle vertical dotted line), a reduction in rat nerve sorbitol of 85% would correspond to an inhibition of AR flux of ∼62%. Therefore, to the extent that the human nerve resembles the rat nerve, e.g., Figs. 12 and 14, this analysis suggests an 85% reduction in human nerve sorbitol corresponds to an inhibition of AR flux of ∼60%.

B. LIMITATIONS OF MODEL

The two-compartment model is only a starting point for trying to understand the role of the polyol pathway in peripheral nerve function. Experimental measurements of polyol pathway flux vs pool size will provide a direct and important challenge to the model. Definitive localization of SDH within the peripheral rat and human nerve will be another key test. Correlation of the enzymatic activity of the polyol pathway enzymes vs immunoreactive proteins will need to be verified. In this regard, there is a discrepancy reported between immunoreactive SDH vis à vis SDH enzymatic activity in diabetic rat liver; the difference is attributed to inactivation of SDH by glycation (Hoshi *et al.*, 1996). Careful, quantitative morphometry needs to be employed to define the actual nerve compartment sizes and the extent of heterogeneity of various nerve fiber types present (Stevens *et al.*, 2000b). In addition, the model needs to be extended in various ways, in particular, to model peripheral nerve fructose levels in rats and humans, and to incorporate new data on the epineurial compartment (Kasajima *et al.*, 2001).

X. Future Research Directions

A. BASIC RESEARCH QUESTIONS

1. Why Is Such Strong Suppression of Nerve Sorbitol Necessary?

Many fundamental questions of fact remain to be answered about the polyol pathway in the diabetic peripheral nerve. In addition to the points mentioned in the preceding paragraph, the detailed basis for the observed need to inhibit the polyol pathway very strongly with ARIs (e.g., Figs. 10–12), demands explanation. Is it simply that the nerve metabolism is exquisitely sensitive to polyol pathway flux? Is there a small, but critical, compartment in the vasa nervorum that is difficult to penetrate (Cameron *et al.*, 1994)? Alternatively, and/or in addition, is overnormalization of polyol pathway flux required to compensate for one or more additional, independent metabolic insults with similar biochemical consequences, such as elevated fatty acid oxidation (Williamson *et al.*, 1993; Edelstein *et al.*, 2000; Inoguchi *et al.*, 2000)?

2. What Is the Link between Polyol Pathway Flux and Oxidative Stress?

A related key strategic issue is to define the quantitative significance of each of the several possible mechanisms that may contribute to the generation of oxidative stress in the peripheral nerve (Figs. 1 and 2).

Particularly important will be to assess the role of "overheated" mitochondrial metabolism (Nishikawa *et al.*, 2000) and the potential link, if any, to the polyol pathway. Similar information needs to be generated for the potential link between the polyol pathway and protein kinase C activation in peripheral nerve. Likewise, the observed normalization of carboxymethyllysine in erythrocytes of diabetic patients by ARI epalrestat (Hamada *et al.*, 2000) suggests that the polyol pathway plays a significant role in the generation of intracellular glycative stress (Fig. 2). Do inhibitors of the polyol pathway reduce other relevant markers of oxidant stress, and to what extent? Will any advantage arise from combination of ARIs and SDIs with one another or with directly acting antioxidants, e.g., (Coppey *et al.*, 2001a,b)? The list of fundamental questions is substantial, even without elaborating on the many questions elicited by the new genetically altered mouse models. This is an important and exciting time for experimental research in this area.

B. CLINICAL RESEARCH DIRECTIONS

Additional work is clearly warranted to establish the extent of possible linkage of human alleles associated with high or low expression levels of AR with the rate of onset, severity, and progression of diabetic neuropathy (Heesom *et al.*, 1998; Oates and Mylari, 1999), as well as the interactions of AR alleles with alleles of other physiologically relevant genes, e.g., GLUT1 (Hodgkinson *et al.*, 2001b). Clinical markers of oxidative stress reported to be linked with diabetic complications, such as urinary 8-oxo,2′-deoxyguanosine, e.g., (Hinokio *et al.*, 1999), merit intense further investigation. Taken together, the currently available experimental, genetic and clinical data, together with the "new" paradigm of polyol pathway metabolic flux and its linkage to glucose-induced oxidative stress, render the rationale for creating and evaluating potent polyol pathway inhibitors stronger than ever. Inhibitors that are at least 10-fold more potent *in vivo* than most previous ARIs are called for. Development of "super-potent" polyol pathway inhibitors with appropriate pharmacokinetic properties (Dvornik, 1996) will at long last allow evaluation of the neurophysiological and clinical consequences of inhibiting metabolic flux through the polyol pathway vis à vis simply lowering nerve sorbitol. It is hoped that the dose-dependent partial efficacy of previous ARIs will be totally eclipsed by a new generation of well tolerated "super-potent" polyol pathway inhibitors, agents that will allow clear demonstration of robust, clinically meaningful efficacy against relevant neuropathic end points. Only then, when clearly efficacious therapy has been proven, will we know that we finally have some real insight into the "occasions and causes why and wherefore" that connect the polyol pathway and diabetic peripheral neuropathy.

Acknowledgments

I express my sincere gratitude to my many co-workers and colleagues for their support and contributions to the work described here. In my laboratory, I thank David A. Beebe, James B. Coutcher, and Craig A. Ellery for generating data in Figs. 4, 7, 8, and 11 and for assisting with the preparation of several of the figures. Special thanks to Peter F. Kador (National Eye Institute, Bethesda) for his kind gift of rat rAR standard and to Jan-Olaf Höög (Karolinska Institute, Stockholm) for his gift of cDNA for rat rSDH. In our Department of General Pharmacology, thanks to John P. Hakkinen, Teresa M. Schelhorn, Matthew W. Miller, and Eric Hammerlund for electrophysiology data shown in Figs. 11 and 13. Special thanks to Virginia L. Rath for providing the beautiful X-ray images of AR and SDH in Figs. 3 and 6. I am also very grateful to the Groton Pfizer Research Library staff, particularly Kathy M. Freidenfelds, Lisa M. Powers, and Gregory C. Aldinger, for the efficient and cheerful help provided in obtaining many of the references cited herein. Heartfelt thanks are especially due to Andrew G. Demaine, Banavara L. Mylari, David R. Tomlinson, and Joseph R. Williamson for critically reviewing this manuscript. For many stimulating discussions on this topic, I am also grateful to colleagues Andrew G. Demaine, Charles E. Grimshaw, Banavara L. Mylari, Ravichandran Ramasamy, and Joseph R. Williamson. I am also very thankful for the inspiration received in the laboratory of my graduate school mentor, Oscar Touster, a coauthor of the first review on the polyol pathway (1962). Finally, my deep appreciation is offered to my wife Nancy, for her astonishing patience and unswerving support.

"[W]hat's past is prologue."

The Tempest, II.1, William Shakespeare

References

Aida, K., Ikegishi, Y., Chen, J., Tawata, M., Ito, S., Maeda, S., and Onaya, T. (2000). Disruption of aldose reductase gene (Akr1b1) causes defect in urinary concentrating ability and divalent cation homeostasis. *Biochem. Biophys. Res. Commun.* **277**, 281–286.

Alvarez, A., Martinez, A., Ibarra, B., Medina, C., Bracamontes, M., Perea, J., and Vaca, G. (1993). Membrane-bound sorbitol dehydrogenase in human red blood cells: Studies in normal subjects and in enzyme-deficient subjects with congenital cataracts. *J. Inherit. Metab. Dis.* **16**, 67–72.

Anand, P., Llewelyn, J. G., Thomas, P. K., Gillon, K. R., Lisk, R., and Bloom, S. R. (1988). Water content, vasoactive intestinal polypeptide and substance P in intact and crushed sciatic nerves of normal and streptozotocin-diabetic rats. *J. Neurol. Sci.* **83**, 167–177.

Angyal, S. J. (1984). The composition of reducing sugars in solution. *Adv. Carbohydr. Chem. Biochem.* **42**, 15–68.

Ao, S., Shingu, Y., Kikuchi, C., Takano, Y., Nomura, K., Fujiwara, T., Ohkubo, Y., Notsu, Y., and Yamaguchi, I. (1991). Characterization of a novel aldose reductase inhibitor, FR74366, and its effects on diabetic cataract and neuropathy in the rat. *Metabolism* **40**, 77–87.

Arezzo, J. C., Klioze, S. S., Peterson, M. J., Lakshminarayanan, M. Y., and the Zopolrestat Phase II Neuropathy Study Group. (1996). Efficacy and safety results of a phase II multicenter study of the aldose reductase inhibitor zopolrestat in patients with peripheral symmetrical diabetic polyneuropathy. *Diabetes* **45**(Suppl. 2), 276A.

Bagnasco, S. M., Uchida, S., Balaban, R. S., Kador, P. F., and Burg, M. B. (1987). Induction of aldose reductase and sorbitol in renal inner medullary cells by elevated extracellular NaCl. *Proc. Natl. Acad. Sci. USA* **84**, 1718–1720.

Bairoch, A. (2000). The ENZYME database in 2000. *Nucleic Acids Res.* **28**, 304–305.

Ballinger, W. E., Day, W. W., Beebe, D. A., Oates, P. J., Zembrowski, W. J., and Mylari, B. L. (1996). The effect of multiple bolus dosing vs. chronic water dosing with CP-166,572, a potent sorbitol dehydrogenase inhibitor, on nerve fructose normalization in diabetic rats. *Proc. 7th North Am. ISSX Meet.* **10**, 295.

Banfield, M. J., Salvucci, M. E., Baker, E. N., and Smith, C. A. (2001). Crystal structure of the NADP(H)-dependent ketose reductase from *Bemisia argentifolii* at 2.3 Å resolution. *J. Mol. Biol.* **306**, 239–250.

Basso, M., Banas, D., Lai, K.-D., Hohman, T. C., Cameron, N. E., and Cotter, M. A. (1998). Oxidative stress, and Na/K ATPase activity in diabetic rats: Aldose reductase inhibitor withdrawal studies. *Diabetologia* **41**, A271.

Bateman, J. B., Kojis, T., Heinzmann, C., Klisak, I., Diep, A., Carper, D., Nishimura, C., Mohandas, T., and Sparkes, R. S. (1993). Mapping of aldose reductase gene sequences to human chromosomes 1, 3, 7, 9, 11, and 13. *Genomics* **17**, 560–565.

Baynes, J. W., and Thorpe, S. R. (1999). Role of oxidative stress in diabetic complications: A new perspective on an old paradigm. *Diabetes* **48**, 1–9.

Bianchi, R., Marelli, C., Marini, P., Fabris, M., Triban, C., and Fiori, M. G. (1990). Diabetic neuropathy in db/db mice develops independently of changes in ATPase and aldose reductase: A biochemical and immunohistochemical study. *Diabetologia* **33**, 131–136.

Biessels, G. J., and VanDam, P. S. (1997). Diabetic neuropathy: Pathogenesis and current treatment perspectives. *Neurosci. Res. Commun.* **20**, 1–10.

Blakley, R. L. (1951). The metabolism and antiketogenic effects of sorbitol: Sorbitol dehydrogenase. *Biochem. J.* **49**, 257–271.

Bohren, K. M., and Grimshaw, C. E. (2000). The sorbinil trap: A predicted dead-end complex confirms the mechanism of aldose reductase inhibition. *Biochemistry* **39**, 9967–9974.

Boulton, A. J. M., and Malik, R. A. (1998). Diabetic neuropathy. *Med. Clin. North Am.* **82**, 909–929.

Brahmachari, S. K., Meera, G., Sarkar, P. S., Balagurumoorthy, P., Tripathi, J., Raghavan, S., Shaligram, U., and Pataskar, S. (1995). Simple repetitive sequences in the genome: Structure and functional significance. *Electrophoresis* **16**, 1705–1714.

Bravi, M. C., Pietrangeli, P., Laurenti, O., Basili, S., Cassone-Faldetta, M., Ferri, C., and De, M. G. (1997). Polyol pathway activation and glutathione redox status in non-insulin-dependent diabetic patients. *Metab. Clin. Exper.* **46**, 1194–1198.

Bril, V. (2001). Status of current clinical trials in diabetic polyneuropathy. *Can. J. Neurol. Sci.* **28**, 191–198.

Brownlee, M. (1997). Advanced products of nonenzymatic glycosylation and the pathogenesis of diabetic complications. *In* "Diabetes Mellitus" (D. Porte, Jr. and R. S. Sherwin, eds.), pp. 229–256. Appleton & Lange, Stamford.

Burg, M. B. (1996). Coordinate regulation of organic osmolytes in renal cells. *Kidney Int.* **49**, 1684–1685.

Burg, M. B., Kwon, D., and Kultz, D. (1997). Regulation of gene expression by hypertonicity. *Annu. Rev. Physiol.* **59**, 437–455.

Burger-Kentischer, A., Mueller, E., Maerz, J., Fraek, M. L., Thurau, K., and Beck, F.-X. (1999). Hypertonicity-induced accumulation of organic osmolytes in papillar interstitial cells. *Kidney Int.* **55**, 1417–1425.

Burnell, J. N., and Holmes, R. S. (1983). Purification and properties of sorbitol dehydrogenase from mouse liver. *Int. J. Biochem.* **15**, 507–511.

Calcutt, N. A., and Dunn, J. S. (1997). Diabetic neuropathy. *Anesthiol. Clin. North Am.* **15**, 429–444.

Calcutt, N. A., Tomlinson, D. R., and Biswas, S. (1990). Coexistence of nerve conduction deficit with increased Na(+)-K(+)-ATPase activity in galactose-fed mice: Implications for polyol pathway and diabetic neuropathy. *Diabetes* **39**, 663–666.

Calcutt, N. A., Willars, G. B., and Tomlinson, D. R. (1988). Statil-sensitive polyol formation in nerve of galactose-fed mice. *Metabolism* **37**, 450–453.

Cameron, N. E., and Cotter, M. A. (1992). Dissociation between biochemical and functional effects of the aldose reductase inhibitor, ponalrestat, on peripheral nerve in diabetic rats. *Br. J. Pharmacol.* **107**, 939–944.

Cameron, N. E., and Cotter, M. A. (1997). Metabolic and vascular factors in the pathogenesis of diabetic neuropathy. *Diabetes* **46**(Suppl. 2), S31–37.

Cameron, N. E., and Cotter, M. A. (2000). Oxidative stress and abnormal lipid metabolism in diabetic complications. *In* "Chronic Complications in Diabetes. Animal Models and Chronic Complications" (A. A. F. Sima, ed.), Vol. 1, pp. 97–130. Harwood Academic, Amsterdam.

Cameron, N. E., Cotter, M. A., Basso, M., and Hohman, T. C. (1997). Comparison of the effects of inhibitors of aldose reductase and sorbitol dehydrogenase on neurovascular function, nerve conduction and tissue polyol pathway metabolites in streptozotocin-diabetic rats. *Diabetologia* **40**, 271–281.

Cameron, N. E., Cotter, M. A., Dines, K. C., and Hohman, T. C. (1996). Reversal of defective peripheral nerve conduction velocity, nutritive endoneurial blood flow, and oxygenation by a novel aldose reductase inhibitor, WAY-121,509, in streptozotocin-induced diabetic rats. *J. Diabet. Compl.* **10**, 43–53.

Cameron, N. E., Cotter, M. A., Dines, K. C., Maxfield, E. K., Carey, F., and Mirrlees, D. J. (1994). Aldose reductase inhibition, nerve perfusion, oxygenation and function in streptozotocin-diabetic rats: Dose-response considerations and independence from a *myo*-inositol mechanism. *Diabetologia* **37**, 651–663.

Carr, I. M., and Markham, A. F. (1995). Molecular genetic analysis of the human sorbitol dehydrogenase gene. *Mamm. Genome* **6**, 645–652.

Carr, I. M., Markham, A. F., and Coletta, P. L. (1997). Identification and characterisation of a sequence related to human sorbitol dehydrogenase. *Eur. J. Biochem.* **245**, 760–767.

Carr, I. M., Whitehouse, A., Coletta, P. L., and Markham, A. F. (1998). Structural and evolutionary characterization of the human sorbitol dehydrogenase gene duplication. *Mamm. Genome* **9**, 1042–1048.

Ceriello, A. (2000). Oxidative stress and glycemic regulation. *Metabolism* **49**(Suppl. 1), 27–29.

Chakrabarti, S., Sima, A. A., Nakajima, T., Yagihashi, S., and Greene, D. A. (1987). Aldose reductase in the BB rat: Isolation, immunological identification and localization in the retina and peripheral nerve. *Diabetologia* **30**, 244–251.

Chandler, C. E., and Miller, L. J. (1986). Studies of aldose reductase using neuronal cell culture and ligated rat sciatic nerve. *Metabolism* **35**(Suppl. 1), 71–77.

Chavers, B. M., Bilous, R. W., Ellis, E. N., Steffes, M. W., and Mauer, S. M. (1989). Glomerular lesions and urinary albumin excretion in type I diabetes without overt proteinuria. *N. Engl. J. Med.* **320**, 966–970.

Cheng, H.-M., and Gonzalez, R. G. (1986). The effect of high glucose and oxidative stress on lens metabolism, aldose reductase, and senile cataractogenesis. *Metabolism* **35**(Suppl. 1), 10–14.

Chida, K., Yamamoto, N., and Yasuda, K. (1975). Histochemical study of polyol dehydrogenase: Localization of sorbitol dehydrogenase in mouse-liver and kidney. *Acta Histochem. Cytochem.* **8**, 234–246.

Chistyakov, D. A., Tourakulov, R. I., Shestakova, M. V., Shamkhalova, M. V., Chugunova, L. A., Milen'kaya, T. M., Debabov, V. G., Dedov, I. I., and Nosikov, V. V. (2000). Polymorphism of aldose reductase gene is associated with diabetic microangiopathy in a Moscow population. *Diabetologia* **43**(Suppl. 1), A241.

Chung, S., and LaMendola, J. (1989). Cloning and sequence determination of human placental aldose reductase gene. *J. Biol. Chem.* **264**, 14775–14777.

Clements, R. S., Jr. (1986). The polyol pathway: A historical review. *Drugs* **32**, 3–5.

Clements, R. S., Jr., Weaver, J. P., and Winegrad, A. I. (1969). The distribution of polyol: NADP oxidoreductase in mammalian tissues. *Biochem. Biophys. Res. Commun.* **37**, 347–353.

Cohen, M. P. (1996). "Diabetes and Protein Glycation," pp. 183–192. JC Press, Philadelphia.

Cohen, R. A. (1997). Glucotoxicity and mediators: Nitric oxide, arachidonic acid, superoxide anion. *Therapie (Paris)* **52**, 387–388.

Cohen, R. B. (1961). Studies on a polyhydric alcohol dehydrogenase system utilizing a histochemical method. *Lab. Invest.* **10**, 459–465.

Colciago, A., Negri-Cesi, P., and Celotti, F. (2002). Pathogenesis of diabetic neuropathy: Do hyperglycemia and aldose reductase inhibitors affect neuroactive steroid formation in the rat sciatic nerves? *Exp. Clin. Endocrinol. Diabet.* **110**, 22–26.

Coppey, L. J., Gellett, J. S., Davidson, E. P., Dunlap, J. A., Lund, D. D., and Yorek, M. A. (2001a). Effect of antioxidant treatment of streptozotocin-induced diabetic rats on endoneurial blood flow, motor nerve conduction velocity, and vascular reactivity of epineurial arterioles of the sciatic nerve. *Diabetes* **50**, 1927–1937.

Coppey, L. J., Gellett, J. S., Davidson, E. P., Dunlap, J. A., Lund, D. D., Salvemini, D., and Yorek, M. A. (2001b). Effect of M40403 treatment of diabetic rats on endoneurial blood flow, motor nerve conduction velocity and vascular function of epineurial arterioles of the sciatic nerve. *Br. J. Pharmacol.* **134**, 21–29.

Costantino, L., Rastelli, G., Vianello, P., Cignarella, G., and Barlocco, D. (1999). Diabetes complications and their potential prevention: Aldose reductase inhibition and other approaches. *Med. Res. Rev.* **19**, 3–23.

Darmanin, C., and El-Kabbani, O. (2001). Modelling studies of the active site of human sorbitol dehydrogenase: an approach to structure-based inhibitor design of the enzyme. *Bioorg. Med. Chem. Lett.* **11**, 3133–3136.

DCCT Research Group: The Diabetes Control and Complications Research Group. (1995). The effect of intensive diabetes therapy on the development and progression of neuropathy. *Ann. Intern. Med.* **122**, 561–568.

Dejgaard, A. (1998). Pathophysiology and treatment of diabetic neuropathy. *Diabet. Med.* **15**, 97–112.

Demaine, A., Cross, D., and Millward, A. (2000). Polymorphisms of the aldose reductase gene and susceptibility to retinopathy in type 1 diabetes mellitus. *Invest. Ophthalmol. Vis. Sci.* **41**, 4064–4068.

Dixit, B. L., Balendiran, G. K., Watowich, S. J., Srivastava, S., Ramana, K. V., Petrash, J. M., Bhatnagar, A., and Srivastava, S. K. (2000). Kinetic and structural characterization of the glutathione-binding site of aldose reductase. *J. Biol. Chem.* **275**, 21587–21595.

Donald, L. J., Wang, H. S., and Hamerton, J. L. (1980). Assignment of the sorbitol dehydrogenase locus to human chromosome 15 pter leads to q21. *Biochem. Genet.* **18**, 425–431.

Dvornik, D. (1987). "Aldose Reductase Inhibition," pp. 1–368. Biomedical Information Corp., New York.

Dvornik, D. (1996). A perspective of aldose reductase inhibitors and diabetic complications. *Croat. Chem. Acta* **69**, 613–630.

Dvornik, D., Millen, J., Hicks, D. R., and Kraml, M. (1994). Tolrestat pharmacokinetics in rat peripheral nerve. *J. Diabetes Compl.* **8**, 18–26.

Dyck, P. J., Zimmerman, B. R., Vilen, T. H., Minnerath, S. R., Karnes, J. L., Yao, J. K., and Poduslo, J. F. (1988). Nerve glucose, fructose, sorbitol, myo-inositol, and fiber degeneration and regeneration in diabetic neuropathy. *N. Engl. J. Med.* **319**, 542–548.

Dyer, P. H., Chowdhury, T. A., Dronsfield, M. J., Dunger, D., Barnett, A. H., and Bain, S. C. (1999). The 5'-end polymorphism of the aldose reductase gene is not associated with diabetic nephropathy in Caucasian type I diabetic patients. *Diabetologia* **42**, 1028–1032.

Edelstein, D., Du, X.-L., and Brownlee, M. (2000). Unbound free fatty acids induce mitochondrial superoxide overproduction in endothelial cells, which activates NFκB. *Diabetologia* **43**(Suppl. 1), A265.

Eklund, H., Horjales, E., Jörnvall, H., Branden, C. I., and Jeffery, J. (1985). Molecular aspects of functional differences between alcohol and sorbitol dehydrogenases. *Biochemistry* **24**, 8005–8012.

El-Kabbani, O., Rogniaux, H., Barth, P., Chung, R. P., Fletcher, E. V., Van Dorsselaer, A., and Podjarny, A. (2000). Aldose and aldehyde reductases: Correlation of molecular modeling and mass spectrometric studies on the binding of inhibitors to the active site. *Proteins* **41**, 407–414.

Engerman, R. L., Kern, T. S., and Larson, M. E. (1994). Nerve conduction and aldose reductase inhibition during 5 years of diabetes or galactosaemia in dogs. *Diabetologia* **37**, 141–144.

Estonius, M., Danielsson, O., Karlsson, C., Persson, H., Jörnvall, H., and Höög, J. O. (1993). Distribution of alcohol and sorbitol dehydrogenases: Assessment of mRNA species in mammalian tissues. *Eur. J. Biochem.* **215**, 497–503.

Fairwell, T., Krutzsch, H., Hempel, J., Jeffery, J., and Jörnvall, H. (1984). Acetyl-blocked N-terminal structures of sorbitol and aldehyde dehydrogenases. *FEBS Lett.* **170**, 281–289.

Fedele, D., and Giugliano, D. (1997). Peripheral diabetic neuropathy: Current recommendations and future prospects for its prevention and management. *Drugs* **54**, 414–421.

Feldman, E. L., Stevens, M. J., and Greene, D. A. (1997). Pathogenesis of diabetic neuropathy. *Clin. Neurosci.* **4**, 365–370.

Finegold, D., Lattimer, S. A., Nolle, S., Bernstein, M., and Greene, D. A. (1983). Polyol pathway activity and myo-inositol metabolism: A suggested relationship in the pathogenesis of diabetic neuropathy. *Diabetes* **32**, 988–992.

Fondelli, C., Signorini, A. M., Borgogni, L., and Gragnoli, G. (1993). Tolrestat and superoxide anion production in type 2 diabetic patients. *Diabetologia* **36**(Suppl. 1), A204.

Fu, D. T. W., Lee, A. Y. W., Lin, C. X. F., Chung, S. S. M., and Chung, S. K. (1996). Downregulation of mouse aldose reductase mRNA in the Schwann cells but not in the endothelial cells of sciatic nerve by hyperglycemia. *Soc. Neurosci. Abs.* **22**, 1498.

Fujisawa, T., Ikegami, H., Kawaguchi, Y., Yamato, E., Nakagawa, Y., Shen, G. Q., Fukuda, M., and Ogihara, T. (1999). Length rather than a specific allele of dinucleotide repeat in the 5' upstream region of the aldose reductase gene is associated with diabetic retinopathy. *Diabet. Med.* **16**, 1044–1047.

Gabbay, K. H. (1973). Role of sorbitol pathway in neuropathy. *Adv. Metabol. Disorders* **2**(Suppl), 417–424.

Gabbay, K. H. (1975). Hyperglycemia, polyol metabolism, and complications of diabetes mellitus. *Annu. Rev. Med.* **26**, 521–536.

Gabbay, K. H., Merola, L. O., and Field, R. A. (1966). Sorbitol pathway: Presence in nerve and cord with substrate accumulation in diabetes. *Science* **151**, 209–210.

Gabbay, K. H., and O'Sullivan, J. B. (1968). The sorbitol pathway: Enzyme localization and content in normal and diabetic nerve and cord. *Diabetes* **17**, 239–243.

Geisen, K. (1992). Substituierte Pyrimidin-derivate, verfahren zu ihrer Herstellung und ihre Verwendung als Reagenzien. *Eur. Pat. Appl.* No. 470, 616.

Geisen, K., Utz, R., Grotsch, H., Lang, H. J., and Nimmesgern, H. (1994). Sorbitol-accumulating pyrimidine derivatives. *Arzneimittelforschung* **44**, 1032–1043.

Gillon, K. R., Hawthorne, J. N., and Tomlinson, D. R. (1983). Myo-inositol and sorbitol metabolism in relation to peripheral nerve function in experimental diabetes in the rat: The effect of aldose reductase inhibition. *Diabetologia* **25**, 365–371.

Graham, A., Brown, L., Hedge, P. J., Gammack, A. J., and Markham, A. F. (1991a). Structure of the human aldose reductase gene. *J. Biol. Chem.* **266**, 6872–6877.

Graham, A., Heath, P., Morten, J. E. N., and Markham, A. F. (1991b). The human aldose reductase gene maps to chromosome region 7q35. *Hum. Genet.* **86**, 509–514.

Greene, D. A., Arezzo, J. C., Brown, M. B., and the Zenarestat Study Group. (1999). Effect of aldose reductase inhibition on nerve conduction and morphometry in diabetic neuropathy. *Neurology* **53**, 580–591.

Greene, D. A., Lattimer, S. A., and Sima, A. A. (1987). Sorbitol, phosphoinositides, and sodium-potassium-ATPase in the pathogenesis of diabetic complications. *N. Engl. J. Med.* **316**, 599–606.

Griffin, B. W., McNatt, L. G., and York, B. M. (1987). Characterization of aldose reductase activities from human and animal sources by a sensitive fluorescence assay. *Prog. Clin. Biol. Res.* **232**, 325–340.

Grimshaw, C. E. (1986). Direct measurement of the rate of ring opening of D-glucose by enzyme-catalyzed reduction. *Carbohydr. Res.* **148**, 345–348.

Grimshaw, C. E. (1992). Aldose reductase: Model for a new paradigm of enzymic perfection in detoxification catalysts. *Biochemistry* **31**, 10139–10145.

Grimshaw, C. E., Bohren, K. M., Lai, C. J., and Gabbay, K. H. (1995). Human aldose reductase: Rate constants for a mechanism including interconversion of ternary complexes by recombinant wild-type enzyme. *Biochemistry* **34**, 14356–14365.

Grimshaw, C. E., and Lai, C. J. (1996). Oxidized aldose reductase: *In vivo* factor not *in vitro* artifact. *Arch. Biochem. Biophys.* **327**, 89–97.

Grimshaw, C. E., and Mathur, E. J. (1989). Immunoquantitation of aldose reductase in human tissues. *Anal. Biochem.* **176**, 66–71.

Gupta, S., Chough, E., Daley, J., Oates, P. J., Tornheim, K., Ruderman, N. B., and Keaney, J. F., Jr. (2001). Hyperglycemia increases superoxide anion production in rabbit endothelium leading to decreased Na^+K^+-ATPase activity. *Am. J. Physiol.* **282**, C560–566.

Hamada, Y., Nakamura, J., Naruse, K., Komori, T., Kato, K., Kasuya, Y., Nagai, R., Horiuchi, S., and Hotta, N. (2000). Epalrestat, an aldose reductase inhibitor, reduces the levels of N-epsilon-(carboxymethyl)lysine protein adducts and their precursors in erythrocytes from diabetic patients. *Diabet. Care* **23**, 1539–1544.

Harrison, D. H., Bohren, K. M., Ringe, D., Petsko, G. A., and Gabbay, K. H. (1994). An anion binding site in human aldose reductase: Mechanistic implications for the binding of citrate, cacodylate, and glucose 6-phosphate. *Biochemistry* **33**, 2011–2020.

Hayward, L. D., and Angyal, S. J. (1977). A symmetry rule for the circular dichroism of reducing sugars, and the proportion of carbonyl forms in aqueous solutions thereof. *Carbohydr. Res.* **53**, 13–20.

Heesom, A. E., Hibberd, M. L., Millward, B. A., and Demaine, A. G. (1997). Polymorphism in the 5'-end of the aldose reductase gene is strongly associated with the development of diabetic nephropathy in type I diabetes. *Diabetes* **46**, 287–291.

Heesom, A. E., Millward, B. A., and Demaine, A. G. (1998). Susceptibility to diabetic neuropathy in patients with insulin dependent diabetes mellitus is associated with a polymorphism at the 5' end of the aldose reductase gene. *J. Neurol. Neurosurg. Psychiat.* **64**, 213–216.

Henry, D. N., Busik, J. V., Brosius, F. C., III, and Heilig, C. W. (1999). Glucose transporters control gene expression of aldose reductase, PKCalpha, and GLUT1 in mesangial cells *in vitro. Am. J. Physiol.* **277**, F97–104.

Hers, H. G. (1956). Le méchnaisme de la transformation de glucose en fructose par les vésicules seminales. *Biochim. Biophys. Acta* **22**, 202–203.

Hers, H. G. (1960). L'Aldose-reductase. *Biochim. Biophys. Acta* **37**, 120–126.

Hinokio, Y., Suzuki, S., Hirai, M., Chiba, M., Hirai, A., and Toyota, T. (1999). Oxidative DNA damage in diabetes mellitus: its association with diabetic complications. *Diabetologia* **42**, 995–998.

Ho, E. C. M., Lam, K. S. L., Chung, S. S. M., and Chung, S. K. (2001). Aldose reductase-deficient mice are alleviated from the depletion of GSH in the peripheral nerve and MNCV deficit associated with diabetes. *Diabetes* **50**(Suppl. 2), A59.

Ho, H. T., Chung, S. K., Law, J. W., Ko, B. C., Tam, S. C., Brooks, H. L., Knepper, M. A., and Chung, S. S. (2000). Aldose reductase-deficient mice develop nephrogenic diabetes insipidus. *Mol. Cell Biol.* **20**, 5840–5846.

Hodgkinson, A. D., Millward, B. A., and Demaine, A. G. (2001b). Polymorphisms of the glucose transporter (GLUT1) gene are associated with diabetic nephropathy. *Kid. Int.* **59**, 985–989.

Hodgkinson, A. D., Sondergaard, K. L., Yang, B., Cross, D. F., Millward, B. A., and Demaine, A. G. (2001a). Aldose reductase expression is induced by hyperglycemia in diabetic nephropathy. *Kid. Int.* **60**, 211–218.

Hohman, T., Meng, X., Tse, S., Beg, M., Neefe, L., Bochenek, W., and the ARI-509 Development Team. (1996). Enhanced biochemical efficacy of the novel aldose reductase inhibitor, ARI-509, in human tissues. *Diabetes* **45**(Suppl. 2), 5A.

Hohman, T. C., Banas, D., Basso, M., Cotter, M. A., and Cameron, N. E. (1997a). Increased oxidative stress in experimental diabetic neuropathy. *Diabetologia* **40**(Suppl. 1), A549.

Hohman, T. C., Bochenek, W. J., Meng, X., Neefe, L., Beg, M., and the ARI-509 Development Team. (1997b). Nerve function and biochemistry in diabetic patients treated with ARI-509, a novel aldose reductase inhibitor. *Diabetologia* **40**(Suppl. 1), A552.

Hollman, S., and Touster, O. (1957). The L-xylulose-xylitol enzyme and other polyol dehydro-genases of guinea pig liver mitochondria. *J. Biol. Chem.* **225**, 87–102.

Holmes, R. S., Duley, J. A., and Hilgers, J. (1982). Sorbitol dehydrogenase genetics in the mouse: A 'null' mutant in a 'European' C57BL strain. *Anim. Blood Groups Biochem. Genet.* **13**, 263–272.

Hoshi, A., Takahashi, M., Fujii, J., Myint, T., Kaneto, H., Suzuki, K., Yamasaki, Y., Kamada, T., and Taniguchi, N. (1996). Glycation and inactivation of sorbitol dehydrogenase in normal and diabetic rats. *Biochem. J.* **318**, 119–123.

Hotta, N. (1997). New concepts and insights on pathogenesis and treatment of diabetic complications: Polyol pathway and its inhibition. *Nagoya J. Med. Sci.* **60**, 89–100.

Hotta, N., Toyota, T., Matsuoka, K., Shigeta, Y., Kikkawa, R., Kaneko, T., Takahashi, A., Sugimura, K., Koike, Y., Ishii, J., Sakamoto, N., and the SNK-860 Diabetic Neuropathy study group. (2001). Clinical efficacy of fidarestat, a novel aldose reductase inhibitor, for diabetic peripheral neuropathy: 52-week multicenter placebo-controlled double-blind parallel group study. *Diabet. Care* **24**, 1776–1782.

Ibarra, B., Gonzalez-Quiroga, G., Vaca, G., Diaz, M., Rivas, F., and Cantu, J. M. (1982). Sorbitol dehydrogenase polymorphism in human seminal plasma. *Ann. Genet.* **25**, 53–55.

Ibarra, B., Vaca, G., Perea, F. J., Rabago, M., Alvarez, C., Ortiz, C., and Medina, C. (1989). Screening for thermostability and electrophoretic red blood cell sorbitol dehydrogenase variants. *Ann. Genet.* **32**, 102–105.

Ido, Y., Chang, K. C., Oates, P. J., Mylari, B. L., and Williamson, J. R. (1999). Inhibitors of sorbitol dehydrogenase (SDI) and aldose reductase (ARI) reverse impaired motor nerve conduction velocity (MNCV) in diabetic rats. *Diabetes* **48**(Suppl. 1), A150.

Ido, Y., Chang, K. C., Woolsey, T. A., and Williamson, J. R. (2001). NADH: Sensor of blood flow need in brain, muscle, and other tissues. *FASEB J.* **15**, 1419–1421.

Ido, Y., Oates, P. J., Mylari, B. L., Burgan, J., Bache, M., Mizisin, A. P., and Williamson, J. R. (1995). Effects of sorbitol dehydrogenase inhibition on peripheral nerve conduction and vascular function in control and diabetic rats. *Diabetes* **44**(Suppl. 1), 66A.

Ido, Y., Oates, P. J., Mylari, B. L., Burgan, J., Bache, M., Mizisin, A. P., and Williamson, J. R. (1996). Inhibitors of sorbitol dehydrogenase and aldose reductase prevent decreased MNCV caused by diabetes. *Diabetologia* **39**(Suppl. 1), A248.

Ikegishi, Y., Tawata, M., Aida, K., and Onaya, T. (1999). Z − 4 allele upstream of the aldose reductase gene is associated with proliferative retinopathy in Japanese patients with NIDDM, and elevated luciferase gene transcription *in vitro*. *Life Sci.* **65**, 2061–2070.

Imperatore, G., Hanson, R. I., Pettitt, D. J., Kobes, S., Bennett, P. H., and Knowler, W. C. (1998). Sib-pair linkage analysis for susceptibility genes for microvascular complications among Pima Indians with type 2 diabetes. *Diabetes* **47**, 821–830.

Inagaki, K., Miwa, I., and Okuda, J. (1982). Affinity purification and glucose specificity of aldose reductase from bovine lens. *Arch. Biochem. Biophys.* **216**, 337–344.

Inoguchi, T., Li, P., Umeda, F., Yu, H. Y., Kakimoto, M., Imamura, M., Aoki, T., Etoh, T., Hashimoto, T., Naruse, M., Sano, H., Utsumi, H., and Nawata, H. (2000). High glucose level and free fatty acid stimulate reactive oxygen species production through protein kinase C-dependent activation of NAD(P)H oxidase in cultured vascular cells. *Diabetes* **49**, 1939–1945.

Iwata, N., Inazu, N., and Satoh, T. (1990). The purification and properties of aldose reductase from rat ovary. *Arch. Biochem. Biophys.* **282**, 70–77.

Iwata, T., Popescu, N. C., Zimonjic, D. B., Karlsson, C., Höög, J.-O., Vaca, G., Rodriguez, I. R., and Carper, D. (1995). Structural organization of the human sorbitol dehydrogenase gene (SORD). *Genomics* **26**, 55–62.

Jedziniak, J. A., Chylack, L. T., Jr., Cheng, H. M., Gillis, M. K., Kalustian, A. A., and Tung, W. H. (1981). The sorbitol pathway in the human lens: Aldose reductase and polyol dehydrogenase. *Invest. Ophthalmol. Vis. Sci.* **20**, 314–326.

Jedziniak, J. A., and Kinoshita, J. H. (1971). Activators and inhibitors of lens aldose reductase. *Invest. Ophthalmol.* **10**, 357–366.

Jeffery, J., Cederlund, E., and Jörnvall, H. (1984a). Sorbitol dehydrogenase: The primary structure of the sheep-liver enzyme. *Eur. J. Biochem.* **140**, 7–16.

Jeffery, J., Chesters, J., Mills, C., Sadler, P. J., and Jörnvall, H. (1984b). Sorbitol dehydrogenase is a zinc enzyme. *EMBO J.* **3**, 357–360.

Jeffery, J., Cummins, L., Carlquist, M., and Jörnvall, H. (1981). Properties of sorbitol dehydrogenase and characterization of a reactive cysteine residue reveal unexpected similarities to alcohol dehydrogenases. *Eur. J. Biochem.* **120**, 229–234.

Jeffery, J., and Jörnvall, H. (1988). Sorbitol dehydrogenase. *Adv. Enzymol. Relat. Areas Mol. Biol.* **61**, 47–106.

Jez, J. M., and Penning, T. M. (2001). The aldo-keto reductase (AKR) superfamily: An update. *Chem. Biol. Interact.* **130–132**, 499–525.

Jiang, Z. Y., Zhou, Q. L., Eaton, J. W., Koppenol, W. H., Hunt, J. V., and Wolff, S. P. (1991). Spirohydantoin inhibitors of aldose reductase inhibit iron- and copper-catalysed ascorbate oxidation *in vitro*. *Biochem. Pharmacol.* **42**, 1273–1278.

Johansson, K., El-Ahmad, M., Kaiser, C., Jörnvall, H., Eklund, H., Höög, J.-O., and Ramaswamy, S. (2001). Crystal structure of sorbitol dehydrogenase. *Chem. Biol. Interact.* **130–132**, 351–358.

Johnson, A. B. (1965). Histochemical demonstration of the soluble enzyme, ketose reductase (sorbitol dehydrogenase), by utilizing a new fixing solution. *J. Histochem. Cytochem.* **13**, 583–594.

Judzewitsch, R. G., Jaspan, J. B., Polonsky, K. S., Weinberg, C. R., Halter, J. B., Halar, E., Pfeifer, M. A., Vukadinovic, C., Bernstein, L., Schneider, M., Liang, K. Y., Gabbay, K. H.,

Rubenstein, A. H., and Porte, Jr. D. (1983). Aldose reductase inhibition improves nerve conduction velocity in diabetic patients. *N. Engl. J. Med.* **308**, 119–125.

Kador, P. F., Inoue, J., Secchi, E. F., Lizak, M. J., Rodriguez, L., Mori, K., Greentree, W., Blessing, K., Lackner, P. A., and Sato, S. (1998). Effect of sorbitol dehydrogenase inhibition on sugar cataract formation in galactose-fed and diabetic rats. *Exp. Eye Res.* **67**, 203–208.

Kao, Y. L., Donaghue, K., Chan, A., Knight, J., and Silink, M. (1999a). A novel polymorphism in the aldose reductase gene promoter region is strongly associated with diabetic retinopathy in adolescents with type 1 diabetes. *Diabetes* **48**, 1338–1340.

Kao, Y. L., Donaghue, K., Chan, A., Knight, J., and Silink, M. (1999b). An aldose reductase intragenic polymorphism associated with diabetic retinopathy. *Diabete. Res. Clin. Pract.* **46**, 155–160.

Karlsson, C. (1994). Mammalian Sorbitol Dehydrogenase. Department of Medical Biochemistry and Biophysics. Stockholm, Karolinska Institutet: 69.

Karlsson, C., and Höög, J.-O. (1993). Zinc coordination in mammalian sorbitol dehydrogenase: Replacement of putative zinc ligands by site-directed mutagenesis. *Eur. J. Biochem.* **216**, 103–107.

Karlsson, C., Jörnvall, H., and Höög, J.-O. (1991). Sorbitol dehydrogenase: cDNA coding for the rat enzyme. Variations within the alcohol dehydrogenase family independent of quaternary structure and metal content. *Eur. J. Biochem.* **198**, 761–765.

Karlsson, C., Maret, W., Auld, D. S., Höög, J.-O., and Jörnvall, H. (1989). Variability within mammalian sorbitol dehydrogenases: The primary structure of the human liver enzyme. *Eur. J. Biochem.* **186**, 543–550.

Karp, G. (1999). "Cell and Molecular Biology, Concepts and Experiments." p. 544, John Wiley & Sons, Inc., New York.

Kasajima, H., Yamagishi, S.-I., Sugai, S., Yagihashi, N., and Yagihashi, S. (2001). Enhanced in situ expression of aldose reductase in peripheral nerve and renal glomeruli in diabetic patients. *Virch. Arch.* **439**, 46–54.

Kato, N., Mizuno, K., Makino, M., Suzuki, T., and Yagihashi, S. (2000). Effects of 15-month aldose reductase inhibition with fidarestat on the experimental diabetic neuropathy in rats. *Diabet. Res. Clin. Pract.* **50**, 77–85.

Kawamura, M., Kopin, I. J., Kador, P. F., Sato, S., Tjurmina, O., and Eisenhofer, G. (1997). Effects of aldehyde/aldose reductase inhibition on neuronal metabolism of norepinephrine. *J. Autonom. Nerv. Sys.* **66**, 145–148.

Kida, Y. (1974). Studies on properties and biologic function of rat sorbitol dehydrogenase (EC 1.1.1.14) (Japanese). *J. Nara Med. Assoc.* **25**, 1–2.

King, G. L., Ishii, H., and Koya, D. (1997). Diabetic vascular dysfunctions: A model of excessive activation of protein kinase C. *Kidney Int.* **60**(Suppl.), S77–S85.

Kinoshita, J. H. (1990). A thirty year journey in the polyol pathway. *Exp. Eye Res.* **50**, 567–573.

Kinoshita, J. H., Datiles, M. B., Kador, P. F., and Robison, Jr., W. G. (1990). Aldose reductase and diabetic eye complications. *In* "Diabetes Mellitus, Theory and Practice" (H. Rifkin and D. Porte, Jr., eds.), pp. 264–278. Elsevier Science, New York.

Kinoshita, J. H., Futterman, S., Satoh, K., and Merola, L. O. (1963). Factors affecting the formation of sugar alcohols in ocular lens. *Biochim. Biophys. Acta* **74**, 340–350.

Ko, B. C., Lam, K. S., Wat, N. M., and Chung, S. S. (1995). An (A-C)n dinucleotide repeat polymorphic marker at the 5′ end of the aldose reductase gene is associated with early-onset diabetic retinopathy in NIDDM patients. *Diabetes* **44**, 727–732.

Ko, B. C. B., Ruepp, B., Bohren, K. M., Gabbay, K. H., and Chung, S. S. (1997). Identification and characterization of multiple osmotic response sequences in the human aldose reductase gene. *J. Biol. Chem.* **272**, 16431–16437.

Lee, A. Y., and Chung, S. S. (1999). Contributions of polyol pathway to oxidative stress in diabetic cataract. *FASEB J.* **13**, 23–30.

Lee, F. K., Cheung, M. C., and Chung, S. (1994). The human sorbitol dehydrogenase gene: cDNA cloning, sequence determination, and mapping by fluorescence in situ hybridization. *Genomics* **21**, 354–358.

Lee, A. Y., Chung, S. K., and Chung, S. S. (1995a). Demonstration that polyol accumulation is responsible for diabetic cataract by the use of transgenic mice expressing the aldose reductase gene in the lens. *Proc. Natl. Acad. Sci. USA* **92**, 2780–2784.

Lee, F. K., Lee, A. Y., Lin, C. X., Chung, S. S., and Chung, S. K. (1995b). Cloning, sequencing, and determination of the sites of expression of mouse sorbitol dehydrogenase cDNA. *Eur. J. Biochem.* **230**, 1059–1065.

Lee, F. K., Chung, S. K., and Chung, S. S. (1997). Aberrant mRNA splicing causes sorbitol dehydrogenase deficiency in C57BL/LiA mice. *Genomics* **46**, 86–92.

Leissing, N. C., and McGuinness, E. T. (1982). Sorbitol dehydrogenase from rat liver. *Methods Enzymol.* **89**, 135–140.

Leissing, N. C., and McGuinness, E. T. (1983). Kinetic analysis of rat liver sorbitol dehydrogenase. *Int. J. Biochem.* **15**, 651–656.

Lindstad, R. I., Hermansen, L. F., and McKinley-McKee, J. S. (1992). The kinetic mechanism of sheep liver sorbitol dehydrogenase. *Eur. J. Biochem.* **210**, 641–647.

Lindstad, R. I., Koll, P., and McKinley-McKee, J. S. (1998). Substrate specificity of sheep liver sorbitol dehydrogenase. *Biochem. J.* **330**, 479–487.

Lindstad, R. I., and McKinley-McKee, J. S. (1993). Methylglyoxal and the polyol pathway: Three-carbon compounds are substrates for sheep liver sorbitol dehydrogenase. *FEBS Lett.* **330**, 31–35.

Lindstad, R. I., and McKinley-McKee, J. S. (1995). Effect of pH on sheep liver sorbitol dehydrogenase steady-state kinetics. *Eur. J. Biochem.* **233**, 891–898.

Lindstad, R. I., and McKinley-McKee, J. S. (1997). Reversible inhibition of sheep liver sorbitol dehydrogenase by the antidiabetogenic drug 2-hydroxymethyl-4-(4-N,N-dimethyl-aminosulfonyl-1-piperazino) pyrimidine. *FEBS Lett.* **408**, 57–61.

Los, J. M., Simposon, L. B., and Wiesner, K. (1956). The kinetics of mutarotation of D-glucose with consideration of an intermediate free-aldehyde form. *J. Am. Chem. Soc.* **78**, 1564–1568.

Lowitt, S., Malone, J. I., Salem, A. F., Korthals, J., and Benford, S. (1995). Acetyl-L-carnitine corrects the altered peripheral nerve function of experimental diabetes. *Metabolism* **44**, 677–680.

Ludvigson, M. A., and Sorenson, R. L. (1980). Immunohistochemical localization of aldose reductase. I. Enzyme purification and antibody preparation: Localization in peripheral nerve, artery, and testis. *Diabetes* **29**, 438–449.

Maeda, S., Haneda, M., Yasuda, H., Tachikawa, T., Isshiki, K., Koya, D., Terada, M., Hidaka, H., Kashiwagi, A., and Kikkawa, R. (1999). Diabetic nephropathy is not associated with the dinucleotide repeat polymorphism upstream of the aldose reductase (ALR2) gene but with erythrocyte aldose reductase content in Japanese subjects with type 2 diabetes. *Diabetes* **48**, 420–422.

Magnani, P., Thomas, T. P., Tennekoon, G., DeVries, G. H., Greene, D. A., and Brosius, F. C. (1998). Regulation of glucose transport in cultured Schwann cells. *J. Peripher. Nerv. Syst.* **3**, 28–36.

Malone, J. I., Leavengood, H., Peterson, M. J., O'Brien, M. M., Page, M. G., and Aldinger, C. E. (1984). Red blood cell sorbitol as an indicator of polyol pathway activity: Inhibition by sorbinil in insulin-dependent diabetic subjects. *Diabetes* **33**, 45–49.

Malone, J. I., Lowitt, S., Korthals, J. K., Salem, A., and Miranda, C. (1996). The effect of hyperglycemia on nerve conduction and structure is age dependent. *Diabetes* **45**, 209–215.

Maret, W. (1991). Novel substrates and inhibitors of human liver sorbitol dehydrogenase. *Adv. Exp. Med. Biol.* **284**, 327–336.

Maret, W. (1996). Human sorbitol dehydrogenase: A secondary alcohol dehydrogenase with distinct pathophysiological roles. pH-dependent kinetic studies. *Adv. Exp. Med. Biol.* **414**, 383–393.

Maret, W., and Auld, D. S. (1988). Purification and characterization of human liver sorbitol dehydrogenase. *Biochemistry* **27**, 1622–1628.

Marini, I., Bucchioni, L., Borella, P., DelCorso, A., and Mura, U. (1997). Sorbitol dehydrogenase from bovine lens: Purification and properties. *Arch. Biochem. Biophys.* **340**, 383–391.

Markus, H. B., Raducha, M., and Harris, H. (1983). Tissue distribution of mammalian aldose reductase and related enzymes. *Biochem. Med.* **29**, 31–45.

Mayer, J. H., and Tomlinson, D. R. (1983). Prevention of defects of axonal transport and nerve conduction velocity by oral administration of myo-inositol or an aldose reductase inhibitor. *Diabetologia* **25**, 433–438.

McCorkindale, J., and Edson, N. L. (1954). Polyol dehydrogenases. I. The specificity of rat liver polyol dehydrogenase. *Biochem. J.* **57**, 518–523.

Medina, C., Vaca, G., Rios, J., Chavez-Anaya, E., Ocampo, R., Garcia-Maravilla, S., Arreola, R., and Cantu, J. M. (1987). Screening for red blood cell sorbitol dehydrogenase deficiency in patients with diabetes or cataracts. *Ophthalmic. Pediatr. Genet.* **8**, 197–202.

Miwa, I., Kanbara, M., and Okuda, J. (1989). Improvement of nerve conduction velocity in mutant diabetic mice by aldose reductase inhibitor without affecting nerve myo-inositol content. *Chem. Pharm. Bull. (Tokyo)* **37**, 1581–1582.

Mizisin, A. P., Li, L., and Calcutt, N. A. (1997). Sorbitol accumulation and transmembrane efflux in osmotically stressed JS1 schwannoma cells. *Neurosci. Lett.* **229**, 53–56.

Mizuno, K., Kato, N., Makino, M., Suzuki, T., and Shindo, M. (1999). Continuous inhibition of excessive polyol pathway flux in peripheral nerves by aldose reductase inhibitor fidarestat leads to improvement of diabetic neuropathy. *J. Diabet. Compl.* **13**, 141–150.

Mizuno, K., Kato, N., Matsubara, A., Nakano, K., and Kurono, M. (1992). Effects of a new aldose reductase inhibitor, (2S, 4S)-6-fluoro-2′,5′-dioxospiro[chroman-4,4′-imidazolidine]-2-carboxamide (SNK-860), on the slowing of motor nerve conduction velocity and metabolic abnormalities in the peripheral nerve in acute streptozotocin-induced diabetic rats. *Metabolism* **41**, 1081–1086.

Moczulski, D. K., Burak, W., Doria, A., Zychma, M., Zukowska-Szczechowska, E., Warram, J. H., and Grzeszczak, W. (1999). The role of aldose reductase gene in the susceptibility to diabetic nephropathy in type II (non-insulin-dependent) diabetes mellitus. *Diabetologia* **42**, 94–97.

Moczulski, D. K., Scott, L., Antonellis, A., Rogus, J. J., Rich, S. S., Warram, J. H., and Krolewski, A. S. (2000). Aldose reductase gene polymorphisms and susceptibility to diabetic nephropathy in type 1 diabetes mellitus. *Diabet. Med.* **17**, 111–118.

Moriyama, T., Nakano, T., Wada, T., *et al.* (1973). Crystallization and properties of liver sorbitol dehydrogenase (EC 1.1.1.14) (Japanese). *J. Nara Med. Assoc.* **24**, 356–362.

Murray, R. K., Gadacz, I., Bach, M., Hardin, S., and Morris, H. P. (1969). Metabolic and electrophoretic studies of rat liver sorbitol dehydrogenase. *Can. J. Biochem.* **47**, 587–593.

Mylari, B. L., Oates, P. J., Beebe, D. A., Brackett, N. S., Coutcher, J. B., Dina, M. S., and Zembrowski, W. J. (2001). Sorbitol dehydrogenase inhibitors (SDIs): A new potent, enantiomeric SDI, 4-[2-1R hydroxy-ethyl)-pyrimidin-4-yl]-piperazine-1-sulfonic acid dimethylamide. *J. Med. Chem.* **44**, 2695–2700.

Nakano, T., and Petrash, J. M. (1996). Kinetic and spectroscopic evidence for active site inhibition of human aldose reductase. *Biochemistry* **35**, 11196–11202.

Ng, D. P., Conn, J., Chung, S. S., and Larkins, R. G. (2001). Aldose reductase (AC)n microsatellite polymorphism and diabetic microvascular complications in Caucasian type 1 diabetes mellitus. *Diabet. Res. Clin. Pract.* **52**, 21–27.

Ng, T. F., Lee, F. K., Song, Z. T., Calcutt, N. A., Lee, A. Y., Chung, S. S., Chung, S. K., Ng, D. T., and Lee, L. W. (1998). Effects of sorbitol dehydrogenase deficiency on nerve conduction in experimental diabetic mice. *Diabetes* **47**, 961–966.

Nishikawa, T., Edelstein, D., Du, X. L., Yamagishi, S., Matsumura, T., Kaneda, Y., Yorek, M. A., Beebe, D. A., Oates, P. J., Hammes, H. P., Giardino, I., and Brownlee, M. (2000). Normalizing mitochondrial superoxide production blocks three pathways of hyperglycaemic damage. *Nature* **404**, 787–790.

Nishimura, C., Furue, M., Ito, T., Omori, Y., and Tanimoto, T. (1993). Quantitative determination of human aldose reductase by enzyme-linked immunosorbent assay: Immunoassay of human aldose reductase. *Biochem. Pharmacol.* **46**, 21–28.

Nishimura-Yabe, C. (1998). Aldose reductase in the polyol pathway: A potential target for the therapeutic intervention of diabetic complications. *Fol. Pharmacol. Jap.* **111**, 137–145.

Oates, P. J. (1997). The polyol pathway: A culprit in diabetic neuropathy? *Neurosci. Res. Commun.* **21**, 33–40.

Oates, P., Schelhorn, T., Miller, M., Hammerlund, E., Ellery, C., Beebe, D., and Hakkinen, J. (1998). Polyol pathway inhibitors dose-dependently preserve nerve function in diabetic rats. *Diabetologia* **41**(Suppl. 1), A271.

Oates, P., Williamson, J., Chang, K., Mylari, B., Beebe, D., Coutcher, J., Siegel, T., and Zembrowski, W. (1994). Attenuation of diabetes-induced vascular dysfunction by an inhibitor of sorbitol dehydrogenase. *Diabetes* **43**(Suppl. 1), 17A.

Oates, P. J., Ellery, C. A., Coutcher, J. B., and Beebe, D. A. (2001). A metabolic model of sorbitol in normal and diabetic peripheral nerve. *Diabetes* **50**(Suppl. 1), A188.

Oates, P. J., and Goddu, K. J. (1987). A sorbitol gradient in the rat renal medulla. *Kidney Int.* **31**, 448.

Oates, P. J., and Mylari, B. L. (1999). Aldose reductase inhibitors: Therapeutic implications for diabetic complications. *Exp. Opin. Invest. Drugs* **8**, 2095–2119.

Oates, P. J., Beebe, D. A., and Mylari, B. L. (1997). Sorbitol dehydrogenase catalysis and inhibition by CP-166,572: Ligand binding and kinetic studies. In "US–Japan Aldose Reductase Workshop Proceedings," p. 71. Kona, HI.

O'Brien, M. M., Schofield, P. J., and Edwards, M. R. (1983). Polyol-pathway enzymes of human brain: Partial purification and properties of sorbitol dehydrogenase. *Biochem. J.* **211**, 81–90.

Obrosova, I. G., Lang, H. J., and Greene, D. A. (1998). Antioxidative defense in diabetic peripheral nerve: Role for aldose reductase and sorbitol dehydrogenase. *Diabetologia* **41**, A270.

Obrosova, I. G., Fathallah, L., Lang, H. J., and Greene, D. A. (1999). Evaluation of a sorbitol dehydrogenase inhibitor on diabetic peripheral nerve metabolism: A prevention study. *Diabetologia* **42**, 1187–1194.

Obrosova, I. G., Greene, D. A., and Fathallah, L. (2000). Relationship between aldose reductase and oxidative stress in diabetic peripheral nerve. *Diabetes* **49**(Suppl. 1), A77–A78.

Oka, M., Matsumoto, Y., Sugiyama, S., Tsuruta, N., and Matsushima, M. (2000). A potent aldose reductase inhibitor, (2S,4S)-6-fluoro-2', 5'-dioxospiro[chroman-4,4'-imidazolidine]-2-carboxamide (fidarestat): Its absolute configuration and interactions with the aldose reductase by X-ray crystallography. *J. Med. Chem.* **43**, 2479–2483.

Olmos, P., Futers, S., Acosta, A. M., Siegel, S., Maiz, A., Schiaffino, R., Morales, P., Diaz, R., Arriagada, P., Claro, J. C., Vega, R., Vollrath, V., Velasco, S., and Emmerich, M. (2000). (AC)23 [Z − 2] polymorphism of the aldose reductase gene and fast progression of retinopathy in Chilean type 2 diabetics. *Diabet. Res. Clin. Pract.* **47**, 169–176.

Orosz, S. E., Townsend, S. F., Tornheim, P. A., and Brownscheidle, C. M. (1981). Localization of aldose reductase and sorbitol dehydrogenase in the nervous system of normal and diabetic rats. *Acta Diabetol. Lat.* **18**, 373–381.

Patel, A., Hibberd, M. L., Millward, B. A., and Demaine, A. G. (1996). Chromosome 7q35 and susceptibility to diabetic microvascular complications. *J. Diabet. Compl.* **10**, 62–67.

Perkins, B. A., Greene, D. A., and Bril, V. (2001). Glycemic control is related to the morphological severity of diabetic sensorimotor polyneuropathy. *Diabet. Care* **24**, 748–752.

Persson, B., Zigler, J. S., and Jörnvall, H. (1994). A super-family of medium-chain dehydrogenases/reductases (mdr): Sub-lines including xi-crystallin, alcohol and polyol dehydrogenases, quinone oxidoreductases, enoyl reductases, vat-1 and other proteins. *Eur. J. Biochem.* **226**, 15–22.

Petrash, J. M., Harter, T. M., Devine, C. S., Olins, P. O., Bhatnagar, A., Liu, S., and Srivastava, S. K. (1992). Involvement of cysteine residues in catalysis and inhibition of human aldose reductase: Site-directed mutagenesis of Cys-80, -298, and -303. *J. Biol. Chem.* **267**, 24833–24840.

Petrash, J. M., Harter, T. M., and Murdock, G. L. (1997). A potential role for aldose reductase in steroid metabolism. *Adv. Exp. Med. Biol.* **414**, 465–473.

Petrash, J. M., Tarle, I., Wilson, D. K., and Quiocho, F. A. (1994). Aldose reductase catalysis and crystallography: Insights from recent advances in enzyme structure and function. *Diabetes* **43**, 955–959.

Pfeifer, M. A., Schumer, M. P., and Gelber, D. A. (1997). Aldose reductase inhibitors: The end of an era or the need for different trial designs? *Diabetes* **46**(Suppl. 2), S82–S89.

Poduslo, J. F., Dyck, P. J., and Berg, C. T. (1985). Regulation of myelination: Schwann cell transition from a myelin-maintaining state to a quiescent state after permanent nerve transection. *J. Neurochem.* **44**, 388–400.

Powell, H. C., Garrett, R. S., Kador, P. F., and Mizisin, A. P. (1991). Fine-structural localization of aldose reductase and ouabain-sensitive, K(+)-dependent p-nitro-phenylphosphatase in rat peripheral nerve. *Acta Neuropathol. (Berl.)* **81**, 529–539.

Ramasamy, R., Oates, P. J., and Schaefer, S. (1997). Aldose reductase inhibition protects diabetic and non-diabetic rat hearts from ischemic injury. *Diabetes* **46**, 292–300.

Rees-Milton, K. J., Jia, Z., Green, N. C., Bhatia, M., El-Kabbani, O., and Flynn, T. G. (1998). Aldehyde reductase: The role of C-terminal residues in defining substrate and cofactor specificities. *Arch. Biochem. Biophys.* **355**, 137–144.

Reiner, J. M. (1969). "Behavior of Enzyme Systems," p. 345. Van Nostrand Reinhold Co., New York.

Robinson, B., Hunsaker, L. A., Stangebye, L. A., and Vander Jagt, D. L. (1993). Aldose and aldehyde reductases from human kidney cortex and medulla. *Biochim. Biophys. Acta* **1203**, 260–266.

Roelfzema, H., Van Den Hurk, J. J. M. A., and Mier, P. D. (1976). Absence of the sorbitol pathway in epidermis. *Br. J. Derm.* **94**, 577–578.

Rondeau, J.-M., Tete-Favier, F., Podjarny, A., Reymann, J.-M., Barth, P., Biellmann, J.-F., and Moras, D. (1992). Novel NADPH-binding domain revealed by the crystal structure of aldose reductase. *Nature* **355**, 469–472.

Rosen, P., Nawroth, P. P., King, G., Moller, W., Tritschler, H. J., and Packer, L. (2001). The role of oxidative stress in the onset and progression of diabetes and its complications: a summary of a Congress Series sponsored by UNESCO-MCBN, the American Diabetes Association and the German Diabetes Society. *Diabetes Metab. Res. Rev.* **17**, 189–212.

Sarges, R., and Oates, P. J. (1993). Aldose reductase inhibitors: Recent developments. *Prog. Drug Res.* **40**, 99–161.

Sato, S. (1992). Rat kidney aldose reductase and aldehyde reductase and polyol production in rat kidney. *Am. J. Physiol.* **263**, F799–F805.

Schmidt, A. M., Yan, S. D., Yan, S. F., and Stern, D. M. (2000). The biology of the receptor for advanced glycation end products and its ligands. *Biochim. Biophys. Acta* **1498**, 99–111.

Schmidt, R. E., Dorsey, D. A., Beaudet, L. N., Plurad, S. B., Parvin, C. A., Yarasheski, K. E., Smith, S. R., Lang, H., Williamson, J. R., and Ido, Y. (2001). Inhibition of sorbitol dehydrogenase exacerbates autonomic neuropathy in rats with streptozotocin-induced diabetes. *J. Neuropathol. & Exp. Neurol.* **60**, 1153–1169.

Schmidt, R. E., Dorsey, D. A., Beaudet, L. N., Plurad, S. B., Williamson, J. R., and Ido, Y. (1998). Effect of sorbitol dehydrogenase inhibition on experimental diabetic autonomic neuropathy. *J. Neuropathol. Exp. Neurol.* **57**, 1175–1189.

Schmidt, R. E., Plurad, S. B., Coleman, B. D., Williamson, J. R. and Tilton, R. G. (1991). Effects of sorbinil, dietary myo-inositol supplementation, and insulin on resolution of neuroaxonal dystrophy in mesenteric nerves of streptozocin-induced diabetic rats. *Diabetes* **40**, 574–582.

Segel, I. H. (1975). "Enzyme Kinetics," pp. 957, John Wiley & Sons, Inc., New York.

Service, F. J., Rizza, R. A., Daube, J. R., O'Brien, P. C., and Dyck, P. J. (1985). Near normoglycaemia improved nerve conduction and vibration sensation in diabetic neuropathy. *Diabetologia* **28**, 722–727.

Sgambato, S., Ceriello, A., Passariello, N., and Giuliano, D. (1979). Plasma sorbitol dehydrogenase in diabetic subjects with or without vascular complications. *Biochem. Exp. Biol.* **15**, 61–64.

Shah, V. O., Scavini, M., Nikolic, J., Sun, Y., Vai, S., Griffith, J. K., Dorin, R. I., Stidley, C., Yacoub, M., Vander Jagt, D. L., Eaton, R. P., and Zager, P. G. (1998). Z − 2 microsatellite allele is linked to increased expression of the aldose reductase gene in diabetic nephropathy. *J. Clin. Endocrinol. Metab.* **83**, 2886–2891.

Shimoshige, Y., Ikuma, K., Yamamoto, T., Takakura, S., Kawamura, I., Seki, J., Mutoh, S., and Goto, T. (2000). The effects of zenarestat, an aldose reductase inhibitor, on peripheral neuropathy in Zucker diabetic fatty rats. *Metabolism* **49**, 1395–1399.

Shin, Y. S., Rieth, M., Endres, W., and Haas, P. (1984). Sorbitol dehydrogenase deficiency in a family with congenital cataracts. *J. Inherit. Metab. Dis.* **7**, 151–152.

Sima, A. A., Bril, V., Nathaniel, V., McEwen, T. A., Brown, M. B., Lattimer, S. A., and Greene, D. A. (1988). Regeneration and repair of myelinated fibers in sural-nerve biopsy specimens from patients with diabetic neuropathy treated with sorbinil. *N. Engl. J. Med.* **319**, 548–555.

Smith, M. G. (1962). Polyol dehydrogenases: Crystallization of the L-iditol dehydrogenase of sheep liver. *Biochem. J.* **83**, 135–144.

Sobajima, H., Aoki, T., Sassa, H., Suzuki, T., Taniko, K., Makino, M., and Mizuno, K. (2001). Pharmacological properties of fidarestat, a potent aldose reductase inhibitor, clarified by using sorbitol in human and rat erythrocytes. *Pharmacology* **62**, 193–199.

Srivastava, S. K., Ansari, N. H., Brown, J. H., and Petrash, J. M. (1982). Formation of sorbitol 6-phosphate by bovine and human lens aldose reductase, sorbitol dehydrogenase and sorbitol kinase. *Biochim. Biophys. Acta* **717**, 210–214.

Srivastava, S. K., Ansari, N. H., Hair, G. A., and Das, B. (1984). Aldose and aldehyde reductases in human tissues. *Biochim. Biophys. Acta* **800**, 220–227.

Stevens, M. J., Feldman, E. L., and Greene, D. A. (1995). The aetiology of diabetic neuropathy: The combined roles of metabolic and vascular defects. *Diabet. Med.* **12**, 566–579.

Stevens, M. J., Killen, P. D., Wang, P., Larkin, D., and Greene, D. A. (2000a). Overexpression of aldose reductase (AR) in human retinal pigment epithelial (RPE) cell lines is associated with a polymorphism in the ar basal promoter region. *Diabetes* **49**(Suppl. 1), A167.

Stevens, M. J., Obrosova, I. G., Cao, X., Van Huysen, C., and Greene, D. A. (2000b). Effects of DL-alpha-lipoic acid on peripheral nerve conduction, blood flow, energy metabolism, and oxidative stress in experimental diabetic neuropathy. *Diabetes* **49**, 1006–1015.

Stewart, M. A., Sherman, W. R., and Anthony, S. (1966). Free sugars in alloxan diabetic rat nerve. *Biochem. Biophys. Res. Commun.* **22**, 4–91.

Sugimoto, K., Kasahara, T., Yonezawa, H., and Yagihashi, S. (1999). Peripheral nerve structure and function in long-term galactosemic dogs: Morphometric and electron microscopic analyses. *Acta Neuropathol.* **97**, 369–376.

Sugimoto, K., Murakawa, Y., and Sima, A. A. (2000). Diabetic neuropathy: A continuing enigma. *Diabet. Metab. Res. Rev.* **16**, 408–433.

Suzuki, T., Mizuno, K., Yashima, S., Watanabe, K., Taniko, K., and Yabe-Nishimura, C. (1999). Characterization of polyol pathway in Schwann cells isolated from adult rat sciatic nerves. *J. Neurosci. Res.* **57**, 495–503.

Szwergold, B. S., Kappler, F., and Brown, T. R. (1990). Identification of fructose 3-phosphate in the lens of diabetic rats. *Science* **247**, 451–454.

Szwergold, B. S., Kappler, F., Brown, T. R., Pfeffer, P., and Osman, S. F. (1989). Identification of D-sorbitol 3-phosphate in the normal and diabetic mammalian lens. *J. Biol. Chem.* **264**, 9278–9282.

Takahashi, M., Fujii, J., Teshima, T., Suzuki, K., Shiba, T., and Taniguchi, N. (1993). Identity of a major 3-deoxyglucosone-reducing enzyme with aldehyde reductase in rat liver established by amino acid sequencing and cDNA expression. *Gene* **127**, 249–253.

Tanimoto, T., Maekawa, K., Okada, S., and Yabe-Nishimura, C. (1998). Clinical analysis of aldose reductase for differential diagnosis of the pathogenesis of diabetic complications. *Anal. Chim. Acta* **365**, 285–292.

Thurston, J. H. (1990). Increased lactate in diabetic rat nerve *in vivo*: Hypoxia/ischemia or inhibition of pyruvate oxidation in ketotic animals? *Clin. Res.* **38**, 840A.

Tilton, R. G., Chang, K., Nyengaard, J. R., Van den Enden, M., Ido, Y., and Williamson, J. R. (1995). Inhibition of sorbitol dehydrogenase. Effects on vascular and neural dysfunction in streptozocin-induced diabetic rats. *Diabetes* **44**, 234–242.

Tilton, R. G., Chang, K., Pugliese, G., Eades, D. M., Province, M. A., Sherman, W. R., Kilo, C., and Williamson, J. R. (1989). Prevention of hemodynamic and vascular albumin filtration changes in diabetic rats by aldose reductase inhibitors. *Diabetes* **38**, 1258–1270.

Tomlinson, D. R., Willars, G. B., and Carrington, A. L. (1992). Aldose reductase inhibitors and diabetic complications. *Pharmacol. Therap.* **54**, 151–194.

Tomlinson, D. R. (1998). Future prevention and treatment of diabetic neuropathy. *Diabet. Metab.* **24**(Suppl. 3), 79–83.

Tomlinson, D. R. (1999). Mitogen-activated protein kinases as glucose transducers for diabetic complications. *Diabetologia* **42**, 1271–1281.

Touster, O., and Shaw, D. R. D. (1962). Biochemistry of the acyclic polyols. *Physiol. Rev.* **42**, 181–225.

Urzhumtsev, A., Tete-Favier, F., Mitschler, A., Barbanton, J., Barth, P., Urzhumtseva, L., Biellmann, J.-F., Podjarny, A. D., and Moras, D. (1997). A 'specificity' pocket inferred from the crystal structures of the complexes of aldose reductase with the pharmaceutically important inhibitors tolrestat and sorbinil. *Structure (London)* **5**, 601–612.

Vaca, G., Alonso, R., Zuniga, P., Gonzalez-Quiroga, G., Chavez, M. A., Medina, C., Wunsch, C., and Cantu, J. M. (1984). Sorbitol dehydrogenase deficiency in several pig tissues: Potential implications for studies of experimental diabetes. *Diabetologia* **27**, 482–483.

Vaca, G., Ibarra, B., Bracamontes, M., Garcia-Cruz, D., Sanchez-Corona, J., Medina, C., Wunsch, C., Gonzalez-Quiroga, G., and Cantu, J. M. (1982). Red blood cell sorbitol dehydrogenase deficiency in a family with cataracts. *Hum. Genet.* **61**, 338–341.

VanDam, P. S., and Bravenboer, B. (1997). Oxidative stress and antioxidant treatment in diabetic neuropathy. *Neurosci. Res. Commun.* **21**, 41–48.

Vander Jagt, D. L., Robinson, B., Taylor, K. K., and Hunsaker, L. A. (1990). Aldose reductase from human skeletal and heart muscle: Interconvertible forms related by thiol-disulfide exchange. *J. Biol. Chem.* **265**, 20982–20987.

Varfolomeev, S. D. (1977). Kinetics of reactions in multienzymesystems. I. Steady-state processes in two-enzyme systems. *Molek. Biol.* **11**, 564–581.

Vasiliou, V., Pappa, A., and Petersen, D. R. (2000). Role of aldehyde dehydrogenases in endogenous and xenobiotic metabolism. *Chem. Biol. Interact.* **129**, 1–19.

Vinik, A. I., Park, T. S., Stansberry, K. B., and Pittenger, G. L. (2000). Diabetic neuropathies. *Diabetologia* **43**, 957–973.

Vlassara, H. (2001). Diabetes and advanced glycation end products. *In* "Diabetes and Cardiovascular Disease" (M. T. Johnstone and A. Veves, eds.), pp. 81–102. Humana Press, Totowa, NJ.

Walsall, E. P., Lyons, S. A., and Metzger, R. P. (1978). A comparison of selected physical properties of hepatic sorbitol dehydrogenases [L-iditol: NAD oxidoreductases] from four mammalian species. *Comp. Biochem. Physiol.* **59B**, 213–218.

Wang, K., Bohren, K. M., and Gabbay, K. H. (1993). Characterization of the human aldose reductase gene promoter. *J. Biol. Chem.* **268**, 16052–16058.

Ward, J. D. (1999). Improving prognosis in type 2 diabetes: Diabetic neuropathy is in trouble. *Diabet. Care* **22**(Suppl. 2), B84–B88.

Wautier, M. P., Chappey, O., Corda, S., Stern, D. M., Schmidt, A. M., and Wautier, J. L. (2001). Activation of NADPH oxidase by AGE links oxidant stress to altered gene expression via RAGE. *Am. J. Physiol.* **280**, E685–E694.

Wen, Y., and Bekhor, I. (1993). Sorbitol dehydrogenase: Full-length cDNA sequencing reveals a mRNA coding for a protein containing an additional 42 amino acids at the N-terminal end. *Eur. J. Biochem.* **217**, 83–87.

Whiting, P. H., and Ross, I. S. (1988). Increased nerve polyol levels in experimental diabetes and their reversal by sorbinil. *Br. J. Exp. Pathol.* **69**, 697–702.

Williamson, J. R., Chang, K., Frangos, M., Hasan, K. S., Ido, Y., Kawamura, T., Nyengaard, J. R., van den Enden, M., Kilo, C., and Tilton, R. G. (1993). Hyperglycemic pseudohypoxia and diabetic complications. *Diabetes* **42**, 801–813.

Williamson, J. R., Kilo, C., and Ido, Y. (1999). The role of cytosolic reductive stress in oxidant formation and diabetic complications. *Diabet. Res. Clin. Pract.* **45**, 81–82.

Wilson, D. K., Tarle, I., Petrash, J. M., and Quiocho, F. A. (1993). Refined 1.8 Å structure of human aldose reductase complexed with the potent inhibitor zopolrestat. *Proc. Natl. Acad. Sci. USA* **90**, 9847–9851.

Winegrad, A. (1973). Discussion. *Adv. Metabol. Disorders* **2** (Suppl.), 430–432.

Wolin, M. S. (2000). Interactions of oxidants with vascular signaling systems. *Arterioscler. Thromb. Vasc. Biol.* **20**, 1430–1442.

Yabe-Nishimura, C. (1998). Aldose reductase in glucose toxicity: A potential target for the prevention of diabetic complications. *Pharmacol. Rev.* **50**, 21–33.

Yagihashi, S. (1997). Pathogenetic mechanisms of diabetic neuropathy: Lessons from animal models. *J. Peripher. Nerv. Syst.* **2**, 113–132.

Yagihashi, S., Yamagishi, S., Wada, R., Sugimoto, K., Baba, M., Wong, H. G., Fujimoto, J., Nishimura, C., and Kokai, Y. (1996). Galactosemic neuropathy in transgenic mice for human aldose reductase. *Diabetes* **45**, 56–59.

Yue, D. K., Hanwell, M. A., Satchell, P. M., and Turtle, J. R. (1982). The effect of aldose reductase inhibition on motor nerve conduction velocity in diabetic rats. *Diabetes* **31**, 789–794.

Ziegler, D., Dannehl, K., Wiefels, K., and Gries, F. A. (1992). Differential effects of near-normoglycaemia for 4 years on somatic nerve dysfunction and heart rate variation in type 1 diabetic patients. *Diabet. Med.* **9**, 622–629.

Zochodne, D. W. (1999). Diabetic neuropathies: Features and mechanisms. *Brain Pathol.* **9**, 369–391.

Zou, X., Lu, J., and Pan, C. (2000). Effect of polymorphism of (AC)n in the 5′-end of aldose reductase gene on erythrocyte aldose reductase activity in patients with type 2 diabetes mellitus. *Chin. J. Endocrinol. Metabol.* **16**, 346–349.

NERVE GROWTH FACTOR FOR THE TREATMENT OF DIABETIC NEUROPATHY: WHAT WENT WRONG, WHAT WENT RIGHT, AND WHAT DOES THE FUTURE HOLD?

Stuart C. Apfel

Albert Einstein College of Medicine, Bronx, New York 10461

Since their discovery in the 1950s, neurotrophic factors have raised expectations that their clinical application to neurodegenerative diseases might provide an effective therapy for what are now untreatable conditions. Nerve growth factor (NGF) was the first neurotrophic factor to be discovered and was one of the earliest to proceed to clinical trials. NGF, which is selectively trophic for small fiber sensory and sympathetic neurons, was selected as a potential therapy for diabetic polyneuropathy because of the serious consequences associated with degeneration of those neuronal populations in this condition. In addition, evidence shows that reduced availability of NGF may contribute to the pathogenesis of diabetic neuropathy, and animal models of neuropathy respond to the exogenous administration of NGF. Two sets of phase II clinical trials suggested that recombinant human NGF (rhNGF) administration was effective at ameliorating the symptoms associated with both diabetic polyneuropathy and HIV-related neuropathy. These early studies, however, revealed that painful side effects were dose limiting for NGF. A large-scale phase III clinical trial of 1019 patients randomized to receive either rhNGF or placebo for 48 weeks failed to confirm the earlier indications of efficacy. Among the explanations offered for the

393

discrepancy between the two sets of trials was a robust placebo effect, inadequate dosage, different study populations, and changes to the formulation of rhNGF for the phase III trial. As a result of the phase III outcome, Genentech has decided not to proceed with further development of rhNGF. © 2002, Elsevier Science (USA).

I. Introduction

A. The Beginnings

In 1953, Rita Levi-Montalcini, working with Victor Hamburger, grafted mouse sarcoma tissue onto chick embryos whose wing buds had been extirpated. She discovered that the mouse sarcoma tissue produced a soluble factor that promoted the growth of nearby sensory and sympathetic ganglia (Levi-Montalcini, 1987). Collaborating with a protein chemist named Stanley Cohen, they isolated the substance responsible and named it nerve growth factor (NGF) (Cohen *et al.*, 1954). Nerve growth factor was the first neurotrophic factor to be discovered, and for decades was the best characterized. Since the discovery of NGF, numerous other neurotrophic factors have been identified, most within the last 10–20 years (Table I). The rapid growth of this field has been stimulated, at least in part, by the prospects of

TABLE I

PARTIAL LIST OF NEUROTROPHIC FACTORS

Neurotrophins: NGF, BDNF, NT-3, NT-4/5, NT-6, NT-7

Hematopoietic cytokines: CNTF, LIF, Onc M, CT-1, interleukin(IL)-1, IL-3, IL-6, IL-7, IL-9, IL-11, GCSF, GRO-β, NNT-1/BSF-3

IGF family: Insulin, IGF-I, IGF-II

Heparin-binding family: aFGF, bFGF, int-2-onc, hst/k-fgf onc, FGF-4, FGF-5, FGF-6, KGF

EGF family: EGF, HB-EGF, TGF-α

TGF-β family: TGF-β1, TGF-β2, TGF-β3, activin A, TGF-ss, GDF-5, BMPs

GDNF family: GDNF, Neurturin, Persephin, Artemin

Tyrosine kinase-associated cytokines: PDGF, CSF-1, SCF

Others: ACTH analogs, gangliosides, APP, interferon-γ, neuroimmunophilins, various small molecules

NGF, nerve growth factor; BDNF, brain derived neurotrophic factor; NT-3, neurotrophin-3; CNTF, ciliary derived neurotrophic factor; LIF, leukemia inhibitory factor; Onc M, Oncogene M; CT-1, cardiotropin-1; GCSF, granulocyte colony stimulating factor; GRO, growth regulated oncogene; IGF, insulin like growth factor; aFGF, acidic fibroblast growth factor; bFGF, basic fibroblast growth factor; KGF, keratinocyte growth factor; EGF, epidermal growth factor; TGF, transforming growth factor; GDF, growth differentiation factor; BMP, bone morphogenic proteins; GDNF, glial-cell derived neurotrophic factor; PDGF, platelet derived growth factor; CSF-1, colony stimulating factor; SCF, stem cell factor; ACTH, adrenocorticotrophic hormone; APP, amyloid precursor protein.

applying these factors clinically. At the same time, each new discovery has raised expectations regarding their potential clinical application. For most of this period, these expectations were fueled primarily by highly promising preclinical data from tissue culture and animal-based studies. In the mid-1990s, clinical trials were initiated for the first time with a wide variety of different neurotrophic factors. The results were by and large disappointing and much of the original enthusiasm in the field has been dampened.

One of the earliest indications sought for the application of neurotrophic factors was the treatment of diabetic polyneuropathy. Although several neurotrophic factors were considered for this application, nerve growth factor was the only one to be tested extensively in clinical trials. The early clinical trials of NGF for the treatment of diabetic polyneuropathy were highly encouraging. Subsequent extensive phase III studies failed to confirm any benefit from this therapy. Consequently, further clinical testing of NGF has been suspended. This chapter reviews the NGF story, examines data both supporting and rejecting its application in the treatment of diabetic neuropathy, and attempts to address the question: "What next?"

B. What are Neurotrophic Factors?

The more we learn about neurotrophic factors the more difficult it has become to provide a simple definition. By conventional use, neurotrophic factors may be thought of as endogenous proteins that promote the survival of specific populations of neurons, stimulate their morphological differentiation, and regulate neuronal gene expression in a variety of ways. Historically, they were thought of as having highly specific and characteristic properties in common; however, experience has taught us that they are quite heterogeneous. Apart from their common ability to promote neuronal survival, the biological properties of the different neurotrophic factors differ vastly from one another. Many of these proteins have significant activity outside the nervous system and, in some cases, their primary physiological role appears to be linked to nonneuronal systems.

Classically, neurotrophic factors were thought to be released by neuronal target tissue, bind to specific receptors on nerve terminals, and be retro-gradely transported back to the cell nucleus where they exert their biological effects via regulation of gene expression. While in many instances this notion still applies, we now know that it is an oversimplification. Neurotrophic factors are frequently produced and released by cells that are not neuronal targets and they may bind to receptors all along the neuron, frequently to receptors on nonneuronal cells. They are active not only in the nucleus, but also at nerve terminals and even on local nonneuronal cells. While they play an important role in promoting the survival and normal development of specific populations of neurons, they may have equally important roles in regulating the physiological activities of mature neurons.

C. The Neurotrophin Gene Family

NGF was the first neurotrophic factor to be discovered and was felt to be the prototype for all neurotrophic factors subsequently discovered. It is now recognized that NGF is a member of a specific gene family of structurally related neurotrophic factors called "neurotrophins." Other members of this gene family include brain-derived neurotrophic factor (BDNF), neurotrophin-3 (NT-3), NT-4/5, NT-6, and NT-7. The last two members of this gene family are not found in mammals and are not discussed further here. Although there is a high degree of structural homology among the neurotrophins, they differ from one another in their specific populations of responsive neurons (Table II). Responsiveness is conveyed by expression of particular high-affinity neurotrophin receptors.

Neurotrophin high-affinity receptors are members of a parallel gene family called tropomyosin-related kinases (trks). Trk A, originally identified as a protooncogene from colon carcinoma cells, selectively mediates the activity of NGF (Klein *et al.*, 1991a). Trk B serves as the high-affinity receptor for both BDNF and NT-4/5 (Klein *et al.*, 1991b), and trk C is the high-affinity receptor for NT-3 (Lamballe *et al.*, 1991). In addition, a low-affinity receptor exists that binds all the neurotrophins and has been shown to mediate some of the biological activities associated with NGF (Dobrowsky *et al.*, 1994). This receptor, called p75^{NTR} (neurotrophin receptor of apparent molecular mass of 75 kDa), is believed to act primarily through sphingomyelin signaling pathways.

TABLE II

NEUROTROPHINS

Neurotrophin	High-affinity receptor	Peripheral populations
NGF	Trk A	Small fiber sensory (pain and temperature) Sympathetic
BDNF	Trk B	Midsize fiber sensory (touch, pressure) Motor
NT-3	Trk C	Large fiber sensory (proprioception and vibration) Motor Sympathetic
NT-4/5	Trk B	Midsize fiber sensory (touch, pressure) Motor

In the peripheral nervous system, the individual neuronal cell types are differentially responsive to specific members of the neurotrophin gene family. NGF is trophic primarily for sympathetic neurons and small diameter sensory neurons that mediate pain and temperature sensation. BDNF and NT-4/5 are trophic for motor neurons and medium size diameter sensory neurons that mediate touch and pressure sensation. NT-3 is trophic for the largest diameter sensory neurons, as well as motor neurons and sympathetic neurons.

In addition to the neurotrophins, another family of neurotrophic factors exists that also appears to play an important role in the development and maintenance of the peripheral nervous system. This is the glial cell-derived neurotrophic factor (GDNF) family, a part of the larger superfamily of transforming growth factor-β (TGF-β)-related growth factors. GDNF is a potent trophic agent for motor neurons (Henderson *et al.*, 1994), populations of sensory neurons (including the 20% or so of dorsal root ganglia neurons that do not respond to any neurotrophin), and both sympathetic and parasympathetic neurons (Buj-Bello *et al.*, 1995).

II. Rationale for the Application of Nerve Growth Factor to Diabetic Polyneuropathy

A. Population of Neurons Involved

In general, the rationale for applying neurotrophic factors to neuro-degenerative diseases of any type is based on the expectation that The neurotrophic factor (1) will promote survival of the affected population of neurons, making them more resistant to the underlying disease process; (2) will stimulate the physiological activity of surviving neurons, enabling them to compensate for the loss of activity due to the disease process; and (3) may stimulate some degree of nerve regeneration or nerve terminal sprouting in affected, but surviving neurons.

The difficulty in implementing this approach begins with the selection of the most appropriate neurotrophic factor for the clinical application. As can be seen previously, a variety of neurotrophic factors are active in the peripheral nervous system, but they are each trophic for only a subset of the neurons affected in diabetic neuropathy. Table III summarizes the signs and symptoms associated with the degeneration of different fiber types in diabetic neuropathy. Motor involvement in polyneuropathy is minor and tends to be a late feature. Sensory involvement can be profound, but most of the more debilitating symptoms are associated with the degeneration of small diameter sensory fibers. These symptoms include the various types

TABLE III
SIGNS AND SYMPTOMS OF DIABETIC NEUROPATHY BY FIBER TYPE

Small fiber sensory	Large fiber sensory	Autonomic
Burning pain	Loss of vibratory sensation	Heart rate abnormalities
Cutaneous hyperesthesia	Loss of proprioception	Postural hypotension
Paresthesiae	Loss of reflexes	Abnormal sweating
Loss of pain and temperature sensation	Slowed nerve conduction velocities	Gastroparesis
Foot ulceration		Neuropathic diarrhea
Lancinating pain		Impotence
Loss of visceral pain		Retrograde ejaculation

of neuropathic pain, loss of normal pain sensation leading to injury, and foot ulceration. Large fiber sensory neuropathy, however, is more easily quantifiable but less bothersome to the patient, at least until the condition becomes highly advanced. Autonomic symptoms and signs are also common and may be debilitating to the patient with diabetic neuropathy.

Given the importance of degeneration of small sensory fibers and autonomic neurons in the clinical setting, a growth factor that supports these two groups was felt to be the most useful for the treatment of diabetic neuropathy. The most logical choice was clearly NGF.

B. AN ETIOLOGICAL ROLE FOR NERVE GROWTH FACTOR

Lending further support to the clinical rationale was a growing body of evidence suggesting that NGF may play a pathogenic role in the etiology of diabetic polyneuropathy. Remember that classically, neurotrophic factors are released by target tissue, bind to specific receptors on the terminals of responsive neurons, and are retrogradely transported back to the cell nucleus where they exert their biological affects. Interference with this process could reduce the availability of neurotrophic factors to the neuron, resulting in degeneration. Several studies suggest that the normal retrograde transport of NGF, released from skin keratinocytes and retrogradely transported through small fiber sensory or sympathetic neurons, is impaired in animal models of diabetes. It was discovered, for example, that following the induction of diabetes in rats through the administration of streptozotocin (STZ), NGF levels increased in the skin (target tissue), but decreased in the sciatic nerve and sympathetic ganglia (Hellweg and Hartung, 1990). This suggested that the normal process of NGF uptake and retrograde transport might be impaired. Others have demonstrated more directly that the retrograde transport of NGF is impaired in experimental diabetes (Jakobsen *et al.*, 1981; Hellweg *et al.*, 1994).

Expression of NGF is also reduced in experimental diabetes. An early study demonstrated that levels of NGF are reduced in submandibular glands and serum of diabetic mice (Ordonez *et al.*, 1994). It is not clear, however, whether reduced levels of systemic NGF would have a significant effect on the pathogenesis of diabetic neuropathy, as it is believed that normally most of the endogenous NGF is provided to the peripheral nerves from the local target tissue. In another study in diabetic rats, it was shown that levels of NGF messenger RNA (mRNA) in the skin of the hindfoot are significantly reduced, and the reduction is proportional to the duration of diabetes (Brewster *et al.*, 1994; Fernyhough *et al.*, 1995). This reduction in neuronal target tissue NGF expression is more likely to be significant for the pathogenesis of neuropathy than a reduction in systemic NGF.

Human diabetic patients have also been shown to have a significant reduction in levels of NGF in skin keratinocytes (Anand *et al.*, 1996). This study was particularly interesting because it showed for the first time a direct correlation between the reduction of NGF levels in skin and the presence of early findings associated with small fiber sensory neuropathy. The authors demonstrated that patients who did not yet have symptomatic neuropathy, but had reduced levels of skin NGF, had early abnormalities of temperature sensitivity and axon reflex vasodilation. This suggested that reduced levels of skin NGF probably preceded the onset of small fiber neuropathy, as opposed to being its consequence. Although these and other studies strongly suggest that reduced availability of endogenous NGF contributes to the pathogenesis of diabetic neuropathy, it is highly unlikely to be the only or even the most important contributing factor. Numerous other possible etiologies have been proposed and many are likely to play a role. Nevertheless, the possibility that reduced availability of NGF contributes to the pathogenesis of diabetic neuropathy reinforces the rationale of administering NGF as a potential treatment.

C. PRECLINICAL EFFICACY STUDIES

Nerve growth factor administration has been shown to be efficacious in a variety of experimental models of peripheral neuropathy. These animal models may be grouped into three major categories, including mechanical injury to the nerve, toxic injury, and metabolic (diabetes) injury. In each preclinical model category, NGF administration was able to prevent the onset of behavioral or biochemical measures associated with peripheral nerve dysfunction.

1. *Mechanical Injury*

The mechanical model studied most extensively was the chronic constriction injury model originally developed by Bennett and Xie (1988). In this model of compressive neuropathy, four loose ligatures are tied around the sciatic nerve, resulting in marked hyperalgesia of the foot and trophic changes within a few days. Local application of NGF to the compressed nerve has been reported to prevent the onset of hyperalgesia (Ren *et al.*, 1995; Ro *et al.*, 1999). Interestingly, in the study of Ro and colleagues (1999) NGF antiserum also prevented hyperalgesia, although with a different time course. We have replicated these findings with systemic (subcutaneous) administration of NGF to rats with chronic constriction injury (unpublished observations). Although we found that NGF administration alleviated the hyperalgesia, we found that it had no effect on slowed conduction velocities, reduced action potential amplitudes, and histological changes to the nerve fibers associated with the chronic constriction injury. How NGF administration prevents hyperalgesia in this model is not yet fully understood.

2. *Toxic Injury*

NGF administration has also proven effective in the treatment of experimental toxic neuropathy models in which predominantly sensory or sympathetic neurons are injured. The earliest demonstration of a beneficial NGF effect was in a neurotoxicity model, where neonatal rats were treated with the vinca alkaloid vinblastine (Johnson, 1978). NGF administration prevented the massive loss of sympathetic neurons resulting from the vinblastine. Johnson postulated that vinblastine interrupted the normal retrograde transport of NGF from peripheral target tissues by disrupting microtubules, which caused the neurons to die. By injecting NGF systemically, the need for retrograde transport was bypassed and the neurons survived.

We have extended these observations to the more subtle neurotoxicities associated with the administration of the chemotherapeutic agents paclitaxel and cisplatin (Apfel *et al.*, 1991, 1992). In the paclitaxel mouse model, we demonstrated that NGF administration prevented the loss of thermal pain sensitivity and a reduction in substance P levels in the dorsal root ganglia. In the cisplatin mouse model, NGF administration prevented the loss of thermal sensitivity as well as a loss of proprioception. It also prevented the loss of both substance P and calcitonin gene-related peptide (CGRP) in the dorsal root ganglia. Substance P and CGRP are neuropeptides present in small diameter sensory neurons and appear to play a role in the transmission and modulation of pain sensation. It is believed that they may also serve as markers of small fiber sensory neuron integrity. NGF has also been reported to prevent neurotoxicity associated with the application of these agents to neurons in tissue culture (Peterson and Crain, 1982; Hayakawa *et al.*, 1994; Konings *et al.*, 1994).

3. *Metabolic Injury*

Neurotoxicity studies demonstrated that NGF administration may prevent neuropathy caused by a limited, but direct, insult to the nerve. It was not possible to reliably extrapolate from these studies and conclude that NGF could prevent the more insidious injury associated with diabetes. Most of the published preclinical data that directly relate to diabetic neuropathy were derived from studies involving the streptozotocin-induced diabetic rat. We found that within 13 weeks of streptozotocin administration to Wistar rats, the rats displayed behavioral, electrophysiological, and biochemical abnormalities suggestive of peripheral nerve dysfunction (Apfel *et al.*, 1994). The tail-flick threshold, which measures the response of the rat to a thermal noxious stimulus, was elevated significantly ($p < 0.05$ by ANOVA) in streptozotocin-treated animals. NGF administration prevented that elevation. The diabetic rats also had approximately 30% reduction in sensory ganglia levels of substance P and CGRP as compared with nondiabetic controls. Coadministration of NGF resulted in neuropeptide levels that were greater than those in the nondiabetic control group ($p < 0.01$ by ANOVA). Similar results were reported by Fernyhough *et al.* (1995), who measured levels of substance P and CGRP in the sciatic nerve of streptozotocin-induced diabetic rats.

The loss of sensitivity to a thermal noxious stimulus and the reductions in levels of substance P and CGRP were all markers of small diameter sensory fiber dysfunction, and not surprisingly, NGF administration was beneficial. Nerve conduction velocities in the caudal nerve of the tail were also reduced significantly ($P < 0.01$ by ANOVA) in the diabetic animals, but this being primarily a measure of large diameter function, there was no significant improvement with NGF administration (Apfel *et al.*, 1994). In a study of transgenic hypoinsulinemic mice, it was observed that administration of recombinant human nerve growth factor (rhNGF) prevented the loss of C-fiber response amplitude and integral measured in untreated diabetic animals (Elias *et al.*, 1998). In contrast, the administration of rhNGF had no effect on the reduction of sensory nerve conduction velocities, demonstrating again a selective beneficial effect on small fiber sensory function.

A report has been published suggesting that administration of excessive amounts of exogenous NGF to diabetic animals may result in the exacerbation of some aspects of neuropathy (Schmidt *et al.*, 2001). This study followed the occurrence of dystrophic axons in prevertebral sympathetic ganglia and illeal mesenteric nerves of streptozotocin-treated rats. They found that treatment with NGF for 2–3 months appeared to increase the frequency of neuroaxonal dystrophy in the superior mesenteric ganglia. The significance of these surprising findings is not yet clear.

III. Early Clinical Studies

A. PHASE I AND II STUDIES IN DIABETIC NEUROPATHY

Given this background, Genentech Inc. (South San Francisco, CA) made the decision to proceed with clinical trials of rhNGF administered subcutaneously. Initial phase I trials in healthy volunteers and patients with small fiber sensory neuropathy revealed that the dose-limiting toxicity of rhNGF appeared to be pain (Perry *et al.*, 1994). Pain was manifest as injection site hyperalgesia in virtually all the volunteers. Patients receiving larger doses frequently experienced diffuse myalgias or arthralgias. The painful side effects, often severe, limited the dose of rhNGF for future clinical trials to less than 1 μg/kg. Other than the transient pain, there were no serious adverse events reported in these early trials.

Following the initial phase I clinical trials, a multicenter, double-blind, placebo-controlled phase II trial was initiated in 250 patients with symptomatic diabetic polyneuropathy (Apfel *et al.*, 1998). Patients were randomized to receive placebo, 0.1, or 0.3 μg/kg rhNGF subcutaneously three times a week for 6 months. The adverse events reported in this trial that were more common in rhNGF-treated groups than placebo-treated groups (see Table IV) were similar to those reported in the phase I trials and were largely pain related. Efficacy was measured with a variety of instruments. These included the Neuropathy Impairment Score for the Lower Limbs (NIS-LL), a quantitative version of the neurological examination focused on the lower extremities (Dyck *et al.*, 1995), quantitative sensory tests using the CASE IV system (Gruener and Dyck, 1994), and subjective symptom questionnaires such as the Neuropathy Symptom Profile (NSP), the Neuropathy Symptom and Change (NSC), and the Global Symptom Questionnaire. The Global Symptom Questionnaire was a single question that asked patients to rate overall whether their neuropathy symptoms improved or worsened on a seven point scale over the course of the study.

TABLE IV
MOST FREQUENT ADVERSE EVENTS REPORTED IN PHASE II AND PHASE III TRIALS

| | | Phase II trial | | | Phase III trial |
	Placebo	0.1 μg/kg rhNGF	0.3 μg/kg rhNGF	Placebo	0.1 μg/kg rhNGF
Injection site hyperalgesia	15%	93%	94%	11.6%	67.2%
Peripheral edema	6%	11%	8%	1.2%	5%
Myalgias	5%	11%	16%	2.2%	6.2%
Arthralgias		7%			
Leg cramps	2%	5%	5%		

Patients receiving rhNGF in this study appeared to benefit to a significantly greater degree than those receiving placebo. Subjectively, the global symptom questionnaire assessing the patient's overall perception of their neuropathic symptoms demonstrated the most robust beneficial effect favoring rhNGF-treated groups (see Table V). Among the quantitative sensory studies, the cooling detection threshold (CDT) and the intermediate heat as pain (HP:5.0) measures were suggestive of a beneficial response to rhNGF administration. These two measures were the most sensitive to small diameter fiber sensory function. The vibratory detection threshold (VDT), which is a quantitative sensory test that measures a large diameter function, did not demonstrate an rhNGF effect. The NIS-LL also indicated a trend toward improvement with rhNGF administration, and this improvement was largely driven by the small fiber components of the

TABLE V
SUMMARY OF EFFICACY RESULTS FROM THE PHASE II CLINICAL TRIAL

rhNGF groups compared with placebo		
Variable	P value	O'Brien's rank sum
NIS-LL	0.17	
CDT	0.049	$P = 0.005$
HP:5.0	0.15	
VDT	0.76	$P = 0.032$
NSP	0.83	
Global questionnaire	0.001	$P = 0.008$
NSC	0.192	

NIS-LL (i.e., pinprick sensation, $p = 0.01$). Not surprisingly, electrophysiological studies, which primarily measure large diameter fiber function, did not show any improvement with rhNGF administration. Prospectively designated multiple end point analyses (using O'Brien's rank sum test) lent strong support to the conclusion that rhNGF was efficacious in this trial (Table V).

Although the positive outcome from this trial generated additional enthusiasm for the potential use of NGF in the treatment of diabetic neuropathy, some concerns were expressed. The most important flaw in the trial was the possible unblinding of the study due to the frequent occurrence of injection site hyperalgesia in patients receiving rhNGF. During the course of the study, 93% of patients receiving 0.1 μg/kg and 94% of patients receiving 0.3 μg/kg rhNGF reported injection site hyperalgesia or allodynia as compared with 15% of placebo patients. When asked at the conclusion of the trial, 81% of the subjects correctly concluded that they were receiving rhNGF and 57% that they were receiving placebo. They indicated both perceived presence or absence of efficacy, and the presence or absence of adverse events as determining their conclusion. Nevertheless, the fact that only small diameter measures showed a beneficial effect of rhNGF, and that relatively objective measures such as the quantitative sensory tests and the neurological examination provided more evidence of efficacy than subjective measures such as the symptom questionnaires, suggested that the study was not unduly compromised by the partial unblinding.

Additional studies conducted following the phase II trial in diabetic neuropathy served to further enhance expectations that rhNGF would prove useful for the treatment of neuropathy. A double-blind, placebo-controlled, phase II study of 270 patients with painful HIV-induced neuropathy demonstrated that administration of rhNGF at doses of 0.1 or 0.3 μg/kg twice a week for 18 weeks was superior to placebo in improving neuropathic pain and sensitivity to pinprick (McArthur *et al.*, 2000). In Europe, several studies have been conducted in which 200 μg murine NGF was applied topically to the eye for the treatment of refractory corneal neurotrophic ulcers (Lambiase *et al.*, 1998; Bonini *et al.*, 2000). These ulcers may result from a variety of causes, including the loss of sensory innervation of the cornea. These studies reported the outcome in 43 patients with corneal ulcers that failed to respond to conventional treatment. All patients were reported to have complete resolution of their corneal ulcers, along with a return of corneal sensitivity, suggesting sensory reinnervation. These studies lacked a placebo control group, but 100% healing for a condition that does not readily improve is nevertheless highly impressive.

IV. Phase III Clinical Trial of Recombinant Human Nerve Growth Factor in Diabetic Polyneuropathy

A. STUDY DESIGN

The encouraging results from the phase II clinical trials prompted Genentech to proceed with plans for a pivotal phase III clinical trial of rhNGF to treat diabetic polyneuropathy (Apfel *et al.*, 2000). The phase III study enrolled 1019 patients with demonstrable diabetic polyneuropathy, ranging in age from 18 to 74 years old. Patients were randomized to receive either rhNGF 0.1 μg/kg or placebo by subcutaneous injection, three times a week for 48 weeks. Assessments were done at baseline, 12, 24, and 48 weeks. A number of changes were included in the protocol to avoid some of the problems that were associated with the phase II clinical studies. For example, the placebo consisted of hypertonic saline to mimic the injection site hyperalgesia seen with rhNGF and to minimize problems with unblinding. In addition, formulation of the rhNGF itself was changed to a more dilute solution than was used in the phase II trials. The purpose of this was to avoid the accidental overdosing that occurred with the phase II trial, where patients were instructed to dilute the rhNGF themselves prior to injection.

The primary outcome measure for this study was the change in the NIS-LL score over 48 weeks. The NIS-LL, a subset of the Neuropathy Impairment Score, was based on a focused neurological examination of the lower extremities using a graded battery of muscle strength, reflexes, and sensory testing. Patients were classified as having improved if their NIS-LL increased by two points or more, worsened if their NIS-LL decreased by two points or more, and was unchanged if the change was less than two points in any direction. Secondary outcome measures included quantitative sensory tests (CDT, VDT, HP:5.0), the neuropathy symptoms and change questionnaire, and a global assessment of neuropathy (overall rating of change in neuropathy symptoms, as baseline was recorded on a seven-point scale).

B. RESULTS

The adverse event profile for the phase III study was similar to the previous studies in that the most frequent adverse events were pain related. Injection site hyperalgesia was reported in 67.2% of patients receiving rhNGF, compared with 11.6% of placebo patients. It is worth noting here that the percentage of patients reporting injection site hyperalgesia was much lower in this study than in the previous phase II clinical trial (where

TABLE VI
PHASE III TRIAL NIS-LL CATEGORICAL ANALYSIS

Change in NIS-LL	Placebo ($n = 515$)	rhNGF ($n = 504$)
Improved	175 (34%)	159 (32%)
Unchanged	212 (41%)	204 (41%)
Worsened	128 (25%)	141 (28%)

it was greater than 90%). Myalgias were the next most commonly reported adverse events and were present in 6.2% of the patients receiving rhNGF versus 2.2% of placebo patients. Serious adverse events and deaths were distributed evenly between the two treatment groups.

The phase III clinical trial was disappointing in that there was little evidence to suggest efficacy in this clinical setting. Efficacy data are summarized in Tables VI and VII. With regard to the primary outcome variable, there were no significant differences between rhNGF and placebo ($p = 0.25$). If anything, there might have been a slight trend suggesting that patients receiving rhNGF did worse. Furthermore, there were no significant differences between the two treatment groups with regard to the secondary efficacy variables, with the exception of the global symptom assessment (Table VIII). There, the rhNGF group was superior to placebo with a $p = 0.03$. Despite this single positive outcome, it was clear that overall, the study failed to demonstrate the efficacy of rhNGF in treating diabetic polyneuropathy.

C. WHAT WENT WRONG?

Given the early success with the phase II clinical trials, the surprising failure of the large phase III trial prompted considerable interest in

TABLE VII
PHASE III TRIAL SECONDARY EFFICACY VARIABLES CHANGE
FROM BASELINE[a]

Measures	Placebo	rhNGF	P value
CDT	0.0 (0.28)	0.0 (0.4)	0.46
HP:5.0	−0.2 (1.1)	−0.3 (1.1)	0.88
NIS-LL sensory	−0.3 (2.7)	−0.2 (2.6)	0.47
NSC sensory	−0.2 (1.7)	−0.3 (1.7)	0.88
NSC severity	−0.8 (4.0)	−0.9 (3.8)	0.93

[a]Values represent the mean change in score from baseline. Standard deviation is represented in parentheses. A negative score signifies improvement from baseline.

TABLE VIII
PHASE III TRIAL GLOBAL ASSESSMENT ($P = 0.03$)

	Placebo	RhNGF
Significantly worse	21 (4%)	17 (4%)
Moderately worse	36 (7%)	25 (5%)
Mildly worse	71 (15%)	61 (13%)
Unchanged	178 (36%)	150 (33%)
Mildly better	75 (15%)	92 (20%)
Moderately better	67 (14%)	63 (14%)
Significantly better	40 (8%)	52 (11%)

determining why the outcome of the two sets of clinical trials was so different. A number of possible explanations have been proposed, and it is likely that many, if not all, of these possibilities may have been important contributors to the negative outcome.

1. *Failure of the Placebo Group to Progress*

One of the difficulties confronting the phase III study was that with virtually every measure of neuropathy, the placebo-treated group failed to demonstrate any progression of neuropathy over the course of the year and, in many instances, showed clear signs of improvement. When nearly 35% of the placebo-treated patients improve by two or more points on the NIS-LL, it is very difficult for any drug to do significantly better. This study was powered based on the outcome of a large epidemiological study, the Rochester Diabetic Neuropathy Study (Dyck *et al.*, 1997), to show significant progression of the neuropathy in the placebo-treated group over the course of a year. This study, and others like it, showed that in the short term, individual measures of peripheral nerve function may have a variable course in diabetic patients and that transient improvements are common. The Rochester study showed that progression of diabetic neuropathy over a 2-year course may be clearly demonstrated, but it is less clear over 1 year. Nevertheless, based on their data and given the large sample size of our study population, the absence of significant progression of neuropathy in the placebo-treated group is surprising.

A possible explanation may be due to the fact that the phase III rhNGF trial population differed in an important way from the Rochester population. The mean serum glycosylated hemoglobin for patients enrolled in the phase III trial was 8.7% compared with 11.2% for the Rochester patients with neuropathy, suggesting the former had more rigorous glucose control. The Rochester population was a large survey population that was screened for an epidemiological study. The rhNGF study population was self-selected by their desire to participate in a clinical trial. It is not surprising that the latter population may have been more highly motivated to maintain tighter control of their serum glucose, particularly with frequent visits to their investigative study sites and routine monitoring of their serum glucose. The tighter glucose control in the rhNGF population may explain why there was no evidence of neuropathy progression over the course of a year. As a result, there was not sufficient separation between the two treatment groups to demonstrate a beneficial effect of rhNGF. The difficulty with this explanation was that in the phase II trial, the placebo group also did well, yet the rhNGF-treated group did better. Other factors must also be responsible for the differences seen.

2. *Doses Were Too Low*

Much of the evidence favoring a beneficial effect of NGF in the treatment of neuropathy is derived from animal models (see earlier discussion). The doses of NGF used in those studies ranged from 1 to 10 mg/kg administered from three times a week to seven times a week. In contrast, rhNGF was administered at a dose of 0.1 μg/kg in the clinical trial. This is at minimum a 10,000-fold difference. The dose of rhNGF used in the clinical trials was limited by the painful side effects, such as injection site hyperalgesia, myalgias, and arthralgias, which became more frequent and severe with larger doses. Although no minimum effective dose was identified in the animal studies, it is possible that the dose of NGF used in the clinical studies was either subtherapeutic or borderline therapeutic. When circumstances favored a demonstration of efficacy, as they did in the phase II trial, it was possible to demonstrate efficacy even with this dose. When conditions did not favor a demonstration of efficacy, as in the case of the phase III trial, the dose was too low to see efficacy.

3. *Different Patient Populations*

Inclusion and exclusion criteria changed between clinical trials, which may have resulted in patients with more advanced neuropathy being included in the phase III clinical trial. The upper limit on age, for example, in the phase II clinical trial was 59 years, whereas for the phase III clinical trial, the age range extended until 75 years. Older patients may have both more advanced neuropathy and be intrinsically less likely to respond to neurotrophic factors. In fact, it was informally observed that most of the patients from the phase II trial who had the most robust response to rhNGF were young (less than 40 years old) and had relatively mild neuropathy to begin with (personal observation). Although age and severity of neuropathy may have been significant issues distinguishing the two trials, it could not be proven from the data. Data from the phase III study were reexamined with patients older than 59 years of age excluded, and there were still no significant differences between the treatment groups. There was, however, one other important variable that distinguished the populations in the two studies. In the phase II clinical trial, patients with proliferative diabetic retinopathy were excluded. This was not the case in the phase III study. As a group, patients with proliferative retinopathy are more likely to have severe neuropathy than patients without retinopathy. Unfortunately, because fundus photographs were only done in the phase II trial and not in the phase III trial, it is impossible to go back and determine whether this was an important distinction between the two trials.

4. *The rhNGF May Have Been Different in the Two Studies*

Although the amino acid sequence and structure of the study drug in the two trials were identical, changes were made in the manufacturing process and in the final formulation prior to the phase III clinical trial. The new rhNGF was in a solution with a slightly different concentration of acetate and sodium chloride than was used in the phase II clinical trials. In addition, the concentration of the formulation was changed significantly. For phase II, the drug was formulated at a concentration of 2 mg/ml, and patients were expected to dilute the drug with normal saline immediately prior to injection. Because several patients forgot to dilute the drug, it was formulated at a concentration of 0.1 mg/ml for phase III. When tested in PC12 cell cultures, the two formulations of rhNGF had similar bioactivity. Unfortunately, they were not compared in animal bioassays, nor were they compared after storage over extended periods of time, as was the case in the phase III clinical trials. It is impossible to say at this point in time whether the difference in formulation had a significant effect on the outcome of the trial. However, if the dose was near threshold for efficacy and the population in the phase III trial had more severe neuropathy, it would be easy to imagine that even a relatively small reduction in bioactivity might result in a negative trial. In support of the possibility that the rhNGF was less active in phase III is the observation that there were far fewer reports of injection site hyperalgesia during the year of treatment with the phase III preparation (67%) than during the 6 months of treatment with the phase II preparation (93%).

5. *NGF Might Not Be Effective in Reversing Diabetic Polyneuropathy*

All of the preclinical data supporting the efficacy of NGF in treating peripheral neuropathy comes from animal models in which NGF was administered from before the onset of neuropathy. These studies demonstrate that NGF administration may be useful to prevent the onset of neuropathy before it starts, but provide no evidence that NGF may reverse neuropathy that is already present. It is possible then that the phase III results are more indicative of the true activity of rhNGF than the phase II studies. The phase II studies may have been falsely positive due to the unblinding of the study from the very high incidence of injection site hyperalgesia.

Personally, I think that this possibility is unlikely. The efficacy of rhNGF was apparent in both the phase II diabetes study and the HIV neuropathy study. Although there was some unblinding, the evidence favoring efficacy was restricted to just small fiber sensory function and not at all to even subjective measures of large diameter sensory function. Nevertheless, there is no way to know for certain unless a new, revised clinical trial is conducted that addresses all of the deficiencies of the prior studies. Currently such a trial is unlikely, as Genentech has decided to discontinue further studies with rhNGF.

V. What Does the Future Hold?

In theory, neurotrophic factors appear to be ideal agents for the treatment of diabetic neuropathy and other neurodegenerative diseases. They promote the survival of selective populations of neurons, they stimulate their growth and regeneration, they protect them from injury, and they stimulate a variety of physiological properties characteristic of the responsive neuronal population. However, as we have learned from the past decade of clinical trials, many problems are associated with their clinical application.

We do not yet have a thorough understanding of their activities outside the nervous system. Some of these activities are responsible for dose-limiting side effects, such as the hyperalgesia associated with NGF. Other neurotrophic factors applied to different clinical conditions have run into similar, although distinct, problems. Large proteins are also difficult to work with from a practical perspective. Access to the nervous system is often limited by the blood–brain or the blood–nerve barrier. It is necessary to administer these factors parenterally as well because large proteins are not absorbed intact from the gastrointestinal tract. Large proteins are also often difficult and expensive to produce in sufficient quantities for clinical application. The manufacturing process is complex and may be prone to variable bioactivity. Their pharmacology is not understood very well either.

Given these difficulties, it is clear in hindsight that many of us involved in the rapid progression of neurotrophic factors to clinical trials were fairly naïve about the difficulty of the journey. Unfortunately, the many clinical trial failures have reduced enthusiasm for the entire approach, and few additional clinical trials are being considered for any growth factor in any application. The field, however, is not quite dead. Scientists continue to study the growth factors themselves, and our understanding of their physiological role is growing steadily. Many are beginning to explore alternative approaches, studying small molecules that either mimic the activity of the larger neurotrophic proteins or stimulate their endogenous overexpression in a more natural way. Examples of potentially promising small molecules include clenbuterol (Riaz and Tomlinson, 1999) and the vitamin D analog CB1093 (Riaz *et al.*, 1999). These small molecules have numerous theoretical advantages over the larger proteins, including ease of manufacture, simpler routes of administration, access to the nervous system, and perhaps greater selectivity for specific neuronal populations with fewer systemic side effects.

References

Anand, P., Terenghi, G., Warner, G., Kopelman, P., Williams-Chestnut, R. E., and Sinicropi, D. V. (1996). The role of endogenous nerve growth factor in human diabetic neuropathy. *Nature Med.* **2**, 703–707.

Apfel, S. C., Arezzo, J. C., Brownlee, M., Federoff, H., and Kessler, J. A. (1994). Nerve growth factor administration protects against experimental diabetic sensory neuropathy. *Brain Res.* **634**, 7–12.

Apfel, S. C., Arezzo, J. C., Lipson, L., and Kessler, J. A. (1992). Nerve growth factor prevents experimental cisplatin neuropathy. *Ann. Neurol.* **31**, 76–80.

Apfel, S. C., Kessler, J. A., Adornato, B. T., Litchy, W. J., Sanders, C., Rask, C. A. (1998). Recombinant human nerve growth factor in the treatment of diabetic polyneuropathy. *Neurology* **51**, 695–702.

Apfel, S. C., Lipton, R. B., Arezzo, J. C., and Kessler, J. A. (1991). Nerve growth factor prevents toxic neuropathy in mice. *Ann. Neurol.* **29**, 87–90.

Apfel, S. C., Schwartz, S., Adornato, B. T., Freeman, R., Biton, V., Rendell, M., Vinik, A., Giuliani, M., Stevens, J. C., Barbano, R., and Dyck, P. J. (2000). Efficacy and safety of recombinant human nerve growth factor in patients with diabetic polyneuropathy: A randomized controlled trial. *JAMA* **284**, 2215–2221.

Bennett, G. J., and Xie, Y.-K. (1988). A peripheral mononeuropathy in the rat produces disorders of pain sensation like those seen in man. *Pain* **33**, 87–107.

Bonini, S., Lambiase, A., Rama, P., Caprioglio, G., and Aloe, L. (2000). Topical treatment with nerve growth factor for neurotrophic keratitis. *Ophthalmology* **107**, 1347–1351.

Brewster, W. J., Fernyhough, P., Diemel, L. T., *et al.* (1994). Diabetic neuropathy, nerve growth factor and other neurotrophic factors. *Trends Neurosci.* **17**, 321–325.

Buj-Bello, A., Buchman, V. L., Horton, A., Rosenthal, A., and Davies, A. M. (1995). GDNF is an age specific survival factor for sensory and autonomic neurons. *Neuron* **15**, 821–828.

Cohen, S., Levi-Montalcini, R., and Hamburger, V. (1954). A nerve growth stimulating factor isolated from sarcomas 37 and 180. *Proc. Natl. Acad. Sci. USA* **40**, 1014–1018.

Dobrowsky, R. T., Werner, M. H., Castellino, A. M., *et al.* (1994). Activation of the sphingomyelin cycle through the low affinity neurotrophin receptor. *Science* **265**, 1596–1599.

Dyck, P. J., Davies, J. L., Litchy, W. J., and O'Brien, P. C. (1997). Longitudinal assessment of diabetic polyneuropathy using a composite score in the Rochester Diabetic Neuropathy Study cohort. *Neurology* **49**, 229–239.

Dyck, P. J., Litchy, W. J., Lehman, K. A., Hokanson, B. A., Low, P. A., and O'Brien, P. C. (1995). Variables influencing neuropathic endpoints: The Rochester diabetic neuropathy study of healthy subjects. *Neurology* **45**, 1115–1121.

Elias, K. A., Cronin, M. J., Stewart, T. A., and Carlsen, R. C. (1998). Peripheral neuropathy I transgenic diabetic mice: Restoration of C-fiber function with human recombinant nerve growth factor. *Diabetes* **47**, 1637–1642.

Fernyhough, P., Diemel, L. T., Brewster, W. J., and Tomlinson, D. R. (1994). Deficits in sciatic nerve neuropeptide content coincide with a reduction in target tissue nerve growth factor messenger RNA in streptozocin-diabetic rat. *Neuroscience* **62**, 337–344.

Fernyhough, P., Diemel, L. T., Hardy, J., *et al.* (1995). Human recombinant nerve growth factor replaces deficient neurotrophic support in the diabetic rat. *Eur. J. Neurosci.* **7**, 1107–1110.

Greuner, G., and Dyck, P. J. (1994). Quantitative sensory testing: Methodology, applications, and future directions. *J. Clin. Neurophysiol.* **11**, 568–583.

Hayakawa, K., Sobue, G., Itoh, T., and Mitsuma, T. (1994). Nerve growth factor prevents neurotoxic effects of cisplatin, vincristine and taxol on adult rat sympathetic explants *in vitro*. *Life Sci.* **55**, 519–525.

Hellweg, R., and Hartung, H. D. (1990). Endogenous levels of nerve growth factor (NGF) are altered in experimental diabetes mellitus: A possible role for NGF in the pathogenesis of diabetic neuropathy. *J Neurosci. Res.* **26**, 258–267.

Hellweg, R., Raivitch, G., Hartung, H. D., *et al.* (1994). Axonal transport of endogenous nerve growth factor (NGF) and NGF receptor in experimental diabetic neuropathy. *Exp. Neurol.* **130**, 24–30.

Henderson, C. E., Phillips, H. S., Pollock, R. A., Davies, A. M., Lemeulle, C., Armanini, M., Simmons, L., Moffet, B., Vandlen, R. A., Koliatsos, V. E., and Rosenthal, A. (1994). GDNF: A potent survival factor for motoneurons present in peripheral nerve and muscle. *Science* **266**, 1062–1064.

Jakobsen, J., Brimijoin, S., Skau, K., *et al.* (1981). Retrograde axonal transport of transmitter enzymes, fucose-labeled protein, and nerve growth factor in streptozocin-diabetic rats. *Diabetes* **30**, 797–803.

Johnson, E. M., Jr. (1978). Destruction of the sympathetic nervous system in neonatal rats and hamsters by vinblastine: Prevention by concomitant administration of nerve growth factor. *Brain Res.* **141**, 105–118.

Klein, R., Jing, S. Q., Nanduri, E., O'Rourke, E., and Barbacid, M. (1991a). The trk proto-oncogene encodes a receptor for nerve growth factor. *Cell* **65**, 189–197.

Klein, R., Nanduri, E., Jing, E., *et al.* (1991b). The trk B tyrosine protein kinase is a receptor for brain-derived neurotrophic factor and neurotrophin-3. *Cell* **66**, 395–403.

Konings, P. N. M., Makkink, W. K., van Delf, A. M. L., and Ruigt, G. S. F. (1994). Reversal by NGF of cytostatic drug reduction of neurite outgrowth in dorsal root ganglia *in vitro*. *Brain Res.* **640**, 195–204.

Lamballe, F., Klein, R., and Barbacid, M. (1991). Trk C, a new member of the trk family of tyrosine protein kinases, is a receptor for neurotrophin-3. *Cell* **66**, 967–979.

Lambiase, A., Rama, P., Bonini, S., Caprioglio, G., and Aloe, L. (1998). Topical treatment with nerve growth factor for corneal neurotrophic ulcers. *N. Engl. J. Med.* **338**, 1174–1180.

Levi-Montalcini, R. (1987). The nerve growth factor: Thirty five years later. *EMBO J.* **6**, 1145–1154.

McArthur, J. C., Yiannoutsos, C., Simpson, D. M., Adornato, B. T., Singer, E. J., Hollander, H., *et al.* (2000). A phase II trial of nerve growth factor for sensory neuropathy associated with HIV infection. *Neurology* **54**, 1080–1088.

Ordonez, G., Fernandez, A., Perez, R., and Sotelo, J. (1994). Low contents of nerve growth factor in serum and submaxillary gland of diabetic mice: A possible etiological element of diabetic neuropathy. *J Neurol. Sci.* **121**, 163–166.

Perry, B. G., Cornblath, D. R., Adornato, B. T., *et al.* (1994). The effect of systemically administered recombinant human nerve growth factor in healthy human subjects. *Ann. Neurol.* **36**, 244–246.

Peterson, E. R., and Crain, S. M. (1982). Nerve growth factor attenuates neurotoxic effects of taxol on spinal cord-ganglion explants from fetal mice. *Science* **217**, 377–379.

Ren, K., Thomas, D. A., and Dubner, R. (1995). Nerve growth factor alleviates a painful peripheral neuropathy in rats. *Brain Res.* **699**, 286–292.

Riaz, S., Malcangio, M., Miller, M., and Tomlinson, D. R. (1999). A vitamin D(3) derivative (CB1093) induces nerve growth factor and prevents neurotrophic deficits in stretozotocin-diabetic rats. *Diabetologia* **42**, 1308–1313.

Riaz, S. S., and Tomlinson, D. R. (1999). Clenbuterol stimulates neurotrophic support in streptozotocin-diabetic rats. *Diabet. Obes. Metab.* **1**, 43–51.

Ro, L.-S., Chen, S.-T., Tang, L.-M., and Jacobs, J. M. (1999). Effect of NGF and anti-NGF on neuropathic pain in rats following chronic constriction injury of the sciatic nerve. *Pain* **79**, 265–274.

Schmidt, R. E., Dorsey, D. A., Beaudet, L. N., Parvin, C. A., and Escandon, E. (2001). Effect of NGF and neurotrophin-3 treatment on experimental diabetic autonomic neuropathy. *J. Neuropathol. Exp. Neurol.* **60**, 263–273.

ANGIOTENSIN-CONVERTING ENZYME INHIBITORS: ARE THERE CREDIBLE MECHANISMS FOR BENEFICIAL EFFECTS IN DIABETIC NEUROPATHY?

Rayaz A. Malik and David R. Tomlinson

Department of Medicine, Manchester Royal Infirmary
Manchester M13 9WL, United Kingdom

I. Introduction

There is no treatment for diabetic neuropathy, and this lack of an effective therapy derives from three principal factors: (1) the complex etiology of the condition, (2) imperfect animal modeling and failure to accurately translate experimental findings to the clinical arena, and (3) imperfect drug candidates and flawed clinical trial design.

Retrospective rationalization offers plausible explanations for the spectacular failures we have witnessed in clinical trials of human diabetic neuropathy. They include inappropriate selection of patients according to neuropathic severity, drug toxicity, inadequate drug potency, inappropriate measures of therapeutic efficacy, and inadequately short periods of study aimed to rapidly reverse a process, which has taken many years to develop.

In 1998, a small clinical trial with the angiotensin-converting enzyme (ACE) inhibitor trandolapril showed a significant improvement in patients with diabetic neuropathy (Malik *et al.*, 1998) of a magnitude comparable to that seen with improved glycemic control (DCCT Trial Research Group, 1995; UK Prospective Diabetes Study Group, 1998). This chapter provides a systematic review of the clinical, pathogenetic, and pharmacological basis for this observation.

415

ACE inhibitors have surpassed all predictions for their widespread use in clinical medicine. Initially deemed useful only in a select group of patients with renovascular hypertension (Di Giulio *et al.*, 1981), they now constitute the panacea for the treatment of diabetes and its complications, ischemic heart and cerebrovascular disease, and nephropathy from a variety of causes. The pharmacology of ACE inhibition is complex and provides for a number of major interactions with pathogenetically relevant pathways resulting in human diabetic neuropathy.

II. Vascular Basis for Diabetic Neuropathy

The potential role of ischemia in the pathogenesis of diabetic neuropathy is reviewed in detail in the chapter by Zochodne (this volume). The simplest explanation for the efficacy of ACE inhibition in human diabetic neuropathy demands acknowledgement of the premise that vascular dysfunction is important in the genesis of this condition and that inhibition of ACE normalizes nerve blood flow. Epidemiological studies define microvascular disease as a crucial determinant for the development and progression of the diabetic triopathy (nephropathy, retinopathy, and neuropathy) (Dyck *et al.*, 1999). Traditional clinical risk factors, such as glycemic control, hypertension, and hyperlipidemia, induce vascular dysfunction and lead not only to retinopathy and nephropathy, but also neuropathy (Forrest *et al.*, 1997; The EURODIAB PCS Group, 1999).

The link between hypertension and nephropathy is firmly established, and treatment of the former markedly improves the prognosis for the latter. Furthermore, blood pressure and microalbuminuria/proteinuria are easily evaluated, enabling early detection and ready assessment of therapeutic efficacy (Malik, 1998; Maschio *et al.*, 1996). For retinopathy, optimal glycemic control limits its development and progression (UK Prospective Diabetes Study Group, 1998; Di Giulio *et al.*, 1981). There are encouraging data demonstrating a significant reduction in retinal permeability (Parving *et al.*, 1989) and a reduction in the progression of background diabetic retinopathy following treatment with an ACE inhibitor (Chaturvedi *et al.*, 1998).

The demonstration that long duration type 1 diabetic patients without nephropathy, retinopathy, or neuropathy have preserved vascular function suggests an intimate relationship between vascular dysfunction and these complications (Giannattasio *et al.*, 1999; Enderle *et al.*, 1998). Indeed, endothelium-dependent and -independent resistance vessel responses are normal in long duration type 1 diabetic patients with minimal evidence of microvascular complications (Malik *et al.*, 1999). The expression of vascular dysfunction may be related to an alteration in nitric oxide (NO) expression or action. In patients with established neuropathy, two studies demonstrate a reduction in the intensity of immunostaining for skin capillary endothelial nitric oxide synthetase (eNOS) (Veves *et al.*, 1998) and eNOS protein levels determined by Western blotting (Jude *et al.*, 1999). Although interestingly, in the latter study, both inducible nitric oxide synthetase (iNOS) and eNOS protein levels were increased markedly in diabetic patients with severe neuropathy and foot ulceration (Jude *et al.*, 1999). Conversely, we have previously demonstrated enhanced vasoconstriction and sensitivity of resistance vessels to angiotensin II (ATII) in type 1 diabetic patients (Malik *et al.*, 1999). It has also been demonstrated that intrabrachial norepinephrine causes an enhanced vasoconstrictor response in conjunction with a worsening of ulnar nerve latencies and sensory conduction velocity in patients with type 2 diabetes (Hogikyan *et al.*, 1999). Kihara and colleagues (1999) have demonstrated increased sensitivity and augmented epineurial arteriolar constriction to angiotensin II in the sciatic nerve of diabetic rats, which can be ameliorated with ACE inhibition. Augmentation of epineurial constriction has been further attributed to an alteration in the redox state of peripheral nerve (Coppey *et al.*, 2001). Thus there appears to be a diminution in vasodilator and an enhancement of vasoconstrictor responses in diabetes. These proximal functional changes may mediate downstream structural changes in the microvasculature.

In peripheral nerve itself, many studies have now demonstrated endoneurial microangiopathy, characterized by basement membrane thickening, endothelial cell hyperplasia and hypertrophy, and pericyte cell degeneration. These structural changes will further compound an impairment in tissue perfusion. In support of this, we have demonstrated previously a capillary microangiopathy of the skin, which correlated significantly with the impaired hyperemic response to thermal and mechanical injury, relating structure to function (Rayman *et al.*, 1995). In peripheral nerve trunks, the nerve fibers reside in the endoneurium and, therefore, one should expect maximal vascular damage in this compartment. Accordingly, we have localized more advanced microangiopathy to endoneurial capillaries compared to muscle, skin (Malik *et al.*, 1989), and epineurial (Malik *et al.*, 1993) capillaries from the same diabetic patients. Although these abnormalities are most prominent in patients with severe (Rayman *et al.*, 1995; Malik *et al.*, 1989; Yasuda and Dyck, 1987; Britland *et al.*, 1990; Dyck *et al.*, 1985) neuropathy, they are also present in diabetic patients with mild neuropathy (Malik *et al.*, 1992) and even without clinically significant neuropathy (Giannini and Dyck, 1995). The functional consequences of such structural abnormalities are those of a significant reduction in sural nerve endoneurial oxygen tension (Newrick *et al.*, 1986) and epineurial oxygen saturation (Ibrahim *et al.*, 1999). Furthermore, *in vivo* sural nerve photography demonstrates active epineurial arteriovenous shunts (Tesfaye *et al.*, 1996), which may explain apparently normal blood flow observed using laser Doppler flux measurements (Theriault *et al.*, 1997), which detects epineurial but not endoneurial blood flow. Simultaneous fluorescein angiography, used to quantify nerve blood flow by defining the fluorescein appearance time, confirms impaired endoneurial nerve blood flow (Tesfaye *et al.*, 1993).

III. Interventions

A. EXPERIMENTAL

Oxygen supplementation (Low *et al.*, 1984) and hyperbaric oxygenation (Low *et al.*, 1988) improve nerve conduction velocity in diabetic rats without ameliorating metabolic abnormalities. A host of vasoactive compounds, including the calcium channel blocker nifedipine (Robertson *et al.*, 1992), α-adrenergic receptor blocker prazosin (Cameron *et al.*, 1991), and the nicotinic acid derivative niceritrol (Hotta *et al.*, 1992), improve nerve conduction velocity in the streptozotocin (STZ)-diabetic rat. However, the calcium channel blocker nimodipine has failed to improve nerve conduction velocity in the BB rat (Ristic *et al.*, 1996). More recent studies have demonstrated improvements in nerve blood flow and conduction velocity with the β_2-adrenoceptor agonist salbutamol and the α_1-adrenoceptor blocker doxazocin (Cotter and Cameron, 1998). The angiotensin-converting enzyme inhibitors, lisinopril (Cameron *et al.*, 1992) and, more recently, cilazipril (Kihara *et al.*, 1999), as well as an angiotensin II receptor blocker (Maxfield *et al.*, 1993), all improve nerve blood flow and conduction velocity in experimental diabetes. Furthermore, a combination of low-dose endothelin-1 and angiotensin II antagonists demonstrate remarkable synergy, producing an improvement in both nerve blood flow and nerve conduction velocity, above and beyond the additive response one would expect with either drug alone (Cameron and Cotter, 1996). The thiazolidinedione troglitazone has also been shown to improve nerve conduction velocity and myelinated nerve fiber morphology in the STZ rat (Qiang *et al.*, 1998), presumably via improved endothelial function (Walker *et al.*, 1999). Protein kinase C-ß inhibitors have been shown to improve a number of parameters, including nerve blood flow and conduction velocity (Cameron *et al.*, 1999; Nakamura *et al.*, 1999) in the STZ-diabetic rat.

B. CLINICAL

In human diabetic neuropathy, large vessel revascularization improves (Young *et al.*, 1992) or, at worst, prevents the progression (Akbari *et al.*, 1997) of deficits in nerve conduction velocity. More recently, intramuscular gene transfer of vascular endothelial growth factor (VEGF)$_{165}$ has been shown to reduce clinical symptom scores, vibration threshold, and electrophysiology in both diabetic and nondiabetic patients with critical lower limb ischemia (Simovic *et al.*, 2001). α-Lipoic acid therapy results in a significant improvement in sensory nerve conduction velocity and amplitude after 2 years (Reljanovic *et al.*, 1999). The mechanisms of benefit of this drug are unclear, but it may well be related to an alteration in oxidative stress (Nishikawa *et al.*, 2000).

ACE inhibitors ameliorate vascular dysfunction and promote vasodilatation by preventing the generation of the active pressor agent angiotensin II and breakdown of the vasodilator agent bradykinin, which in turn increases the production of prostaglandins (Curzen and Timmis, 1997). They also mediate an increased flow-dependent release of nitric oxide (Vanhoutte *et al.*, 1996) and mediate endothelium-dependent vessel relaxation. In addition to its vasoconstrictor properties, angiotensin II stimulates aldosterone release, promotes sympathetic outflow, and is capable of stimulating arteriolar smooth muscle cell growth by inducing protooncogenes and growth factors. Thus ACE inhibitors have been shown to improve vascular hypertrophy (Cooper *et al.*, 1990) and reduce basement membrane thickening (Cooper *et al.*, 1994) in animal models of diabetes.

Large clinical trials demonstrate a significant benefit in outcomes following myocardial infarction, particularly in diabetic patients, if treated with an ACE inhibitor (Zuanetti *et al.*, 1997). With regard to neuropathy, it is paradoxical that for a group of compounds that were initially feared to cause a neuropathy (Hormigo and Alves, 1992), we are now arguing for their efficacy in the treatment of human diabetic neuropathy.

The earliest study to show a potential benefit assessed the acute effects of 25 mg of captopril in eight diabetic subjects with moderately elevated blood pressure and evidence of mild asymptomatic autonomic neuropathy (impairment of forced sinus arrhythmia). The bradycardic response to apnoeic face immersion was enhanced significantly, but there was no significant change in resting heart rate, sinus arrhythmia, blood pressure response to standing, or the cold pressor test. Ingestion of a single dose of captopril increased vagal but not sympathetic nerve function (Moore *et al.*, 1987). While some of these neurological changes may be the result of acute hemodynamic changes secondary to ACE inhibition, angiotensin II also has direct effects on the adrenergic nervous system. Thus ATII enhances adrenergic tone by (1) promoting the release of noradrenaline from adrenergic terminal neurons; (2) enhancing central activation of vagal outflow; (3) facilitating ganglionic transmission; and (4) amplifying α_1-receptor-mediated vasoconstriction (Antonaccio and Kerwin, 1981). Furthermore, acute ACE inhibition has a significant sympatholytic effect in the human forearm (Lyons *et al.*, 1997) and skeletal muscle (Noll *et al.*, 1997). Chronic ACE inhibition in patients with diabetic autonomic neuropathy (DAN) has shown benefit within 3 months, which was maintained at both 6 (Kontopoulos *et al.*, 1997) and 12 (Athyros *et al.*, 1998) months. Quinapril increased high frequency (HF) and decreased low frequency (LF), as well as their ratio, all indicative of a reduction in sympathetic predominance.

With regard to somatic neuropathy, 13 hypertensive diabetic patients were studied in an open label, dose escalation study to assess the effects of 20 mg lisinopril daily on nerve function. After 12 weeks of treatment, there was a significant improvement in median motor (mean change $\pm$ SEM $2.7 \pm 0.6 \text{ ms}^{-1}$, $p < 0.0001$), median sensory ($2.1 \pm 0.9 \text{ ms}^{-1}$, $p = 0.03$), peroneal motor ($1.0 \pm 0.4 \text{ ms}^{-1}$, $p = 0.03$), and sural sensory ($1.9 \pm 0.7 \text{ ms}^{-1}$, $p = 0.01$) nerve conduction velocities. There were also significant improvements in warm thermal discrimination threshold (TDT) and vibration perception threshold (VPT). Diastolic BP decreased significantly, but there was no significant change in HbA1c (Reja *et al.*, 1995).

We have carried out the only double-blind, placebo-controlled clinical trial in 41 diabetic patients randomized to receive the ACE inhibitor trandalopril over 12 months. Peroneal motor nerve conduction velocity ($p = 0.03$) and M-wave amplitude ($p = 0.03$) increased, the F-wave latency ($p = 0.03$) decreased, and the sural nerve action potential amplitude increased ($p = 0.04$) significantly after 12 months of treatment. Vibration-perception threshold, autonomic function, and the neuropathy symptom and deficit score remained unchanged (Malik *et al.*, 1998).

The Appropriate Blood Pressure Control in Diabetes (ABCD) trial, a prospective randomized-blinded trial, has assessed the effects of intensive versus moderate blood pressure control comparing nisoldipine against enalapril. The incidence and progression of nephropathy, retinopathy, and neuropathy were assessed in 470 type 2 diabetic patients over 5.3 years (Estacio *et al.*, 2000). While the mean blood pressure achieved was 132/78 mm Hg in the intensive group and 138/86 mm Hg in the moderate control group, there was no difference between intensive and moderate groups regarding the progression of diabetic nephropathy, retinopathy, and neuropathy. Thus, after the first year of antihypertensive treatment, creatinine clearance stabilized in both the intensive and the moderate blood pressure control groups in those patients with baseline normo- or microalbuminuria. In contrast, patients starting with overt albuminuria demonstrated a steady decline in creatinine clearance of 5–6 ml·min^{-1}·1.73 m^{-2} per year throughout the follow-up period whether they were on intensive or moderate therapy. There was also no difference between the interventions regarding individuals progressing from normoalbuminuria to microalbuminuria (25% intensive therapy vs 18% moderate therapy, $p = 0.20$) or microalbuminuria to overt albuminuria (16% intensive therapy vs 23% moderate therapy, $p = 0.28$). These findings are contrary to most previously published studies, which show a definite benefit of antihypertensive therapy, at least in nephropathy. This therefore questions the validity of the disappointing results in neuropathy and retinopathy.

Whether or not ACE inhibitors or similar compounds will become the mainstay of therapy in diabetic neuropathy will be decided after carefully controlled trials have shown efficacy. However, one should consider the basis on which predicted benefits might be observed.

IV. Physiology of Angiotensin II

Angiotensin-converting enzyme or kininase II is a metalloprotease, denoting the presence of two zinc groups. It is found chiefly in the vascular endothelium of the lung, but is also distributed widely in the endothelium of other vascular beds, as well as myocardium. Its two main function are (a) to convert the deca peptide angiotensin I to the octa peptide and potent vasoconstrictor angiotensin II and (b) to convert the potent vasodilator bradykinin to its inactive peptides.

The principal responses to angiotensin II are mediated via the AT$_1$ receptor, which is itself subdivided into AT-1a and AT-1b. The AT$_2$ receptor is thought to play a relatively insignificant role in normal vascular physiology, but has been show to be upregulated in heart failure (Haywood *et al.*, 1997). It has been postulated that it may counteract the harmful effects of ATII-mediated vasoconstriction and vascular hypertrophy.

Occupation of the ATII receptor results in acute and chronic effects on the vasculature. The major mechanism by which contraction is achieved is by calcium mobilization in the smooth muscle. This is achieved by activation of the phosphodiesterase phospholipase C, leading to the hydrolysis of phosphatidylinositol bisphosphate to inositol trisphosphate (IP$_3$) and 1,2-diacylglycerol (DAG). IP$_3$ enters the cytosolic space and acts on its receptor to liberate calcium from the intracellular sarcoplasmic reticulum (Berridge, 1993). DAG acts within the plane of the cell wall to activate protein kinase C, which phosphorylates the contractile proteins, leading to sustained contraction and maintenance of vascular tone (Osicka *et al.*, 2001). Additionally,

ATII enhances the activation of voltage-dependent calcium channels via an interaction with a guanosine triphosphatase (GTP)-binding protein (Brown and Birnbaumer, 1988).

The chronic vascular growth effects are mediated as follows.

a. ATII stimulates rapid activation of extracellular signal-regulated kinase (ERK) 1/2, c-Jun N-terminal kinase (JNK), and p38 in rat aortic smooth muscle cells (Yoshizumi *et al.*, 2001).
b. ATII activation of SAPK, JNK, and ERK in turn stimulates protooncogenes (c-fos, c-jun, c-myc), resulting in cellular growth and proliferation (Lyall *et al.*, 1992).
c. ATII mediates increased growth of collagen I, collagen III, and fibronectin (Nishimoto *et al.*, 2001).
d. ATII mediates increased production of endothelin, a powerful mitogen (Moreau *et al.*, 1997).

V. Nonvascular Effects of Angiotensin II

The reasonable doubts concerning a primary vascular causation of diabetic neuropathy (see Zochodne, this volume) prompt consideration of nonvascular mechanisms that might contribute to the obvious efficacy of ACE inhibitors. Work in this area is in its infancy, but there are clear indicators that activation of the ATII receptor in peripheral nervous tissues may have effects that are independent of the vascular supply. The most potentially interesting of these are the effects on cellular apoptosis. Demyelination is a critical feature of diabetic neuropathy. As is described elsewhere in this volume, the failure of the Schwann cell to maintain the myelin sheath and the death of Schwann cells cause internodal demyelination. This triggers Schwann cell proliferation and remyelination, which may not form a competent sheath. Even if it does, the internodal distances are shortened, with an increase in the number of nodes. This is the major contributor to the slowing of nerve conduction seen at advanced stages of the disease (see Arezzo, this volume). Thus, switches in Schwann cell phenotype away from myelination and the death of Schwann cells are pivotal features of irreversible degradative changes. Evidence suggests that Schwann cells die by apoptosis in diabetes and that the MAP kinases are implicated in this process (Cheng and Feldman, 1998; Cheng *et al.*, 2001). Studies show that the ATII receptor can mediate activation of MAP kinases (Stroth *et al.*, 2000) and that angiotensin can promote apoptosis (Shenoy *et al.*, 1999; Fiordaliso *et al.*, 2000). Neurons also express the ATII receptor and its expression increases after nerve damage, indicating some role in pathophysiological responses. There are clear interactions between angiotensin and nerve growth factor (NGF), and although angiotensin II stimulates NGF expression (Jeffreson *et al.*, 1995; Tuttle *et al.*, 1993), it appears to oppose the positive neurotrophic effects of NGF (Stroth *et al.*, 1998). This latter effect may represent functional antagonism because angiotensin II inhibits formation and/or assembly of critical proteins of the axonal endoskeleton (Gallinat *et al.*, 1997; Stroth *et al.*, 1998). Thus, although reversal of ischemic hypoxia of neurons is an obvious and likely mechanism for the beneficial effects of ACE inhibitors in diabetic neuropathy, neurotrophic and neurochemical effects cannot be discounted. This is important given the very equivocal evidence for the vascular causation of diabetic neuropathy.

VI. Pharmacology of Angiotensin-Converting Enzyme Inhibition

ACE inhibitors deactivate ACE by interacting with the terminal negatively charged carboxyl group on the positively charged ions of the zinc atom on the ACE molecule. Thus, by acting on circulating and endogenous tissue ACE levels, ACE inhibitors can potentially have a powerful net vasodilator effect. Another major axis of the action of ACE inhibitors is the kallikrein–kinin system and bradykinin. Bradykinin is a nonapeptide, generated in the endothelium, which is formed from kininogen by the actions of kallikrein. It is inactivated by two kininases (kininase 1 or ACE, and kininase 2). Bradykinin formation is stimulated by shear stress or endothelial damage and acts on bradykinin receptors (BK_2 subtype), which mediate the formation and release of NO (Farhy *et al.*, 1993) and also increase the conversion of arachidonic acid to vasodilatory prostaglandins, such as PGE_2 and prostacyclin (Schror, 1992), promoting vasodilatation. The systemic levels of bradykinin are not increased during ACE inhibitor therapy, and because the actions of bradykinin are local, it may be that local bradykinin levels are increased. The importance of bradykinin is highlighted by administering aprotonin, which inhibits the conversion of prekallikrein to kallikrein, essential for the formation of bradykinin. It reduces the hypotensive effect of ACE inhibition (Mimran *et al.*, 1980). Furthermore, indomethacin, which presumably blocks prostaglandin synthesis and therefore interferes with bradykinin-mediated prostaglandin synthesis, blunts the hypotensive effect of ACE inhibitors.

VII. Tissue Renin/Angiotensin

The presence of a local tissue renin/angiotensin system (RAS), which regulates the production of ATII, adds another dimension to the role of

ACE inhibitor therapy (Miyazaki and Takai, 2001). If local RAS activity varies in different tissues, then the relative efficacy of the ACE inhibitor will vary. In diseases such as diabetes, therefore, the relative efficacy of ACE inhibition may depend very much on the extent of RA activation in the retina, kidney, or nerve, thus determining the beneficial effects in retinopathy, nephropathy, and neuropathy, respectively. Furthermore, the local tissue concentration, and hence action of different ACE inhibitors, varies and may explain the observed differences in efficacy arguing against a "class effect" (Anderson *et al.*, 2000).

VIII. Alternative Pathways of Angiotensin II Formation

Not all ATII is generated by ACE but by serine and cysteine proteases. Although the contribution of these alternative pathways is minimal under normal physiological conditions, during pathophysiological states the role of these pathways may become more important and the relative benefits of ACE inhibition will diminish (Miyazaki and Takai, 2001). In normal vascular tissues, ACE plays a crucial role in angiotensin II production, whereas chymase is stored in mast cells in an inactive form. Chymase acquires the ability to form angiotensin II following mast cell activation.

In conclusion, the renin/angiotensin system may be implicated in the pathogenesis of peripheral nerve dysfunction and eventual degeneration. This axis may be inhibited at various levels: renin itself may be inhibited, angiotensin I conversion to angiotensin II, or binding of ATII at the ATII type 1 (ATII1) receptor. Angiotensin-converting enzyme inhibitors and ATII1 receptor antagonists are now clinically established. Because ACE is a relatively nonspecific peptidase, which catalyzes the breakdown of ATI, bradykinin, and neuropeptides such as substance P and neurotensin, the effects of ACE inhibitors go far beyond the prevention of ATII production (Ertl and Hu, 2001). Alternatively, in certain tissues such as vascular and cardiac tissue, and perhaps peripheral nerve, ATII is produced by other enzymes, e.g., chymase, and ACE inhibitors do not consistently prevent ATII production (Miyazaki and Takai, 2001). The action of ATII1 receptor antagonists may also not be confined to the prevention of binding of ATII at the ATII1 receptor, as by rebound, more ATII may bind at the ATII type 2 (ATII2) receptor and thus mediate, until now, not well-defined effects. Conversely, the beneficial vasodilator effects of ACE inhibition are thought to be mediated principally by bradykinin-mediated NO release, which would not be influenced by ATII receptor antagonism, as demonstrated by the lack of improvement in flow-mediated vasodilatation in patients treated with the ATII antagonist Losartan (Anderson *et al.*, 2000). Thus, the actions of these drugs may be related to multiple mechanisms resulting in net vasodilatation and a limitation of vascular cell growth. The systemic circulating renin/angiotensin system is probably responsible for the control of tissue blood flow and hence oxygenation, whereas local RAS seem to control vascular tissue cell growth. Interrupting the RAS appears to be a logical treatment that does not require a protracted clinical trial program to develop new but as yet untested therapies for patients with diabetic neuropathy.

References

Akbari, C. M., Gibbons, G. W., Habershaw, G. M., LoGerfo, F. W., and Veves, A. (1997). The effect of arterial reconstruction on the natural history of diabetic neuropathy. *Arch. Surg.* **132**, 148–152.

Anderson, T. J., Elstein, E., Haber, H., and Charbonneau, F. (2000). Comparative study of ACE-inhibition, angiotensin II antagonism, and calcium channel blockade on flow-mediated vasodilation in patients with coronary disease (BANFF study). *J. Am. Coll. Cardiol.* **35**, 60–66.

Antonaccio, M. J., and Kerwin, L. (1981). Pre- and postjunctional inhibition of vascular sympathetic function by captopril in SHR. Implication of vascular angiotensin II in hypertension and antihypertensive actions of captopril. *Hypertension* **3**, 154–162.

Athyros, V. G., Didangelos, T. P., Karamitsos, D. T., Papageorgiou, A. A., Boudoulas, H., and Kontopoulos, A. G. (1998). Long-term effect of converting enzyme inhibition on circadian sympathetic and parasympathetic modulation in patients with diabetic autonomic neuropathy. *Acta Cardiol.* **53**, 201–209.

Berridge, M. J. (1993). Inositol trisphosphate and calcium signalling. *Nature* **361**, 315–325.

Britland, S. T., Young, R. J., Sharma, A. K., and Clarke, B. F. (1990). Relationship of endoneurial capillary abnormalities to type and severity of diabetic polyneuropathy. *Diabetes* **39**, 909–913.

Brown, A. M., and Birnbaumer, L. (1988). Direct G protein gating of ion channels. *Am. J. Physiol.* **254**, H401–H410.

Cameron, N. E., and Cotter, M. A. (1996). Effects of a nonpeptide endothelin-1 ET_A antagonist on neurovascular function in diabetic rats: Interaction with the renin-angiotensin system. *J. Pharmacol. Exp. Ther.* **278**, 1262–1268.

Cameron, N. E., Cotter, M. A., Ferguson, K., Robertson, S., and Radcliffe, M. A. (1991). Effects of chronic α-adrenergic receptor blockade on peripheral nerve conduction, hypoxic resistance, polyols, Na^+-K^+-ATPase activity, and vascular supply in STZ-D rats. *Diabetes* **40**, 1652–1658.

Cameron, N. E., Cotter, M. A., Jack, A. M., Basso, M. D., and Hohman, T. C. (1999). Protein kinase C effects on nerve function, perfusion, Na^+,K^+-ATPase activity and glutathione content in diabetic rats. *Diabetologia* **42**, 1120–1130.

Cameron, N. E., Cotter, M. A., and Robertson, S. (1992). Angiotensin converting enzyme inhibition prevents development of muscle and nerve dysfunction and stimulates angiogenesis in streptozotocin-diabetic rats. *Diabetologia* **35**, 12–18.

Chaturvedi, N., Sjolie, A. K., Stephenson, J. M., Abrahamian, H., Keipes, M., Castellarin, A., Rogulja-Pepeonik, Z., and Fuller, J. H. (1998). Effect of lisinopril on progression of retinopathy in normotensive people with type 1 diabetes. The EUCLID Study Group. EURODIAB Controlled Trial of Lisinopril in Insulin-Dependent Diabetes Mellitus. *Lancet* **351**, 28–31.

Cheng, H.-L., and Feldman, E. L. (1998). Bidirectional regulation of p38 kinase and c-Jun N-terminal protein kinase by insulin-like growth factor-I. *J. Biol. Chem.* **273**, 14560–14565.

Cheng, H. L., Steinway, M. L., Xin, X., and Feldman, E. L. (2001). Insulin-like growth factor-I and Bcl-X(L) inhibit c-jun N-terminal kinase activation and rescue Schwann cells from apoptosis. *J. Neurochem.* **76**, 935–943.

Cooper, M. E., Allen, T. J., O'Brien, R. C., Papazoglou, D., Clarke, B. E., Jerums, G., and Doyle, A. E. (1990). Nephropathy in model combining genetic hypertension with experimental diabetes: Enalapril versus hydralazine and metoprolol therapy. *Diabetes* **39**, 1575–1579.

Cooper, M. E., Rumble, J., Komers, R., He-Cheng, D., Jandeleit, K., and Sheung-To, C. (1994). Diabetes-associated mesenteric vascular hypertrophy is attenuated by angiotensin-converting enzyme inhibition. *Diabetes* **43**, 1221–1228.

Coppey, L. J., Gellett, J. S., Davidson, E. P., Dunlap, J. A., Lund, D. D., and Yorek, M. A. (2001). Effect of antioxidant treatment of streptozotocin-induced diabetic rats on endoneurial blood flow, motor nerve conduction velocity, and vascular reactivity of epineurial arterioles of the sciatic nerve. *Diabetes* **50**, 1927–1937.

Cotter, M. A., and Cameron, N. E. (1998). Correction of neurovascular deficits in diabetic rats by β_2-adrenoceptor agonist and α_1-adrenoceptor antagonist treatment: Interactions with the nitric oxide system. *Eur. J. Pharmacol.* **343**, 217–223.

Curzen, N. C., and Timmis, A. (1997). Endothelial dysfunction in chronic heart failure. *In* "Controversies in the Management of Heart Failure" (A. Coats, and J. G. F. Cleland, eds.), pp. 25–40. Churchill Livingstone, Edinburgh.

Di Giulio, S., Perez-Ramirez, J. L., Jamoun, P., Grunfeld, J. P., and Meyer, P. (1981). Treatment of severe arterial hypertension with captopril. *Nouv. Presse Med.* **10**, 1551–1555.

Dyck, P. J., Davies, J. L., Wilson, D. M., Service, F. J., Melton, L. J., III, and O'Brien, P. C. (1999). Risk factors for severity of diabetic polyneuropathy: Intensive longitudinal assessment of the Rochester Diabetic Neuropathy Study cohort. *Diabet. Care* **22**, 1479–1486.

Dyck, P. J., Hansen, S., Karnes, J., O'Brien, P. C., Yasuda, H., Windebank, A. J., and Zimmerman, B. R. (1985). Capillary number and percentage closed in human diabetic sural nerve. *Proc. Natl. Acad. Sci. USA* **82**, 2513–2517.

Enderle, M. D., Benda, N., Schmuelling, R. M., Haering, H. U., and Pfohl, M. (1998). Preserved endothelial function in IDDM patients, but not in NIDDM patients, compared with healthy subjects. *Diabet. Care* **21**, 271–277.

Ertl, G., and Hu, K. (2001). Anti-ischemic potential of drugs related to the renin–angiotensin system. *J. Cardiovasc. Pharmacol.* **37**(Suppl. 1), S11–S20.

Estacio, R. O., Jeffers, B. W., Gifford, N., and Schrier, R. W. (2000). Effect of blood pressure control on diabetic microvascular complications in patients with hypertension and type 2 diabetes. *Diabet. Care* **23**(Suppl. 2), B54–B64.

Farhy, R. D., Carretero, O. A., Ho, K. L., and Scicli, A. G. (1993). Role of kinins and nitric oxide in the effects of angiotensin converting enzyme inhibitors on neointima formation. *Circ. Res.* **72**, 1202–1210.

Fiordaliso, F., Li, B. S., Latini, R., Sonnenblick, E. H., Anversa, P., Leri, A., and Kajstura, J. (2000). Myocyte death in streptozotocin-induced diabetes in rats is angiotensin II-dependent. *Lab. Invest.* **80**, 513–527.

Forrest, K. Y. Z., Maser, R. E., Pambianco, G., Becker, D. J., and Orchard, T. J. (1997). Hypertension as a risk factor for diabetic neuropathy: A prospective study. *Diabetes* **46**, 665–670.

Gallinat, S., Csikos, T., Meffert, S., Herdegen, T., Stoll, M., and Unger, T. (1997). The angiotensin AT2 receptor down-regulates neurofilament M in PC12W cells. *Neurosci. Lett.* **227**, 29–32.

Giannattasio, C., Failla, M., Piperno, A., Grappiolo, A., Gamba, P., Paleari, F., and Mancia, G. (1999). Early impairment of large artery structure and function in type I diabetes mellitus. *Diabetologia* **42**, 987–994.

Giannini, C., and Dyck, P. J. (1995). Basement membrane reduplication and pericyte degeneration precede development of diabetic polyneuropathy and are associated with its severity. *Ann. Neurol.* **37**, 498–504.

Haywood, G. A., Gullestad, L., Katsuya, T., Hutchinson, H. G., Pratt, R. E., Horiuchi, M., and Fowler, M. B. (1997). AT1 and AT2 angiotensin receptor gene expression in human heart failure. *Circulation* **95**, 1201–1206.

Hogikyan, R. V., Wald, J. J., Feldman, E. L., Greene, D. A., Halter, J. B., and Supiano, M. A. (1999). Acute effects of adrenergic-mediated ischemia on nerve conduction in subjects with type 2 diabetes. *Metabolism* **48**, 495–500.

Hormigo, A., and Alves, M. (1992). Peripheral neuropathy in a patient receiving enalapril. *Br. Med. J.* **305**, 1332.

Hotta, N., Kakuta, H., Fukasawa, H., Koh, N., Sakakibara, F., Komori, H., and Sakamoto, N. (1992). Effect of niceritrol on streptozocin-induced diabetic neuropathy in rats. *Diabetes* **41**, 587–591.

Ibrahim, S., Harris, N. D., Radatz, M., Selmi, F., Rajbhandari, S., and Brady, L. (1999). A new minimally invasive technique to show nerve ischaemia in diabetic neuropathy. *Diabetologia* **42**, 737–742.

Jeffreson, S., Rush, R., Zettler, C., Frewin, D. B., and Head, R. J. (1995). The influence of the renin angiotensin system on abnormal expression of nerve growth factor in the spontaneously hypertensive rat. *Clin. Exp. Pharmacol. Physiol.* **22**, 478–480.

Jude, E. B., Boulton, A. J. M., Ferguson, M. W. J., and Appleton, I. (1999). The role of nitric oxide synthase isoforms and arginase in the pathogenesis of diabetic foot ulcers: Possible modulatory effects by transforming growth factor beta 1. *Diabetologia* **42**, 748–757.

Kihara, M., Mitsui, M. K., Mitsui, Y., Okuda, K., Nakasaka, Y., Takahashi, M., and Schmelzer, J. D. (1999). Altered vasoreactivity to angiotensin II in experimental diabetic neuropathy: Role of nitric oxide. *Muscle Nerve* **22**, 920–925.

Kontopoulos, A. G., Athyros, V. G., Didangelos, T. P., Papageorgiou, A. A., Avramidis, M. J., Mayroudi, M. C., and Karamitsos, D. T. (1997). Effect of chronic quinapril administration on heart rate variability in patients with diabetic autonomic neuropathy. *Diabet. Care* **20**, 355–361.

Low, P. A., Schmelzer, J. D., Ward, K. K., Curran, G. L., and Poduslo, J. F. (1988). Effect of hyperbaric oxygenation on normal and chronic streptozotocin diabetic peripheral nerves. *Exp. Neurol.* **99**, 201–212.

Low, P. A., Tuck, R. R., Dyck, P. J., Schmelzer, J. D., and Yao, J. K. (1984). Prevention of some electrophysiologic and biochemical abnormalities with oxygen supplementation in experimental diabetic neuropathy. *Proc. Natl. Acad. Sci. USA* **81**, 6894–6898.

Lyall, F., Dornan, E. S., McQueen, J., Boswell, F., and Kelly, M. (1992). Angiotensin II increases proto-oncogene expression and phosphoinositide turnover in vascular smooth muscle cells via the angiotensin II AT1 receptor. *J. Hypertens.* **10**, 1463–1469.

Lyons, D., Roy, S., O'Byrne, S., and Swift, C. G. (1997). ACE inhibition: Postsynaptic adrenergic sympatholytic action in men. *Circulation* **96**, 911–915.

Malik, R. A. (1998). Microalbuminuria: Pathophysiology, relevance and treatment. *Heart Fail. Hypertens.* **10**, 4–6.

Malik, R. A., Newrick, P. G., Sharma, A. K., Jennings, A., Ah-See, A. K., Mayhew, T. M., Jakubowski, J., Boulton, A. J. M., and Ward, J. D. (1989). Microangiopathy in human diabetic neuropathy: Relationship between capillary abnormalities and the severity of neuropathy. *Diabetologia* **32**, 92–102.

Malik, R. A., Paniagua, O., Shaw, L., Austin, C., and Heagerty, A. M. (1999). Resistance vessel function and structure in normotensive patients with type I diabetes. *Diabetologia* **42**, A75.

Malik, R. A., Tesfaye, S., Thompson, S. D., Veves, A., Sharma, A. K., Boulton, A. J. M., and Ward, J. D. (1993). Endoneurial localisation of microvascular damage in human diabetic neuropathy. *Diabetologia* **36**, 454–459.

Malik, R. A., Veves, A., Masson, E. A., Sharma, A. K., Ah-See, A. K., Schady, W., Lye, R. H., and Boulton, A. J. M. (1992). Endoneurial capillary abnormalities in mild human diabetic neuropathy. *J. Neurol. Neurosurg. Psychiat.* **55**, 557–561.

Malik, R. A., Williamson, S., Abbott, C., Carrington, A. L., Iqbal, J., Schady, W., and Boulton, A. J. M. (1998). Effect of angiotensin-converting-enzyme (ACE) inhibitor trandolapril on human diabetic neuropathy: Randomised doubleblind controlled trial. *Lancet* **352**, 1978–1981.

Maschio, G., Alberti, D., Janin, G., Locatelli, F., Mann, J. F., Motolese, M., Ponticelli, C., Ritz, E., and Zucchelli, P. (1996). Effect of the angiotensin-converting-enzyme inhibitor benazepril on the progression of chronic renal insufficiency. The Angiotensin-Converting-Enzyme Inhibition in Progressive Renal Insufficiency Study Group. *N. Engl. J. Med.* **334**, 939–945.

Maxfield, E. K., Cameron, N. E., Cotter, M. A., and Dines, K. C. (1993). Angiotensin II receptor blockade improves nerve function, modulates nerve blood flow and stimulates endoneurial angiogenesis in streptozotocin-diabetic rats and nerve function. *Diabetologia* **36**, 1230–1237.

Mimran, A., Targhetta, R., and Laroche, B. (1980). The antihypertensive effect of captopril: Evidence for an influence of kinins. *Hypertension* **2**, 732–737.

Miyazaki, M., and Takai, S. (2001). Local angiotensin II-generating system in vascular tissues: The roles of chymase. *Hypertens. Res.* **24**, 189–193.

Moore, M. V., Jeffcoate, W. J., and MacDonald, I. A. (1987). Apparent improvement in diabetic autonomic neuropathy induced by captopril. *J. Hum. Hypertens.* **1**, 161–165.

Moreau, P., d'Uscio, L. V., Shaw, S., Takase, H., Barton, M., and Luscher, T. F. (1997). Angiotensin II increases tissue endothelin and induces vascular hypertrophy: Reversal by ET(A)-receptor antagonist. *Circulation* **96**, 1593–1597.

Nakamura, J., Kato, K., Hamada, Y., Nakayama, M., Chaya, S., Nakashima, E., Naruse, K., Kasuya, Y., Mizubayashi, R., Miwa, K., Yasuda, Y., Kamiya, H., Ienaga, K., Sakakibara, F., Koh, N., and Hotta, N. (1999). A protein kinase C-β-selective inhibitor ameliorates neural dysfunction in streptozotocin-induced diabetic rats. *Diabetes* **48**, 2090–2095.

Newrick, P. G., Wilson, A. J., Jakubowski, J. A., Boulton, A. J. M., and Ward, J. D. (1986). Sural nerve oxygen tension in diabetes. *Br. Med. J.* **293**, 1053–1054.

Nishikawa, T., Edelstein, D., Du, X. L., Yamagishi, S., Matsumura, T., Kaneda, Y., Yorek, M. A., Beebe, D., Oates, P. J., Hammes, H. P., Giardino, I., and Brownlee, M. (2000). Normalizing mitochondrial superoxide production blocks three pathways of hyperglycaemic damage. *Nature* **404**, 787–790.

Nishimoto, M., Takai, S., Kim, S., Jin, D., Yuda, A., Sakaguchi, M., Yamada, M., Sawada, Y., Kondo, K., Asada, K., Iwao, H., Sasaki, S., and Miyazaki, M. (2001). Significance of chymase-dependent angiotensin II-forming pathway in the development of vascular proliferation. *Circulation* **104**, 1274–1279.

Noll, G., Wenzel, R. R., de Marchi, S., Shaw, S., and Luscher, T. F. (1997). Differential effects of captopril and nitrates on muscle sympathetic nerve activity in volunteers. *Circulation* **95**, 2286–2292.

Osicka, T. M., Yu, Y. X., Lee, V., Panagiotopoulos, S., Kemp, B. E., and Jerums, G. (2001). Aminoguanidine and ramipril prevent diabetes-induced increases in protein kinase C activity in glomeruli, retina and mesenteric artery. *Clin. Sci. (Colch.)* **100**, 249–257.

Parving, H. H., Larsen, M., Hommel, E., and Lund-Andersen, H. (1989). Effect of antihypertensive treatment on blood-retinal barrier permeability to fluorescein in hypertensive type 1 (insulin-dependent) diabetic patients with background retinopathy. *Diabetologia* **32**, 440–444.

Qiang, X., Satoh, J., Sagara, M., Fukuzawa, M., Masuda, T., Sakata, Y., Muto, G., Muto, Y., Takahashi, K., and Toyota, T. (1998). Inhibitory effect of troglitazone on diabetic neuropathy in streptozotocin-induced diabetic rats. *Diabetologia* **41**, 1321–1326.

Rayman, G., Malik, R. A., Sharma, A. K., and Day, J. L. (1995). Microvascular response to tissue injury and capillary ultrastructure in the foot skin of type I diabetic patients. *Clin. Sci. (Colch.)* **89**, 467–474.

Reja, A., Tesfaye, S., Harris, N. D., and Ward, J. D. (1995). Is ACE inhibition with lisinopril helpful in diabetic neuropathy? *Diabet. Med.* **12**, 307–309.

Reljanovic, M., Reichel, G., Rett, K., Lobisch, M., Schuette, K., Möller, W., Tritschler, H. J., Mehnert, H., and ALADIN II Study Group. (1999). Treatment of diabetic polyneuropathy with the antioxidant thioctic acid (α-lipoic acid): A two year multicenter randomized double-blind placebo-controlled trial (ALADIN II). *Free Radic. Res.* **31**, 171–179.

Ristic, H., Wiley, J. W., Hall, K. E., and Sima, A. A. F. (1996). Failure of nimodipine to prevent or correct the long-term nerve conduction defect and increased neuronal Ca^{2+}-currents in the diabetic BB/W-rat. *Diabet. Res. Clin. Pract.* **32**, 135–140.

Robertson, S., Cameron, N. E., and Cotter, M. A. (1992). The effect of the calcium antagonist nifedipine on peripheral nerve function in streptozotocin-diabetic rats. *Diabetologia* **35**, 1113–1117.

Schror, K. (1992). Role of prostaglandins in the cardiovascular effects of bradykinin and angiotensin-converting enzyme inhibitors. *J. Cardiovasc. Pharmacol.* **20**(Suppl. 9), S68–S73.

Shenoy, U. V., Richards, E. M., Huang, X. C., and Sumners, C. (1999). Angiotensin II type 2 receptor-mediated apoptosis of cultured neurons from newborn rat brain. *Endocrinology* **140**, 500–509.

Simovic, D., Isner, J. M., Ropper, A. H., Pieczek, A., and Weinberg, D. H. (2001). Improvement in chronic ischemic neuropathy after intramuscular phVEGF165 gene transfer in patients with critical limb ischemia. *Arch. Neurol.* **58**, 761–768.

Stroth, U., Blume, A., Mielke, K., and Unger, T. (2000). Angiotensin AT(2) receptor stimulates ERK1 and ERK2 in quiescent but inhibits ERK in NGF-stimulated PC12W cells. *Mol. Brain Res.* **78**, 175–180.

Stroth, U., Meffert, S., Gallinat, S., and Unger, T. (1998). Angiotensin II and NGF differentially influence microtubule proteins in PC12W cells: Role of the AT_2 receptor. *Mol. Brain Res.* **53**, 187–195.

Tesfaye, S., Harris, N., Jakubowski, J. J., Mody, C., Wilson, R. M., Rennie, I. G., and Ward, J. D. (1993). Impaired blood flow and arterio-venous shunting in human diabetic neuropathy: A novel technique of nerve photography and fluorescein angiography. *Diabetologia* **36**, 1266–1274.

Tesfaye, S., Malik, R., Harris, N., Jakubowski, J. J., Mody, C., Rennie, I. G., and Ward, J. D. (1996). Arterio-venous shunting and proliferating new vessels in acute painful neuropathy of rapid glycaemic control (insulin neuritis). *Diabetologia* **39**, 329–335.

The Diabetes Control and Complications Trial Research Group. (1995). The effect of intensive diabetes therapy on the development and progression of neuropathy. *Ann. Intern. Med.* **122**, 561–568.

The EURODIAB Prospective Complications Study (PCS) Group. (1999). Cardiovascular risk factors predict diabetic peripheral neuropathy in type I subjects in Europe. *Diabetologia* **42**, A50.

Theriault, M., Dort, J., Sutherland, G., and Zochodne, D. W. (1997). Local human sural nerve blood flow in diabetic and other polyneuropathies. *Brain* **120**, 1131–1138.

Tuttle, J. B., Etheridge, R., and Creedon, D. J. (1993). Receptor-mediated stimulation and inhibition of nerve growth factor secretion by vascular smooth muscle. *Exp.Cell Res.* **208**, 350–361.

UK Prospective Diabetes Study Group. (1998). Intensive blood glucose control with sulphonylureas or insulin compared with conventional treatment and risk of complications in patients with type II diabetes (UK PDS 33). *Lancet* **352**, 837–853.

Vanhoutte, P. M., Feletou, M., Boulanger, C. M., Hoffner, U., and Rubanyi, G. M. (1996). Existence of multiple endothelium-derived relaxing factors. *In* "Endothelium-Derived Hyperpolarizing Factor" (P. M. Vanhoutte, ed.), pp. 1–10. Harwood Academic, Reading, UK.

Veves, A., Akbari, C. M., Primavera, J., Donaghue, V. M., Zacharoulis, D., Chrzan, J. S., DeGirolami, U., LoGerfo, F. W., and Freeman, R. (1998). Endothelial dysfunction and the expression of endothelial nitric oxide synthetase in diabetic neuropathy, vascular disease, and foot ulceration. *Diabetes* **47**, 457–463.

Walker, A. B., Chattington, P. D., Buckingham, R. E., and Williams, G. (1999). The thiazolidinedione rosiglitazone (BRL-49653) lowers blood pressure and protects against impairment of endothelial function in Zucker fatty rats. *Diabetes* **48**, 1448–1453.

Yasuda, H., and Dyck, P. J. (1987). Abnormalities of endoneurial microvessels and sural nerve pathology in diabetic neuropathy. *Neurology* **37**, 20–28.

Yoshizumi, M., Tsuchiya, K., Kirima, K., Kyaw, M., Suzaki, Y., and Tamaki, T. (2001). Quercetin inhibits Shc- and phosphatidylinositol 3-kinase-mediated c-Jun N-terminal kinase activation by angiotensin II in cultured rat aortic smooth muscle cells. *Mol. Pharmacol.* **60**, 656–665.

Young, M. J., Veves, A., Walker, M. G., and Boulton, A. J. M. (1992). Correlations between nerve function and tissue oxygenation in diabetic patients: Further clues to the aetiology of diabetic neuropathy? *Diabetologia* **35**, 1146–1150.

Zuanetti, G., Latini, R., Maggioni, A. P., Franzosi, M., Santoro, L., and Tognoni, G. (1997). Effect of the ACE inhibitor lisinopril on mortality in diabetic patients with acute myocardial infarction: Data from the GISSI-3 study. *Circulation* **96**, 4239–4245.

CLINICAL TRIALS FOR DRUGS AGAINST DIABETIC NEUROPATHY: CAN WE COMBINE SCIENTIFIC NEEDS WITH CLINICAL PRACTICALITIES?

Dan Ziegler

German Diabetes Research Institute at the Heinrich Heine University
German Diabetes Clinic, 40225 Düsseldorf, Germany

Dieter Luft

4th Medical Department, Eberhard Karls University
72076 Tübingen, Germany

Diabetic neuropathy is a chronic progressive disease accounting for considerable morbidity and reduced quality of life among patients with diabetes. Accumulating evidence suggests that the clinical and neurophysiological markers used to assess neuropathy not only predict the development

of neuropathic foot ulceration, one of the most common causes for hospital admission and lower limb amputations, but are also predictors of increased mortality in diabetic patients. In addition to metabolic control, drug treatment of both incipient and clinically manifest diabetic neuropathy will be necessary for the years to come. Because 1–2% of the whole population in western societies may be affected, the search for effective drug treatment is not only a very important goal for the patient suffering from diabetic neuropathy and for the practicing physician, but also an economic task for both the health care systems and the pharmaceutical companies. The validity of inferences about the clinical consequences of the use of any given agent to induce a specific pharmacologic effect will depend not only on the extent to which it affects the targeted biological phenomenon, but also on the extent to which all of the actions of the agent have been defined and the extent to which all affect the entire organism, alone and in concert. The ultimate test of the usefulness of a drug or device depends on the determination of outcomes, ideally in randomized clinical trials (RCTs) of sufficient scope and duration. The efficacy and safety of a variety of drugs based on the different pathogenetic hypotheses proposed have been evaluated in RCTs since the 1970s. However, the quality of RCTs published between 1981 and 1992 that evaluated the effects of medical treatment in diabetic polyneuropathy was poor. Adequate designs for RCTs in diabetic neuropathy must consider the following criteria: type and stage of neuropathy, homogeneity of the study population, outcome measures (neurophysiological markers, intermediate clinical end points, ultimate clinical outcomes, quality of life), natural history, sample size, study duration, reproducibility of neurophysiological and intermediate end points, nonspecific effects of treatment, measures of treatment effect, the extent to which the overall trial result applies to individual patients (external validity), and the reporting of RCTs. Trials focusing preferentially on patients with mild or moderate early stages of neuropathy over long periods of 3–5 years aimed at slowing or prevention, rather than reversal, using end point measures that have clinical and prognostic significance are most likely to produce meaningful results.
© 2002, Elsevier Science (USA).

I. Clinical Impact of Diabetic Polyneuropathy

Diabetic neuropathy has been defined as a demonstrable disorder, either clinically evident or subclinical, that occurs in the setting of diabetes mellitus without other causes for peripheral neuropathy. It includes manifestations in the somatic and/or autonomic parts of the peripheral nervous system (San Antonio Agreement, 1988). Distal symmetrical sensory or sensorimotor polyneuropathy (DSP) represents the most important clinical manifestation, affecting approximately 30% of the hospital-based population and 20% of community-based samples of diabetic patients. The incidence of DSP is approximately 2% per year (Shaw and Zimmet, 1999). DSP is related to both lower-extremity impairments, such as diminished position sense, and functional limitations, such as walking ability (Resnick *et al.*, 2000). Accumulating evidence suggests that not only surrogate markers of microangiopathy, such as albuminuria, but also those used for polyneuropathy, such as nerve conduction velocity (NCV) and vibration perception threshold (VPT), may predict mortality in diabetic patients (Forsblom *et al.*, 1998; Coppini *et al.*, 2000). Elevated VPT also predicts the development of neuropathic foot ulceration, one of the most common causes for hospital admission and lower limb amputations among diabetic patients (Abbott *et al.*, 1998). Neuropathic symptoms are present in 15–20% of diabetic patients, 7.5% of whom experience chronic neuropathic pain (Chan *et al.*, 1990). Pain is a subjective symptom of major clinical importance, as it is often this complaint that motivates patients to seek health care. Pain associated with diabetic neuropathy exerts a substantial impact on the quality of life, particularly by causing considerable interference in sleep and enjoyment of life (Galer *et al.*, 2000).

II. Role of Drug Treatment in Diabetic Polyneuropathy

In addition to metabolic control, drug treatment of both incipient and clinically manifest diabetic neuropathy will be necessary for the years to come because (1) it is not yet possible to reach near normoglycemic control in all type 1 diabetic patients with currently available methods of treatment, which are of proven efficacy to reduce the incidence of diabetic neuropathy, and, (2) type 2 diabetes mellitus is sometimes only diagnosed when neuropathy is the presenting complication. Moreover, in type 2 diabetes, we are far from reaching normoglycemic control in most patients. Because 1–2% of the whole population in western societies may be affected, the search for effective drug treatment is not only a very important goal for the patient suffering from diabetic neuropathy and for the practicing physician, but also an economic task for both the health care systems and the pharmaceutical companies.

The efficacy and safety of a variety of drugs based on the different pathogenetic hypotheses proposed have been evaluated in randomized clinical trials (RCTs) (Table I) since the 1970s. RCTs are a widely accepted means of applying experimental methods to a clinical setting and have been advocated as the gold standard for comparing and evaluating different treatments. However, the quality of RCTs that evaluated the effects of medical

TABLE I
TREATMENT OPTIONS FOR DIABETIC POLYNEUROPATHY BASED ON
PATHOGENETIC CONSIDERATIONS[a]

Near-normal glycemic control
Aldose reductase inhibitors[b–d]
α-Lipoic acid (thioctic acid)
Vitamin E
γ-Linolenic acid[e]
Vasodilators, e.g., prostaglandins, ACE inhibitors
Nucleosides
Vitamin B mixtures (B_1, B_6, B_{12})
Nerve growth factor[b]
myo-Inositol[b]
Gangliosides[c]
Aminoguanidine[f]
Acetyl-L-carnitine[b]
C peptide[f]
Protein kinase C ß inhibitor[f]
Vascular endothelial growth factor[f]

[a]Several agents have been withdrawn due to various reasons.
[b]Lack of efficacy.
[c]Side effects.
[d]Except for epalrestat.
[e]Failed to be licensed in the United Kingdom.
[f]No clinical trials in diabetic polyneuropathy yet available.

treatment in diabetic polyneuropathy was poor. Cavaliere *et al.* (1994) assessed the quality of scientific evidence for the efficacy of various pathogenetically oriented treatment approaches for diabetic polyneuropathy, which were examined in RCTs published between 1981 and 1992. They used a quality system covering the internal (scientific) validity (the ability to demonstrate a treatment effect if it really exists) and external validity (the possibility of generalizing the study results to patients seen in clinical practice) (Chalmers *et al.*, 1981). The analysis, based on 38 RCTs in total, revealed a devastating picture: the methods of randomization were unspecified and a detailed *a priori* estimate of the sample size needed to detect a treatment difference was not reported in 95% of the RCTs, respectively. Only 11% of the RCTs had sufficient statistical power to detect a clinically meaningful difference (Cavaliere *et al.*, 1994). Thus, as stated by Altman and Bland (1995), absence of evidence is not evidence of absence. In other words, to interpret the majority of published RCTs as providing evidence of an ineffectiveness of the respective treatment in diabetic neuropathy would be clearly misleading, as the likelihood of detecting differences in the parameters of nerve function between the treatment groups was too low in view of the small sample sizes. Furthermore, an *a posteriori* estimate of study power was provided in only 17% of those studies in which a significant difference between arms was not reached. The median treatment duration in the 38 RCTs was only 24 weeks, and in only 11% the duration of treatment was 2 or more years. Finally, only 5% of the RCTs focused on major complications (ulcers) (Coppini *et al.*, 2000). These disappointing results, therefore, do not allow any conclusion as to whether the missing success was due to a lack of efficacy of the drug tested or an inappropriate study design. Several reviews dealing with these methodological problems, which have to be solved before any judgement about the efficacy of a study drug is possible, have been published (Pfeifer and Schumer, 1995; Sima and Laudadio, 1996; Ziegler, 1997; Luft, 1998). This review focuses on the various aspects to be considered in designing clinical trials for the treatment of the distal symmetrical predominantly sensory neuropathy (Table II), as this type is the clinically most relevant in terms of prevalence, morbidity, and mortality in diabetic patients (Shaw and Zimmet, 1999; Resnick *et al.*, 2000; Forsblom *et al.*, 1998; Coppini *et al.*, 2000; Abbott *et al.*, 1998).

III. Classification, Diagnosis, and Staging

Because a classification of diabetic neuropathy based on pathogenetic grounds is not possible, the different manifestations must be classified

TABLE II

ASPECTS CONSIDERED IN DESIGNING RANDOMIZED CLINICAL
TRIALS IN DIABETIC NEUROPATHY

Type and stage of neuropathy
Homogeneity of study population
Natural history and risk factors
Sample size
Study duration
Outcome measures
    Neurophysiological markers
    Intermediate clinical end points
    Ultimate clinical outcomes
    Quality of life
Reproducibility
Measures of treatment effect
Nonspecific effects of treatment
    True and perceived placebo effects
    Regression to the mean
Generalizability (external validity)
Reporting of RCTs (CONSORT statement)

on clinical criteria. However, due to the variety of the clinical syndromes with possible overlaps, not even a generally accepted classification exists. The most widely used approach is the one proposed by Thomas (1973), who differentiates between a reversible "hyperglycemic neuropathy," possibly reflecting nerve hypoxia, and more persistent manifestations, the commonest being DSP, as well as focal and multifocal neuropathy, including cranial, thoraco-abdominal and focal limb neuropathies and diabetic amyotrophy. For research purposes, DSP has been classified into two main classes. Subclinical neuropathy (class I) refers to patients without demonstrable signs or symptoms, whereas clinical neuropathy (class II) refers to those with symptoms, signs, or both. Further subclasses (A,B,C) are defined by neurophysiological tests, including electrodiagnosis and quantitative sensory and autonomic testing (San Antonio Agreement, 1998). Because neuropathy may be attributed to a number of causes other than diabetes, several conditions need to be excluded, e.g., alcoholism, uremia, vitamin B_{12} deficiency, chronic liver diseases, intoxications, and malignancies, as well as hereditary, drug-induced, or autoimmune-mediated neuropathies.

Due to the increasing recognition of diabetic neuropathy as a major contributor to morbidity and the recent burst of clinical trials in this field on one hand, but lack of agreement on the definition and diagnostic assessment of neuropathy on the other hand, several consensus conferences were convened to overcome these barriers (San Antonio Conference, 1988; Consensus Statement, 1998; Albers *et al.*, 1995; Peripheral Neuropathy Association, 1993; American Academy of Neurology, 1996). The Consensus Development Conference on Standardized Measures in Diabetic Neuropathy (San Antonio Agreement, 1998) recommended the following five measures for the diagnosis of diabetic neuropathy after reviewing their diagnostic validity: (a) clinical measures, (b) morphological and biochemical analyses, (c) electrodiagnostic assessment, (d) quantitative sensory testing, and (e) autonomic nervous system testing. In the Rochester Diabetic Neuropathy Study (RDNS), Dyck *et al.* (1993) employed the following criteria for polyneuropathy: (1) neuropathic symptoms [neuropathy symptom score (NSS)], (2) neuropathic deficits [neuropathy impairment score (NIS)], (3) motor/sensory nerve conduction velocity (M/SNCV), (4) quantitative sensory testing (QST: vibration or cold detection thresholds), and (5) autonomic function testing [AFT: R-R variation to deep breathing (DB) or Valsalva ratio]. Minimal criteria for the diagnosis of polyneuropathy required ≥ 2 abnormalities among criteria 1–5 with at least one being 3 or 5. The following staging approach was used: no neuropathy (N0: minimal criteria unfulfilled); asymptomatic neuropathy (N1a: minimal criteria fulfilled, NSS = 0, normal ankle dorsiflection; N1b:

minimal criteria fulfilled, NSS = 0, abnormal ankle dorsiflection); symptomatic neuropathy (N2a: minimal criteria fulfilled, NSS $\geq$ 1, normal ankle dorsiflection; N2b: minimal criteria fulfilled, NSS $\geq$ 1, abnormal ankle dorsiflection); and disabling neuropathy (N3: minimal criteria fulfilled, disabling features).

IV. Outcome Measures

Phase III RCTs, which evaluate the effect that new interventions have on the clinical outcomes of particular relevance to the patient (e.g., foot ulcers, amputations), may require many participants to be followed for a long time. There has been a great interest in using surrogate end points to reduce the cost and duration of RCTs. A surrogate end point is a laboratory measurement or a physical sign used as a substitute for a clinically meaningful end point that measures directly how a patient feels, functions, or survives. Changes induced by a therapy on a surrogate end point are expected to reflect changes in a clinically meaningful end point. In theory, for a surrogate end point to be an effective substitute for the clinical outcome, effects of the intervention on the surrogate must reliably predict (not only correlate with) the overall effect on the clinical outcome (Fleming and DeMets, 1996). In practice, however, this requirement frequently fails as there are examples from several disease areas showing how surrogate end points have been misleading about the actual effects that treatments have on the health of patients [e.g., in cardiology, drugs (encainide, flecainide, moricizine, sotalol) that suppressed cardiac arrhythmia, but unexpectedly increased the risk of mortality]. Surrogate end points can be useful in phase II screening trials for identifying whether a new intervention is biologically active and for guiding decisions about whether the intervention is promising enough to justify a large definitive trial with clinically meaningful outcomes. In definitive phase III trials, except for rare circumstances in which the validity of the surrogate end point has already been rigorously established, the primary end point should be the true clinical outcome (Fleming and DeMets, 1996; Sobel and Furberg, 1997). However, effectiveness in phase II studies by no means predicts clinical usefulness in phase III trials. According to Malik (2000), at least 30 different substances have been shown to be effective in phase II trials and reached phase III trials, but none eventually proved to be effective in clinical practice.

According to the model of the consequences of disease developed by the World Health Organization in 1980, outcome measures can be classified into measures of impairment, disability, and handicap. Impairment refers to organ dysfunction or abnormalities of body structure (e.g., numbness, absent reflexes, slowing in NCV), disability to the patient's functional performance (e.g., walking or eating), and handicap to the social disadvantages resulting from impairment and disability (e.g., the ability to work). A review of the medical literature between 1978 and 1993 (Molenaar *et al.*, 1995) revealed that only 4% of 54 studies in patients with diabetic neuropathy have used disability or handicap measures. It has been emphasized that treatment recommendations should be based on trials in which disability and handicap are assessed, not on trials in which impairment only is assessed (Molenaar *et al.*, 1995). As the focus of health care researchers is increasingly directed at physical health and functional ability, global and disease-specific quality of life measures must be included in any large-scale trial of diabetic neuropathy. Examples for end point variables in RCTs of DSP are listed in Table III.

The Peripheral Nerve Society (Albers *et al.*, 1995) has reassessed the suitability of the possible end points employed in controlled clinical trials of diabetic polyneuropathy. Because neuropathic symptoms may reduce the quality of life, they can be used as meaningful end points, but their change should be associated with at least the minimal change in the NIS that physicians can detect in the clinical examination, i.e., two points. Linear regression analyses from population-based epidemiological data indicate that a change in the NIS of two points corresponds, for example, to a change in peroneal MNCV of 2.2 m/s, median SNCV of 1.9 m/s, and sural sensory amplitude of 3.8 μV. These differences are considered as clinically meaningful minimal degrees of improvement, prevention, or

TABLE III

END POINT VARIABLES IN CLINICAL TRIALS OF DIABETIC NEUROPATHY

Dependent variable	Example
Mechanistic studies	
Biochemical	Decrease in nerve or (red blood cell) sorbitol
Pathophysiological	Improvement in nerve conduction velocity, quantitative sensory testing, autonomic function tests
Morphological	Sural nerve-myelinated fiber density
Clinical studies	
Symptomatic	Symptomatic relief; improvement in quality of life
Functional	Reduction in neuropathy impairment score
Hospitalization	Reduction in number of hospitalizations
Disease	Prevention of neuropathy
Event	Reduction in new foot ulcers or amputations

deterioration of neuropathic findings (Dyck and O'Brien, 1989). Electrodiagnostic measures have the advantage of being the most objective, sensitive, specific, and reproducible methods, which are available in many neurophysiological laboratories worldwide. Their limitations are that they measure only function in the largest, fastest conducting myelinated fibers; have relatively low specificity in detecting diabetic neuropathy; show relatively high intraindividual variability for certain parameters (amplitudes); are vulnerable to external factors, such as electrode locations or limb temperature; and may be influenced by conditions other than neuropathy, such as electrolyte changes (San Antonio Agreement, 1998).

The Peripheral Neuropathy Association (1993) has recommended that detection thresholds of touch-pressure, vibration, coolness, warmth, heat pain, cold pain, and mechanical pain be used to characterize cutaneous sensation. The procedures that are being used for QST include (1) the method of limits (continuous increase or decrease in intensity to appearance or disappearance threshold), (2) threshold tracking (combination of appearance or disappearance threshold), (3) titration method (graded steps to appearance and disappearance threshold), and (4) the two-alternative forced-choice method (pairs of stimulus and null-stimulus phases).

QST has the advantages that it is highly sensitive, relatively simple, noninvasive, and nonaversive; affords precise control over stimulus intensity and testing algorithms; contributes to differentiation of the relative deficit in small fibers (cooling and heat-pain perception thresholds) vs large fibers (vibration and touch-pressure thresholds); and is particularly valuable in screening large populations or in longitudinal trials. The limitations to QST include that they constitute psychophysical methods vulnerable to the effects of alertness, mood, concentration, ambient noise, etc.; show a relatively high intraindividual variability; have not been standardized adequately; and may be time-consuming (forced-choice methods), which may lead to a decline in concentration or boredom in the person tested, thereby resulting in measurement errors (San Antonio Agreement, 1998).

Among the parameters obtained from sural nerve biopsy, the myelinated fiber density and the diameter histogram are accurate and reproducible. However, because this technique is invasive and there is little information as to whether neuropathological measures predict the severity and course of diabetic polyneuropathy, their further validation is needed (Albers *et al.*, 1995). Given the high prevalence of long-term impairment in diabetic patients after nerve biopsy, this method should be used only in exceptional circumstances.

The changes in the aforementioned measures are considered meaningful if the degree of change is associated with at least two points of NIS derived from regression analyses of population-based epidemiological data (Albers *et al.*, 1995; Dyck and O'Brien, 1989).

Measures of cardiovascular AFT based on changes in heart rate variability (HRV) may be used as markers for vagal and sympathetic dysfunction, and there is increasing evidence that abnormalities in these tests are associated with increased mortality, not only in diabetic autonomic neuropathy, but also other disorders, such as coronary heart disease, heart failure, or chronic liver diseases (American Academy of Neurology, 1996).

V. Natural History and Risk Factors

An adequate design for a RCT must take into account the natural history of the disease. Various problems are encountered in assessing the progression of DSP: (1) Various nerve fiber populations might be affected at different rates; (2) expected changes may take place over several years; (3) minor changes may not be detected due to the suboptimal reproducibility of some measures of neuropathy; (4) nerve function may deteriorate with age in nondiabetic subjects (NCV: 0.5–2 m/s per decade), emphasizing the need for a control comparator; and (5) glycemic control or the risk factor profile may change over time. Overwhelming evidence now suggests that diabetic polyneuropathy develops in strong association with long-term poor glycemic control (Diabetes Control and Complications Trial Research Group, 1993; Ziegler *et al.*, 1996; Forrest *et al.*, 1997; Sands *et al.*, 1997). In population-based studies, the overall incidence rate of DSP was 6.1 per 100 person years in type 2 (Sands *et al.*, 1997) and 2.8 per 100 person years in type 1 diabetic patients (Forrest *et al.*, 1997) derived from a 6-year incidence rate of 15%, with a relative risk of 2.6 in poorly controlled patients vs those who were fairly controlled (Forrest *et al.*, 1997). We have shown that over the first decade of type 1 diabetes, the annual rate of slowing of peroneal MNCV is 0.6 m/s/year and that of sural SNCV is 0.7 m/s/year in poorly controlled vs well-controlled patients (Ziegler *et al.*, 1996). A similar rate of nerve conduction slowing was shown in type 2 patients (Partanen *et al.*, 1995). Macleod *et al.* (1991) estimated the annual rate of change for the vibration perception threshold (VPT) on the great toe to be 0.4 V in healthy subjects and 2.5 V in those with diabetic neuropathy. In RCTs that evaluated the effects of the aldose reductase inhibitor tolrestat on distal-symmetric polyneuropathy, placebo-treated patients had an 82 and 70% increased risk for a decrease in MNCV of 1 and 2 m/s in at least two of four nerves, respectively (Nicolucci *et al.*, 1996). The annual losses in nerve conduction observed in several studies are shown in Table IV. These relatively slowly

TABLE IV

ANNUAL LOSS IN NERVE CONDUCTION IN POORLY CONTROLLED VS WELL-CONTROLLED DIABETIC PATIENTS[a]

	Peroneal MNCV (m/s/year)	Sural SNCV (m/s/year)	Poor vs good glycemic control
Type 1 first decade (Ziegler)	0.6	0.7	9.9 vs 7.8%[b]
Type 2 first decade (Partanen)	0.3 vs baseline	0.8 vs baseline	9.0%[c]
DCCT PP PNP−/PNP+	0.8/0.7	0.4/0.6	9.1 vs 7.2%[c]
DCCT SI PNP−/PNP+	0.7/0.6	0.7/0.1	9.1 vs 7.2%[c]

[a]MNCV, motor nerve conduction velocity; SNCV, sensory nerve conduction velocity; PP, primary prevention cohort; SI, secondary intervention cohort; PNP−, polyneuropathy absent; PNP+, polyneuropathy present.
[b]HbA1a-c.
[c]HbA1c.

developing small changes underline the need for long-term trials using techniques of high precision. The risk factors and risk indicators that need to be considered in the natural history of diabetic neuropathy and their relative impact are summarized in Table V.

TABLE V

RISK FACTORS AND MARKERS OF DIABETIC POLYNEUROPATHY[a]

	Type 1 diabetes	Type 2 diabetes
Age	+	+
Sex	—	—
Height	+	(+)
Weight	—	(+)
Hyperglycemia	++	++
Hypoinsulinemia	Not applicable	+
Duration of diabetes	++	++
Smoking	+	(+)
Alcohol	(+)	(+)
Hyperlipidemia	(+)	(+)
Hypertension	++	(+)
Nephropathy	++	+
Retinopathy	++	+
Cardiovascular autonomic neuropathy	++	++
Macroangiopathy	(+)	(+)

[a]Association strong, ++; moderate, +; controversial; (+); not found, —.

VI. Sample Size and Duration of Trials

The number of patients needed in pharmacological studies to reach meaningful and significant results at the completion of a study depends on the following criteria.

1. The rates of success in both active drug and placebo-treated groups.
2. The absolute difference of success rates between both groups, which depends on the duration of treatment, the severity of neuropathy to be defined for inclusion in a trial, the natural course of end points, and the rate of improvement ever possible, e.g., morphological changes may need a very long time to show any difference between placebo and active treatment arms. Prospective studies of intensified insulin therapy in type 1 diabetic patients (Diabetes Control and Complications Trial Research Group, 1993; Amthor *et al.*, 1994; Reichard *et al.*, 1993), as well as trials evaluating the effects of combined pancreas–kidney transplantation (Navarro *et al.*, 1990; Solders *et al.*, 1992; Muller-Felber *et al.*, 1993), suggest that the longer the diabetes duration and the more advanced the neuropathy, the longer will be the treatment time needed to identify any significant difference between groups. Possibly, stages of predominantly structural destruction may not be reversible at all. Pfeifer and Schumer (1995), therefore, recommended the inclusion of patients with only mild to moderate degrees of neuropathy and to observe these patients over 3–5 years to find clinically relevant changes.
3. The *a priori* specified statistical power, i.e., the probability to detect a difference between both arms of a study at a certain *a priori* specified level of significance.

Negative results from a large number of small trials by no means prove the inefficacy of the treatment under study due to the low number of patients investigated so that the power of the statistical tests used was too low to detect small, but nonetheless real treatment effects (Altman and Bland, 1995). A meta-analysis of randomized parallel group, placebo-controlled clinical trials with negative results showed that only 16% of the studies had a sufficient statistical power of more than 80% to detect a relative difference of 25% between placebo-treated and verum-treated patients. Thirty-six percent of all studies had sufficient power to detect a difference of 50% between the two groups (Moher *et al.*, 1994). Therefore, it is absolutely mandatory to calculate the sample size needed to detect significant differences. Formulas and examples have been published by Campbell *et al.* (1995).

One of the prerequisites of the calculation of the number of patients needed in such studies is the knowledge of the size of the effect, which may be clinically meaningful, implicating that the clinical impact of the measured variable is known, which is not the case using surrogate variables. Cavaliere and co-workers (1994) showed that the median number of patients in studies to treat diabetic neuropathy is 30 (range 9–259), which is too small to demonstrate significant effects. Dyck *et al.* (1997) calculated the sample size needed in a clinical trial based on the 2-year follow-up of the RDNS. The estimates were derived from changes of a composite score, including NIS of the lower limbs plus seven tests [NIS(LL)+7: VPT great toe, R-R variation to DB, peroneal CMAP, MNCV, and MNDL, tibial MNDL, and sural SNAP), which, among other measures, performed best at showing monotone worsening over time. Assuming that a treatment effect of two NIS points is clinically meaningful, a 2-year study would need 68 patients in each treatment arm to have a power of 0.90 at the two-sided 0.05 level. If the effect of treatment is to halt the progression of neuropathy without improving it, a study of 3.7 years would be required. A 4-year study would require 45 patients per arm to achieve power of 0.90 to detect a treatment effect that inhibits the progression of neuropathy, and a clinically meaningful effect could be expected after approximately 2.4 years. Thus, a conservative estimate would yield 70–100 patients per arm for a period of at least 3 years to achieve a high probability of detecting a clinically meaningful effect (Dyck *et al.*, 1997). A 4-year RCT evaluating the effects of α-lipoic in diabetic polyneuropathy has been designed on the basis of these estimates.

Among patients with no/mild impairment of VPT, the annual incidence rate of neuropathic foot ulcers is about 0.8% (Abbott *et al.*, 1998). Nicolucci *et al.* (1996) estimated that if a medication could reduce such a risk by 50%, about 3000 patients should be enrolled in a 5-year RCT ($\alpha = 0.05$, $\beta = 0.80$).

VII. Reproducibility of Outcome Measures

Numerous studies have reported the day-to-day reproducibility of electrophysiologic measures, QST, and AFTs. In large-scale multicenter studies the coefficients of variation were lowest for electrophysiologic measurements, slightly higher for vibration perception threshold, and highest for AFT (ranging from about 5 to 25%). Thus, while the intraindividual variability may be reasonably low, the intercenter variability may be rather high (Valensi *et al.*, 1993; Bril *et al.*, 1998; Santiago *et al.*, 1993; Sundkvist *et al.*, 1992). This in turn leads to an additional increase in the number of patients needed to obtain significant treatment results. Given an equal probability and statistical power in a clinical trial, the number of patients to be included in a study must increase if the expected treatment difference will either be small or the variability of the measurement is high. Hence, at a given probability of $\alpha = 0.05$ and a power of $1-\beta = 0.90$, the numbers in each group of patients must be 42 to detect a difference in NCV of 2.5 m/s, 175 patients are needed to detect a significant difference between amplitudes, and 550 patients are necessary to detect significant differences in VPT (Bril *et al.*, 1998). With regard to the magnitude of the treatment effect, it is obvious that the number of patients in each group needed to detect a difference in NCV must be higher if the difference is smaller, i.e., to detect a difference of 2 and 3 m/s, respectively, these numbers have been calculated to be 262 and 116, respectively (Cavaliere *et al.*, 1994). Regrettably, the number of trials fulfilling these criteria was 2 (5%) and 0, respectively (Cavaliere *et al.*, 1994).

To calculate the number of patients it is therefore mandatory to know exactly the reproducibility of measurements used. Maser and colleagues (1989) have shown that the interobserver agreement varied, depending on what was investigated, e.g., sensory symptoms, sensation, or reflexes. Accordingly, the reproducibility of the NSS is lower than that of the NDS with an intraclass correlation coefficient of 0.95 for the NDS but only of > 0.75 for the NSS (Dyck *et al.*, 1991). These results are corroborated by an Italian group: the coefficient of variation for symptom assessment with a questionnaire was 8–32%, for the neurological examination it was 0–6.5%, and for VPT on the great toe it was 4.4–28% (Gentile *et al.*, 1995). If a tuning fork (Rydel-Seiffer, 128 Hz) was used by the same investigator, the intraindividual coefficient of variation in diabetic patients was 8.4% (Thivolet *et al.*, 1990) in one study and 24% in another (Liniger *et al.*, 1990). Therefore, clinical and psychophysical tests should always be performed at the same site of the patient's body by the same investigator. The difference between both great toes was more than 30% in 24% of all diabetic patients studied (Williams *et al.*, 1988). Generally, intraindividual variabilities either cross-sectionally between contralateral sites or longitudinally are more pronounced in diabetic patients than in normal subjects. In diabetic patients the coefficients of variation of AFTs may also be higher, but this was not observed in all published studies (Kronert *et al.*, 1986). The large variability in psychophysical testing can partially be explained by both changing attention and capability to cooperate, but very large intraindividual variances may point to feigned results (Yarnitsky *et al.*, 1994).

Factors influencing the variation between centers may be differences in patients' characteristics, the equipment used for measurement, the ability to use the equipment, varying skills and experience of technicians involved in measuring, changing the operators during a study, and differing evaluation procedures used in different centers. Ideally, the bias due to all these factors can be reduced by a rigid adherence to central training, identical equipment, the prohibition to change the operators, and centralized evaluation of test results. When interpreting individual values over time, imprecision of measurements, which increases with decreasing absolute readings, must be taken into account. Two values consecutively measured in the same patient can be judged with reasonable certainty to be biologically different only if the difference is larger than the coefficient of variation of this method in the given range of measurements multiplied by 2.6 (Hanseler and Keller, 1994), e.g., if NCV is 37 m/s at the first measurement and 43 m/s at the second measurement, the difference of 6 m/s can be interpreted as being a real change because it exceeds the product of the coefficient of variation 0.05×37 m/s $\times 2.6$, which is 4.8 m/s. The absolute difference of mean conduction velocities in groups of patients over time representing a relevant change is still under discussion. The aforementioned formula does not seem to be appropriate to solve this problem. Dyck and O'Brien (1989) argued that the mean differences between treated and untreated groups of patients do not need to be larger than the reproducibility of the measurement, assuming that the variability in parallel group trials is identical in both groups.

VIII. Nonspecific Effects of Treatment

A. REGRESSION TO THE MEAN

Patients with chronic conditions such as diabetic neuropathy seek medical care and enroll in research studies when symptoms are at their worst. If the value of this first measurement lies at an extreme position of the whole distribution, it is more probable that the next measurement will be more to the center of the distribution than to the more extreme, which may be interpreted as an apparent improvement. Thus, the next change is likely to be an improvement. This tendency of extreme symptoms or findings to return toward the individual's more typical state is known as regression to the mean (Bland and Altman, 1994).

B. PLACEBO EFFECTS

Placebo responses vary greatly and are frequently much higher than the often-cited one-third. Individuals are not consistent in their placebo responses, and a placebo-responder personality has not been identified (Turner *et al.*, 1994). The true placebo effect has to be differentiated from the perceived placebo effect. The latter is a function of several factors, including the true placebo effect and nonspecific effects, including the natural history, regression toward the mean, and other time effects (e.g., increased skill of the investigator) and unidentified parallel interventions (e.g., sensitization of the patient to the problem after inclusion in a trial). In order to obtain the true placebo effect in clinical trials, nonspecific effects can be identified by including an untreated control group (Ernst and Resch, 1995) or by comparing the changes with those of patients on a waiting list.

Placebo effects on pain have been shown repeatedly to be greater than those on other symptoms (Richardson, 1994). However, placebo effects may not only affect subjective variables but also objectively quantifiable ones (Ernst and Resch, 1995). Placebo effects in conjunction with the natural history of the disease and regression to the mean can result in high rates of good outcomes, which may be misattributed to specific treatment effects (Turner *et al.*, 1994; Ernst and Resch, 1995).

Some authors have argued that placebo-controlled trials should no longer be part of the "gold standard" for assessing the efficacy of a new drug. As medical knowledge accumulates, these trials should become infrequent because when an efficacious treatment already exists, it is unethical to assign placebo treatment to patients (Rothman and Michels, 1994). In such situations, one solution is to use an existing drug for the same disease as an active comparator in an equivalence trial. Such trials generally need to be larger than placebo-controlled trials, their standard of conduct needs to be especially high, the handling of withdrawals, losses, and protocol deviations needs more care than usual, and different approaches to analysis and interpretation are appropriate. For example, analysis strategies dealing with unavoidable problems should not center on an intention-to-treat (as randomized) analysis (in contrast to placebo-controlled RCTs), but should seek to show the similarity from a range of approaches (Rothman and Michels, 1994).

When investigating pain treatment, the use of placebo may depend on the pain intensity and the duration of the trial. Problems may arise if manifest neuropathy is treated in pharmacological trials over years. In most studies in the past, metabolic control was maintained constant during the duration of the trial. This, however, may not be possible in the future given the results of the DCCT showing that improved metabolic control is efficient in reducing the incidence and progression of neuropathy in type 1 diabetic patients. These results will not allow to follow patients over years with less than improved metabolic control. This again will further increase the number of patients needed in studies to gain meaningful results after years of treatment. These issues have been discussed very vividly (Clark, 2001).

IX. Measures of Treatment Effects

Problems arising from the definition of treatment effects fall into at least two distinct areas: the definition of treatment effects and adequate presentation, which allows translating results to patients treated outside of studies.

A. DEFINITIONS OF CLINICALLY RELEVANT TREATMENT EFFECTS

The clinically relevant success of drug treatment in studies of diabetic neuropathy is difficult to define and, therefore, discussed controversially. The most frequently used technique, i.e., to investigate surrogate end points, is understandable but problematic (Boissel *et al.*, 1992). A meaningful change in NCV is defined as one that correlates with the minimum unequivocally detectable and relevant change of neuropathic symptoms and deficits. A change in MNCV in the ulnar nerve by 4.6 m/s, median nerve by 2.5 m/s, and peroneal nerve by 2.2 m/s, on average by 2.9 m/s, was equivalent to a change in the neuropathy disability score (NDS) by two points. In a group of type 1 diabetic patients these changes were smaller: a change in peroneal MNCV by 2.0 m/s and, on average, by 2.3 m/s, combining ulnar, median, and peroneal nerves, equaled a change in the NDS by two points. Similar equations can be formulated for changes in the neuropathy symptom score (Dyck *et al.*, 1987). NCV measurements, however, may fluctuate and reflect only the function of the remaining, or hopefully regenerating, large myelinated fibers.

Internal consistency is an important criterion to check for plausibility of treatment-related changes, i.e., changes observed in different nerves do not need to show exactly the same magnitude, but the direction should be concordant. It is unclear how to interpret discordant changes of electrophysiological measurements, e.g., data from an individual subject showing an increase in NCV in one out of four nerves, no change in two others, and a decrease in the fourth. Drawing reliable conclusions on treatment effects in different studies using the same drug may be examined by looking for the patterns of changes that should be similar. Two studies with the aldose reductase inhibitor tolrestat showed discordant results regarding the development of pain and paresthesia. In both studies, pain decreased in the placebo as well as the drug-treated groups, whereas paresthesia improved in one study during drug treatment and during placebo treatment in the other (Macleod *et al.*, 1992; Boulton *et al.*, 1990). A meta-analysis of aldose reductase inhibitor studies (Nicolucci *et al.*, 1996) revealed a large variability between study results regarding the composition of improved nerves and the magnitude of improvement in the same nerve. Internal consistency was lacking in 50% of all studies. A large variability of treatment effects within the same nerve in different studies may point to different populations of patients included, i.e., different stages of neuropathy with different chances to get any improvement. The variability of improvement between nerves in the same study may be due not only to measurement errors and changes by chance in small groups of patients, but also to different stages of neuropathy in different regions of the body. Thus, nerves in the upper part of the body may well improve during therapy, whereas those in the lower part may not.

B. PRESENTATION OF TREATMENT EFFECTS

The relative benefit of an active treatment over a control is usually expressed as the relative risk, the relative risk reduction, or the odds ratio. However, it has been suggested that for clinical decision making, it is more meaningful to use the measure "number needed to treat" (NNT) (Cook and Sackett, 1995). This measure is expressed as the reciprocal of the absolute risk reduction. Laupacis *et al.* (1988) introduced this approach to summarizing the effect of treatment in terms of the number of patients a clinician needs to treat with a particular therapy to expect to prevent one adverse event.

Some drawbacks must be addressed that compromise the usefulness of this number to compare or combine treatment effects of different studies.

1. The mean NNT of a sample of studies is not the average of all weighted NNTs of these studies but is derived as the reciprocal of the arithmetic mean of the weighted absolute risk reductions.
2. In most instances, differing periods of time and different end points or surrogate variables used do not allow for a direct comparison of studies. It is not possible to extrapolate NNTs beyond the time point investigated and to "normalize" NNTs of different studies to one common duration of treatment.
3. In all cases, the 95% confidence interval for the NNT should be given because in studies with small numbers of patients, NNT is a rather crude estimate of the efficacy of the drug used. However, the formula for calculation of the confidence interval given by Sackett and colleagues (1997) may not be appropriate for crossover studies.

4. The absolute number of NNT does not answer the question whether a treatment with proven efficacy will be justified in clinical practice. This can only be decided if prevalence, severity, prognosis, efficacy, side effects, cultural influences on therapy, and cost of treatment are taken into account.

5. NNTs may be overestimated if placebo and drug effects change in the same direction. To use NNTs correctly, it is important to look for the appropriate comparator. In most cases, this will not be a placebo treatment but another drug less effective than the new one. Thus, it has to be computed how many patients must be transferred from the old to the new treatment and which effect can be expected from this change (de Craen *et al.*, 1998).

To judge whether treatment results may justify the transfer to daily practice, it may be necessary to estimate the degree of heterogeneity between studies, which should not exceed that expected by chance alone (L'Abbe *et al.*, 1987). Groups of patients treated in different studies may in fact be suspected to be heterogeneous if the relative risk reduction differs by more than 20%. As an example, it may be 40% in the first study and less than 20% in the second study, or if the difference between the confidence intervals lying farest apart is larger than 5%, such that the lowest risk reduction in study 1 may be 30% and the highest risk reduction in study 2 less than 25%. The confidence intervals can be computed according to Morris and Gardner (1989). If there is heterogeneity between studies, it will be more difficult to generalize results from meta-analyses. Thompson and Pocock (1991) have proposed methods to reduce this problem of heterogeneity.

X. Generalizability of Overall Results of Randomized Clinical Trials (External Validity)

The applicability of study results is generally restricted because, for example, all patients with concurrent or earlier disorders or certain concomitant medications are excluded due to safety reasons. Strictly speaking, with the exception of studies in which $n = 1$, the results of RCTs cannot be applied to individuals. A single patient cannot experience a percentage reduction in death, which can only be calculated from an analysis of groups of similar patients. However, patients included in a RCT are heterogeneous, i.e., they differ in the severity of neuropathy and consequently in the absolute risk of poor outcome. In general, the aim of treatment is to target individuals who are at high risk of a poor outcome without treatment but who are also at low risk of a poor outcome with treatment. A treatment that produces an overall risk reduction, but also has significant morbidity or mortality, may be harmful in patients at low risk. Generalization of the overall trial result to all future individual patients similar to those in the whole trial assumes that groups at high and low risk cannot be identified at baseline. The applicability of trial results to individual patients can be improved by knowledge of the association between relative treatment effect and absolute baseline risk to identify those patients in whom treatment is ineffective and those who are most likely to benefit (e.g., by prognostic models based on multiple regression analysis) (Rothwell, 1995).

The method of recruitment of patients (through newspaper advertisements, from a specialized or a general out-patient clinic, from an unselected population-based group) may influence extrapolation to wider patient groups in a covert manner. Presumably, groups of study patients are more compliant than the "general" patient, which led Haynes and Dantes (1987) to conclude that "it is clearly not reasonable to generalize the results of a study among compliant volunteers to all people with similar disorders since the majority of people will be noncompliant under usual conditions," which implies that study results are generally better than what might be possible in general practice. Even if a clinical outcome was investigated with adequate methods and found to be beneficially influenced, it has to be decided what size of effect is necessary to use an agent in general practice. This threshold above which treatment will be both indicated and beneficial and below which it should not be used can be constructed, at least from an economical point of view, by the formulation of two relevant end points: (1) the clinical outcome that is to be prevented by treatment and (2) the unwanted side effects that may happen during treatment. Then it may be possible to calculate and compare the costs, which are either spared by preventing clinical outcomes or necessary to treat unwanted side effects. This approach, however, denies that the quality of treatment effect and side effects cannot be counted only in costs.

If drugs are tested in studies for the treatment of diabetic neuropathy that are thought to act via a hypothetical metabolic pathway that may be disturbed in diabetic neuropathy, sometimes a dilemma emerges because treatment is aimed not on symptomatic improvement but on structural improvement, which may take years to become detectable. On the one hand, these agents would act best in diabetic patients with the least structural defects, i.e., in very early stages of diabetic neuropathy, e.g., stage 0 or stage 1 in which the probability to develop symptomatic neuropathy may be about 30 to 40%. On the other hand, the fact that the spectrum, frequency, and severity of side effects are only insufficiently known may preclude the use of a new drug in asymptomatic patients who will not inevitably become ill over time. New drugs may, therefore, be investigated only in patients already suffering from diabetic neuropathy, which itself may prevent a meaningful treatment effect due to its advanced stage. This would increase the duration of the study, the number of patients needed to find smaller but nonetheless relevant effects, and the study costs. Although it would be better to test new drugs in patients who would benefit most, it should be kept in mind that compounds that have been tested during the last few years have shown unwanted, sometimes disastrous, adverse reactions (Spielberg *et al.*, 1991; Raschetti *et al.*, 1995; Freedman *et al.*, 1999), whereas the beneficial effects were of only minor significance.

XI. Reporting of Randomized Clinical Trials

In response to increasing evidence that reporting of RCTs is imperfect, the Standards of Reporting Trials (SORT) group and the Asilomar Working Group on Recommendations for Reporting of Clinical Trials in the Biomedical Literature convened a meeting that resulted in the Consolidated Standards of Reporting Trials (CONSORT) statement (Moher *et al.*, 2001). This statement includes a checklist and a flow diagram. The checklist consists of 21 items that should be included in a report, and the flow chart describes patient progress through the trial. It is expected that in addition to improved reporting, the CONSORT statement will improve the conduct of future research by increasing awareness of the requirements for a good trial (Altman, 1996).

XII. Current State of Pharmacological Treatments Based on Pathogenetic Concepts

A. ALDOSE REDUCTASE INHIBITORS

Detailed consideration of the biochemistry underlying these drugs is given in the chapter by Oates (this volume). Only a brief background is given here. An increased flux through the polyol pathway, resulting in multiple biochemical abnormalities in the diabetic nerve, is thought to play a major role in the pathogenesis of diabetic neuropathy. Aldose reductase inhibitors (ARIs) block the increased activity of aldose reductase, the rate-limiting enzyme that converts glucose to sorbitol. The first trials of ARIs in diabetic neuropathy were published in the 1980s. The various compounds that have been evaluated are alrestatin, sorbinil, ponalrestat, tolrestat, epalrestat, zopolrestat, zenarestat, and fidarestat. Except for epalrestat, which is marketed in Japan, none of these agents could be permanently licensed due to serious adverse events (sorbinil, tolrestat, zenarestat) or lack of efficacy (ponalrestat, zopolrestat). A meta-analysis of 13 clinical trials with ARIs revealed a marginal effect on peroneal motor NCV of 1.24 m/s and an even weaker effect on median motor NCV of 0.69 m/s after 1 year (Nicolucci *et al.*, 1996a). Data of 738 subjects from three trials of tolrestat showed a benefit equal to 1 m/s in a pooled analysis of NCV in all the nerves studied (Nicolucci *et al.*, 1996b). The following degrees of changes in motor and sensory NCV that are associated with a change in the neuropathy impairment score of two points have been considered to be clinically meaningful in controlled clinical trials: median motor NCV, 2.5 m/s; ulnar MNCV, 4.6 m/s; peroneal MNCV, 2.2 m/s; median SNCV, 1.9 m/s; and sural SNCV, 5.6 m/s (Dyck and O'Brien, 1989). According to this suggestion, changes in NCV obtained from the ARI trials so far do not appear to reflect a meaningful magnitude of a treatment effect. In a 1-year phase II trial of zenarestat including 208 patients with diabetic polyneuropathy, a dose-dependent improvement in small myelinated fiber loss and peroneal NCV was observed (Greene *et al.*, 1999), but subsequent large phase III trials of zenarestat had to be prematurely terminated due to a significant deterioration in renal function in some patients.

A 52-week controlled multicenter trial of fidarestat (1 mg/day) including 279 patients with diabetic polyneuropathy showed an improvement in F-wave conduction velocity and a reduction in neuropathic symptoms (Hotta *et al.*, 2000). No significant adverse reactions to fidarestat were observed in this trial. The results of ongoing phase III trials remain to be seen.

B. γ-LINOLENIC ACID

Two multicenter trials have demonstrated improvement in neuropathic deficits and NCV after 1 year of treatment with γ-linolenic acid (GLA) in diabetic peripheral neuropathy (Keen *et al.*, 1993; Horrobin, 1997). However, because GLA could not be licensed on the basis of these data in the United Kingdom, no further trials have been initiated.

C. α-LIPOIC ACID (THIOCTIC ACID)

Accumulating evidence suggests that free radical-mediated oxidative stress is implicated in the pathogenesis of diabetic neuropathy by inducing neurovascular defects that result in endoneurial hypoxia and subsequent nerve dysfunction (Cameron and Cotter, 1997). Antioxidant treatment with α-lipoic acid has been shown to prevent these abnormalities in experimental diabetes (Cameron *et al.*, 1998; Hounsom *et al.*, 1998), thus providing a rationale for a potential therapeutic value in diabetic patients. In Germany, α-lipoic acid is licensed and used for the treatment of symptomatic diabetic neuropathy since the 1960s. Thus far, five randomized, placebo-controlled clinical trials have been published, suggesting the following: (1) Short-term treatment for 3 weeks using 600 mg of thioctic acid iv per day appears to reduce the chief neuropathic symptoms, including pain, paresthesia, and numbness. A 3-week pilot study of 1800 mg per day indicates that the therapeutic effect may be independent of the route of administration, but this needs to be confirmed in a larger sample size. (2) Three-week treatment also improves neuropathic deficits, and subsequent oral treatment for 4–7 months tends to reduce neuropathic deficits and improves cardiac autonomic neuropathy. (3) Preliminary data over 2 years indicate possible long-term improvement in motor and sensory NCV in the lower limbs. (4) Clinical and postmarketing surveillance studies have revealed a highly favorable safety profile of the drug (Ziegler *et al.*, 1999). Another short-term study has been completed that used 600 mg of α-lipoic acid iv over 3 weeks in type 1 and type 2 diabetic inpatients (SYDNEY study: SYDNEY Study Group, personal communication). Two large multicenter trials are being conducted in North America and Europe to verify the results of the ALADIN studies (NATHAN 2 study) and to evaluate the efficacy and safety of long-term treatment with α-lipoic acid over 4 years on neuropathic deficits (NATHAN 1 study). The design of the NATHAN 1 study is summarized in Table VI. At present, this is the only long-term study ever conducted over several years to evaluate whether pathogenetically oriented drug treatment may slow the progression of DSP.

Regarding other antioxidants, a preliminary study including 21 patients with symptomatic polyneuropathy showed that vitamin E may improve motor but not sensory NCV after 6 months, but it was not reported whether neuropathic symptoms were influenced (Tütüncü *et al.*, 1998).

D. VASODILATORS

Microvascular changes of the vasa nervorum and reduced endoneurial blood flow resulting in hypoxia are thought to be important factors in the pathogenesis of diabetic neuropathy (Cameron and Cotter, 1997). Thus, there is solid theoretical background to support treatment with vasodilating drugs. In a 1-year trial including 41 normotensive patients with mild neuropathy, several attributes of NCV, but not neuropathic symptoms

TABLE VI

DESIGN OF THE NEUROLOGICAL ASSESSMENT OF THIOCTIC ACID (α-LIPOIC ACID) IN DIABETIC NEUROPATHY (NATHAN) 1 STUDY[a]

Design Randomized, double-blind, placebo-controlled, prospective, parallel group, multicenter trial

Subjects Two parallel groups of type 1 or type 2 diabetic patients ($n = 500$ enrolled)

Medication Thioctic acid 600-mg or placebo tablets once daily orally

Duration Screening: 2 weeks, placebo run in: 6 weeks, treatment: 192 weeks, follow-up: 4 weeks (interim analysis: 96 weeks)

Inclusion criteria Stage 1 or 2a polyneuropathy [NIS(LL)+7 tests score ≥ 97.5 centile; TSS≤ 5]

Primary outcome measure NIS(LL)+7 tests score (VDT, HBDB, peroneal CMAP, MNCV, and MNDL, tibial MNDL, sural SNAP)

Secondary outcome measures NSC, TSS, CDT, HP, other NIS and NC

[a]NIS(LL), neuropathy impairment score (lower limbs); VDT, vibration detection threshold; HBDB, heart beat deep breathing; CMAP, compound muscle action potential; MNCV, motor nerve conduction velocity; MNDL, motor nerve distal latency; SNAP, sensory nerve action potential; NSC, neuropathy symptoms and changes; TSS, total symptom score; HP, heat pain; NC, nerve conduction.

and deficits, were improved after 1 year of treatment with the ACE inhibitor trandolapril (Malik *et al.*, 1998). Further studies are clearly needed to define the therapeutic role of ACE inhibitors in diabetic neuropathy.

Several open-label trials from Japan reported pain relief after treatment with vasodilating agents such as the prostacyclin (PGI_2) analogs iloprost or beraprost and prostaglandin E_1 (PGE_1) reported relief of pain or dysesthetic symptoms after 2,12, and 4 weeks, respectively. Due to uncontrolled study designs, these effects are uninterpretable. However, a large controlled multicenter trial including 170 patients with symptomatic polyneuropathy or foot ulcers showed a $> 50\%$ improvement in pain or other neuropathic symptoms in 56% of the patients treated with an iv infusion of PGE_1 incorporated in lipid microspheres (lipo-PGE_1) for 4 weeks compared to 28% on placebo. In a second trial comparing lipo-PGE_1 with PGE_1 in 194 patients, the corresponding rates were 51 and 35%. Side effects were observed in 7% of the patients treated with lipo-PGE_1 (Toyota *et al.*, 1993). Further studies are needed to confirm these findings.

E. NERVE GROWTH FACTOR

The reader is referred to the chapter by Apfel in this volume for further information on NGF and its clinical trials. NGF selectively promotes the survival, differentiation, and maintenance of small fiber sensory and sympathetic neurons in the peripheral nervous system. It is expressed in the skin and other target tissues of its responsive neuronal populations, binds to its high-affinity receptor (trk A) on nerve terminals, and exerts its trophic effects after being retrogradely transported back to the neuronal perikaryon (Fernyhough and Tomlinson, 1999). A 6-month phase II trial including 250 patients with symptomatic diabetic neuropathy showed an improvement of the sensory component of the neurologic examination and both cooling detection and heat as pain threshold, but no effect on neuropathic symptoms could be observed following treatment with recombinant human NGF (Apfel *et al.*, 1998). In contrast, a subsequent large 12-month phase III trial failed to demonstrate a favorable effect of rhNGF on subjective and objective variables of diabetic neuropathy (Apfel *et al.*, 2000). The reasons for the latter disappointing result could be the following: (1) the DSP did not progress during the trial in the placebo group, (2) the dose chosen may have been below the threshold to produce an effect, (3) the most distal testing site (big toe) was selected for assessment, where the most advanced neuropathic changes are expected, which are less susceptible to intervention than more proximal sites, (4) the primary outcome measure [neuropathy impairment score at the lower limbs (NIS-LL)] is not sensitive to small fiber sensory dysfunction, (5) the drug did not get to the target tissue, and (6) the manufacturing process for NGF has been altered after the phase II trial prior to the phase III trial, leaving the possibility that the drug was not identical (Apfel *et al.*, 2000).

F. Protein Kinase C ß Inhibitor

Increased activity of protein kinase C (PKC), a family of serine-threonine kinases that regulate various vascular functions, including contractility, hemodynamics, and cellular proliferation, has been implicated in the pathogenesis of diabetic complications, including neuropathy (Koya and King, 1998). Treatment with a PKC-ß-selective inhibitor ameliorated several neuropathic deficits in experimental diabetic neuropathy (Nakamura *et al.*, 1999). Clinical trials using this agent are currently underway.

G. C Peptide

Studies suggest that C peptide shows specific binding to cell membrane-binding sites and augments skin microcirculation in type 1 diabetic patients (Forst *et al.*, 1998) possibly via an increase in both nitric oxide production and Na^+/K^+-ATPase activity (Forst *et al.*, 2000). In experimental diabetic neuropathy, C-peptide administration prevented the NCV deficit, axonal atrophy, and paranodal swelling and demyelination and produced an increase in Na^+/K^+-ATPase activity and phosphorylation of the insulin receptor (Sima, 2000). A pilot study showed an improvement in small fiber sensory and autonomic function in type 1 diabetic patients (Johansson *et al.*, 2000). Phase II and phase III trials in diabetic neuropathy are needed to confirm these preliminary data.

H. VASCULAR ENDOTHELIAL GROWTH FACTOR

Based on the experimental concept of endoneurial microvascular abnormalities and reduced nerve blood flow resulting in ischemia and hypoxia, it has been hypothesized that destruction of the vasa nervorum can be reversed by administration of vascular endothelial growth factor (VEGF), an endothelial cell mitogen that promotes angiogenesis in several animal models and in humans. Intramuscular gene transfer of plasmid DNA encoding VEGF-1 or VEGF-2 reversed experimental neuropathy after 4 weeks in diabetic rats (Schratzberger *et al.*, 2001). Preliminary data in patients with chronic ischemic neuropathy and critical limb ischemia indicate neurologic improvement in four out of six diabetic patients after 6 months following intramuscular phVEGF165 gene transfer (Simovic *et al.*, 2001). However, caution has been expressed regarding possible adverse effects of VEGF such as retinal neovascularization and increased retinal vascular permeability, induction of peripheral edema, activation of the PKC pathway, and the possible mitogenic effects in tumor development (Veves and King, 2001). Thus, provided that VEGF will be evaluated in larger scale clinical trials, a close monitoring of these and other possible consequences will be mandatory.

XIII. Conclusions

Problems in one or more of the aforementioned areas may explain the relatively scarce information on the effective treatment of diabetic neuropathies. Most important appear to be the lack of homogeneity of patients studied with regard to both the form of neuropathy and the degree of metabolic control, different pathogenetic pathways, the relative importance of which may vary intraindividually, advanced stages of neuropathy, which may preclude any significant improvement, the use of end points with rather large variability between individuals and between centers, the unknown relevance of end points used, study durations too short to allow for a significant morphological improvement, and potentially the reliance on pathogenetic hypotheses derived from animal studies (Cameron and Cotter, 1993), the significance of which for the pathogenesis of human diabetic neuropathy remains a matter of debate (Thomas, 1986; Tomlinson, 1989; Hounsom and Tomlinson, 1997; Malone *et al.*, 1996).

References

Abbott, C. A., Vileikyte, L., Williamson, S., Carrington, A. L. and Boulton, A. J. M. (1998). Multicenter study of the incidence of and predictive risk factors for diabetic neuropathic foot ulceration. *Diabet. Care* **21**, 1071–1075.

Albers, J. W., Andersen, H., Arezzo, J. C., Asbury, A., Bolton, C., Boulton, A. J. M., Bril, V., Brown, M., Brownlee, M., Canal, N., Chalk, C., Donaghy, M., Dyck, P. J., Feasby, T., Giannini, C., Goto, Y., Grant, I., Greene, D., Griffin, J., Hahn, A., Harati, Y., Hoffman, P., Hughes, R., Jakobsen, J., Jaradeh, S., Kennedy, W. R., Litchy, W., Llewellyn, G., Low, P. A., McLean, W. G., Malik, R., Mayer, R. F., Midroni, G., Mendell, J. R., Mohiuddin, L., Nagamatsu, M., Nukada, H., O'Brien, P., Ohi, T., Ohnishi, A., Parry, G., Pascoe, M., Porte, D., Pollock, M., Powell, H., Rizza, R. A., Rubenstein, A. H., Said, G., Saida, K., Schaumburg, H. H., Schroder, J. M., Service, F. J., Sidenius, P., Sima, A. A. F., Suarez, G., Thomas, P. K., Tomlinson, D. R., Toyka, K. V., Ward, J. D., Watkins, P. J., Windebank, A. J., Wright, A., Yagihashi, S., Yamamoto, T., Ziegler, D., Zimmerman, B. R., and Zochodne, D. W. (1995). Diabetic polyneuropathy in controlled clinical trials: Consensus report of the peripheral nerve society. *Ann. Neurol.* **38**, 478–482.

Altman, D. G. (1996). Better reporting of randomised controlled trials: The CONSORT statement. *Br. Med. J.* **313**, 570–571.

Altman, D. G., and Bland, J. M. (1995). Absence of evidence is not evidence of absence. *Br. Med. J.* **311**, 485.

American Academy of Neurology. (1996). Assessment: Clinical autonomic testing report of the Therapeutics and Technology Assessment Subcommittee of the American Academy of Neurology. *Neurology* **46**, 873–880.

Amthor, K.-F., Dahl-Jorgensen, K., Berg, T. J., Skard Heier, M., Sandvik, L., Aagenæs, O., and Hanssen, K. F. (1994). The effect of 8 years of strict glycaemic control on peripheral nerve function in IDDM patients: The Oslo study. *Diabetologia* **37**, 579–584.

Apfel, S. C., Kessler, J. A., Adornato, B. T., Litchy, W. J., Sanders, C., Rask, C. A., and NGF, S. G. (1998). Recombinant human nerve growth factor in the treatment of diabetic polyneuropathy. *Neurology* **51**, 695–702.

Apfel, S. C., Schwartz, S., Adornato, B. T., Freeman, R., Biton, V., Rendell, M., Vinik, A., Giuliani, M., Stevens, J. C., Barbano, R., Dyck, P. J., and RhNGF Clin Investigator Group. (2000). Efficacy and safety of recombinant human nerve growth factor in patients with diabetic polyneuropathy: A randomized controlled trial. *JAMA* **284**, 2215–2221.

Bland, J. M., and Altman, D. G. (1994). Some examples of regression towards the mean. *Br. Med. J.* **309**, 780.

Boissel, J. P., Collet, J. P., Moleur, P., and Haugh, M. (1992). Surrogate endpoints: A basis for a rational approach. *Eur. J. Clin. Pharmacol.* **43**, 235–244.

Boulton, A. J. M., Levin, S., and Comstock, J. (1990). A multicentre trial of the aldose-reductase inhibitor, tolrestat, in patients with symptomatic diabetic neuropathy. *Diabetologia* **33**, 431–437.

Bril, V., Ellison, R., Ngo, M., Bergstrom, B., Raynard, D., Gin, H., and Roche, N. S. G. (1998). Electrophysiological monitoring in clinical trials. *Muscle Nerve* **21**, 1368–1373.

Cameron, N. E., and Cotter, M. A. (1993). Potential therapeutic approaches to the treatment or prevention of diabetic neuropathy: Evidence from experimental studies. *Diabet. Med.* **10**, 593–605.

Cameron, N. E., and Cotter, M. A. (1997). Metabolic and vascular factors in the pathogenesis of diabetic neuropathy. *Diabetes* **46**, S31–S37.

Cameron, N. E., Cotter, M. A., Horrobin, D. H., and Tritschler, H. J. (1998). Effects of α-lipoic acid on neurovascular function in diabetic rats: Interaction with essential fatty acids. *Diabetologia* **41**, 390–399.

Campbell, M. J., Julious, S. A., and Altman, D. G. (1995). Estimating sample sizes for binary, ordered categorical, and continuous outcomes in two group comparisons. *Br. Med. J.* **311**, 1145–1148.

Cavaliere, D., Belfiglio, M., Carinci, F., Cubasso, D., Labbrozzi, D., Mari, E., Massi Benedetti, M., Pontano, C., Scorpiglione, N., Tognoni, G., and Nicolucci, A. (1994). Quality assessment of randomized clinical trials on medical treatment of diabetic neuropathy. *Diabet. Nutr. Metab.* **7**, 287–294.

Chalmers, T. C., Smith, H. J., Blackburn, B., Silverman, B., Schroeder, B., Reitman, D., and Ambroz, A. (1981). A method for assessing the quality of a randomized control trial. *Control Clin. Trials* **2**, 31–49.

Chan, A. W., MacFarlane, I. A., Bowsher, D., Wells, J. C., Bessex, C., and Griffiths, K. (1990). Chronic pain in patients with diabetes mellitus: Comparison with a non-diabetic population. *Pain Clin.* **3**, 147–159.

Clark, C. M., Jr. (2001). Introduction to commentaries on the use of placebo-controlled trails of new therapies in the treatment of type 2 diabetes. *Diabet. Care* **24**, 768–768.

Consensus Statement. (1998). Standardized measures in diabetic neuropathy. *Diabet. Care* **15**, 1080–1107.

Cook, R. J., and Sackett, D. L. (1995). The number needed to treat: A clinically useful measure of treatment effect. *Br. Med. J.* **310**, 452–454.

Coppini, D. V., Bowtell, P. A., Weng, C., Young, P. J., and Sönksen, P. H. (2000). Showing neuropathy is related to increased mortality in diabetic patients: A survival analysis using an accelerated failure time model. *J. Clin. Epidemiol.* **53**, 519–523.

de Craen, A. J., Vickers, A. J., Tijssen, J. G., and Kleijnen, J. (1998). Number-needed-to-treat and placebo-controlled trials. *Lancet* **351**, 310.

Diabetes Control and Complications Trial Research Group. (1993). The effect of intensive treatment of diabetes on the development and progression of long-term complications in insulin-dependent diabetes mellitus. *N. Engl. J. Med.* **329**, 977–986.

Dyck, P. J., Bushek, W., Spring, E. M., Karnes, J. L., Litchy, W. J., O'Brien, P. C., and Service, F. J. (1987). Vibratory and cooling detection thresholds compared with other tests in diagnosing and staging diabetic neuropathy. *Diabet. Care* **10**, 432–440.

Dyck, P. J., Davies, J. L., Litchy, W. J., and O'Brien, P. C. (1997). Longitudinal assessment of diabetic polyneuropathy using a composite score in the Rochester diabetic neuropathy study cohort. *Neurology* **49**, 229–239.

Dyck, P. J., Kratz, K. M., Karnes, J. L., Litchy, W. J., Klein, R., Pach, J. M., Wilson, D. M., O'Brien, P. C., and Melton, L. J., III (1993). The prevalence by staged severity of various types of diabetic neuropathy, retinopathy, and nephropathy in a population-based cohort: The Rochester Diabetic Neuropathy Study. *Neurology* **43**, 817–824.

Dyck, P. J., Kratz, K. M., Lehman, K. A., Karnes, J. L., Melton, L. J., III, O'Brien, P. C., Litchy, W. J., Windebank, A. J., Smith, B. E., Low, P. A., Service, F. J., Rizza, R. A., and Zimmerman, B. R. (1991). The Rochester diabetic neuropathy study: Design, criteria for types of neuropathy, selection bias, and reproducibility of neuropathic tests. *Neurology* **41**, 799–807.

Dyck, P. J., and O'Brien, P. C. (1989). Meaningful degrees of prevention or improvement of nerve conduction in controlled clinical trials of diabetic neuropathy. *Diabet. Care* **12**, 649–652.

Ernst, E., and Resch, K. L. (1995). Concept of true and perceived placebo effects. *Br. Med. J.* **311**, 551–553.

Fernyhough, P., and Tomlinson, D. R. (1999). The therapeutic potential of neurotrophins for the treatment of diabetic neuropathy. *Diabet. Rev.* **7**, 300–311.

Fleming, T. R., and DeMets, D. L. (1996). Surrogate end points in clinical trials: Are we being misled? *Ann. Intern. Med.* **125**, 605–613.

Forrest, K. Y., Maser, R. E., Pambianco, G., Becker, D. J., and Orchard, T. J. (1997). Hypertension as a risk factor for diabetic neuropathy: A prospective study. *Diabetes* **46**, 665–670.

Forsblom, C. M., Sane, T., Groop, P. H., Totterman, K. J., Kallio, M., Saloranta, C., Laasonen, L., Summanen, P., Lepantalo, M., Laatikainen, L., Matikainen, E., Teppo, A. M., Koskimies, S., and Groop, L. (1998). Risk factors for mortality in type II (non-insulin-dependent) diabetes: Evidence of a role for neuropathy and a protective effect of HLA-DR4. *Diabetologia* **41**, 1253–1262.

Forst, T., De La Tour, D. D., Kunt, T., Pfutzner, A., Goitom, K., Pohlmann, T., Schneider, S., Johansson, B. L., Wahren, J., Lobig, M., Engelbach, M., Beyer, J., and Vague, P. (2000). Effects of proinsulin C-peptide on nitric oxide, microvascular blood flow and erythrocyte Na^+,K^+-ATPase activity in diabetes mellitus type I. *Clin. Sci. (Colch.)* **98**, 283–290.

Forst, T., Kunt, T., Pohlmann, T., Goitom, K., Engelbach, M., Beyer, J., and Pfutzner, A. (1998). Biological activity of C-peptide on the skin microcirculation in patients with insulin-dependent diabetes mellitus. *J. Clin. Invest.* **101**, 2036–2041.

Freedman, B. I., Wuerth, J. P., Cartwright, K., Bain, R. P., Dippe, S., Hershon, K., Mooradian, A. D., and Spinowitz, B. S. (1999). Design and baseline characteristics for the aminoguanidine clinical trial in overt type 2 diabetic nephropathy (ACTION II). *Control Clin. Trials* **20**, 493–510.

Galer, B. S., Gianas, A., and Jensen, M. P. (2000). Painful diabetic polyneuropathy: Epidemiology, pain description, and quality of life. *Diabet. Res. Clin. Pract.* **47**, 123–128.

Gentile, S., Turco, S., Corigliano, G., and Marmo, R. (1995). Simplified diagnostic criteria for diabetic distal polyneuropathy: Preliminary data of a multicentre study in the Campania region. S.I.M.S.D.N. Group. *Acta Diabetol.* **32**, 7–12.

Greene, D. A., Arezzo, J. C., Brown, M. B., and the Zenarestat Study Group. (1999). Effect of aldose reductase inhibition on nerve conduction and morphometry in diabetic neuropathy. *Neurology* **53**, 580–591.

Hanseler, E., and Keller, H. (1994). Rational evaluation of laboratory data: Problems and limits. *Internist (Berl.)* **35**, 609–618.

Haynes, R. B., and Dantes, R. (1987). Patient compliance and the conduct and interpretation of therapeutic trials. *Control Clin. Trials* **8**, 12–19.

Horrobin, D. F. (1997). Essential fatty acids in the management of impaired nerve function in diabetes. *Diabetes* **46**, S90–S93.

Hotta, N., Ishii, J., and Sakamoto, N. (2000). Effects of fidarestat, an aldose reductase inhibitor, on diabetic peripheral neuropathy: A 52-week placebo-controlled double-blind study. *Diabetes* **49**, A35–A35. [Abstract]

Hounsom, L., Horrobin, D. F., Tritschler, H., Corder, R., and Tomlinson, D. R. (1998). A lipoic acid-gamma linolenic acid conjugate is effective against multiple indices of experimental diabetic neuropathy. *Diabetologia* **41**, 839–843.

Hounsom, L., and Tomlinson, D. R. (1997). Does neuropathy develop in animal models? *Clin. Neurosci.* **4**, 381–389.

Johansson, B. L., Borg, K., Fernqvist-Forbes, E., Kernell, A., Odergren, T., and Wahren, J. (2000). Beneficial effects of C-peptide on incipient nephropathy and neuropathy in patients with type 1 diabetes mellitus. *Diabet. Med.* **17**, 181–189.

Keen, H., Payan, J., Allawi, J., Walker, J., Jamal, G. A., Weir, A. I., Henderson, L. M., Bissessar, E. A., Watkins, P. J., Sampson, M., Gale, E. A. M., Scarpello, J., Boddie, H. G., Hardy, K. J., Thomas, P. K., Misra, P., and Halonen, J.-P. (1993). Treatment of diabetic neuropathy with γ-linolenic acid. *Diabet. Care* **16**, 8–15.

Koya, D., and King, G. L. (1998). Protein kinase C activation and the development of diabetic complications. *Diabetes* **47**, 859–866.

Kronert, K., Luft, D., Baumann, B., Muller, P. H., and Eggstein, M. (1986). Reduced intraindividual variability of repeated cardiovascular reflex tests: An additional marker of autonomic neuropathy in insulin-dependent (type I) diabetes mellitus. *Acta Diabetol. Lat.* **23**, 279–289.

L'Abbe, K. A., Detsky, A. S., and O'Rourke, K. (1987). Meta-analysis in clinical research. *Ann. Intern. Med.* **107**, 224–233.

Laupacis, A., Sackett, D. L., and Roberts, R. S. (1988). An assessment of clinically useful measures of the consequences of treatment. *N. Engl. J. Med.* **318**, 1728–1733.

Liniger, C., Albeanu, A., Bloise, D., and Assal, J. P. (1990). The tuning fork revisited. *Diabet. Med.* **7**, 859–864.

Luft, D. (1998). Interpretation of clinical trials for the treatment of diabetic neuropathy. *Drugs Today* **34**, 157–175.

Macleod, A. F., Boulton, A. J., Owens, D. R., Van Rooy, P., Van Gerven, J. M., Macrury, S., Scarpello, J. H., Segers, O., Heller, S. R., and Van, D., V (1992). A multicentre trial of the aldose-reductase inhibitor tolrestat, in patients with symptomatic diabetic peripheral neuropathy. North European Tolrestat Study Group. *Diabet. Metab.* **18**, 14–20.

Macleod, A. F., Till, S., and Sönksen, P. H. (1991). Discussion of the clinical trials of aldose reductase inhibitor tolrestat. *Int. Proc. J.* **4**, 17–24.

Malik, R. A. (2000). Can diabetic neuropathy be prevented by angiotensin-converting enzyme inhibitors? *Ann. Med.* **32**, 1–5.

Malik, R. A., Williamson, S., Abbott, C., Carrington, A. L., Iqbal, J., Schady, W., and Boulton, A. J. M. (1998). Effect of angiotensin-converting-enzyme (ACE) inhibitor trandolapril on human diabetic neuropathy: Randomised doubleblind controlled trial. *Lancet* **352**, 1978–1981.

Malone, J. I., Lowitt, S., Korthals, J. K., Salem, A., and Miranda, C. (1996). The effect of hyperglycemia on nerve conduction and structure is age dependent. *Diabetes* **45**, 209–215.

Maser, R. E., Nielsen, V. K., Bass, E. B., Manjoo, Q., Dorman, J. S., Kelsey, S. F., Becker, D. J., and Orchard, T. J. (1989). Measuring diabetic neuropathy: Assessment and comparison of clinical examination and quantitative sensory testing. *Diabet. Care* **12**, 270–275.

Moher, D., Dulberg, C. S., and Wells, G. A. (1994). Statistical power, sample size, and their reporting in randomized controlled trials. *JAMA* **272**, 122–124.

Moher, D., Schulz, K. F., Altman, D. G., and Lepage, L. (2001). The CONSORT statement: Revised recommendations for improving the quality of reports of parallel-group randomised trials. *Lancet* **357**, 1191–1194.

Molenaar, D. S., de Haan, R., and Vermeulen, M. (1995). Impairment, disability, or handicap in peripheral neuropathy: Analysis of the use of outcome measures in clinical trials in patients with peripheral neuropathies. *J. Neurol. Neurosurg. Psychiat.* **59**, 165–169.

Morris, J. A., and Gardner, M. J. (1989). Calculating confidence intervals for relative risks, odds ratios, and standardised ratios and rates. *In* "Statistics with Confidence: Confidence Intervals and Statistical Guidelines" (M. J. Gardner and D. G. Altman, eds.), pp. 50–63 *British Medical Journal*, London.

Muller-Felber, W., Landgraf, R., Scheuer, R., Wagner, S., Reimers, C. D., Nusser, J., Abendroth, D., Illner, W. D., and Land, W. (1993). Diabetic neuropathy 3 years after successful pancreas and kidney transplantation. *Diabetes* **42**, 1482–1486.

Nakamura, J., Kato, K., Hamada, Y., Nakayama, M., Chaya, S., Nakashima, E., Naruse, K., Kasuya, Y., Mizubayashi, R., Miwa, K., Yasuda, Y., Kamiya, H., Ienaga, K., Sakakibara, F., Koh, N., and Hotta, N. (1999). A protein kinase C-ß-selective inhibitor ameliorates neural dysfunction in streptozotocin-induced diabetic rats. *Diabetes* **48**, 2090–2095.

Navarro, X., Kennedy, W. R., Loewenson, R. B., and Sutherland, D. E. R. (1990). Influence of pancreas transplantation on cardiorespiratory reflexes, nerve conduction, and mortality in diabetes mellitus. *Diabetes* **39**, 802–806.

Nicolucci, A., Carinci, F., Cavaliere, D., Scorpiglione, N., Belfiglio, M., Labbrozzi, D., Mari, E., Benedetti, M. M., Tognoni, G., and Liberati, A. (1996a). A meta-analysis of trials on aldose reductase inhibitors in diabetic peripheral neuropathy. The Italian Study Group. The St. Vincent Declaration. *Diabet. Med.* **13**, 1017–1026.

Nicolucci, A., Carinci, F., Graepel, J. G., Hohman, T. C., Ferris, F., and Lachin, J. M. (1996b). The efficacy of tolrestat in the treatment of diabetic peripheral neuropathy: A meta-analysis of individual patient data. *Diabet. Care* **19**, 1091–1096.

Partanen, J., Niskanen, L., Lehtinen, J., Mervaala, E., Siitonen, O., and Uusitupa, M. (1995). Natural history of peripheral neuropathy in patients with non-insulin-dependent diabetes mellitus. *N. Engl. J. Med.* **333**, 89–94.

Peripheral Neuropathy Association. (1993). Quantitative sensory testing: A consensus report from the Peripheral Neuropathy Association. *Neurology* **43**, 1050–1052.

Pfeifer, M. A., and Schumer, M. P. (1995). Clinical trials of diabetic neuropathy: Past, present, and future. *Diabetes* **44**, 1355–1361.

Raschetti, R., Maggini, M., Popoli, P., Caffari, B., Da Cas, R., Menniti-Ippolito, F., Spila-Alegiani, S., and Traversa, G. (1995). Gangliosides and Guillain-Barre syndrome. *J. Clin. Epidemiol.* **48**, 1399–1405.

Reichard, P., Nilsson, B.-Y., and Rosenqvist, U. (1993). The effect of long-term intensified insulin treatment on the development of microvascular complications of diabetes mellitus. *N. Engl. J. Med.* **329**, 304–309.

Resnick, H. E., Vinik, A. I., Schwartz, A. V., Leveille, S. G., Brancati, F. L., Balfour, J., and Guralnik, J. M. (2000). Independent effects of peripheral nerve dysfunction on lower-extremity physical function in old age: The Women's Health and Aging Study. *Diabet. Care* **23**, 1642–1647.

Richardson, P. H. (1994). Placebo effects in pain management. *Pain Rev.* **1**, 15–32.

Rothman, K. J., and Michels, K. B. (1994). The continuing unethical use of placebo controls. *N. Engl. J. Med.* **331**, 394–398.

Rothwell, P. M. (1995). Can overall results of clinical trials be applied to all patients? *Lancet* **345**, 1616–1619.

Sackett, D. L., Richardson, W. S., Rosenberg, W., and Haynes, R. B. (1997). "Evidence-Base Medicine: How to Practice and Teach EBM". Churchill Livingstone, London.

San Antonio Conference. (1988). Consensus statement: Report and recommendations of the San Antonio conference on diabetic neuropathy. American Diabetes Association American Academy of Neurology. *Diabet. Care* **11**, 592–597.

Sands, M. L., Shetterly, S. M., Franklin, G. M., and Hamman, R. F. (1997). Incidence of distal symmetric (sensory) neuropathy in NIDDM: The San Luis Valley Diabetes Study. *Diabet. Care* **20**, 322–329.

Santiago, J. V., Sönksen, P. H., Boulton, A. J., Macleod, A., Beg, M., Bochenek, W., Graepel, G. J., and Gonen, B. (1993). Withdrawal of the aldose reductase inhibitor tolrestat in patients with diabetic neuropathy: Effect on nerve function. The Tolrestat Study Group. *J. Diabet. Complicat.* **7**, 170–178.

Schratzberger, P., Walter, D. H., Rittig, K., Bahlmann, F. H., Pola, R., Curry, C., Silver, M., Krainin, J. G., Weinberg, D. H., Ropper, A. H., and Isner, J. M. (2001). Reversal of experimental diabetic neuropathy by VEGF gene transfer. *J. Clin. Invest.* **107**, 1083–1092.

Shaw, J. E., and Zimmet, P. Z. (1999). The epidemiology of diabetic neuropathy. *Diabet. Rev.* **7**, 245–252.

Sima, A. A. F. (2000). C-peptide deficiency plays a prominent pathogenetic role in type 1 diabetic neuropathy. *Acta Neuropathol.* **99**, 459–459. [Abstract]

Sima, A. A. F., and Laudadio, C. (1996). Design of controlled clinical trails for diabetic neuropathy. *Semin. Neurol.* **16**, 187–191.

Simovic, D., Isner, J. M., Ropper, A. H., Pieczek, A., and Weinberg, D. H. (2001). Improvement in chronic ischemic neuropathy after intramuscular phVEGF165 gene transfer in patients with critical limb ischemia. *Arch. Neurol.* **58**, 761–768.

Sobel, B. E., and Furberg, C. D. (1997). Surrogates, semantics, and sensible public policy. *Circulation* **95**, 1661–1663.

Solders, G., Tydén, G., Persson, A., and Groth, C.-G. (1992). Improvement of nerve conduction in diabetic neuropathy: A follow-up study 4 yr after combined pancreatic and renal transplantation. *Diabetes* **41**, 946–951.

Spielberg, S. P., Shear, N. H., Cannon, M., Hutson, N. J., and Gunderson, K. (1991). In-vitro assessment of a hypersensitivity syndrome associated with sorbinil. *Ann. Intern. Med.* **114**, 720–724.

Sundkvist, G., Armstrong, F. M., Bradbury, J. E., Chaplin, C., Ellis, S. H., Owens, D. R., Rosen, I., and Sonksen, P. (1992). Peripheral and autonomic nerve function in 259 diabetic patients with peripheral neuropathy treated with ponalrestat (an aldose reductase inhibitor) or placebo for 18 months: United Kingdom/Scandinavian Ponalrestat Trial. *J. Diabet. Complicat.* **6**, 123–130.

Thivolet, C., el Farkh, J., Petiot, A., Simonet, C., and Tourniaire, J. (1990). Measuring vibration sensations with graduated tuning fork. Simple and reliable means to detect diabetic patients at risk of neuropathic foot ulceration. *Diabetes Care,* **13**, 1077–1080.

Thomas, P. K. (1973). Metabolic neuropathy. *J. R. Coll. Phys. Lond.* **7**, 154–160.

Thomas, P. K. (1986). Diabetic neuropathy: Human and experimental. *Drugs* **32**, 36–42.

Thompson, S. G., and Pocock, S. J. (1991). Can meta-analyses be trusted? *Lancet* **338**, 1127–1130.

Tomlinson, D. R. (1989). Polyols and *myo*-inositol in diabetic neuropathy—of mice and men. *Mayo Clin. Proc.* **64**, 1030–1033.

Toyota, T., Hirata, Y., Ikeda, Y., Matsuoka, K., Sakuma, A., and Mizushima, Y. (1993). Lipo-PGE1, a new lipid-encapsulated preparation of prostaglandin E1: Placebo- and prostaglandin E1-controlled multicenter trials in patients with diabetic neuropathy and leg ulcers. *Prostaglandins* **46**, 453–468.

Turner, J. A., Deyo, R. A., Loeser, J. D., Von Korff, M., and Fordyce, W. E. (1994). The importance of placebo effects in pain treatment and research. *JAMA* **271**, 1609–1614.

Tütüncü, N. B., Bayraktar, M., and Varli, K. (1998). Reversal of defective nerve conduction with vitamin E supplementation in type 2 diabetes: a preliminary study. *Diabet. Care* **21**, 1915–1918.

Valensi, P., Attali, J. R., and Gagant, S. (1993). Reproducibility of parameters for assessment of diabetic neuropathy: The French Group for Research and Study of Diabetic Neuropathy. *Diabet. Med.* **10**, 933–939.

Veves, A., and King, G. L. (2001). Can VEGF reverse diabetic neuropathy in human subjects? *J. Clin. Invest.* **107**, 1215–1218.

Williams, G., Gill, J. S., Aber, V., and Mather, H. M. (1988). Variability in vibration perception threshold among sites: A potential source of error in biothesiometry. *Br. Med. J. (Clin. Res. Ed.)* **296**, 233–235.

World Health Organization. (1980). "International Classification of Impairments, Disabilities, and Handicaps: A Manual of Classification Relating to the Consequences of Disease." World Health Organization, Geneva.

Yarnitsky, D., Sprecher, E., Tamir, A., Zaslansky, R., and Hemli, J. A. (1994). Variance of sensory threshold measurements: Discrimination of feigners from trustworthy performers. *J. Neurol. Sci.* **125**, 186–189.

Ziegler, D. (1997). The design of clinical trials for treatment of diabetic neuropathy. *Neurosci. Res. Commun.* **21**, 83–91.

Ziegler, D., Piolot, R., and Gries, F. A. (1996). The natural history of diabetic neuropathy is governed by the degree of glycaemic control: A 10-year prospective study in IDDM. *Diabetologia* **39**, A35–A35. [Abstract]

Ziegler, D., Reljanovic, M., Mehnert, H., and Gries, F. A. (1999). α-Lipoic acid in the treatment of diabetic polyneuropathy in Germany: Current evidence from clinical trials. *Exp. Clin. Endocrinol. Diabet.* **107**, 421–430.

I

Immunoreactivity, AGE, localization, 41
Injury
    changes in nerve blood flow with,
        173–175
    free radical, diabetes-associated, 4–5
    impaired neurotrophic response to, 312
    models for nerve growth factor efficacy,
        399–401
    reversibility, in Schwann cells, 313
    Schwann cell response to, 298–300
myo-Inositol, in diabetic nerve dysfunction,
    304
Insulin-like growth factor-I
    in diabetic pathology, 275–276
    effect on apoptosis, 153–154
    etiologic role in diabetic neuropathy,
        307–308
Internodal distance, role in maximal nerve
    conduction velocity, 239
$[^{14}C]$Iodoantipyrine, nerve blood flow
    measurement, 169–170
Ion channel distribution, role in maximal
    nerve conduction velocity, 239–240
Ischemia
    changes in nerve blood flow with, 171–173
    in diabetic pathology, 280
    endoneurial, causing diabetic neuropathy,
        72–73

J

JNK
    activation, 85, 94–95
        long-term, 132
    cell death and cell survival functions,
        102–103
    distinct pools, with different functions,
        104–106
    response to extracellular stress, 119
    in sensory neuron perikarya, 96–97

K

80K-H protein, localization, 44
Kinetic mechanism
    aldose reductase, 335
    sorbitol dehydrogenase, 343–344
Knockout mouse, $AR^{-/-}$, 346–347, 357, 362

L

Lactoferrin-like protein, complex with
    RAGE, 45
Laser Doppler flowmetry, nerve blood flow
    measurement, 168–169
γ-Linolenic acid, clinical trials, 452
Lipid peroxidation
    ARI effect, 9
    diabetes-induced changes, 4
    in peripheral nerve, aldose reductase role,
        20
α-Lipoic acid, clinical trials, 452–453
Luminal size, diabetic microvessels, 184
LY333531
    bisindolylmaleimide-based PKC inhibitor,
        67–68
    PKCβ-selective inhibitor, 72
Lymphocytic infiltrates, in diabetic ganglia,
    261

M

MAP kinases
    activation in
        primary sensory neurons, 90–95
        sural nerve, 101
    in etiology of diabetic complications, 87,
        101–102
    groups, 84–86
    and neuropathy, in STZ diabetes model,
        95–101
    oxidative stress effects, 7
Mechanical injury, nerve growth factor
    efficacy, 399–400
Metabolic flux hypothesis, polyol pathway,
    332–333
Metabolic injury, nerve growth factor
    efficacy, 401
Methylglyoxal, formation, suppression of,
    49–50
Microangiopathy, peripheral nerve trunk,
    188–189, 417–418
Microelectrode hydrogen clearance
    polarography, nerve blood flow
    measurement, 167
Microscopy, in detection of apoptosis,
    149–150
Microsphere approach, in analyses of nerve
    blood flow, 177–178
Mitochondrial pathway, apoptosis, 148–152

Phase III clinical trial: rhNGF (*continued*)
  nerve growth factor effectiveness, 409–410
  placebo group progress, 407–408
  results, 405–406
  study design, 404–405
Phosphoinositide, deranged metabolism,
  272–273
Phospholipase A_2, PKC-activated, 71
Phosphorylation, neurofilament
  aberrant, 131–132
  protein kinases controlling, 118–119
Pimagedine, *see* Aminoguanidine
Placebo effect, 445–447
PNS, *see* Peripheral nervous system
$p75^{NTR}$
  death domains, 147
  in diabetic pathology, 275
  expression on Schwann cells, 299–300
Poly(ADP-ribose)synthetase, inhibition, 8
Polymorphisms
  ALD2, 336–339
  SORD gene, 344–345
Polyol pathway
  aldose reductase, 333–340
  in diabetic autonomic neuropathy,
    272–273
  enzymes
    genetic alterations in, 362
    inhibitors, 352–354
    physiological role, 345–348
  hyperglysolia and, 328–330
  inhibition, in models of diabetic
    neuropathy, 354–363
  inhibitors
    effects in diabetic neuropathy, 363–367
    super-potent, 377
  link with oxidative stress, 376–377
  metabolic flux hypothesis, 332–333
  osmotic hypothesis, 331–332
  in peripheral nerve, 348–349
  sorbitol dehydrogenase, 340–345
Protein kinase C
  activation and regulation, 65–66
  isoform structural design, 63–64
  molecular features, 62–65
  and Na^+,K^+-ATPase, 70–71
  neural *vs.* neurovascular actions, 72–73
  and nonneural diabetic complications,
    66–68

  in normal and diabetic nerve, 68–70
  in pathogenesis of diabetic peripheral
    neuropathy, 12–13
  PKC-β-selective inhibitor, 455
  roles in nerve, 73–75
Protein kinases, control of neurofilament
  phosphorylation, 118–119

Q

Quantitative Sensory Testing, 439
Quiescent nerve, Schwann cell role, 296–298

R

RAGE
  activation of NF-κB mediated by, 46–47
  amphoterin-binding, 45–46
Randomized clinical trials
  adequate design of, 440–441
  generalizability of overall results, 449–451
  phase III, outcome measures, 437–440
  reporting of, 451
  reproducibility of outcome measures,
    443–445
  sample size and trial duration, 442–443
Reactive oxygen species
  elevated glucose effect, 74
  high glucose levels and, 301
  H_2O_2 as source, 88
  hyperglycemia-induced generation, 20–22
  in pathogenesis of diabetic peripheral
    neuropathy, 6–7
Receptor for advanced glycation end
  products, *see* RAGE
Redox cycles
  disturbances, 305
  glutathione in, 89–90
Regeneration
  abnormal, in diabetic neuropathy,
    311–313
  axonal, and collateral sprouting, 133–134
  fibers, 232
  sympathetic axons, 277–278
Renin/angiotensin system, 423–424
Repetitive stimulus paradigm, 247–248
Risk analysis, diabetic neuropathy,
  glycation-related correlates in, 43–44

CONTENTS OF RECENT VOLUMES

Volume 41